Dynamics of Offshore Structures

Minoo H Patel,
BSc, PhD, MIMechE, MRINA, CEng
Reader in Mechanical Engineering, University College, London

Butterworths
London Boston Singapore Sydney Toronto Wellington

Dedicated to Irene, Doly and Homi
for their loving support through the years

PART OF REED INTERNATIONAL P.L.C.

First published 1989

British Library Cataloguing in Publication Data
Patel, Minoo H.
Dynamics of offshore structures.
1. Offshore structures. Dynamics
I. Title
627'.98
ISBN 0-408-01074-6

Library of Congress Cataloging in Publication Data
Patel, Minoo H.
Dynamics of offshore structures/Minoo H. Patel
p. cm.
Includes index.
ISBN 0-408-01074-6
1. Offshore structures—Hydrodynamics. I. Title
TC1665.P38 1989
627'.98—dc 19

Typeset by KEYTEC, Bridport, Dorset
Printed and bound in Great Britain by Butler & Tanner, Frome, Somerset

Preface

Exploitation of oil and gas reserves from hydrocarbon reservoirs below the sea bed has increased rapidly since the early 1960s and in doing so has stimulated a wide-ranging base of theoretical analysis, model testing and practical experience in the scientific disciplines that contribute to the design and operation of offshore structures. These disciplines are, however, spread out over the traditional boundaries of the established physical sciences. The design, construction and operation of fixed and floating offshore structures requires expertise in subject areas ranging from meteorology, oceanography, hydrodynamics, naval architecture, structural and fatigue analysis, corrosion metallurgy, sea bed soil mechanics, mechanical and process engineering, diving physiology and even marine biology. These disciplines are often combined within the descriptive title of 'ocean engineering'.

The wide range of subject areas within ocean engineering has posed problems in the teaching of both undergraduate and postgraduate students because most existing text books are written within traditional discipline boundaries. Teaching material in ocean engineering is, therefore, distributed between many text books and research papers and is not generally available within a suitable context or with an eye to application in ocean engineering.

This text book has been written to help remedy this situation by presenting an integrated treatment of the main subject areas that contribute to the design, construction, installation and operation of fixed and floating offshore structures. The book has been prepared with particular emphasis on the fundamentals of oceanography, basic fluid mechanics, wave theory, hydrodynamics, naval architecture and structural analysis to meet the needs of students reading ocean engineering or naval architecture, at both undergraduate and postgraduate levels. Basic equations and theoretical results are derived in a rigorous manner but sections on model testing, full-scale measurements, design and certification are also induced to ensure that the book is of value to professional engineers seeking a balanced treatment of fundamental and practical issues.

A limited number of original source references are cited for further reading but considerable selectivity has been exercised to ensure that only those research results relevant to current design practice are introduced in the text or through the references. Nevertheless, some recent research advances are included where it is believed that they will have a strong impact on offshore structures design practice in the near future.

I am indebted to a number of my colleagues for their help and advice during the preparation of the manuscript. In particular, the assistance of Dr David Brown, Dr John Harrison, Mr Farman Seyed and Dr Joel Witz was invaluable, as was the infinite patience of Miss Vanessa Smith during the typing of the manuscript. I am especially grateful to Professor Tom Lambert for his encouragement and guidance over many years.

Minoo Patel

Contents

Chapter 1

Introduction

The oceans of planet Earth exercise a dominant influence on its weather systems and ecology. The oceans are believed to have been the medium in which life first appeared on the planet and gradually evolved into land based plant and animal species. The continued existence of land based life is still dependent on the oceans through their effect on the planet's weather and the crucial role that they play in global ecology. Since prehistoric times and up to the early part of this century, however, man's use of the oceans has been largely restricted to near-shore transportation and fisheries, although there is evidence to suggest migrations of people across very large oceanic distances. Since the fifteenth century, the pace of ocean transportation and deep water fishing has gradually increased but man's utilization of the oceans has still been restricted to these two activities.

Over the last five decades, however, traditional uses of the oceans have expanded to include the exploitation of hydrocarbons below the sea bed and the potential of large-scale mineral gathering and energy extraction. The development of offshore oil and gas has led the way with land based exploration and production for crude oil being extended to the sea bed as early as 1887, in California, when the first ever exploration drilling of more than a few feet of water was carried out. Since then the pace of oil exploration and production in shallow water has gradually increased, with the next major development occurring in 1910, when wells were sunk in Ferry Lake, Louisiana. This was followed in 1929 by the exploration and production of a large oil reservoir on Lake Maracaibo in Venezuela. The development of the Gulf of Mexico as an offshore area began in the 1930s, with oil first being produced in 1938 from a timber platform in 4.3 m (14 ft) of water, on the Creole Field, which was located 1.6 km (1 mile) off the coast of Louisiana. Following the Second World War, offshore activity in the Gulf of Mexico rapidly evolved, with oil exploration and production in deeper waters until, by 1959, an oil production platform had been installed by Shell on a block off Grand Isle in 30.5 m (100 ft) of water. At the same time British Petroleum were developing shallow water oil finds in the Persian Gulf.

Such oil and gas exploration and development spread to most of the continental shelf areas of the world from the mid-1950s, commencing

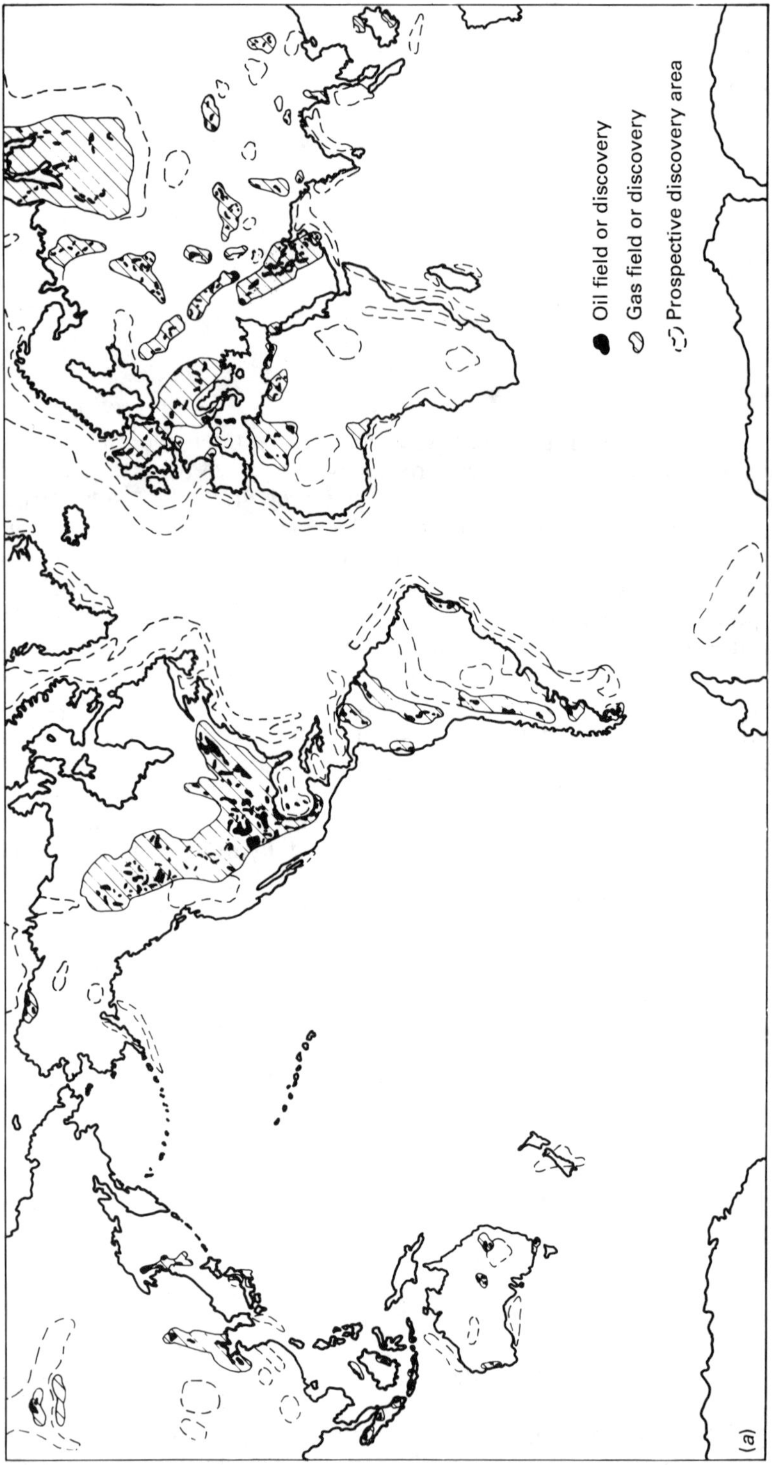

Figure 1.1.(*a*) Worldwide offshore development areas. (*b*) Distribution of water depths of the Earth's oceans. *Key:* □ 500 m; ▲ 1000 m; ○ 3000m; ■ 5000 m; A – Pacific; B – Atlantic; C – Indian Ocean

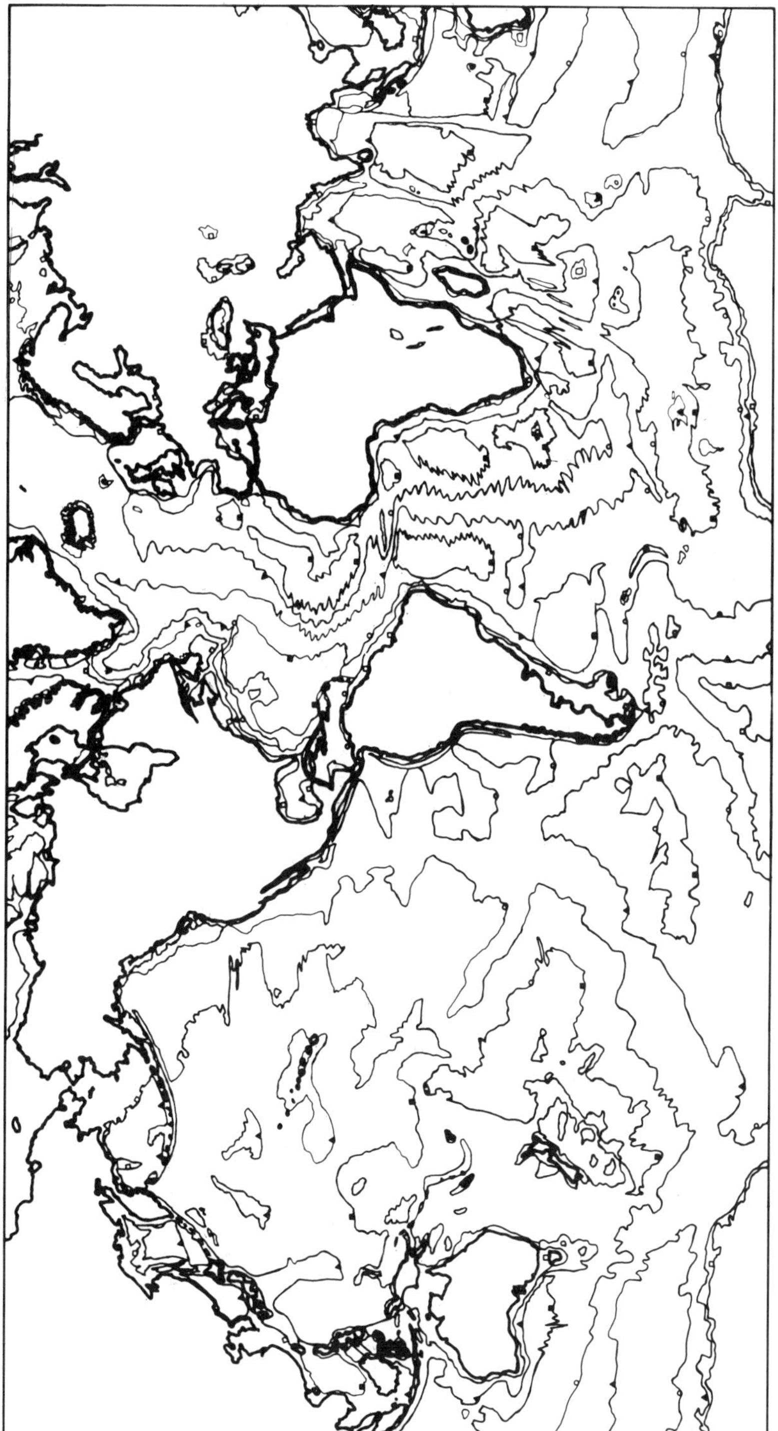

(b)

first in the Gulf of Mexico and then spreading to the waters off Mexico, Brazil and Brunei. The map illustrated at Figure 1.1(*a*) presents an overview of this worldwide offshore exploration and production activity, whereas Figure 1.1(*b*) uses sea bed contours to show the extent of the continental shelves around all the major land masses. The continental shelves cover only a small fraction of the total ocean area but they are potential sites for exploration, or are already known to have large reserves. The offshore industry embarked upon a technically more challenging phase when the North Sea was first explored as a potential offshore area in the early 1960s. Figure 1.2 presents an overview of the fields developed in the North Sea since the mid-1960s. These fields range from large gas reservoirs in the relatively calm waters off East Anglia in south east England to harsh northerly environments in the vicinity of 60° north where fields such as Snorre, Troll, Oseberg and Frigg are exposed to some of the harshest wind and wave conditions of any maritime area in the world. The North Sea is typical of many offshore areas where the development of technology and structures to exploit oil and gas fields has had a significant positive impact on the economies of surrounding countries.

The development of land based and offshore oil and gas deposits has played an essential role in laying down the foundations of the modern world with its high living standards. Table 1.1 shows the rapid growth of total (both land based and offshore) oil production over the last 45 years and its distribution across major regions of the Earth. The ten-fold increase in production over this period reflects the increasing demand for energy throughout the world, although produced oil and gas is also used as feedstock, and in the chemical industry, leading ultimately to a multitude of products including drugs, man made fibres, plastics and industrial chemicals. Table 1.1 shows that production in the Middle East dominated world oil supplies during the 1970s. The resultant economic and political leverage derived by Middle Eastern oil producing countries has played a significant political role in modern times. In particular, oil price rises caused by excess oil demand over supply in the 1970s has prompted the development of offshore oil and gas throughout the world in order to reduce dependence on the Middle East. Figure 1.3 illustrates the relative volumes of onshore and total oil production over the last two decades and shows that offshore oil production is now approaching 30% of total worldwide production.

In the long term, however, man's utilization of the oceans as a resource will stimulate engineering developments in other areas. The oceans have the potential to provide worth while amounts of renewable energy, either from ocean waves of by exploiting thermal gradients between the surface and deeper layers of the ocean. Surface-deployed wave energy extraction devices and ocean thermal energy conversion plants are potentially able to extract such renewable energy. The oceans can also make a worthwhile long-term contribution to the mineral wealth of the world. It is known that sea bed material in the form of nodules is rich in several metals which are in short supply from land based sources – manganese being one example. Sea bed mining of such mineral-rich nodules could contribute to the world's mineral supplies.

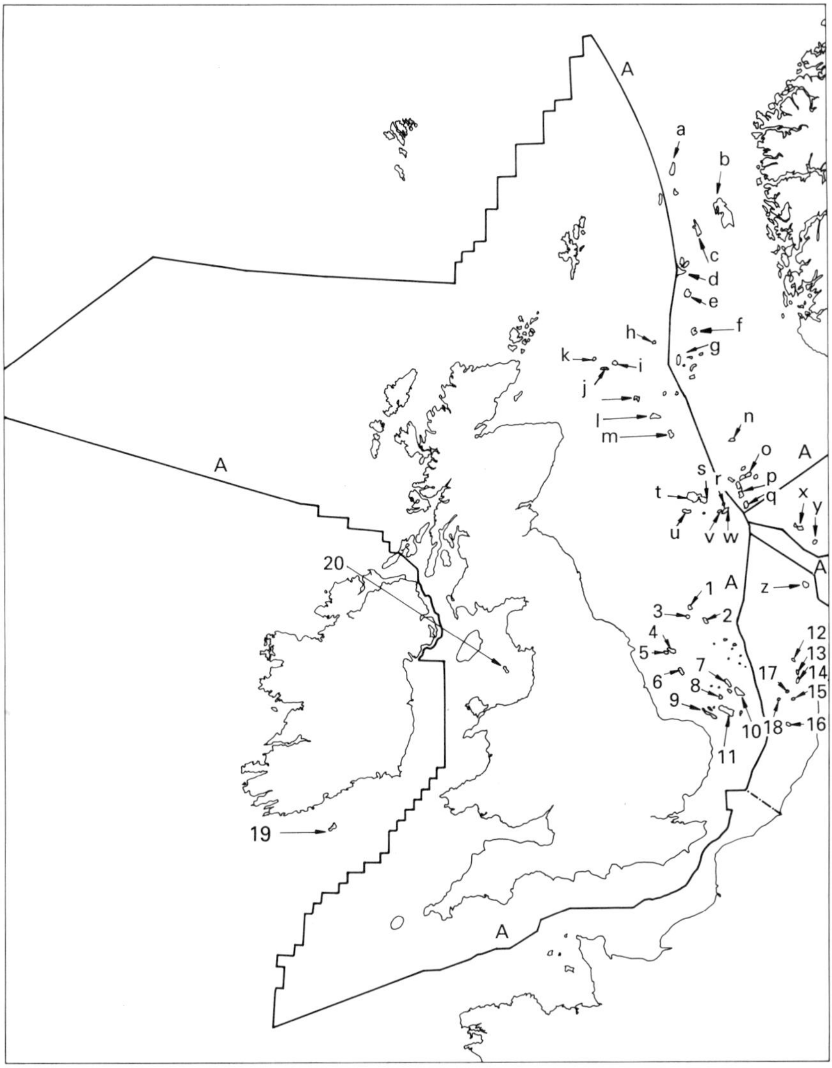

Figure 1.2. Distribution of fields in a typical offshore location – The North Sea. Only a representative selection of fields are illustrated for clarity. *Key*: A – median lines marking limits of neighbouring country's offshore sectors. Names of oil and gas fields: a – Snorre; b – Troll; c – Oseberg; d – Frigg; e – Heimdal; f – Balder; g – Sleipner; h – South Brae; i – Piper; j – Tartan; k – Claymore; l – Forties; m – Montrose; n – Ula; o – Ekofisk; p – Eldfisk; q – Valhall; r – Innes; s – Clyde; t – Fulmar; u – Auk; v – Duncan; w – Argyll; x – Tyra; y – Dan; z – F2–3; 1 – Forbes; 2 – Gordon; 3 – Esmond; 4 – Ravenspurn; 5 – Cleeton; 6 – West Sole; 7 – Viking; 8 – Vulcan; 9 – Hewett; 10 – Indefatigable; 11 – Leman; 12 – L4-A; 13 – L7-C; 15 – K15-FA; 16 – P6; 17 – K14; 18 – K-17, 19 – Kinsale Head; 20 – Morecambe (from the Department of Energy, UK, 1986)

There is also the potential for extracting the extremely low levels of various metallic compounds that are dissolved in sea water. Furthermore, future development of the world's food resources is going to require increasingly stringent conservation of fisheries and a higher level

Table 1.1 Worldwide oil production growth (1000 b/d) (*International Petroleum Encyclopaedia*, 1986)

Region	*1940*	*1950*	*1960*	*1965*	*1970*	*1977*	*1978*	*1979*	*1980*	*1981*	*1982*	*1983*	*1984*	*1985*
North America	3731	5487	7580	8606	10894	9566	10025	10040	9981	9875	9888	10065	10180	10372
Middle East	275	1756	5313	8360	13348	22307	21120	21497	18379	15694	12464	11711	11458	10307
Africa	18	46	285	2212	5969	6277	6002	6638	6032	4360	4455	4333	4673	4839
Europe	30	67	267	353	313	1255	1684	2206	2394	2555	2825	3206	3500	3682
Asia–Pacific	223	238	546	670	1361	2778	2777	2884	2732	2774	2618	2705	3084	3110
Latin America	855	1963	3728	4619	5163	4512	4758	5259	5585	6018	6102	6078	6225	6333
Communist areas	625	872	3370	5462	8012	13323	14043	14321	14580	14645	14764	14908	14975	14849
World total	5757	10428	21088	30280	45059	60018	60409	62845	59683	55922	53116	53007	54095	53493

Note: figures may not add up due to rounding.

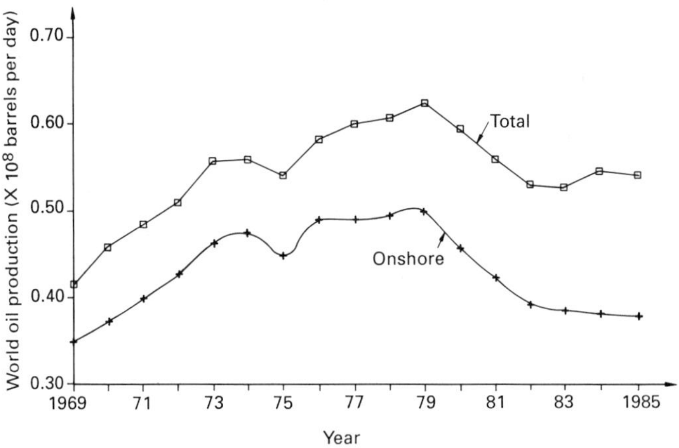

Figure 1.3. Growth of onshore and total oil production (from the Institute of Petroleum, 1986)

of engineering intervention in the oceans, aimed at fish farming and conservation of stocks. In the very long term, the oceans may also be used to increase man's living and working space on the planet by providing room for process plant sites, floating airports and even sea-borne communities living on man made islands or large floating structures. All these activities will pose technically challenging problems for scientists and engineers in the development of materials, structures and equipment for use in the harsh environment of the oceans.

The design of offshore structures used for oil and gas production offers some indication of these problems. A typical steel framed jacket structure used for offshore oil drilling and production will encounter larger steady environmental forces from wind and current flows, and from the structure's self-weight. The jacket will also be subjected to high levels of cyclic load due to gravity waves; its sea bed equipment will be working in the presence of very high sea water pressure; and the steel in its structure will have to survive sea water corrosion for long periods since large parts of the jacket structure will be difficult to access and maintain. At the same time the physical processes that govern interactions between the atmosphere and the ocean surface, and the effects of the structure on the fluid around it and on the behaviour of the sea bed foundation are not completely understood in scientific terms. These problems are compounded by the uncertainties of predicting the most extreme environment likely to be encountered by the structure over its lifetime, which is measured in decades, and the uncertainty of categorizing the geology as well as the recoverable resources of the hydrocarbon reservoir that the steel structure is designed to develop. All these interacting problems offer unique challenges for advanced scientific analysis and engineering design.

The technical evolution of the modern offshore industry can be measured by the depth at which it has been able to carry out exploration drilling and by the structures or vessels that have made such drilling possible. Initially, exploration drilling was carried out from shallow water fixed platforms which were made of timber and were piled to the sea bed. This was later followed by drilling from steel barges or drilling tenders mounted in sheltered locations. The water depth capability of drilling has gradually increased, firstly by the use of jack-up rigs – a typical modern jack-up rig is shown in Figure 1.4. Jack-up rigs are generally of triangular construction containing steel framed legs at each corner which can be jacked up or down by electric or hydraulic machinery. The jack-up hull is transported to the drilling site either by loading on to a barge or by towing the self-floating hull. The legs are jacked down at the drill site, and the platform hull is raised a sufficient distance above mean water level to prevent waves hitting the

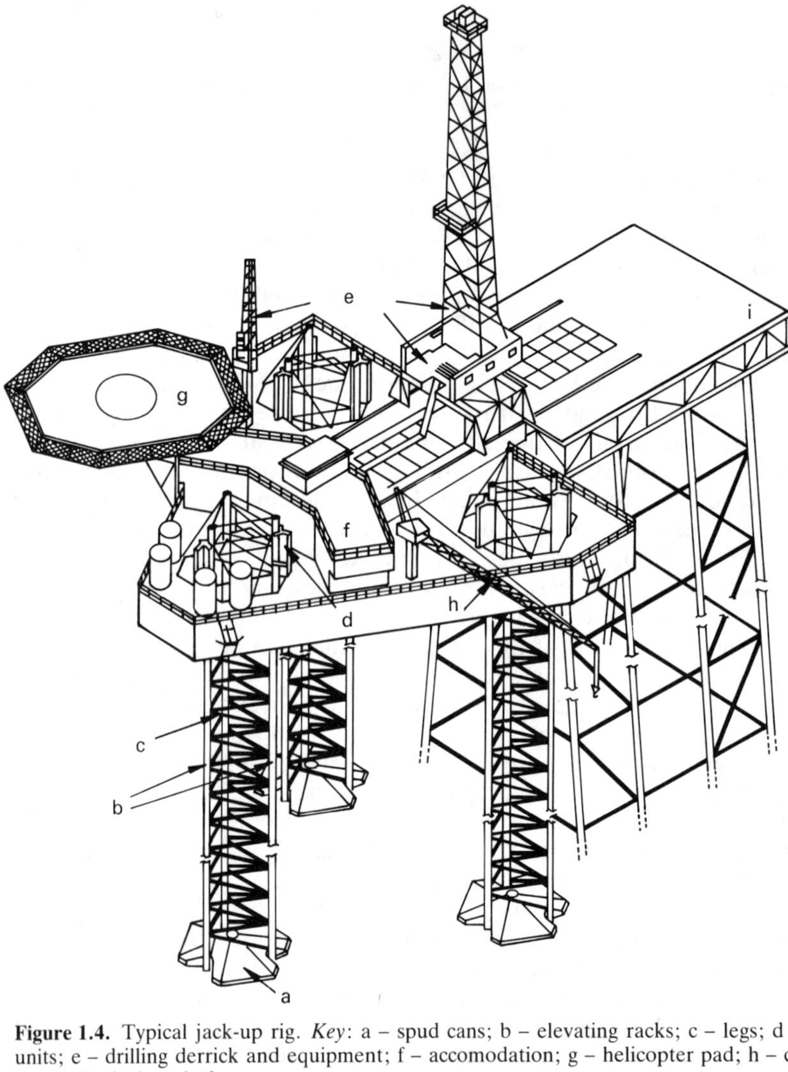

Figure 1.4. Typical jack-up rig. *Key*: a – spud cans; b – elevating racks; c – legs; d – gear units; e – drilling derrick and equipment; f – accomodation; g – helicopter pad; h – cranes; i – nearby jacket platform

underside of the platform. Drilling then commences, usually for a work programme of between 2 and 4 months. When the drilling programme has been completed, the hull is lowered down to the sea surface or on to a transporation barge and the rig is moved to the next drilling location. Some jack-ups have also been used for oil production in shallow water, with modern jack-ups being able to operate in severe weather at water depths of up to 107 m (350 ft).

Exploration drilling in waters of up to 18 m (60 ft) depth has also been carried out from floating vessels with one or more pontoons or caissons supporting a deck with vertical columns. Such platforms were floated on to the drilling site and then ballasted down on to the sea bed to carry out a drilling programme. During the flotation phase of deployment, it was noticed that such column stabilized platforms had exceptionally low motion response to waves. This feature was utilized to develop so-called semisubmersible floating platforms which did not have to be ballasted down to the sea bed in order to carry out drilling. Figure 1.5 shows a perspective view of one such modern vessel with two parallel submerged pontoons at keel level supporting a large deck area

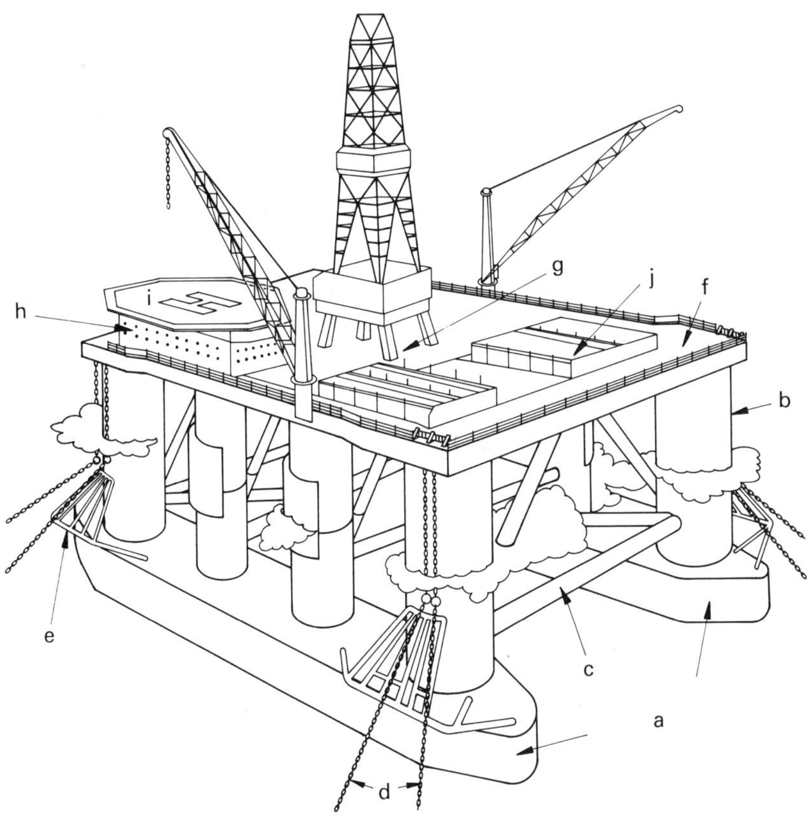

Figure 1.5. Typical semisubmersible vessel. *Key*: a – submerged pontoons; b – surface piercing deck support columns; c – bracing members; d – mooring lines; e – anchor racks; f – deck structure; g – moonpool; h – accomodation; i – helicopter pad; j – drill pipe racks

using eight vertical surface piercing columns. The columns, pontoons and deck are braced to form a space-frame structure which can be moored on location by a spread of catenary mooring ropes or chains. The deck contains a central opening, called the moonpool, through which drilling activities take place. Most semisubmersible vessels are equipped with accommodation, helicopter pads for crew transport, propulsion engines, pipe handling equipment, cranes and a variety of marine services to support crew and drilling operations.

Drilling into the sea bed is carried out through a vertical pipe called the marine riser which connects the area under the moonpool to a subsea well head. A drill bit on the end of the drill pipe is passed through the riser to penetrate the sea bed and carry out the actual drilling. A lubricating fluid, called drilling mud, is pumped to the drill face at high pressure through the hollow drill pipe and returned through the annulus between the drill pipe and riser. The drilling mud pressure is used to drive rotating cones on the drill bit, with the drill bit and pipe itself rotating at up to 250 rpm. The drilling mud serves several purposes. It lubricates the drill bit, drives the drill bit cones, sweeps out drill cuttings to the surface and provides a hydrostatic pressure head to contain high-pressure oil or gas that may be encountered. The small, but significant, heave motions of the floating drilling rig due to ocean waves are compensated for by a slip joint at the base of the drilling derrick. A tensioner also provides an upward force on the riser to keep it in tension and prevent buckling of its slender structure. The moorings of a drilling semisubmersible have to maintain the vessel's horizontal position above a subsea well head within 7% of the water depth, in order to continue drilling. The maximum water depth and weather conditions that a semisubmersible vessel can drill in is governed by its mooring system, the amount of drill pipe and riser pipe that it can carry and by heave compensator limits imposed by the vessel's motion response and prevailing weather.

A catenary line mooring system is usually impracticable at water depths of more than 457 m (1500 ft), although such moorings have been used for water depths of up to 1128 m (3700 ft). At greater depths, a dynamically positioned vessel is required. Dynamic positioning requires a vessel, whether a ship or a semisubmersible, to be installed with rotating thruster pods and propulsion machinery which is able to work against and balance the environmental disturbing forces due to wind, current and wave drift. Dynamic positioning has been most commonly employed on drill ships which utilize position signals from sensors, such as a taut vertical wire to the sea bed, or from acoustic beacons on the sea bed around the well head, to obtain an error signal representing horizontal offset from above a subsea well head. The vessel is fitted with thrusters controlled from an automated computer based algorithm which is programmed to reduce this offset and maintain the vessel as close to directly above the well head as possible. Figure 1.6 shows a schematic view of this arrangement with the drill ship connected by a riser to the blow out preventor (BOP) stack, prior to that being connected to the well head. Acoustic beacons provide a position reference to the drill ship – the position data is then used by a computer to control the main propulsion system and lateral thrusters to resist wind, current and wave drift forces on the vessel. In order to continue drilling it is necessary to maintain position within a 'watch' circle that has a radius of less than about 7% of the water depth.

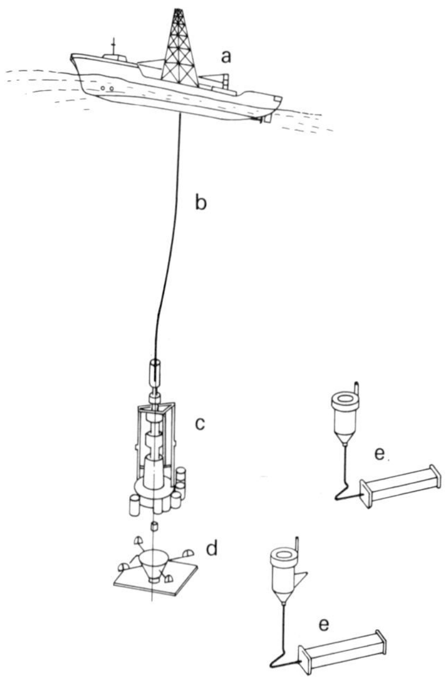

Figure 1.6. Dynamically positioned drill ship. *Key*: a – drill ship; b – riser; c – blow out preventer; d – well head on sea bed; e – acoustic positioning beacons

Several modern semisubmersibles have also been equipped with dynamic positioning systems. The maximum water depth to which a dynamically positioned drilling vessel can work is theoretically unlimited, although in practice limitations are imposed by the length of marine riser that can be carried on the vessel and by tensioner and heave compensator performance. Nevertheless, dynamically positioned vessels have to disconnect the riser in extreme weather conditions, to ride out storms and then to reconnect in calmer weather.

There are approximately 500 offshore drilling rigs worldwide, made up of about 250 jack-up rigs, 130 semisubmersible vessels, 90 drill ships and 30 bottom standing submersible rigs – the latter being of the type which can be floated out on to a shallow water location and then ballasted down to rest on the sea bed prior to commencement of drilling. Worldwide, the active jack-up rig fleet began in 1954, increasing to 27 rigs by 1960 and 103 by 1970, whereas the first semisubmersible drilling vessel started operation in 1962, with 25 active vessels being in use by 1970. The first drill ship commenced operation in 1953, with 16 being developed by 1960 and 39 by 1970. The expansion of the drilling fleet worldwide has been matched by a significant increase in the water depths at which drilling can be accomplished. This is illustrated most vividly by Figure 1.7 which presents water depth against year and curves showing the maximum drilling depth capability, as well as the actual water depth records that were achieved in the years between 1952 and 1985. It can be seen that in the period 1970–1985 the water depth at which drilling was carried out increased from around 1366 m (1200 ft) to 2128 m (6981 ft). At the end of 1988, the water depth achieved for drilling stood at 2328 m (7638 ft). Some drilling in very deep water has been carried out for geological research into the structure of the land mass below the oceans.

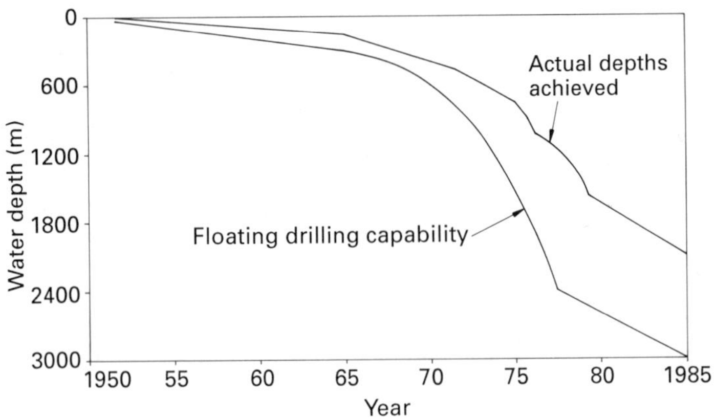

Figure 1.7. Deep water drilling capability and achievements

The water depths at which exploration drilling is carried out is an indicator of future requirements for oil production. In drilling programmes where significant discoveries of hydrocarbons are made, a decision on oil production is dependent upon the prevailing price of oil and the economics of platform construction and operation. Unlike exploration drilling, production platforms must be capable of carrying large amounts of development drilling and oil processing equipment. The processing equipment has to separate water, gas and sand from crude oil emerging from the reservoir, prepare the crude oil for transportation and provide a transportation pumping station. At the same time, the hydrocarbon reservoir depletion procedure may require gas lift, which involves the use of gas to assist in raising the oil if the reservoir pressure is insufficient to do so. Water injection may be required to increase reservoir pressure and thus permit the extraction of more oil. It is common for one platform to be used to deplete a large plan area of the reservoir by using directional drilling or employing step-out satellite wells.

All of these and many other requirements mean that the complexity of equipment on board a production platform can resemble a chemical plant or refinery. At any one point in time, therefore, the depth capability of production platforms tends to be considerably lower than the potential drilling depth capability. Furthermore, the engineering equipment demands of development drilling, oil production, enhanced recovery and oil export are such that the water depth capability with time for production platforms has not increased as steeply as it has for drilling vessels. This is illustrated by Figure 1.8 which shows the evolution of timber or steel framed bottom standing oil production structures – usually called jackets. The diagram illustrates jacket size and water depth capability in 1947 through to a large deep water jacket in use in 1978. In general terms, the 1930s were marked by the use of timber platforms, both in the swamps of Louisiana and the Gulf of Mexico, 1947 being the first year in which a steel platform was installed in 6 m (20 ft) water depth off the Louisiana coast and out of sight of land. By the mid-1950s, jacket platforms were being installed in water depths of about 30 m (98 ft) with the framed structures becoming more open. The 1950s to the early 1970s saw platform water depths extend to approximately 120 m (394 ft). Between 1970 and 1987, fixed platform water depth capability increased substantially from around 120 m (394 ft) through to the largest depth to date for a fixed jacket structure – the Cognac platform in the Gulf of Mexico which stands in 311 m (1020 ft) of water.

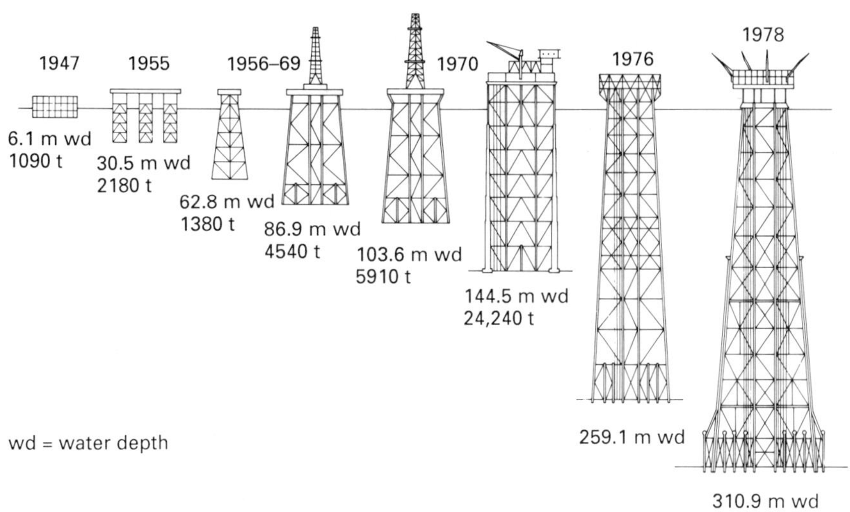

Figure 1.8. Evolution of deep water production capability (from Lee, G. C., 1980)

It is instructive to examine the structure of a typical jacket for deployment at moderate water depth in the North Sea, as illustrated by Figure 1.9. The jacket consists of a steel framed tubular structure connected to the sea bed by piles which are driven through pile guides on the outer members of the jacket. The jacket topsides consist of a series of modules which house drilling equipment, production equipment, living quarters, gas flare stack and revolving cranes. The topsides also have facilities for survival craft, hotel and catering, and a helicopter pad for transfer of crews to and from a shore base. The topside modules are supported on a module support frame, which itself rests on a jacket structure that pierces the wider surface and is piled to the sea bed. The drilling and production tubulars are brought up from the sea bed to the topsides through conductor guides located within the jacket framing. Crude oil and gas are brought up to the surface from the reservoir for processing and then pumped back down through an export pipe line,

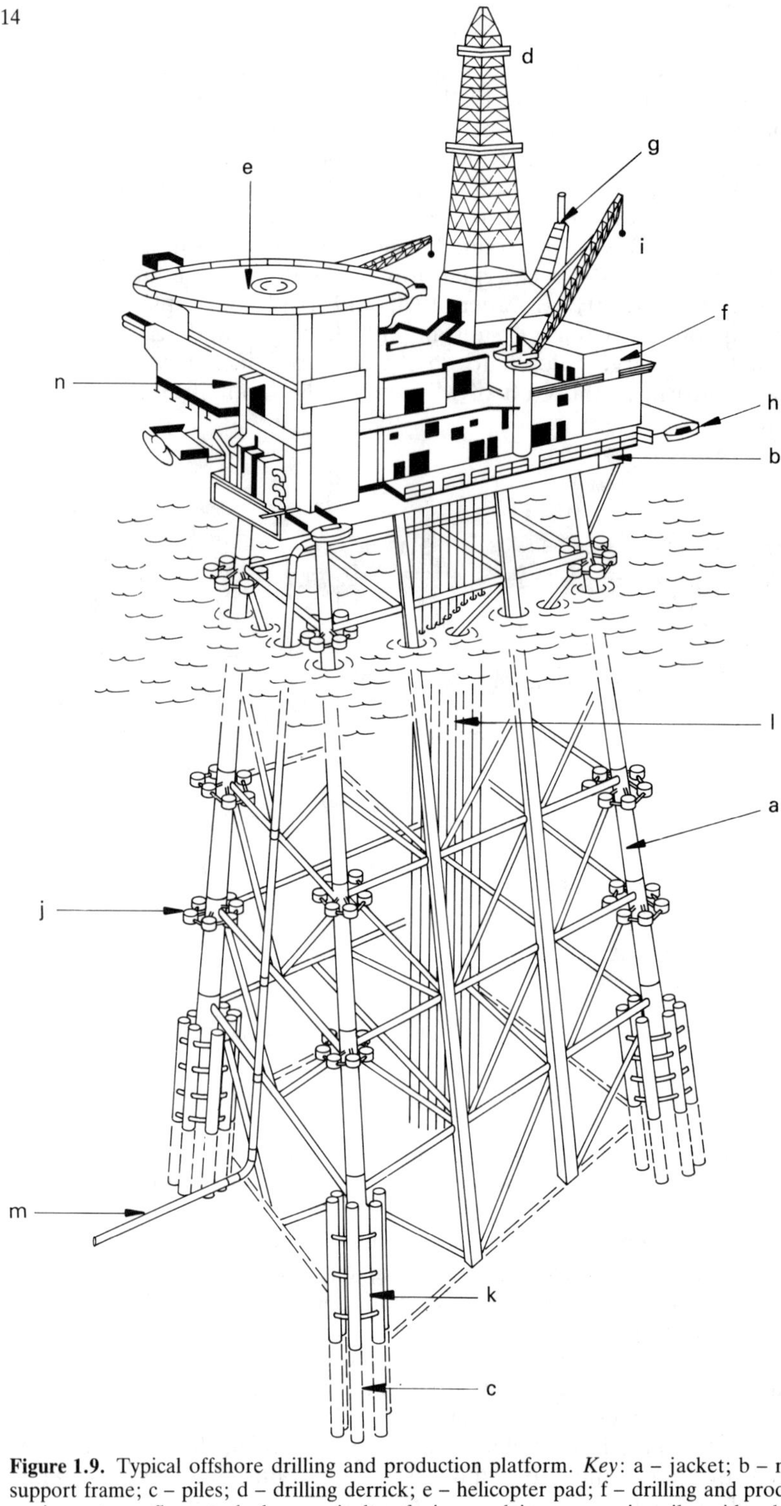

Figure 1.9. Typical offshore drilling and production platform. *Key*: a – jacket; b – module support frame; c – piles; d – drilling derrick; e – helicopter pad; f – drilling and production equipment; g – flare stack; h – survival craft; i – revolving cranes; j – pile guides; k – pile sleeves; l – drilling and production risers; m – export pipeline; n – accomodation

either to a tanker loading buoy or by subsea pipeline to a shore terminal.

The detailed design of the framing for a jacket structure can vary considerably, depending on requirements of strength, fatigue and launch procedure. The main structural members and bracing can be connected by 'X-' or 'K-'shaped joints with sizes ranging from an extensive space frame of small-diameter 'X'-braced or 'K'-braced members to more sparse framing with larger diameter members. Figure 1.8 gives some indication of the variety of design options that are employed.

A jacket designer has to cope with an extensive list of constraints imposed on a jacket structure throughout its life. The life cycle consists of design, construction, load out, launch, installation, piling and hook up phases of the platform before it comes into oil production service. This is followed by a 10–25-year operational life through which the platform has to be maintained, followed by the ecologically desirable requirement to remove and dispose of the platform after the reservoir has been depleted.

The complexity of any one of these phases is illustrated by considering the procedure for launching and installing a jacket. Figure 1.10 illustrates two ways in which this has been done in the past. Figure 1.10(*a*) illustrates a jacket which has been loaded out from a fabrication yard onto a barge and then towed out to the installation site. Steps 1–4 of Figure 1.10(*a*) show the launching phase, during which the jacket is subjected to large dynamic loads. The jacket must then be up-ended into an upright position, either by using a floating crane, as in step 5, or by differential ballasting of jacket structural members and then sunk down on the sea bed in preparation for piling. Figure 1.10(*b*) illustrates another approach to the installation problem by utilizing a self-floating jacket with main members which are large enough that their buoyancy can support the jacket weight, thus dispensing with a launch barge. Such a jacket is towed out on to the site and then up-ended using differential buoyancy in structural members, as shown in steps 2 through to 5. The jacket can then be ballasted down in preparation for piling. There are several pros and cons for a designer to consider in choosing between self-floating and barge launched jackets. Most large jackets constructed for the North Sea in the last few years have been of the barge launch type. They often have to be installed with removable auxiliary buoyancy chambers in order to facilitate launch and up-ending. The only self-floating jacket that has been installed recently in the North Sea was for the Magnus field.

Since barge launched jackets require less buoyancy, they utilize less steel and are in general easier to fabricate. However, such jackets require an appropriate transport barge and are subjected to large dynamic loads in the launch process. On the other hand, self-floaters do not require a launch barge, but do need additional steel to provide extra buoyancy. The larger diameter of the buoyancy members also induces additional unnecessary wave forces during the life of the structure. A compromise between self-floaters and barge launch jackets can be achieved by using a structure which has removable buoyancy chambers. These chambers need to be installed for the load out, launch and up-ending phases only and can be removed after installation is complete,

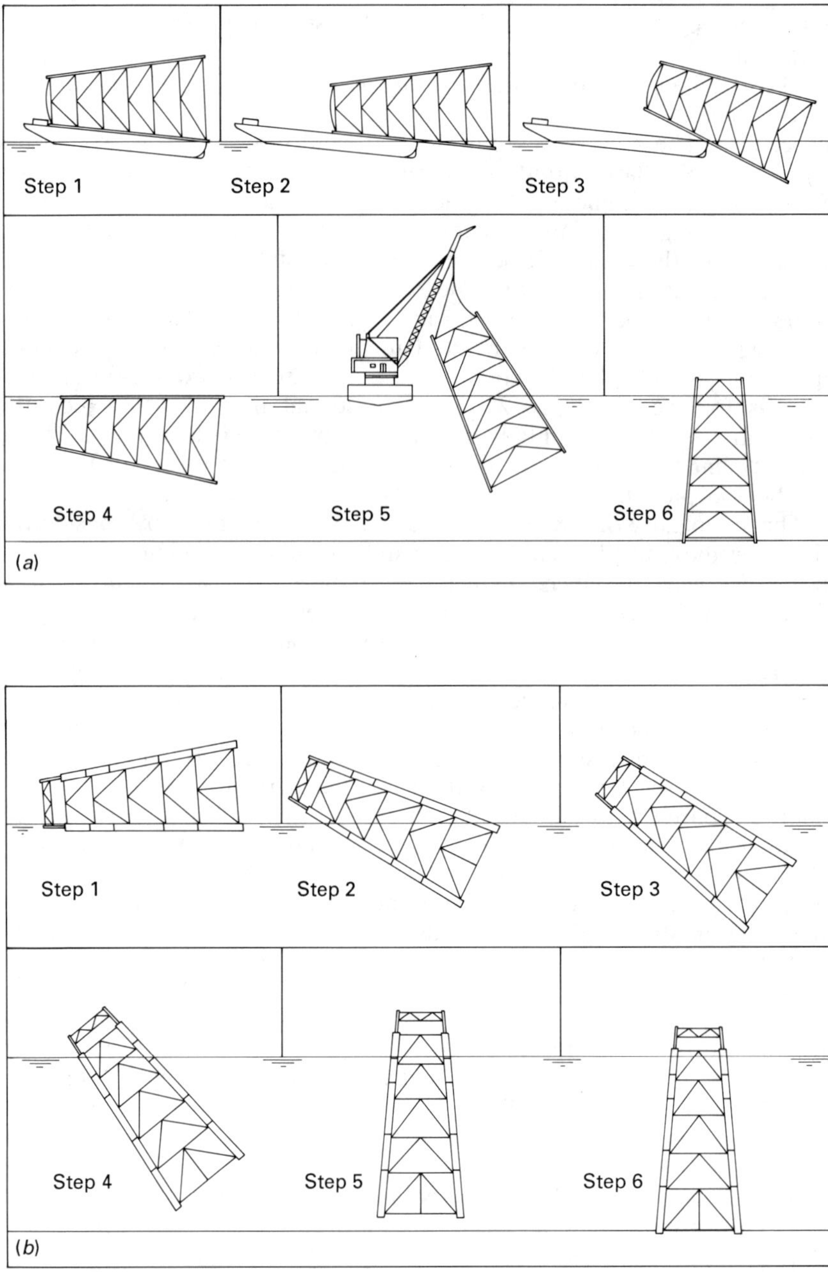

Figure 1.10. Installation of barge launched and self-floating jackets: (*a*) barge launched jacket; (*b*) self-floating jacket

thus largely eliminating the disadvantage of higher wave forces during operation. A typical barge launched jacket structure, such as the jacket on the Cerveza Field weighs 24 500 tonnes. The capabilities of an available fabrication yard for load out or launch of a self-floater as well as the availability of launch barges influences this design decision.

Assessment of jacket foundations and pile design offer another technical challenge. Prior to commencing jacket design, the sea bed soil has to be surveyed and its characteristics evaluated. The designers then determine the number of piles, the required depth of penetration and whether the piles should be drilled or driven by pile hammers. Drilled piling implies that holes for the piles are drilled before the piles are installed and grouted *in situ*. However, recent developments in underwater pile hammers have been so rapid that driven piles are now almost invariably used. Continuing improvements in pile hammer capability have also influenced jacket design and piling installation.

Jacket topside weights in the central and northern North Sea vary between 17 400 tonnes for the North Cormorant jacket, and 35 000 tonnes for the Statfjord and Brae platforms. This topside weight has to be broken down into a number of modules which can be installed offshore by crane barges onto the module support frame on top of the jacket. The design of the jacket and module arrangement is greatly influenced by the availability of crane barges and by their maximum lift capacities. Recent construction of several semisubmersible crane vessels with large, load lifting capacities is beginning to have a significant influence on offshore installation practice. The semisubmersible configuration offers a low down time to weather and the very large size of the vessel means that loads of up to 12 000 tonnes can currently be lifted. It is, therefore, feasible to construct and lift fewer modules, each of larger tonnage and thus to reduce the complexity and cost of module installation and hook-up operations. The availability of larger module sizes also reduces hook-up times and enables earlier oil or gas production to be achieved.

The use of steel frame jacket structures in deeper waters and harsher environments requires the designer to ensure that inspection and maintenance of the underwater parts of the platform are planned for during the design process. Such inspection and consequent repair can be a major and costly activity since it has to be carried out underwater by divers or with remotely operated equipment. The emphasis in engineering design and economic evaluation of modern jacket structures is shifting away from the historical approach of considering initial capital expenditure only, and towards the situation where considerably more attention is paid to designing for economic inspection and maintenance of the structure during service.

Offshore oil developments in the North Sea have also pioneered a completely different design of fixed platform based on the use of concrete to create a large heavy structure which can rest on the sea bed and remain stable under its own weight. The first such platform was installed in the North Sea on the Ekofisk field in 1973, and by 1982 over 17 platforms had been installed. Table 1.2 gives a summary of concrete platforms that have been installed worldwide.

Although concrete gravity structures are considered relatively expensive as compared to steel frame jackets, the structures do offer an attractive alternative to jackets in hostile waters such as those of the

Table 1.2 Summary of concrete platforms (from Furnes and Loset, 1981)

Type of design and location	*Main function*	*Design wave height (m)*	*Depth (m)*
1* Doris Ekofisk 1 (N)	Storage	24.0	70
2* Condeep Beryl A (UK)	Drilling, production, storage	29.5	120
3* Condeep Brent B (UK)	Drilling, production, storage	30.5	142
4* Doris Frigg CDP 1 (UK)	Drilling, compression, production	29.0	96
5* Sea Tank Brent C (UK)	Drilling, production, storage	30.5	142
6* Sea Tank Frigg TP1 (UK)	Production	29.0	104
7* Sea Tank Cormorant A (UK)	Drilling, production, storage	30.5	152
8* Condeep Brent D (UK)	Drilling, production, storage	30.5	142
9* Andoc Dunlin A (UK)	Drilling, production, storage	30.5	152
10* Condeep Statfjord A (N)	Drilling, production, storage	30.5	149
11* Condeep Frigg TCP2 (N)	Treatment, compression, production	29.0	104
12* Doris Frigg MP2 (UK)	Compression station	29.0	94
13* Doris Ninian (UK)	Drilling and production	31.2	139
14* Pub 3 Petrobras	Drilling, production, storage	11.0	15
15* Pub 2	Drilling, production, storage	11.0	15
16* Pag 3 Petrobras	Drilling, production, storage	11.0	15
17* Condeep Statfjord B (N)	Drilling, production, storage	30.5	149

* Supervised by Det Norske Veritas.

North Sea and in situations where the uncertainty of a tanker export system demands a certain amount of oil storage on the platform. Other advantages of concrete gravity platforms over jackets are that the structures can be constructed onshore or in sheltered waters, with all the topsides installed, hooked up and tested prior to floating out and towing the structure to its offshore location. Installation then only requires ballasting the platform down onto the sea bed and consolidating the foundation below the platform by pumping grout into the spaces between the platform base and foundation.

The elimination of steel piling and of having a concrete structure tolerant to overloading and to degradation due to exposure to sea water offers other advantages over the use of steel. It has been demonstrated that concrete used in coastal installations in the 1930s and 1940s has survived up to the present day, essentially unaffected by exposure to sea

Table 1.2 Summary of concrete platforms (*cont.*) (from Furnes and Loset, 1981)

Type of design and location	*Approx. concrete volume (m^3)*	*Base diameter (m)*	*Storage capacity (million barrels)*	*Installation year*
1* Doris Ekofisk 1 (N)	90 000	92	1.0	1973
2* Condeep Beryl A (UK)	55 000	100	0.93	1975
3* Condeep Brent B (UK)	65 000	100	1.0	1975
4* Doris Frigg CDP 1 (UK)	60 000	101		1975
5* Sea Tank Brent C (UK)	105 000	100	0.65	1978
6* Sea Tank Frigg TP1 (UK)	70 000	72		1976
7* Sea Tank Cormorant A (UK)	115 000	100	1.0	1978
8* Condeep Brent D (UK)	65 000	100	1.0	1976
9* Andoc Dunlin A (UK)	89 000	104	0.85	1977
10* Condeep Statfjord A (N)	88 000	110	1.3	1977
11* Condeep Frigg TCP2 (N)	50 000	100		1977
12* Doris Frigg MP2 (UK)	60 000	101		1976
13* Doris Ninian (UK)	142 000	140		1978
14* Pub 3 Petrobras	15 000	52	0.125	1977
15* Pub 2 Petrobras	15 000	52	0.125	1978
16* Pag 3 Petrobras	15 000	52	0.125	1978
17* Condeep Statfjord B (N)	130 000	169	1.5	1981

* Supervised by Det Norske Veritas.

water, whereas conventional steel structures would have been susceptible to sea water corrosion and would require substantial levels of maintenance and protection. These advantages have to be set against the fact that concrete gravity structures are relatively expensive. They often actually employ a greater mass of steel in their reinforcing members than would be required by an equivalent steel frame jacket structure. Concrete gravity structures are likely to suffer from foundation settlement during their working lives which can reduce the air gap between mean water level and the underside of the structure. Another disadvantage with concrete gravity structures is that no feasible means of removing the structure have been defined at present.

Figure 1.11 shows a typical design for a Condeep concrete gravity structure which consists of a cellular concrete caisson or base with three

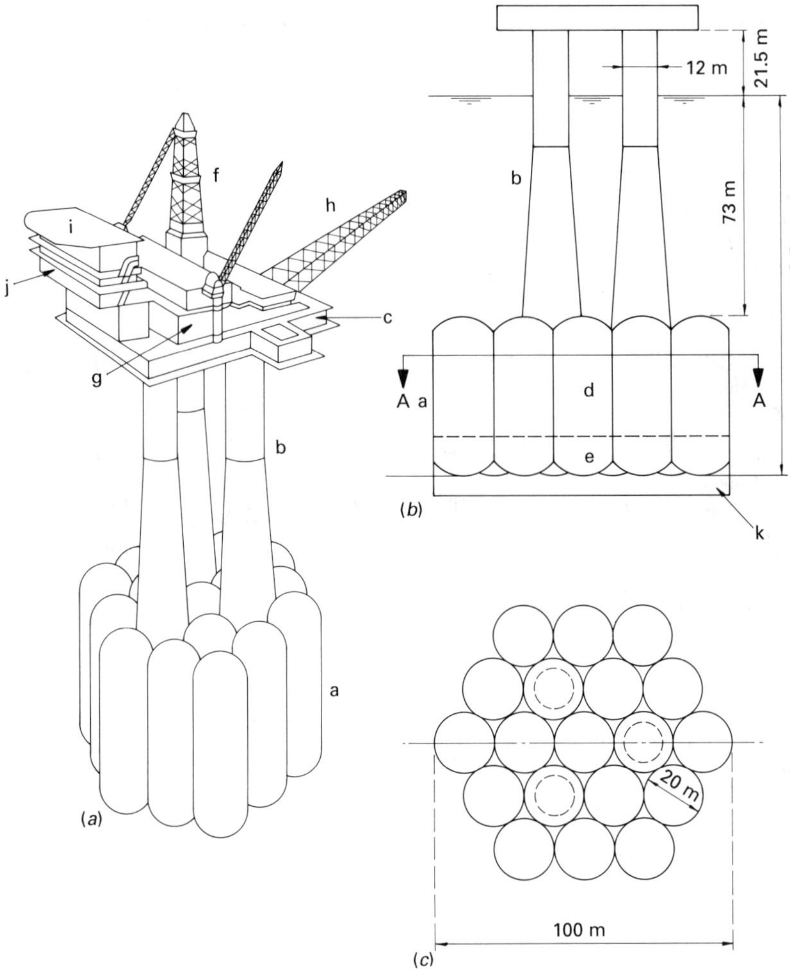

Figure 1.11. Typical Condeep concrete gravity structure: (*a*) perspective view; (*b*) side view; (*c*) section AA. *Key*: a – sea bed caisson; b – surface piercing columns; c – module support structure; d – oil storage; e – sand ballast; f – drilling derrick; g – drilling and production equipment; h – flare stack; i – helicopter pad; j – accomodation; k – steel skirt (from Furnes and Loset, 1980)

columns supporting a topside structure. The base of the platform contains steel skirts which penetrate a short distance into the foundation during installation of the platform. The concrete caisson at the platform base is used for oil storage. Figure 1.11 shows a Condeep design for the Beryl A platform which was installed in June 1975. This platform consists of a base caisson of 100 m (305 ft) maximum width with circular cellular oil storage tanks of 20 m (65.6 ft) diameter and water surface piercing support towers that are 94.5 m (310 ft) high. These towers are used for drilling and oil production conductors (pipelines) and for oil transfer piping. Some concrete production platforms are designed with transverse support frames between their vertical columns to support exposed conductor pipes passing through the base caisson into the sea bed.

Concrete gravity platforms are constructed by a technique known as slip-forming, which uses a structure surrounding the cross-section of the platform in plan view to contain concrete as it is poured and while it sets. As the concrete sets and more is poured, the slip-forming structure is raised and the platform is built or slip-formed from the ground up. Steel reinforcing members are set into the concrete and pre-stressed to give appropriate overall structural properties. Once the entire platform has been constructed, it is towed out to a deep water site, ballasted down, and the pre-prepared topsides are mated to the top. The topsides can be pre-tested and hooked up in the module fabrication yard, thus ensuring that very little offshore hook-up or testing is required. Once the topsides have been installed, the platform is de-ballasted up and towed out to the installation site where it is ballasted down on to the sea bed. Skirt piles of approximately 4–6 m (13–20 ft) height at the base of the platform cut into the foundation under the weight of the platform. The seal between the foundation and the platform is consolidated by pumping grout into the interface region.

There are many technical arguments for and against the selection of steel frame jacket or concrete gravity platforms for a particular location and oil field. Although concrete gravity platforms are expensive to build, they do offer the advantages of low maintenance and high deck payload growth capability during their lifetime. Nevertheless, the construction of new concrete gravity platforms has slowed down since the early 1980s and, compared to steel jacket structures, very few are being considered or are likely to be used for future developments.

The total cost of developing an offshore oil or gas field clearly plays a major part in the selection of the production platform. In the early growth years of the offshore industry when very large oil deposits were being found, the economics of each development permitted large expensive structures to be built. However, as more of the larger reservoirs have been discovered and developed, the remaining reservoirs have a higher probability of containing smaller oil deposits and being in deeper waters and harsher environments. At the same time, the cost of fixed steel frame jackets or gravity structures tends to increase in an approximately exponential fashion with water depth, primarily due to the larger platform structure and the greater volume of material required for structural integrity in deeper water. Figure 1.12 illustrates a typical variation of jacket weight with water depth for two locations – the Gulf of Mexico and offshore California as well as for the North Sea. The jacket weight on the vertical axis can be considered to be representative of the platform cost, whereas the curves denoting North Sea and Gulf of Mexico and California are illustrative of the influence of environmental conditions on jacket weight. The solid portions of the curves denote current developments and the dashed portions of the curves are estimated future projections using the same kind of technology. Figure 1.12 shows that the effect of more extreme environmental conditions for the North Sea in the 100–200 m (328–656 ft) water depth range increases jacket weight and thus cost, by a factor of between 3 and 6. The diagram also illustrates that a jacket weighing 30 000 tonnes in the Gulf of Mexico or offshore California could be used in up to 320 m (1050 ft) water depth, whereas that jacket weight would only suffice for approximately 175 m (574 ft) water depth in the North Sea.

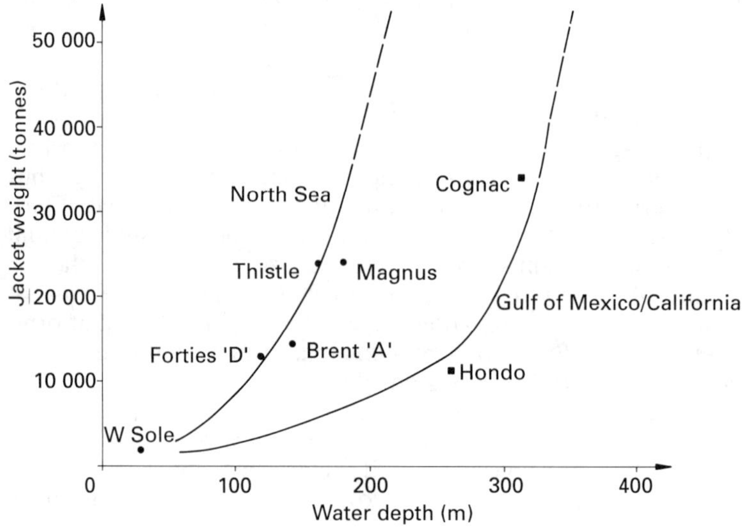

Figure 1.12. Variation of platform jacket weight with waterdepth (from Thornton, 1979)

It is clear that the approximately exponential increase of jacket weight, and hence cost, with increasing water depth cannot be sustained for the North Sea sector since new discoveries are tending to have smaller recoverable reserves and are, therefore, of marginal value economically. This has set the scene for the development of new technology in oil production to try to reduce the rate of increase of cost with water depth. The need to reduce costs has prompted feasibility studies on a wide variety of new types of production platforms based on compliant structures which are fixed to the sea bed but which respond with some degree of flexibility to the ocean environment. The consequent reduction in the volume of steel or concrete material that connects the surface facility to the sea bed can lead to a potential reduction in costs. This is illustrated by Figure 1.13 which presents the cost of design, fabrication and installation for different platform types plotted against water depth. The cost curve for a typical North Sea jacket is also plotted on the same axis for comparison, with the shaded areas around each line denoting the cost range that might be obtained with different design and installation options.

It can be seen that new platform concepts such as the tensioned bouyant platform TBB and semisubmersible or tanker based floating production systems – all of which are described later – offer significant reductions in the rate of increase of cost with water depth. It is now becoming increasingly obvious in the industry that up to the end of this

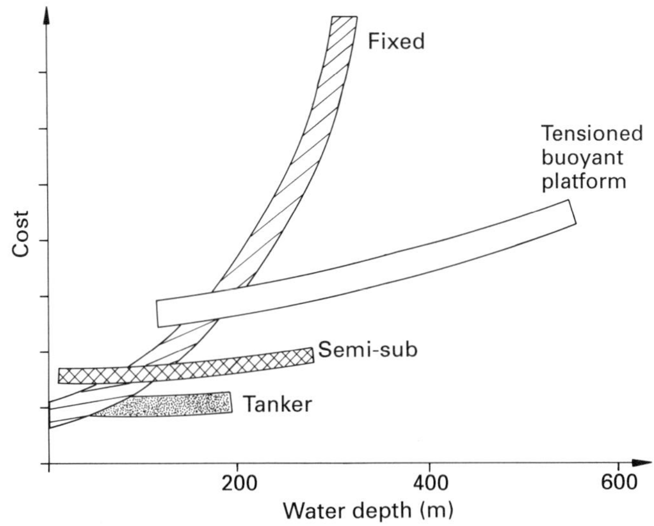

Figure 1.13. Cost comparisons for deep water structures (from Hamilton and Perrett, 1986)

century, oil production in deep water is likely to be carried out by compliant or floating production platforms in all cases except for the very largest of fields or for fields in shallow water. In deep waters above 457 m (1500 ft), a compliant or floating production system may in fact be the only means of producing a field. There are a number of existing and proposed alternatives for cheaper compliant or floating production systems. Table 1.3 gives a summary of developments in this area since the mid-1970s.

One of the most popular floating production platforms is shown in Figure 1.14 and is based around a converted or new build semisubmersible vessel which is catenary moored to the sea bed and is connected to a subsea manifold or well head either by rigid tensioned vertical multi-tube risers or flexible risers. Such systems are technically suitable for depth ranges of approximately 70–250 m (230–820 ft). They are, however, limited by a number of factors and technical risks. Semisubmersibles only have a relatively small deck payload capacity and thus the range of available oil process options is restricted compared to what is possible on fixed platforms. The compliancy of the platform and the catenary mooring system poses the risk of the vessel losing station and damaging its risers during severe environmental conditions or after damage to a mooring line. The platform cannot directly offer oil storage facilities and its heave, roll and pitch motions to waves can limit or degrade processing operations. Nevertheless, catenary moored semisubmersibles offer an economic means of exploiting small reservoirs which have sufficiently good hydrocarbon field characteristics as to require only a small amount of processing on the surface facility.

A technically more advanced form of compliant structure is offered by the tensioned or tethered buoyant platform, shown in Figure 1.15. The

Table 1.3 Floating production installations

Field	*Location*	*Water depth (m)*	*Start-up*		*Operator*	*Equip-ment*
Argyll	UK	76	June	1975	Hamilton Bros	Semi
Enchova	Brazil	122	August	1977	Petrobras	Semi
Castellon	Spain	119	August	1978	Shell	SALS
Dorado	Spain	94	May	1978	Eniepsa	Semi
Garoupa	Brazil	120	June	1979	Petrobras	Semi
Enchova	Brazil	189	December	1979	Chevron	Semi
Casablanca	Spain	134	December	1979	Chevron	Semi
Nilde	Sicily	96	April	1980	Agip	SALS
Buchan	UK	117	May	1981	BP	Semi
Hutton	UK	148	September	1984	Conoco	TBP
Duncan	UK	79	December	1984	Hamilton Bros	Semi
Balmoral	UK	128	June	1986	Sun Oil	Semi
Cyrus	UK	122	March	1988	BP	SWOPS

Notes:
Installations – Semi = semisubmersible surface platform.
SALS = single anchor leg system with tanker.
SWOPS = single well oil production system.
TBP = tensioned buoyant platform.

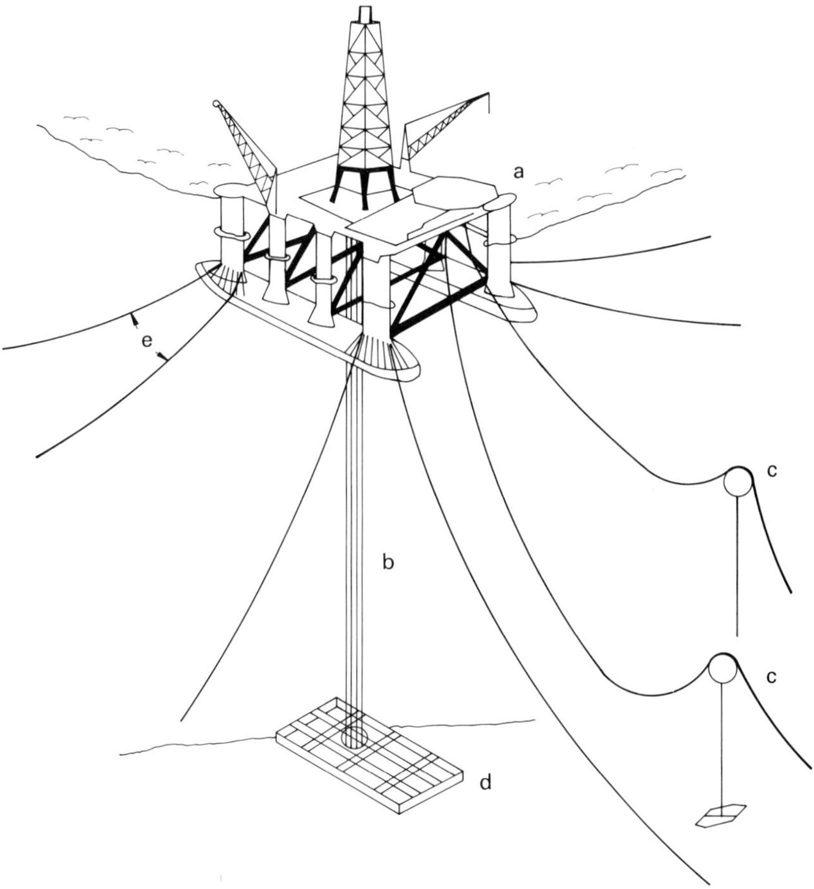

Figure 1.14. Typical semisubmersible based floating production system. *Key*: a – surface platform; b – multitube vertical drilling and production riser; c – flexible production risers; d – sea bed template; e – catenary moorings

surface platform configuration resembles that of a semisubmersible but is tethered to the sea bed by vertical legs, which are kept in tension by excess buoyancy in the platform. The excess buoyancy is of the order of 15–25% of the platform displacement and tensions the tethers to such an extent that heave, roll and pitch motions of the platform induced by ocean waves are virtually eliminated. The platform does, however, experience surge, sway and yaw due to wave action but the vertical tethers are a reliable means of keeping the platform permanently on station above subsea wells. These wells are connected to the surface facility by an array of risers. Oil processing is carried out on the platform and the hydrocarbon products are pumped back down to the sea bed to an export pipe line. In 1984, Conoco Ltd installed the world's first tethered buoyant platform in 148 m (485 ft) of water on the Hutton field in the North Sea. Due to the success of this development, it is likely that tethered buoyant platforms will be considered for a number

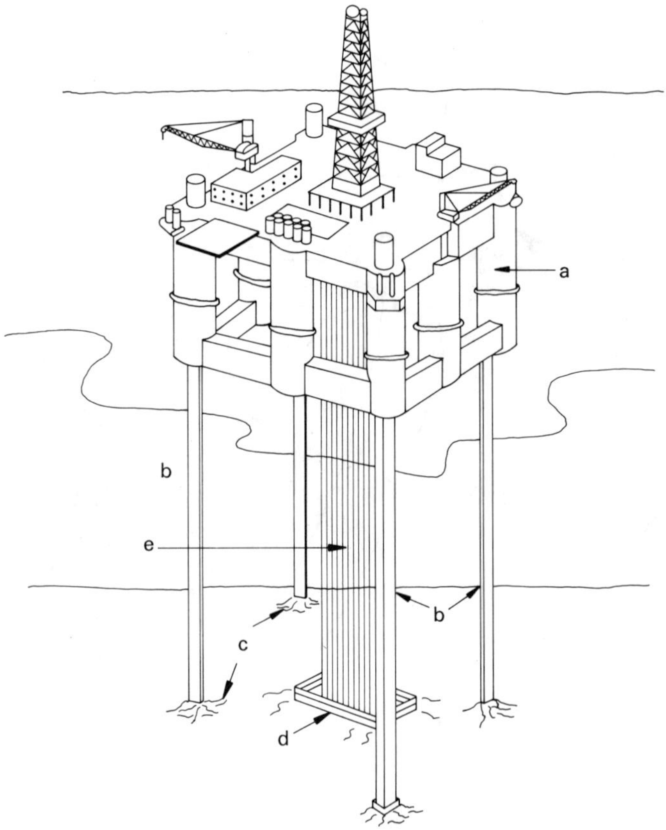

Figure 1.15. A typical tensioned buoyant platform. *Key*: a – surface platform; b – tensioned tethers; c – tether foundations; d – template on seabed; e – marine risers

of future field developments. The probable depth range of tensioned buoyant platforms is from approximately 120 m (394 ft) to 1500 m (4921 ft).

It is also possible to develop small isolated fields in water depths of up to 170 m (558 ft) in calm weather areas by the use of articulated column or single anchor leg storage and tanker systems. Figures 1.16 and 1.17 illustrate two such systems – the first of which is an articulated oil production and loading column. Crude oil is extracted from the reservoir and transferred up the articulated tower to the tethered tanker for processing and storage. The processed hydrocarbons can then be transported by a shuttle tanker mooring up to the production tanker shown in the diagram. An alternative scenario would entail the process equipment being mounted on a larger articulated column and the processed crude pumped to the tanker which shuttles to and from an onshore loading terminal. Figure 1.17 shows a single anchor leg system which uses a yoke structure, a buoyancy tank and a tensioned riser connection to the sea bed to moor a tanker. The tanker houses crude oil processing and storage facilities, with oil export again taking place by

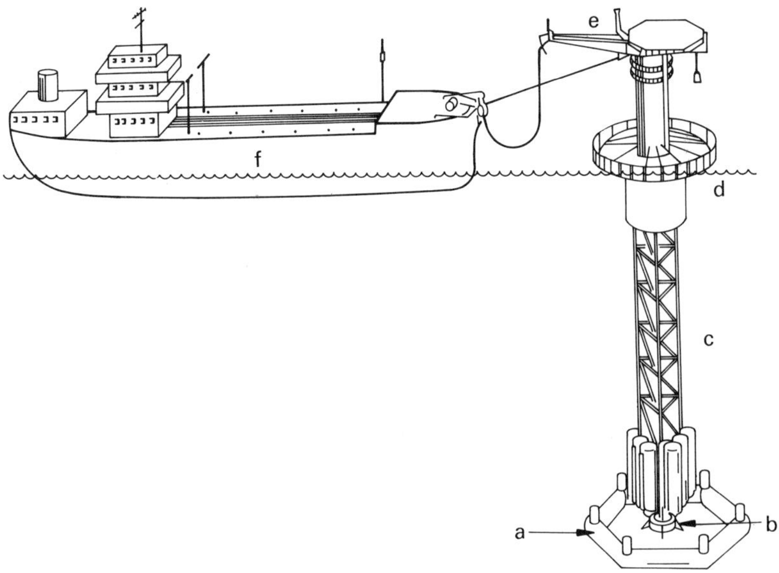

Figure 1.16. Articulated production and loading column. *Key*: a – gravity or piled base; b – universal joint; c – tower; d – buoyancy tanks; e – loading and mooring arm; f – tanker with production equipment and storage

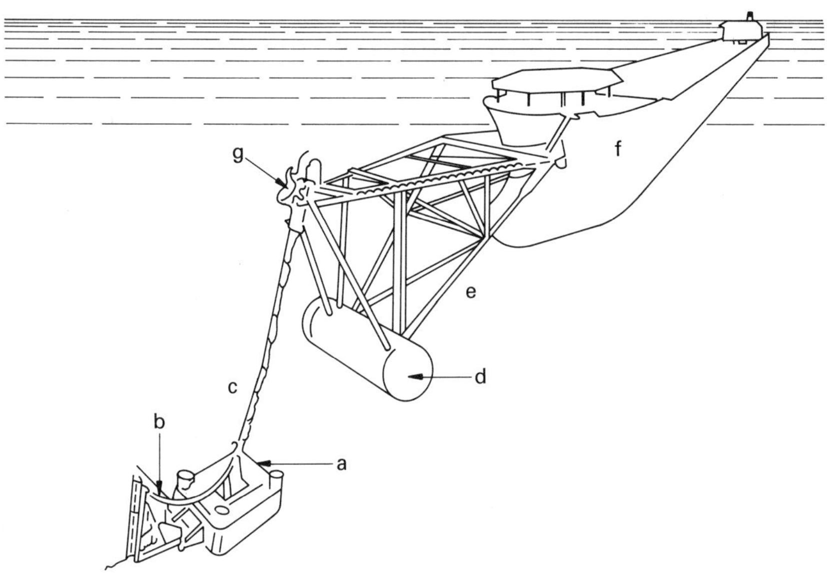

Figure 1.17. Single anchor leg production and storage system. *Key*: a – riser base; b – jumper hose; c – tension and production risers; d – buoyancy tank; e – yoke; f – tanker with production equipment and storage; g – universal joint

shuttle tanker or oil export pipeline. Table 1.3 shows that two systems of the single anchor leg type are in operation at present – the Castellon Field in offshore Spain and the Nilde Field in offshore Sicily.

Another form of economically viable compliant structure is offered by the guyed tower, a slender truss-steel structure, as shown in Figure 1.18, which is supported on the sea bed at its base by a spud can and foundation and held upright by multiple guy lines made up of wire ropes or chains. The guy lines are connected to anchor piles with heavy clump weights attached to the lines between the anchor and the tower to modify the horizontal stiffness of the tower–guy-line system. In normal or operating weather conditions, the guy lines restrain the platform and keep it upright without lifting the clump weights off the sea bed. In more extreme weather, the guy lines are designed such that the clump weights are lifted off the sea bed and a larger restoring force is available to the tower to resist increased wave forces. The guyed tower offers all the advantages of a fixed jacket in that there is a continuous structure between the sea bed and surface facility, but at significantly lower cost. One guyed tower was installed in 1983 in the Gulf of Mexico on the Lena Field. This tower has a steel weight of 24 100 tonnes, a topside weight of 19 640 tonnes and is installed in 305 m (1000 ft) of water. It utilizes galvanized steel spiral wound wire ropes as guy lines and 179-tonne anchor weights with 20 guy lines laid symmetrically around the platform.

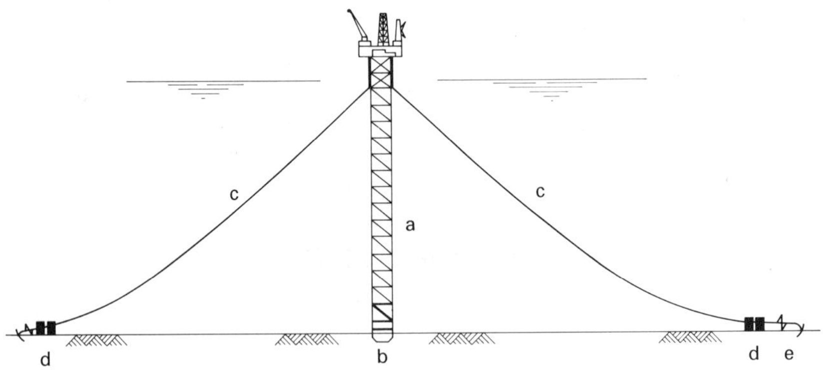

Figure 1.18. Guyed tower production platform. *Key*: a – slender tower; b – spud can foundation; c – mooring lines; d – sea bed clump weights; e – anchors

Apart from the floating production platform types that have already been installed, there are a large number of design concepts which have been evaluated to varying degrees of sophistication. As an example, Figure 1.19 shows one such concept – a concrete tripod-shaped gravity platform in the form of a single water surface piercing tower with a tripod substructure. This platform uses the tripod support structure to provide the necessary resistance to overturning without the cost of a massive gravity structure. Other options that have been proposed are shown in Figures 1.20 and 1.21. Figure 1.20 shows a concrete floating production platform, which uses an array of catenary anchor lines and clump weights for station keeping. Figure 1.21 shows the design of a concrete articulated tower, which consists of a cellular concrete base

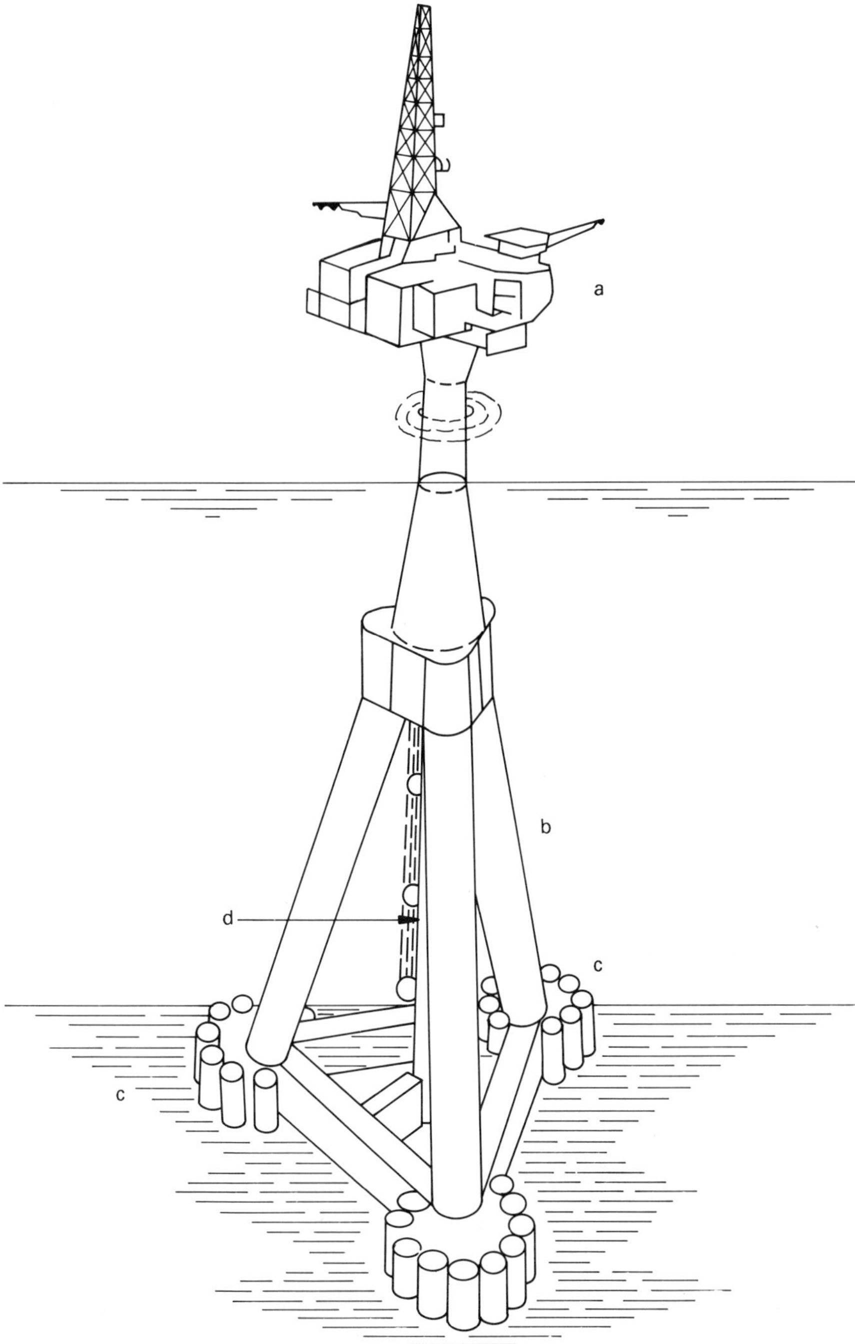

Figure 1.19. Concrete tripod structure. *Key:* a – surface platform; b – concrete tripod structure; c – gravity bases; d – drilling and production risers

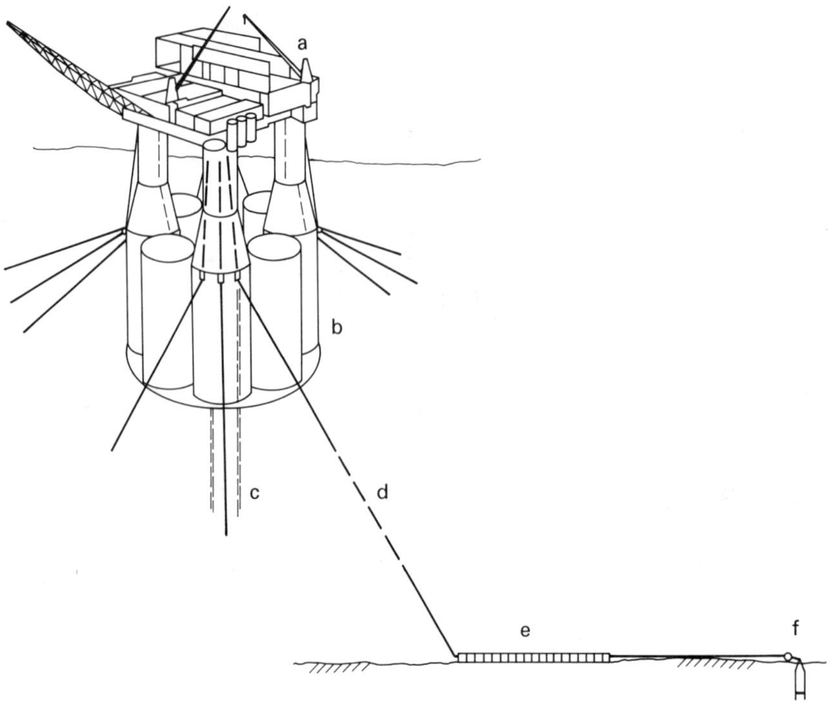

Figure 1.20. Concrete floating production platform. *Key*: a – production facilities; b – oil storage and floatation chambers; c – production risers; d – mooring lines; e – clump weights; f – anchors

piled to the sea bed. An articulated tower is sited above the concrete base, with a universal joint at the base of the tower. The tower, kept upright by excess buoyancy in the structure, can undergo small angular motions due to wave action in a compliant manner. It is probable that such an articulated tower can be used in a depth range of 100–250 m (328–820 ft). Many other novel production platform alternatives have been proposed in recent years and considered in various research papers.

The development of an offshore oil field also requires a variety of other floating vessels and engineering equipment. An example of such vessels is the exposed location single buoy mooring, the ELSBM, shown in Figure 1.22. This buoy enables shuttle tankers to moor up to it and to load crude oil, which is transported to the ELSBM from a nearby oil production platform. Exploration and development of an offshore oil field also requires a range of ship shape and semisubmersible vessels. These include survey vessels for the initial geophysical mapping, drilling vessels to carry out exploration drilling, supply boats to transport equipment and sometimes crew on and off drilling vessels. Once a production platform has been installed, or just prior to its installation, diver support vessels are required to carry out subsurface and sea bed work. During the operating life time of a platform or a group of platforms, maintenance and fire fighting vessels are required.

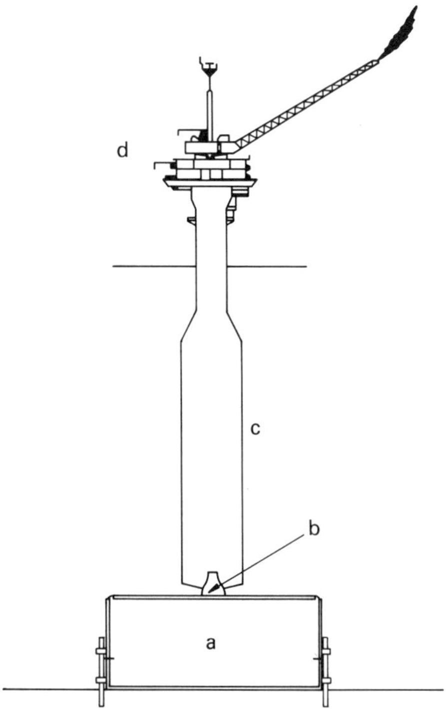

Figure 1.21. Concrete articulated tower. *Key*: a – piled oil storage base; b – universal joint; c – tower; d – production facilities

The installation phase of a conventional production platform requires crane vessels to install topside modules. Typical ship shape and semisubmersible crane vessels are shown in Figures 1.23 and 1.24. The design and operation of these specialized vessels raises a number of unusual engineering analysis requirements. Offshore oil fields also utilize a network of pipelines and control cables on the sea bed, either joining wells within a field or joining a field to an onshore crude oil reception facility. All these pipelines and control cables have to be installed, maintained and repaired. A fleet of semisubmersible and ship shape pipe laying vessels, survey vessels and associated pipe laying methods and repair techniques are part of the technology of offshore field development.

The design, installation and maintenance of subsea well heads, controls, manifolds and protection frames are carried out within a separate subject area called subsea engineering. Some subsea equipment presents difficult design problems since it operates in a harsh, high-pressure environment and is extremely difficult to access and repair. High levels of redundancy and well-thought-out maintenance and repair procedures have to be used with such equipment. Marine riser pipes that connect compliant platform surface facilities to subsea well heads offer another area of interesting analysis and physical problems. Such riser pipes carry crude oil between the well head and surface platform and yet have to accommodate the movement of the compliant platform relative to the sea bed. Under extreme environmental conditions, the riser may have to be disconnected to prevent its failure. The design of

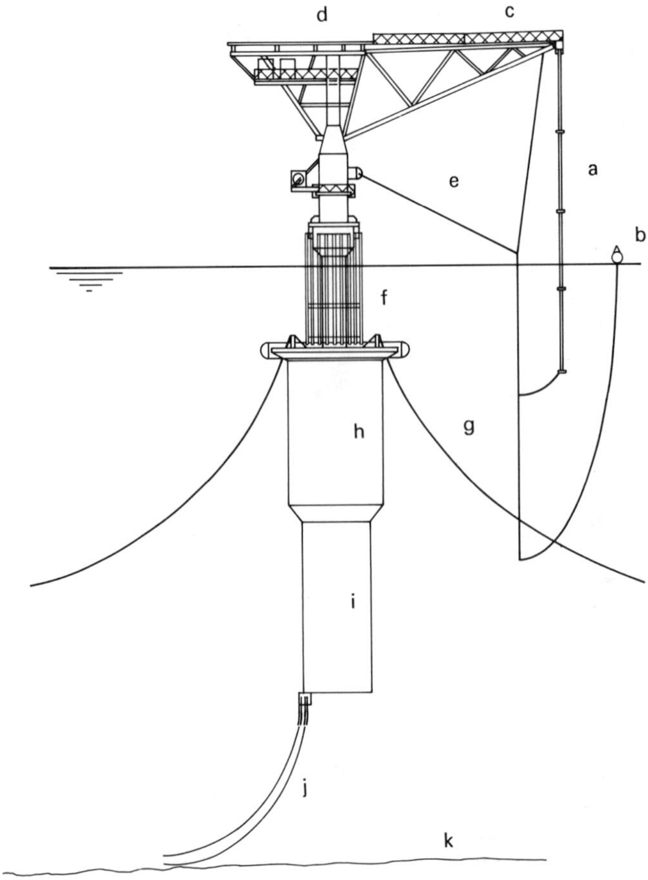

Figure 1.22. Typical exposed location single buoy mooring (ELSBM). *Key*: a – oil loading hose; b – mooring hawser and loading hose pick up buoy; c – revolving loading arm; d – helicopter pad; e – mooring hawser; f – fenders; g – buoy mooring chains; h – buoyancy compartments; i – ballast compartments; j – oil transfer hose

such marine risers requires application of the physics of fluid–structure interaction, which is itself not fully understood. The riser analyst and engineer are thus faced with technically challenging problems where faulty design decisions can have expensive consequences.

When equipment installed on the sea bed does fail or needs to be maintained, manned or unmanned intervention is necessary in order to replace components or carry out repairs. The offshore industry has stimulated development of a range of manned and unmanned vehicles for subsea intervention, survey and repair work. Figure 1.25 shows three examples of manned vehicles, ranging from one-atmosphere diving suits to manned submersibles. All these enable engineers to work underwater in a one-atmosphere environment. One-atmosphere diving suits give single operatives the ability to work close to subsea equipment using tools which are designed as extensions of the operator's arms. A submersible offers one or two men the possibility of carrying out work on subsea equipment or of surveying platforms.

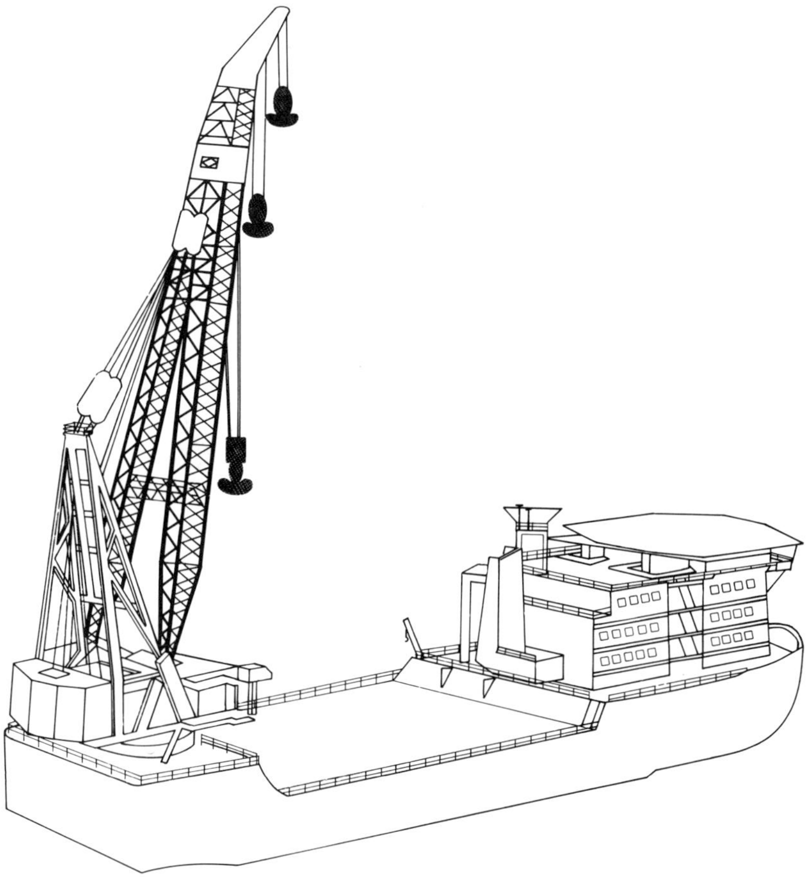

Figure 1.23. Ship shape crane vessel

There is also a demand for unmanned vehicles, which are primarily used to carry out routine survey and simple repair tasks. Figure 1.26 shows a schematic view of the equipment which may be used for such a vehicle and which consists of a control station, a winch, and an A-frame from which the underwater vehicle and its restraint frame are launched. The underwater vehicle operates at the end of a tethered cable with the operator on the surface controlling the vehicle and responding to television camera signals from the unmanned vehicle.

Despite the current emphasis on engineering and equipment related to the offshore oil and gas industry, future needs for exploitation of the oceans must also be borne in mind. The potential of renewable energy from oceans through devices that can collect energy from surface waves or that can make use of thermal gradients between the surface and subsurface layers of the oceans may be exploited in the future. Figures 1.27 show some proposed energy extraction devices. Figure 1.27(*a*) shows the cross-section of a tidal power station that utilizes a low head turbine to extract energy from water level differences between the open ocean and an inland basin separated by a tidal barrage. The

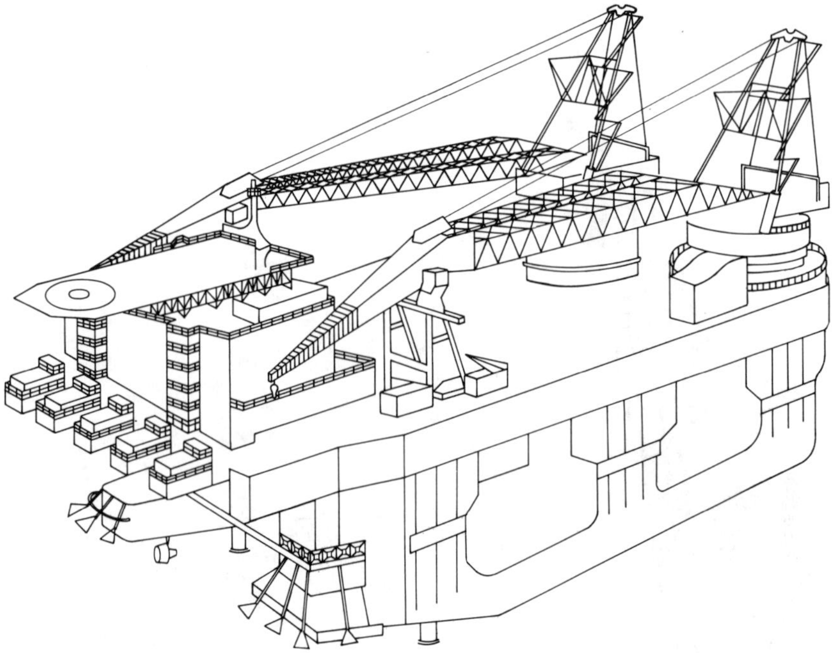

Figure 1.24. Semisubmersible crane vessel

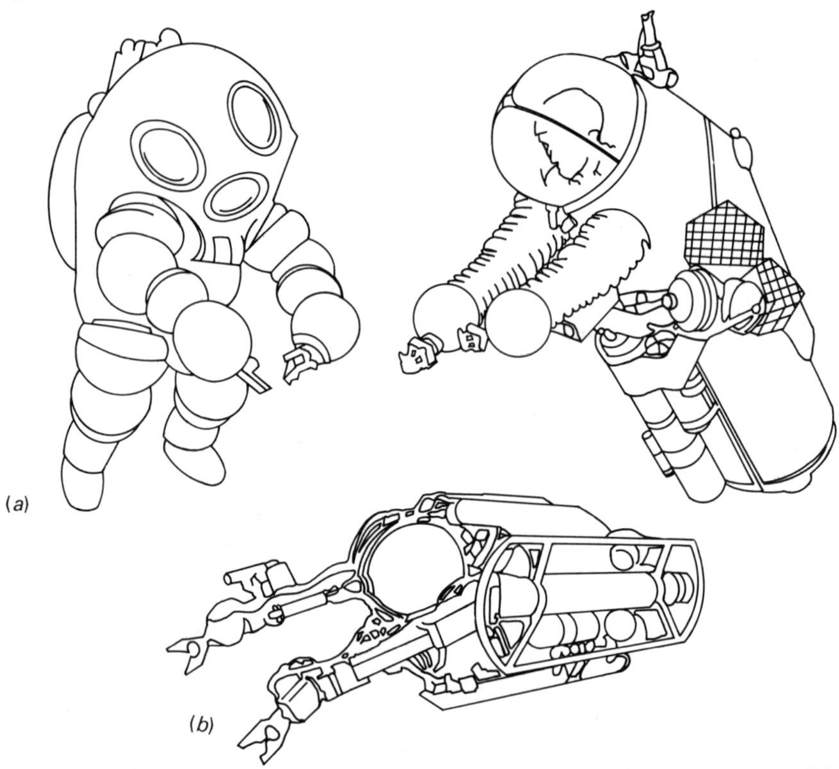

Figure 1.25. Typical 1-atmosphere submersible vehicles: (*a*) atmosphere diving suits; (*b*) manned submersible

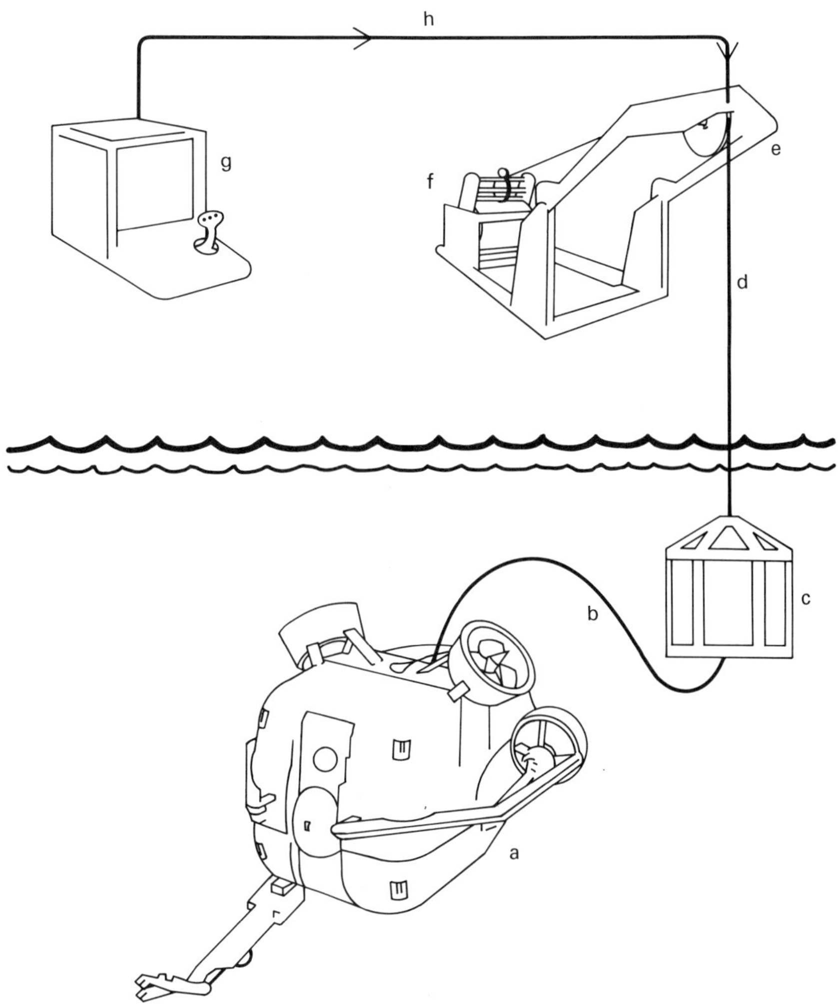

Figure 1.26. Unmanned underwater operations. *Key*: a – remotely operated vehicle; b – umbilical; c – launching restraint; d – lifting and control cable; e – launching A frame; f – winch; g – control station; h – control signals

mechanical energy extracted from the turbine can be converted into pneumatic or electrical energy for utilization. Figure 1.27(*b*) shows a rocker vane device that can extract energy from surface gravity waves by using vane motions to drive hydraulic pumps and electrical generators. An alternative means of extracting wave energy (not shown in Figure 1.27) uses a water column resonating in waves to drive air through a bidirectional turbine – the mechanical energy from the turbine then being converted to electrical energy. Wave energy from the oceans may also be collected by using a sea bed profile to refract the energy into a coastal location where the more concentrated energy is used to drive a localized wave energy device such as an oscillating water column. An ocean thermal energy convertor, as shown in Figure 1.27(*c*), uses a floating surface vessel from which is hung a long

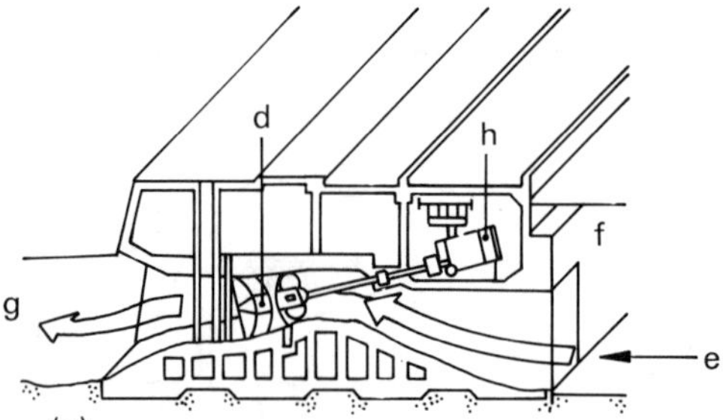

(a)

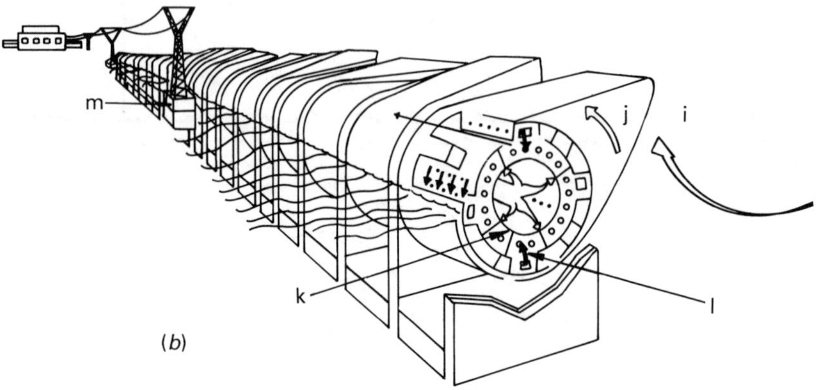

(b)

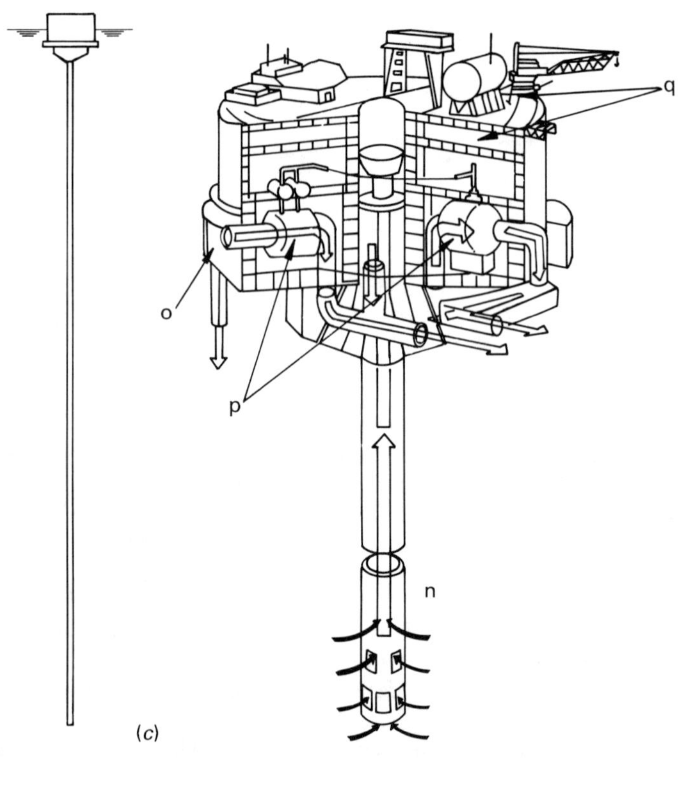

(c)

vertical pipe. The vertical pipe is used to collect cool subsurface ocean water, which is at a temperature of about 21.6 °C (39 °F) lower than that of the surface water. This cool water is brought up to the platform and its temperature difference with the surface water is used within a thermodynamic cycle to generate electrical power. An ocean thermal energy device is really an indirect method of utilizing solar energy collected by the surface water of the ocean. A prototype ocean thermal energy device rated at a power of 1.0 MW has been tested in the Pacific off Hawaii and designs exist for commercial plants delivering up to 40 MW of power.

It has also been proposed that the mining of minerals from nodules on the sea bed be carried out by large surface vessels connected to long slender riser pipes with bottom sea bed crawling vehicles to pick up and pump the nodules up to the surface.

It is difficult to predict exactly how the resources of the oceans will be exploited in the future, but the design, construction, installation, operation and maintenance of structures to carry out activities in the oceans will continue to provide challenging engineering problems. Such developments will need to be supported by a better physical understanding of the mechanics of the environment, aerodynamic and hydrodynamic loadings on structures, their consequent static and dynamic response, the material behaviour of the structure and a multitude of other aspects. This text book gives an introduction to some of the technologies that will need to be developed further, in order to design effective structures for working in the oceans.

References

American Petroleum Institute (1986), *Basic Petroleum Databook*, Volume VI, No 3, September, API, 1220 L Street NW, Washington DC 20005, USA.

Department of Energy (1986), *Development of the Oil and Gas Resources of the United Kingdom*, Appendix 15, Department of Energy, HMSO.

Furnes, O. and Loset, O. (1980), Shell structures in offshore platforms: design and application, *Engineering Structures*, **3**, July, 140–152

Hamilton, J. and Perrett, G. R. (1986), Deep water tension leg platform designs, *Proceedings of the Royal Institution of Naval Architects International Symposium on Developments in Deeper Waters*, 6–7 October, Paper number 10.

International Petroleum Encyclopaedia (1986), Pennwell Publications, 1421 S. Sheriden, PO Box 1260, Tulsa, OK 74101, USA.

Lee, G. C. (1980), Recent advances in design and construction of deep water platforms, Part I, *Ocean Industry*, November, pp. 71–80.

Meteorological Office (1986), *Meteorology for Mariners*, 3rd edn, London, HMSO, ISBN 011 400 311 4.

Thornton, D. (1979), A general review of future problems and their solution, *Proceedings of the Second International Conference on Behaviour of Offshore Structures*, 28–31 August, Paper 88, BHRA Fluid Engineering, Cranfield, Bedford, England.

Figure 1.27. Wave and thermal energy extraction devices: (*a*) tidal power extraction; (*b*) wave energy extraction; (*c*) thermal energy extraction. *Key*: d – low head turbine; e – water flow; f – open ocean; g – inland basin; h – power converter; i – wave action; j – rocker vane; k – central spine; l – hydraulic pumps; m – power converter; n – cold water pipe; o – warm water inlet; p – heat exchangers and generators; q – chemical process plant

Chapter 2

The ocean environment

The design and operation of offshore structures require a thorough understanding of the physical character and behaviour of the ocean environment and also of the atmosphere above it. The oceans of planet Earth form a continuous sheet of salt water that surrounds continental land masses and fills the great depressions on the Earth's surface. The oceans occupy more than 70% of the Earth's surface area and have a dominant influence on the global climate and on its plant, animal and human life. The oceans are also an essential part of man's environment, not only because they provide food and a means of communication, but especially because the oceans are the ultimate source of all water on the globe and supply the moisture which reaches the land masses as rain.

This chapter summarizes the major physical features of the oceans, the composition and properties of its constituent sea water and of air, and also presents an overview of most environmental phenomena in the atmosphere and oceans.

2.1 Physical structure

Planet Earth is a globe with equatorial and polar diameters of 12756 km (7926 mile) and 12714 km (7900 mile), respectively, and a total surface area of 509.859×10^6 km^2 (19.686×10^7 mile^2). A major proportion of this surface area is covered by three large ocean areas which in their order of size are the Pacific, Atlantic and Indian Oceans. These oceanic areas also extend into the waters off the Antarctic, although this region can be regarded as a distinct southernmost ocean. The Arctic ocean is another body of water of significant size but it is not comparable to the three major oceans of the globe.

Figure 1.1(*b*) gives a map of planet Earth and illustrates the extent of all the major and minor sea areas and continents. Table 2.1(a) presents a summary of the physical properties of the planet, whereas Table 2.1(b) gives an interesting summary of physical data for all the world's oceans and adjacent seas in terms of plan areas, volumes and average depths. It can be seen that the average depths of all the oceans, including adjacent seas, is 3795 m (12451 ft) and, therefore, on a global length scale the oceans are very widely spread but thin sheets of water with an average vertical dimension of about 4 km (2.49 mile) compared to the Earth's equatorial circumference of 40 074 km (24 901 mile).

Nevertheless, the depths of water in the world's oceans exhibit considerable variation ranging from the deepest point – Challenger Deep in the Marianas Trench of the Pacific Ocean at a depth of 11 033 m (17 756 ft) – to very shallow water areas bordering the continental land masses. This variation in water depth is illustrated by selected sea bed contours plotted on the world map of Figure 1.1(*b*). The ocean bottom has topographical features that are very similar to those found on land, as is illustrated by Figure 2.1 which shows a schematic of a typical cross-section through an ocean basin bounded by land masses. The areas of relatively shallow water at the edges of the land masses are called continental shelves. These areas are bordered on the seaward side by regions of relatively high slope, called continental slopes, which lead off to substantially deeper water. The global extent of these continental shelves and slopes are indicated by some of the water depth contours on the map shown in Figure 1.1(*b*).

There are a number of possible geological origins of the continental shelves – although most of these are related to local deposition of sediments coupled, in some cases, with a change in level of adjacent land masses. The sediments could have been deposited in the lee of large offshore islands or of submerged sea bed prominences. Some continental shelf areas have been built up from silt and debris deposited by large rivers running into the sea, whereas others could have been generated by sediments scoured out from bed rock by continental glaciers. Ocean currents also have a part to play in sediment transport and deposition for continental shelf formation. All these processes are conducive to trapping organic matter and, over geological timescales, to

Table 2.1 Physical characteristics of the Earth and its major oceans (from Sverdrup, Johnson and Fleming, 1942)

Polar radius = 6357 km (3950 mile)
Equatorial radius = 6378 km (3963 mile)
Equatorial circumference = 40 074 km (24 901 mile)
Mean radius = 6371 km (3959 mile)
Mass = 5980 × 10^{21} t (5.866 × 10^{21} tons)
1° of latitude at equator = 110.5 km (68.66 mile)
1° of arc at poles = 111.7 km (69.41 mile)
1° of longitude at equator = 111.3 km (66.16 mile)
Inclination of equator to ecliptic = 23°27′
Greatest height (Mt Everest) = 8848 m (29 028 ft)
Greatest depth (Challenger Deep, Marianas Trench) = 11 033 m (36 198 ft)
Land area = 148.8 × 10^6 km^2 (5.745 × 10^7 mile^2)
Ocean area = 361.059 × 10^6 km^2 (13.94 × 10^7 mile^2)

Principal elements in the Earth's crust (% by weight)
Oxygen 49.13%, Silicon 26.0%, Aluminium 7.45%, Iron 4.2%, Calcium 3.25%, Sodium 2.4%, Potassium 2.35%, Magnesium 2.35%, Hydrogen 1%, All others 1.87%.

Acceleration of gravity (m/s^2)
At any latitude λ (in degrees) and height h (in metres) above sea level

$$g = 9.806\,16 - 0.025\,928 \cos 2\lambda + 0.000\,069 \cos^2 2\lambda - 0.000\,003h$$

Table 2.1 Physical characteristics of the Earth and its major oceans (from Sverdrup, Johnson and Fleming, 1942)
(*b*) Statistical properties of the oceans

Body	*Area* (10^6 km^2)	*Volume* (10^6 km^3)	*Mean depth* (*m*)
Atlantic Ocean (excluding adjacent seas)	82.441	323.613	3926
Pacific Ocean (excluding adjacent seas)	165.246	707.555	4282
Indian Ocean (excluding adjacent seas)	73.443	291.030	3963
All Oceans (excluding adjacent seas)	321.130	1322.198	4117
Arctic Mediterranean	14.090	16.980	1205
American Mediterranean	4.319	9.573	2216
Mediterranean Sea and Black Sea	2.966	4.238	1429
Asiatic Mediterranean	8.143	9.873	1212
Large Mediterranean seas	29.518	40.664	1378
Baltic Sea	0.422	0.023	55
Hudson Bay	1.232	0.158	128
Red Sea	0.438	0.215	491
Persian Gulf	0.239	0.006	25
Small Mediterranean seas	2.331	0.402	172
All Mediterranean seas	31.849	41.006	1289
North Sea	0.575	0.054	94
English Channel	0.075	0.004	54
Irish Sea	0.103	0.006	60
Gulf of St Lawrence	0.238	0.030	127
Andaman Sea	0.798	0.694	870
Bering Sea	2.268	3.259	1437
Okhostsk Sea	1.528	1.279	838
Japan Sea	1.008	1.361	1350
East China Sea	1.249	0.235	188
Gulf of California	0.162	0.132	813
Bass Strait	0.075	0.005	70
Marginal seas	8.079	7.059	874
All adjacent seas	39.928	48.125	1205
Atlantic Ocean (including adjacent seas)	106.463	354.679	3332
Pacific Ocean (including adjacent seas)	179.679	723.699	4028
Indian Ocean (including adjacent seas)	74.917	291.945	3897
All Oceans (including adjacent seas)	361.059	1370.323	3795

form potential hydrocarbon reservoirs. Despite the continental shelves forming only about 8% of the total area of the oceans, these locations are primary sites for the development of offshore oil and gas because of their potential hydrocarbon deposits and their accessible shallow water depths.

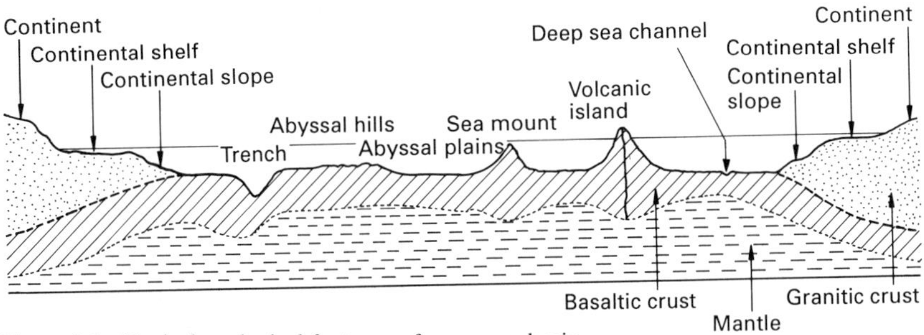

Figure 2.1. Typical geological features of an ocean basin

The continental shelves are bordered at their oceanic side by continental slopes, which have an average steepness of 4°, but this can rise to 45° at some locations. The continental slopes are found over a depth range of 200 m (656 ft) to 3500 m (11 483 ft), and cover approximately 14–17% of the total area of the oceans. Increasingly, the continental slopes are being considered as future locations for oil and gas production, with some exploration drilling presently being carried out to investigate the extent and types of hydrocarbon reservoirs that may be present. However, the continental slopes are likely to be more difficult to develop because of their combination of high water depth, high slope and strong currents, with most continental slope areas also being exposed to more severe weather conditions than adjacent continental shelves.

Figure 2.1 shows that continental shelves and slopes lead to the main ocean basins which have topographical features that include abyssal hills, abyssal plains, undersea mounts and volcanic islands. The mid-ocean basins also have deep trenches with depths that can be substantially greater than those of adjacent plains. Figure 2.1 and the depth contours of Figure 1.1(*b*) show that a substantial portion of the sea bed lies below depths of 3500 m (11 483 ft) and can be classified as a deep ocean floor. Most of the area of such deep ocean floors consists of abyssal plains, which are made up of broad flat areas of fine sediment. These are punctuated by a variety of sea bed features such as ridges, sea mounts and mountain chains. Abyssal hills are a widespread topographical feature of the sea bed and are found over large areas of the Pacific basin, but are less common in the other oceans. These abyssal hills are between 45 m (148 ft) and 900 m (2953 ft) high, with slopes that are relatively steep and often have very irregular local surface features. Mid-ocean basins are also punctuated by more pronounced steep sided sea mounts, which are nearly always volcanic in origin. These sea mounts can be divided into three types – the first which pierce the sea surface make up mid-ocean islands, such as those found in the Pacific in cases, such submerged mountains have flattened tops, hence they are called table mounts or guyots. Sea mounts have steep surfaces which are swept clear of fine sediment and are studded with regular outcroppings of volcanic rock or coral. Other generalized features of sea bed topography are mountain chains or ridges, such as those found in the Mid-Pacific or Mid-Altantic. These subsea features are often of very great length, with typical mountainous irregularities, steep slopes and hard rock outcrops. The largest water depths encountered in the oceans are obtained in trenches, which are narrow deep steep walled slits in the sea floor. The trench walls are made up of igneous rock and the trench bottoms appear to have a covering of sediment on a flat floor. These trenches are generally located in siesmically active areas in the Pacific Ocean.

All these varied sea bottom features can be quantified as a percentage distribution of depth in the oceans. Figure 2.2 presents land mass elevation above the sea surface and sea bed depth below it, plotted against the percentage of land elevation or ocean depth that is less than the indicated elevation or depth. A frequency or histogram distribution of land elevations and ocean depths is also given on the left-hand vertical axis. The diagram shows a highly variable distribution of land mass elevation and ocean depth, with a total range (from Mount Everest at 8848 m (29 028 ft) elevation to Challenger Deep at 11 033 m (36 198 ft) depth) of only 19.881 km (12.353 miles) compared to the Earth's equatorial diameter of 12 756 km (7926 miles). It is instructive to note that unlike the land masses, only the upper 15% or so of the ocean's depths are accessed by man on a routine basis, the rest remaining

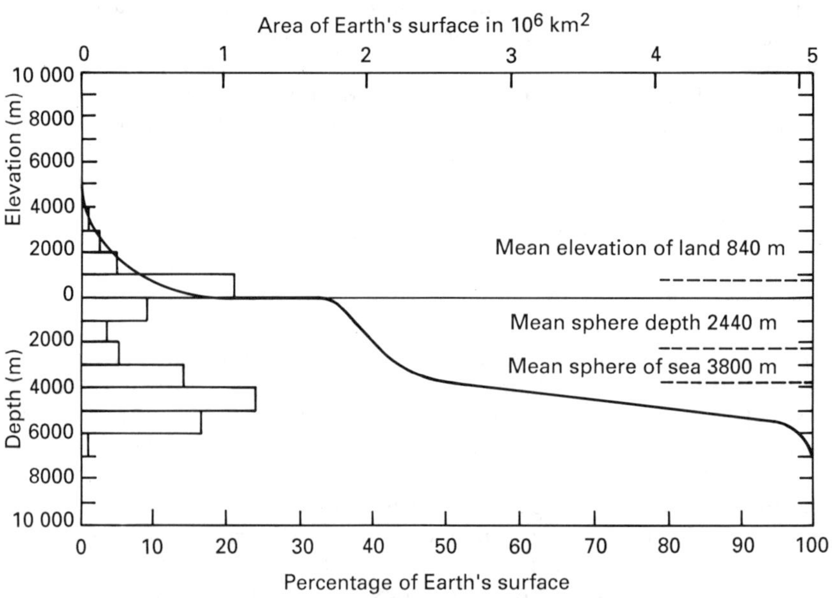

Figure 2.2. Distribution of land height and ocean depths on the Earth's surface (from Sverdrup, Johnson and Fleming, 1942)

largely unworked from the point of view of engineering exploitation and development. The search for hydrocarbons has currently reached water depths of 2 328 m (7 688 ft), although oil production is only being carried out up to water depths of just over 310 m (1020 ft). Engineering activities related to sea bed mining or manganese nodule harvesting are likely to be carried out in deeper waters than those utilized for oil exploration and production.

2.2 Composition and properties

A quantitative knowledge of the composition and physical properties of sea water and of the atmosphere above it is necessary for engineering analysis on offshore structures.

Figure 2.3 shows that 97% of the total Earth's water content is found in the oceans, with polar ice caps, snow fields and glaciers holding 2.25%, rivers, lakes, marshes and ground water taking up 0.75% and water vapour in the atmosphere being only a very small 0.001% of the total. Water (H_2O) is the normal oxide of hydrogen, which in its pure form is colourless, odourless and tasteless. It freezes at 0 °C, boils at 100 °C and has its maximum density at 4 °C.

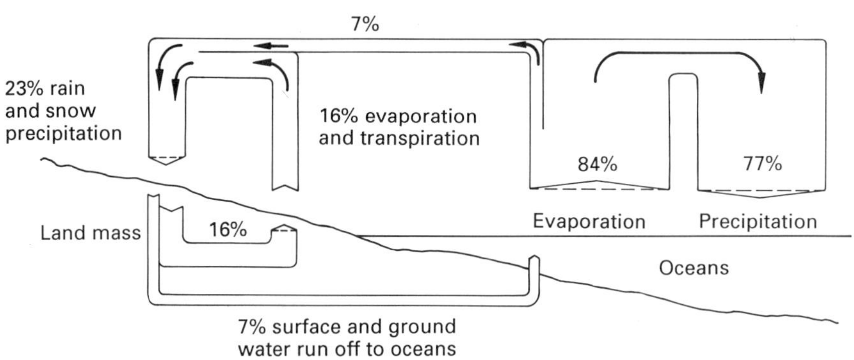

Figure 2.3. Hydrological cycle. Breakdown of the Earth's water content: oceans – 97%; ice caps, snow fields and glaciers – 2.25%; rivers, lakes, marshes and groundwater – 0.75%; water in atmosphere – 0.001%

Water is a remarkable substance for its high heat capacity, its exceptional chemical stability and strong solvent action, compared to many liquid compounds which are present naturally on Earth. Water is one of the few substances that is denser in the liquid than in the solid state and is also unique because in the temperature range 0–4 °C there is an increase in density before it begins to drop again for temperatures greater than 4 °C. Water owes most of these remarkable properties to the fact that it has a polar molecule, that is, the molecule possesses a dipole moment. Such moments arise in molecules where the centres of positive charge and negative charge are different – the centre of negative charge being associated with the more electro-negative element. Thus, one part of the molecule has a partial negative charge, whereas another part has a partial positive charge to form an electric dipole, the magnitude or dipole moment of which is given by the product of the magnitude of either charge and the effective distance between them. The separation of positive and negative charge in a polar molecule tends to give rise to greater forces of attraction between molecules. Therefore, since more energy is required to separate them, polar molecules have higher boiling points. Some of the other unusual properties of water are also due to the extensive hydrogen bonds between its molecules. Weak bonds between a hydrogen atom in one molecule and an atom in an adjacent molecule only form in certain circumstances, but have a profound influence on the bulk chemical properties of the substance. Hydrogen bonds between water molecules also contribute to its stronger solvent action.

The circulation of water between the oceans, the atmosphere and the Earth's land surface makes a predominant contribution to the climate and life supporting processes of the planet. This water exchange is called the hydrological cycle and is schematically illustrated in Figure 2.3. The relative quantities of water involved in various parts of the cycle are expressed as a percentage of the total exchange of water between the Earth's surface and the atmosphere, and are graphically illustrated by the thickness of the bars denoting water movement. Thus 84% of this total exchange evaporates from the oceans, with 77% returning to the oceans through precipitation. The remaining 7% is transported by wind systems on to the land masses, during which the water vapour content is increased by a further 16% given off by evaporation and vegetation transpiration from the land masses themselves. Thus 23% of the total water exchange then precipitates as rain and snow on the land masses, out of which 7% eventually returns to the oceans through surface and ground water run off and the remaining 16% participates in the hydrology of the land masses – through rivers, subsurface water, surface water, lakes and so on – and eventually returns to the atmosphere by direct evaporation or by transpiration from vegetation. This highly simplified explanation of the hydrological cycle masks the complexities of meteorology and geography through which evaporated water from the oceans sustains vegetation and animal life on the planet.

The oceans, atmosphere and land masses are also influenced by the reception, transmission and storage of solar radiation by the Earth's surface. The incoming radiation is of three types – visible light waves, ultra-violet waves of shorter wave length and infra-red rays which are of longer wave length than visible light. Figure 2.4 illustrates the flux of solar radiation into different parts of the Earth's surface layers and also shows how the flux of terrestrial radiation contributes to the total energy budget. The solar radiation is at short wave lengths in the range 0.2–4 μm, whereas terrestrial radiation has longer wave lengths in the range 4–100 μm. As shown by Figure 2.4(*a*), about 31% of all incident solar radiation is reflected back – 23% by clouds and the remaining 4% by reflection from within the atmosphere and 4% by reflection from the Earth's surface; 45% of the incident solar radiation is eventually absorbed by the surface and the remaining 24% is absorbed in the atmosphere and by clouds.

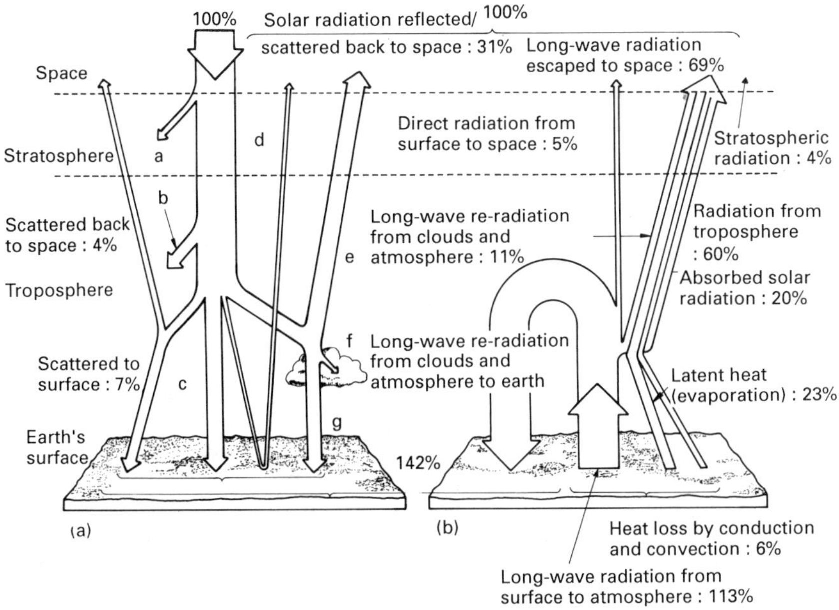

Figure 2.4. Energy flux at the Earth's surface (from Fuchs, 1985): (*a*) solar radiation absorbed by surface = 45%; (*b*) absorbed by surface = 97%. *Key*: a – absorbed 4%; b – absorbed 17%; c – direct to surface 24%; d – reflected from surface 4%; e – reflected by clouds 23%; f – absorbed by clouds 3%; g – reflected diffused by clouds 14%

The 45% of solar radiation absorbed by the surface is re-radiated as long waves (see Figure 2.4(*b*). This re-radiation occurs from a thermal energy cycle in which 142% of the incident solar energy flux is actually radiated out from the surface with 97% re-absorbed after reflection in clouds and from the atmosphere – the difference being the long-wave radiation. Furthermore, 24% of the incident solar energy flux is also re-radiated out directly from the atmosphere, making a total of 69% re-radiated into space. The net result of all of these energy exchanges is that 31% of incident solar radiation is reflected back as short waves, with the remaining 69% re-radiated back into space as long waves. There are variations with latitude within this global energy budget, with smaller levels of solar energy being received near the poles than at the equator. Consequently, latitudes between the poles and 40° north and south experience a net energy loss, while lower latitudes experience a net gain. This imbalance and the resultant transfer of energy from the equatorial region towards the poles drives winds in the planet's atmosphere.

The chemical composition of sea water can be regarded as a weak solution of a variety of salts, gases dissolved from the atmosphere and nutrient biological compounds, either in solution or in suspension. The salts found in sea water are due to the effect of the hydrological cycle over geological timescales in leaching out soluble compounds from land masses and depositing them into the sea, where they remain since they are not transported past the evaporation phase of the cycle. The consequent salinity of the oceans is defined as the proportion by weight of the major ions in parts per thousand

These salt constituents of sea water are listed in Table 2.2. Table 2.2(*a*) presents the composition of ions by weight in parts per thousand, showing clearly that the chlorine and sodium ions predominate, with the total salinity being 35.156 parts per thousand. Table 2.2(*b*) shows the composition of the salts by weight after evaporation. The sodium chloride and magnesium chloride salts form over 88% of the total salts found in sea water. Table 2.3, on the other hand, presents the percentage by volume and by weight of the gaseous compounds in clean dry air at sea level and at atmospheric pressure. The composition of the atmosphere is predominately nitrogen and oxygen in the ratio of 78 : 21 approximately by volume. There are significant levels of argon and carbon dioxide, but the remaining trace gases, neon, helium, methane, krypton and others are present only in very small quantities. This situation is mirrored in sea water, where up to 60 elements are known to occur in trace quantities, that is, in concentrations of less than one part per million. Although proportions such as these are very small, there is such a large amount of sea water that the total amount of trace elements is a significant quantity, although the trace elements would be difficult to extract economically.

Table 2.2 Chemical composition of salts in sea water
(*a*) Composition of ions (From ESDU, 1979)

Common name	*Chemical form in ocean*	*Concentration (g/kg)*
Chloride	Cl^-	19.344
Sodium	Na^+	10.773
Sulphate	$SO_4^=$	2.712
Magnesium	Mg^{++}	1.294
Calcium	Ca^{++}	0.412
Potassium	K^+	0.399
Bicarbonate	HCO_3^-	0.142
Bromide	Br^-	0.067
Strontium	Sr^+	0.008
Boron	H_3BO_3	0.004
Fluoride	F^-	0.0013

Table 2.2 Chemical composition of salts in sea water
(*b*) Composition of salts by weight after evaporation (from Myers *et al*., 1969)

Chemical compound	*Formula*	*%*	*Total salts (%)*	*Common name*
Sodium chloride	$NaCl$	27.21	77.74	Table salt
Magnesium chloride	$MgCl_2$	3.81	10.88	
Magnesium sulphate	$MgSO_4$	1.66	4.74	Epsom salts
Calcium sulphate	$CaSO_4$	1.26	3.60	Gypsum, anhydrite
Potassium sulphate	K_2SO_4	0.86	2.46	
Calcium carbonate	$CaCO_3$	0.12	0.34	Limestone calcite, marl chalk, etc.
Magnesium bromide	$NgBr_2$	0.08	0.23	

Table 2.3 Composition of clean dry air at sea level (from Howatson *et al.*, 1972)

Gas	*Percentage*	
	by volume	*by weight*
Nitrogen	78.08	75.52
Oxygen	20.95	23.14
Argon	0.93	1.28
Carbon dioxide	0.03	0.05
Neon	0.002	0.001
Helium, methane, krypton	0.000 8	0.000 47
Hydrogen, nitrous oxide and xenon	0.000 109	0.000 123

Notes:
Average molecular weight of dry air = 28.96.
Gas constant R = 0.2871 kJ/kg K.

Due to contact between the atmosphere and sea water at the sea surface, all gases present in the atmosphere are also found in solution in the oceans. Of these only nitrogen, oxygen and carbon dioxde are considered here since they are the predominant constituents of the atmosphere. These gases are absorbed by sea water in direct proportion to their partial pressures in the atmosphere and as a function of the absorption coefficient for each gas. The level of dissolved nitrogen in sea water varies from about 8 to 18 ml/l, to the extent that sea water can be regarded as being saturated with nitrogen at all depths. This nitrogen is not necessary for the majority of marine life but compounds of nitrogen, such as nitrates, formed mainly by decomposition of organic matter in the sea, provide an essential source of nutrient for microscopic plants and animals, which form the lower parts of the marine food chain.

Oxygen is essential for both plant and animal life in the sea. It dissolves readily in sea water and levels vary from 5 to 10 ml/l, although these concentrations can go much lower if oxygen is being extracted from sea water by marine organisms or during the occurrence of oxidizing processes such as decomposition. Furthermore, marine plants also release oxygen into sea water in large quantities through the action of photosynthesis. The only other major constituent of the atmosphere, carbon dioxide, can react chemically with the water itself and with some of the salts that are dissolved in it. Thus carbon dioxide dissolves in, and combines with, sea water in complex ways with its concentration in sea water being in a relatively narrow range of 35–60 ml/l. Carbon dioxide itself is, however, essential to life in the sea, since marine plants require it for photosynthesis. The solubility of atmospheric gases in sea water rises as their pressure increases and also as temperature and, to a lesser extent, salinity decrease.

The combination of marine life and dissolved gases in sea water leads to a number of chemical processes – three of which are described here. The first of these processes, photosynthesis, is one in which marine plants manufacture carbohydrates from dissolved carbon dioxide and water in the oceans through the action of sunlight and chlorophyll in plant cells. Photosynthesis produces oxygen as a by-product which enters into solution in sea water and may be utilized by the second process called respiration. During respiration, marine plants extract oxygen from sea water for use in their metabolic processes, and release carbon dioxide wastes. A balance between photosynthesis and respiration can sometimes be achieved. The third chemical process occurs during the decomposition of dead marine creatures during which simple carbohydrates react with oxygen (oxidize) to form carbon dioxide and water. This process has the opposite effect to that of photosynthesis in that oxygen is used up and carbon dioxide is liberated.

These three chemical processes can on occasion become imbalanced, where an excess of photosynthesis can lead to reduced carbon dioxide concentrations with increased oxygen concentrations, and vice versa for a preponderance of respiration. For ocean engineering purposes a predominance of respiration and decomposition can have a deleterious effect on ocean borne structures, since the loss of dissolved oxygen allows anaerobic chemical reactions to occur, hydrogen sulphide to evolve and the water to accumulate an excess of hydrogen ions. These conditions can lead to dramatically increased corrosion.

As explained in Chapter 3, both air and water can be treated as Newtonian fluids whose behaviour can be characterized by well-known sets of governing equations. The behaviour of air as a gas can be predicted by treating it as a perfect gas and using the appropriate gas laws. These prediction methods and the equations they utilize require accurate information on the physical properties of the fluids involved. The following section presents information on these physical properties.

The salinity and density of sea water are two of its principal properties. Figure 2.5 presents a carpet plot of sea water density as a function of temperature and salinity. It shows that for a sea water

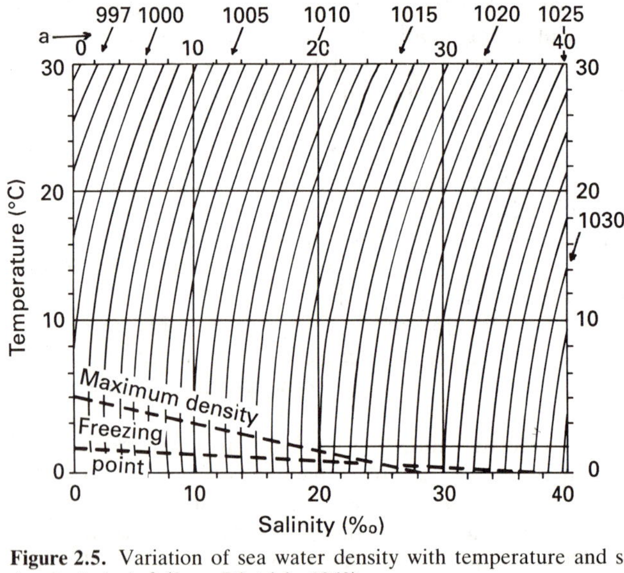

Figure 2.5. Variation of sea water density with temperature and salinity. Labels a denote density in kg/m³ (from Dietrich, 1963)

salinity of 35 parts per thousand and a temperature of 0 °C, the water density can be taken as 1028.1 kg/m^3. The plot also shows that the density varies strongly with salinity, but only weakly with temperature. Water samples from river mouths, estuaries and open oceans will have salinities varying from 0 to greater than 35 parts per thousand. The graph in Figure 2.5 can be used to derive sea water density if its salinity is known. Sea water salinity is also a very weak function of water depth, as illustrated in Figure 2.6. At the surface and up to a depth of 5500m (18 045 ft), the salinity varies by a small amount around an average of approximately 35 parts per thousand. The causes of this variation in salinity are primarily the interaction of fresh water discharges from major rivers and land water outlets, combined with current systems and the balance between evaporation and precipitation in particular ocean areas.

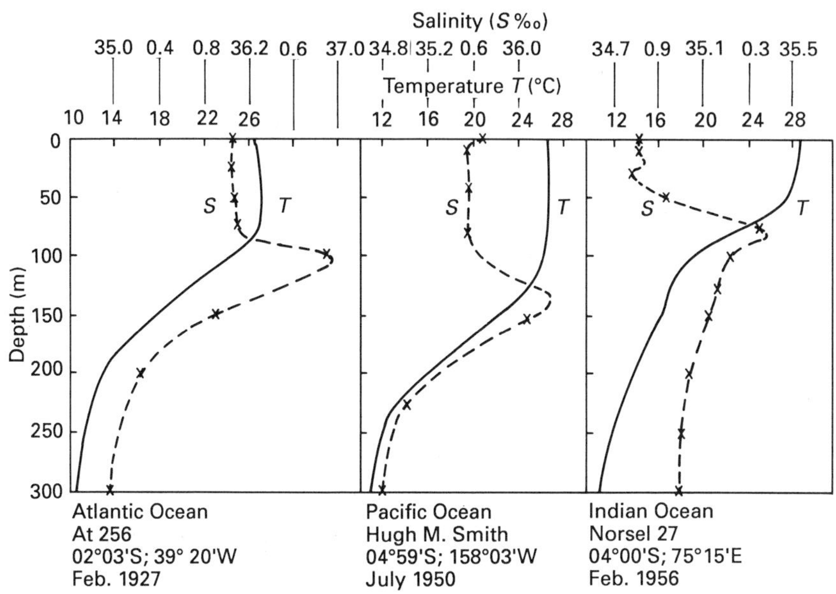

Figure 2.6. Variations of temperature and salinity with depth (from Neumann and Pierson, 1966)

Some remarks about the determination of salinity are worth making here. Salinity is defined as the total amount of dissolved solid material by weight in sea water and may be closely approximated by determining the concentration of one major ion and multiplying it by a constant. It is common practice to use the chloride and bromide ion contents for this purpose, with the bromide content usually converted to its chloride equivalent. The total chloride plus bromide equivalent contents are then known as the chlorinity. Salinity may also be determined by other means, some of which are associated with measuring electrical conductivity and relating it to the ion content and temperature of the sea water. Other techniques that are sometimes used to deduce salinity depend on the refractive index of sea water or its density.

Apart from its salinity, the variation of density, pressure and temperature with vertical distance from the sea bed up through the sea surface into the upper atmosphere, offer an interesting description of the physical properties of the oceans and the atmosphere above them. Figure 2.7(*a, b,* and *c*) presents these density, pressure and temperature profiles with vertical distance. The discontinuity in density and the related rate of pressure change discontinuity shown in Figure 2.7(*a*) and (*b*) are fundamental to many of the physical processes that induce environmental forces on offshore structures. Figure 2.7(*a*) shows that the density of sea water changes only very slightly with depth, although the density of air does change significantly with height above the sea surface. Nevertheless, the density at the free surface between air and sea water changes by a factor of 837. This has an impact on the variation of pressure with vertical distance, illustrated in Figure 2.7(*b*). The total mass of air above the sea surface exerts a pressure of 1.013 25 bar (14.696 lbf/in^2), whereas the much higher density of sea water ensures that an increment of pressure of 1.013 25 bar is achieved for every 10.077 m (33.06 ft) of sea water height. This implies that for a depth of 10 000 m (32 808 ft), a total ambient pressure of 994 atmospheres exists. These enormous hydrostatic pressures with depth below the sea surface have a strong impact on the ability of man made structures and of man himself to work below the ocean surface.

Figure 2.7(*c*) presents a profile of temperature with vertical distance above the sea surface. The temperature below the sea surface has a relatively small range from −1.8 °C to 35 °C, with deep water in the ocean being at a constant cold temperature. The bulk of the oceans forms a thermally strongly stratified structure, with the colder denser water forming the vast bulk of the oceans and a thinner layer of warmer and less dense water at the surface. In polar regions of the Earth, subsurface and surface waters are at nearly the same temperature and form a virtually isothermal body of water. In temperate and tropical waters, however, warm surface waters are separated from cooler subsurface layers by a temperature gradient or thermocline, which can vary in depth both through the annual seasons and daily. Thus seasonal thermoclines can build up each summer and be destroyed by cold winds during the following winter. Furthermore, because of the large heat capacity of water, the surface water temperature lags the seasons, with the surface temperature usually coldest in early March and warmest in August in the northern hemisphere. In temperate and tropical latitudes the thermocline can extend down to depths of 1000 m (3280 ft) or more. Figure 2.7(*c*) shows that the temperature of the air with vertical distance above the sea surface varies much more than over the same vertical distance within the sea. Temperatures in the atmosphere vary over a range of 65 °C compared to 31.8 °C for the oceans. This is consistent with the fact that the large thermal capacity of the oceans enables them to behave like a large heat store and regulate the seasonal heating of the Earth by the Sun.

The variations of density, pressure and temperature of air plotted in Figure 2.7 are taken from a definition of the International Standard

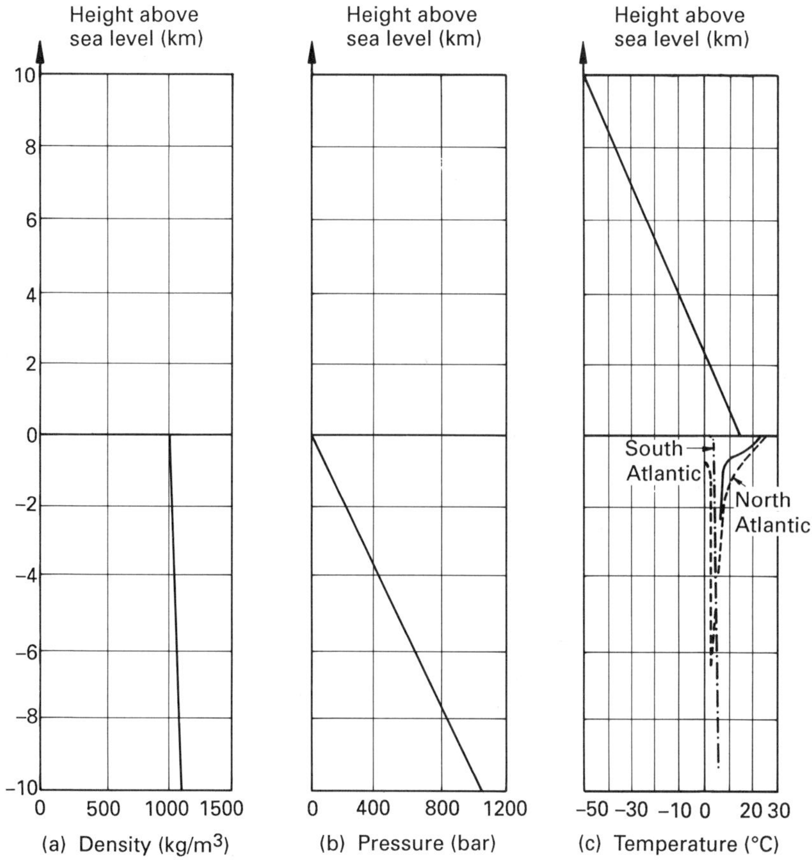

Figure 2.7. Variations of air and seawater properties with vertical distance above sea level: (*a*) density (kg/m^3); (*b*) pressure (bar); (*c*) temperature (°C)

Atmosphere. These variations in relative terms are not clear in the graphs of Figure 2.7, due to the disparity in scales required for air and sea water properties. Thus Table 2.4 lists the properties of the International Standard Atmosphere for completeness. The negative values of distance above sea level presented in this table apply to areas of the world where a local land mass actually lies below the Earth's mean sea level.

In order to calculate the effects of both air and sea water flows, it is necessary to have some knowledge of their viscous, thermal and elastic properties. In general, these properties are not constant and vary with fluid temperature. Figure 2.8 presents the variation of density, dynamic viscosity and kinematic viscosity of sea water over a temperature range of 0–50 °C. The variation of density with temperature is not significant since the scale used is very large, with density change of only 0.67% over a temperature range of 0–30 °C. The dynamic and kinematic viscosities, however, change significantly with temperature. The kinematic viscosity is of particular interest, since it commonly occurs in the

Table 2.4 Properties of the International Standard Atmosphere (from Howatson *et al.*, 1972)

Distance above sea level (km)	*Absolute pressure (bar)*	*Absolute temp. (K)*	*Density ratio (r)*	*Kinematic viscosity (mm^2/s)*	*Thermal conductivity (W/m K)*	*Mean free path (μm)*	*Velocity of sound (m/s)*
− 2.5	1.352	304.4	1.263	12.07	0.0266	0.053	349.8
− 2.0	1.278	301.2	1.207	12.53	0.0264	0.055	347.9
− 1.5	1.207	297.9	1.152	13.01	0.0261	0.058	346.0
− 1.0	1.139	294.7	1.100	13.52	0.0259	0.060	344.1
− 0.5	1.075	291.4	1.049	14.05	0.0256	0.063	342.2
0	1.013	288.2	1.000	14.61	0.0253	0.066	340.3
0.5	0.955	284.9	0.953	15.20	0.0251	0.070	338.4
1.0	0.899	281.7	0.907	15.81	0.0248	0.073	336.4
1.5	0.846	278.4	0.864	16.46	0.0246	0.077	334.5
2.0	0.795	275.2	0.822	17.15	0.0243	0.081	332.5
2.5	0.747	271.9	0.781	17.87	0.0241	0.085	330.6
3.0	0.701	268.7	0.742	18.63	0.0238	0.089	328.6
3.5	0.658	265.4	0.705	19.43	0.0235	0.094	326.6
4.0	0.617	262.2	0.669	20.28	0.0233	0.099	324.6
4.5	0.578	258.9	0.634	21.17	0.0230	0.105	322.6
5.0	0.540	255.7	0.601	22.11	0.0228	0.110	320.5
6.0	0.472	249.2	0.539	24.16	0.0222	0.123	316.5
7.0	0.411	242.7	0.482	26.46	0.0217	0.138	312.3
8.0	0.357	236.2	0.429	29.04	0.0212	0.155	308.1
9.0	0.308	229.7	0.381	31.96	0.0206	0.174	303.8
10.0	0.265	223.3	0.338	35.25	0.0201	0.196	299.5

Notes:
r = density at height given to density at sea level ρ_0.
ρ_0 = 1.225 kg/m³.
For air at 0 °C, specific heat capacity at constant pressure = 1.004 kJ/kg K and Prandtl number = 0.72.

equations describing fluid motion. Figure 2.8 demonstrates that the kinematic viscosity drops by 54% over the temperatures of 0–30 °C. This change in kinematic viscosity with temperature should be accounted for in fluid motion calculations. Table 2.5 presents a comparison of some of the physical properties of pure water, sea water and air at atmospheric pressure for temperatures of 0 °C and 20 °C. The dynamic and kinematic viscosities, the thermal conductivity, Prandtl number, bulk modulus, specific heat capacity and density are given. It is instructive to note that the kinematic and dynamic viscosities, thermal conductivity and Prandtl number do not vary significantly between pure water and sea water, although the bulk modulus of sea water is approximately 9% greater than that of pure water at both 0 °C and 20 °C.

A comparison of these physical properties of air at atmospheric pressure with either pure or sea water shows a number of differences. The dynamic viscosity of air is substantially smaller than that of water, although the kinematic viscosities of both are roughly equivalent due to the division by densities in the transformation from dynamic to kinematic viscosity. The relative thermal properties of air and water are illustrated by values of thermal conductivity, specific heat capacity and Prandtl number presented in Table 2.5. The thermal conductivity and

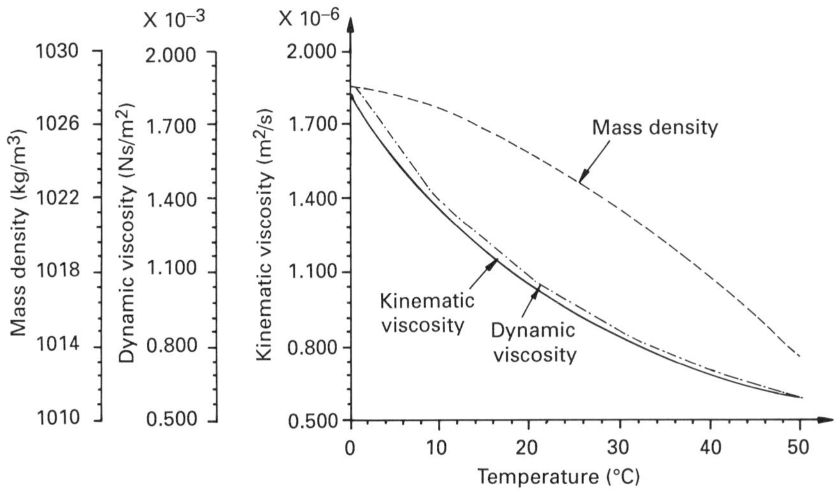

Figure 2.8. Variation of the physical properties of seawater with temperature

specific heat capacity of water are substantially higher than for air, indicating that a water mass is a good conductor of heat when compared to air and also has the capacity to store considerably more heat. This relative property has a crucial impact on diving practice and on the design of manned submersible craft. It also leads to the fact that the survival time of a human being immersed in cold sea water without a protective suit is very short indeed. The equivalent bulk modulus of air at atmospheric pressure is also very small when compared to that of either pure or sea water. In these terms, water can be regarded as virtually incompressible in comparison with air. All the physical properties of pure water, sea water and air presented in Table 2.5 do vary significantly with temperature. However, in general, it is sufficiently accurate for engineering purposes to obtain values of these parameters for a temperature range of 0–30 °C by linear interpolation from the values presented in Table 2.5.

The velocity of propagation and the attenuation of sound waves, light and electromagnetic radiation through sea water is also of interest to engineering activities in the oceans. Sound waves are most commonly utilized in offshore engineering for echo sounding, surveying, acoustic positioning systems and for communication purposes. The properties of visible light transmission in sea water are crucial to subsea tasks, where visual information is necessary – either through the eyes of a manual operative or for remote imaging through a television camera. The very high attenuation of electromagnetic radiation at usable frequencies through sea water prevents it from being used for communication and detection purposes, as it can be in the atmosphere. The smaller attenuation of very low frequency electromagnetic radiation in sea water does permit it to be used for long distance communication with submarines.

The speed of sound in sea water of salinity 35 parts per thousand at atmospheric pressure (1.013 bar) and temperature 0 °C, is 1449.2 m/s. In

Table 2.5 Physical properties of pure water, sea water and air

Quantity and units	*Pure water*		*Sea water (Salinity 35‰)*		*Air at standard temperature and pressure*
	0 °C	*20 °C*	*0 °C*	*20 °C*	*0 °C and 1.013 bar*
Dynamic viscosity (kg/m s)	1.793×10^{-3}	1.002×10^{-3}	1.877×10^{-3}	1.072×10^{-3}	1.709×10^{-5}
Thermal conductivity (W/m K)	0.561	0.599	0.566	0.600	0.0241
Kinematic viscosity (m^2/s)	1.793×10^{-6}	1.004×10^{-6}	1.826×10^{-6}	1.046×10^{-6}	1.322×10^{-5}
Prandtl number	13.2	6.94	13.2	7.13	0.71
Bulk modulus (GN/m^2)	1.993	2.190	2.172	2.388	0.000 101 3
Specific heat capacity (kJ/kg K)	4.218	4.182	3.992	3.993	1.004*
Density (kg/m^3)	999.839	998.205	1028.1	1024.5	1.293

* At constant pressure.

sea water of constant properties, the speed of sound is independent of frequency. The speed of sound does, however, exhibit variations with sea water temperature, pressure and salinity. It increases with increasing hydrostatic pressure, temperature and salinity and is reduced by the presence of bubbles or of excessive concentrations of dissolved gas in the sea. These variations are quantitifed in Figure 2.9. Since hydrostatic pressure and temperature vary much more strongly in the vertical direction, the speed of sound also varies with vertical distance. However, horizontal variations in temperature and salinity can also occur, and horizontal variations of sound speed mirror these changes. The speed of sound is most sensitive to changes in temperature and thus the variation of sound speed with vertical direction, in the upper regions of the ocean, is largely determined by the temperature gradient in these regions. Below the ocean's surface layers, temperature and salinity variations are small, but increasing hydrostatic pressure then becomes the dominant factor in modifying the speed of sound. The sound speed variations with temperature, pressure and salinity presented in Figure 2.9 are for constant sea water properties. However, the speed of sound and its direction of propagation in a large volume of ocean water is strongly influenced by the vertical stratification of sea water properties and by the variation of these properties with horizontal distance. The determination of an accurate value of the average speed and direction of a sound wave in the open ocean is, therefore, quite difficult.

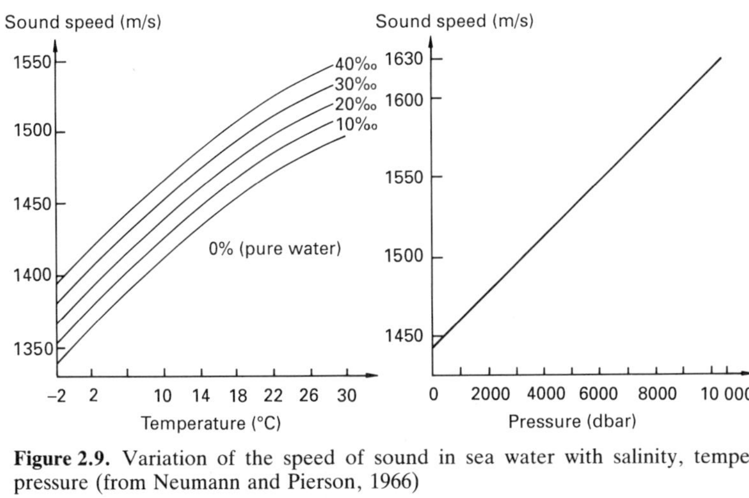

Figure 2.9. Variation of the speed of sound in sea water with salinity, temperature and pressure (from Neumann and Pierson, 1966)

The attenuation or absorption of sound waves during transmission through sea water is relatively low and is strongly dependent on frequency and weakly dependent on salinity, temperature and pressure. Figure 2.10 illustrates the variation of attenuation with frequency at four temperatures with the sea water at atmospheric pressure and with a salinity of 35 parts per thousand. Note that the attenuation in decibels per kilometre is very small for frequencies below 100 kHz. The physical mechanism by which sound is absorbed in sea water tends to be viscous in character for frequencies above 1 MHz, whereas losses below frequencies of 100 kHz are governed by a chemical relaxation process primarily of dissolved magnesium sulphate salts.

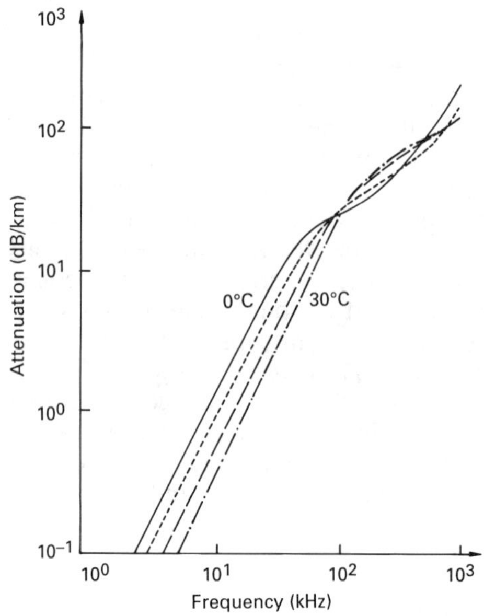

Figure 2.10. Attenuation of sound waves in sea water as a function of frequency (atmospheric pressure and salinity of 35‰) (from Myers *et al.*, 1969)

The transmission of sound waves through sea water is complicated by refraction induced by changes in speed of sound due to pressure, temperature and salinity, as well as by reflections of sound waves from the water surface and sea bed. Reflections from man made or natural free swimming objects also play a part in sound transmission in the oceans. These natural sources of reflections and refractions impart complex characteristics to the sound transmission between two points in the oceans. Nevertheless sound waves (sonar) are used for a variety of detection, communication and measurement tasks in offshore engineering.

The detection of unknown objects by underwater sound depends on a sound projector first emitting a sound pulse and then being able to distinguish the echo returning to the receiver by reflection from the unknown object. A substantial body of literature exists on the use of sound through sonar devices to detect submarines and fish shoals or to characterize sea bed topography, to map the geology of rock layers below the sea bed and also to study a variety of marine species, which emit acoustic detection or communication signals over the entire frequency range between 10 Hz and 100 kHz. Readers are referred to Tucker and Gazey (1966) and Camp (1970) for further information on the applications of acoustics in the oceans.

Visible light has a very limited range in sea water and is attenuated by a combination of scattering and absorption. The scattering loss is primarily due to suspended particles in sea water deflecting rays of light from their normal straight paths. This scattering leads to decreased light transmission because of the background blurring effects of back scattering and the image distortion of forward scattering. Light is also absorbed by sea water in three ways – absorption by molecules of water, by coloured suspended matter, and by coloured dissolved matter. This absorption varies with the wave length of light and it is this selective absorption which gives water its characteristic colour. Absorption of light by water molecules is due to the atomic structure of the water itself with absorption being greatest at the red end of the spectrum. Suspended sediment particles in water also attenuate light, both by scattering, as described above, and by selective absorption. Dissolved organic matter makes a contribution to this kind of absorption and, together with scattering and absorption by suspended matter, can give certain types of water a characteristic colour.

The relative level of absorption by water molecules, and by suspended or dissolved matter is a function of the source of the sample of water. Figure 2.11 presents variation of an extinction coefficient with wave length of light in samples of pure water, water taken from an open ocean location, and water from a coastal location. In all three cases, there is a definite minimum in the extinction coefficient for wave lengths around the blue and green part of the spectrum. The increase of attenuation with decreasing wave length below this blue–green minimum is caused by scattering from small particles, that is, particles having dimensions which are small in comparison to the wave length. The increase in attenuation with increasing wave length to the right-hand side of the minimum is caused by molecular absorption. It is interesting to note that the wave length region of minimum attenuation coincides closely with the wave length of maximum sensitivity of the human eye.

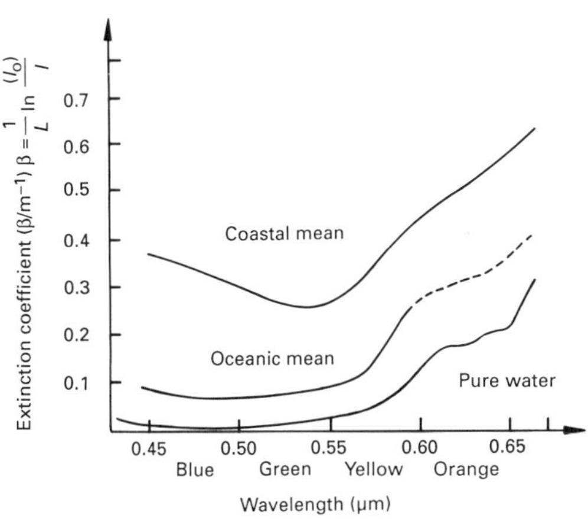

Figure 2.11. Attenuation of sound waves in sea water and pure water: I_0 = light intensity at source; I = light intensity a distance L away (from Sverdup *et al.*, 1942)

In practical terms, the extinction coefficient shown in Figure 2.11 means that sufficient sunlight for the eye to see normally penetrates sea water to depths of only 30 m (98 ft) or so. In very clear water, virtual darkness as far as the human eye is concerned is reached at depths of about 45–75 m (148–246 ft), although in silty water light may only penetrate to a depth of about 10 m (33 ft).

The electromagnetic properties of sea water are also described here for completeness. The dielectric conductivity of sea water can be taken to be approximately 4 mhos/m, which is approximately 4000 times greater than that of pure water. This is illustrated by Figure 2.12(*a*), which shows that the variation of conductivity with salinity is nearly linear. Conductivity also varies with temperature, as illustrated in Figure 2.12(*b*). The permittivity (dielectric constant) of sea water relative to a vacuum is in the range of 78–80, with absolute permittivity of a vacuum being 8.854×10^{-12} F/m. The relative permittivity of sea water does vary slightly with ocean temperature, water depth and salinity, with the higher value generally being for colder and purer water. The value of absolute permeability for sea water is very close to the permeability of all non-ferrous magnetic materials and has a value of $4\pi \times 10^{-7}$ H/m. These electromagnetic properties can be readily translated to the velocity of propagation and attenuation of electromagnetic radio waves in sea water. This data is presented in Figure 2.13(*a*) and (*b*). Figure 2.13(*b*) demonstrates the very high attenuation of radio waves in sea water. There are only two regions where the attenuation is low enough to be of practical significance – one is at low frequency (below 1 kHz) and the other is at frequencies greater than about 10^8 MHz where the polar characteristics of water molecules causes the conductivity to increase and a low attenuation band to occur in the visible light freqency range. This

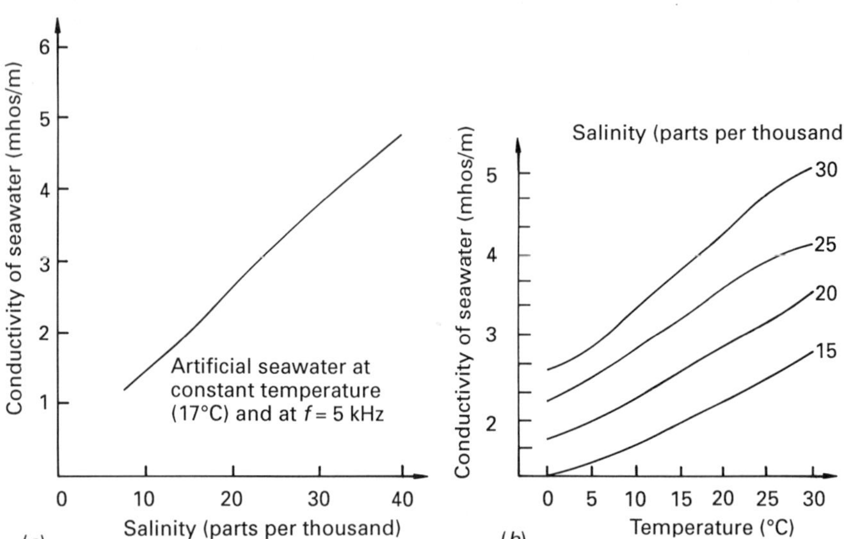

Figure 2.12. Effect of temperature and salinity on the conductivity of sea water (from Myers *et al.*, 1969)

range is mirrored in terms of extinction coefficient by the much larger frequency scale expressed in Figure 2.11.

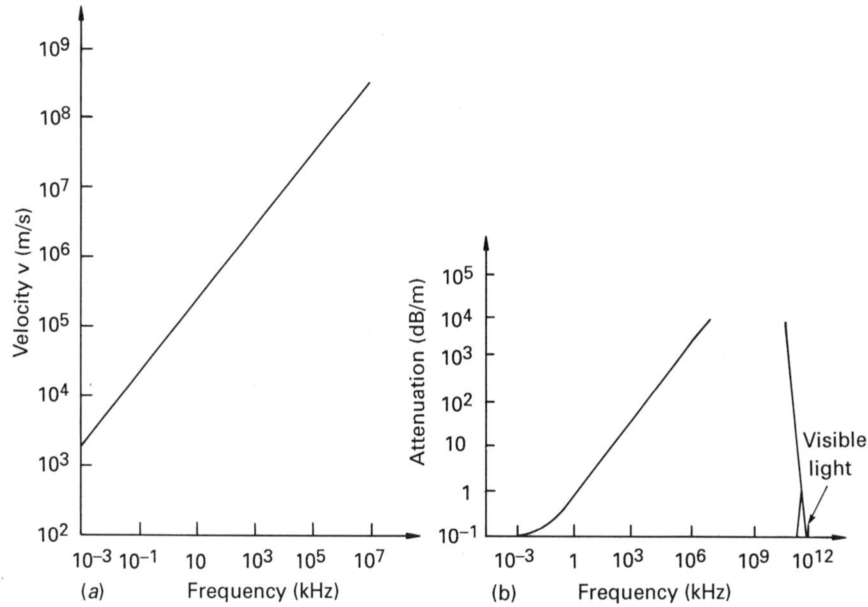

Figure 2.13. Velocity of propagation and attenuation of electromagnetic waves in sea water (for conductivity = 4 mhos/m) (from Myers *et al.*, 1969)

2.3 Environmental phenomena

The large-scale environmental phenomena of wind, current, tides and gravity waves associated with the oceans are considered next. These phenomena serve to directly influence the environmental loading on offshore structures.

Although, globally, the total solar energy incident on the Earth is balanced by energy radiated back out to space, there is a latitude dependent imbalance in that regions of the Earth at higher latitudes radiate more energy than they receive, with the opposite being true for tropical and subtropical regions. The consequent re-distribution of energy that must take place within the Earth's atmosphere is the primary cause for a general world wide circulation pattern of atmospheric winds. This circulation can be defined in terms of four broad latitude bands with similar weather patterns. The tropics lie between the Equator and 5° north and south, with the sub-tropics covering the range between 5° and 35° north and south and the mid-latitudes being the region between 35° and 55° north and south. The polar latitudes lie above 55° north and below 55° south lines.

Wind conditions in the tropics are variable in direction, but seldom exceed speeds of 10 m/s (19.4 knots). The sub-tropical band, however, is typified by winds with a persistent component from the East, with speeds ranging from 3 to 13 m/s (5.8–25.3 knots). These are often described as Trade Winds or Easterlies. The mid-latitude bands, on the other hand, are marked by winds that usually have a component from the west (known as Westerlies), although these winds are highly variable in direction and their speeds may vary from calm to 31 m/s (60.3 knots) or more. Polar regions often have winds with a component from the east, with wind speeds from 3 to 10 m/s (5.8–19.4 knots), although the polar regions can also be subjected to extremely high wind speeds of 52 m/s (101 knots) or more. These speeds may be maintained for periods of days or weeks in interior regions during the winter season. In global terms, the mid-latitudes exhibit broad bands of atmospheric mixing between the warm sub-tropics and cold polar band. The sub-tropical and tropical bands display more constant weather conditions which, nevertheless, occasionally include severe phenomena such as hurricanes, typhoons and water spouts.

Figure 2.14 presents a world chart showing prevailling winds over the oceans in the month of January. The magnitude and direction of the winds is denoted by the length and directions of the repesentative arrows.

There is also some variation in weather patterns with vertical distance above the Earth's ocean and land surfaces. This variability is characterized by two distinct levels of the atmosphere called the troposphere and the stratosphere, separated by a thin layer called the tropopause. The tropopause layer is approximately 8 km (4.97 miles) above the Earth's surface at the poles and 15 km (9.32 miles) at the Equator, with an average height of 12 km (7.46 miles) during the northern hemisphere winter. The tropopause has a slightly higher level in the northern hemisphere summer. The vast majority of variability in atmospheric conditions and the reduction of temperature with height is confined to the troposphere. The stratosphere, at heights of between 12 and 50 km (7.46 and 31.07 miles) above the Earth's surface, contains thin rarified air at temperatures lower than −50 °C at the tropopause and rising up to approximately 20 °C further away. The stratosphere contains the ozone layer, which shields the Earth from ultra-violet light. Atmospheric circulation patterns in the stratosphere are often quite different from those found in the troposphere and are characterized by much less variability, with wind patterns oriented in the general west to east direction. These winds may flow at speeds of between 52 and 130 m/s (101–253 knots) and may sometimes be accompanied by clear air turbulence.

In the troposphere, however, the variability of atmospheric circulation decreases with height where it is less affected by surface topography and convection. The tropospheric general circulation is made up of two types of wind pattern building blocks: cyclones and anti-cyclones, both of which are made up of generalized circulating winds as large as 1600 km (994 miles) to as small as 100 km (61 miles) in diameter. Anti-cyclones, generally, have low wind speeds in a high-pressure region at their centre and high wind speeds at the periphery, with the wind direction being clockwise in the northern hemisphere and anti-clockwise in the southern

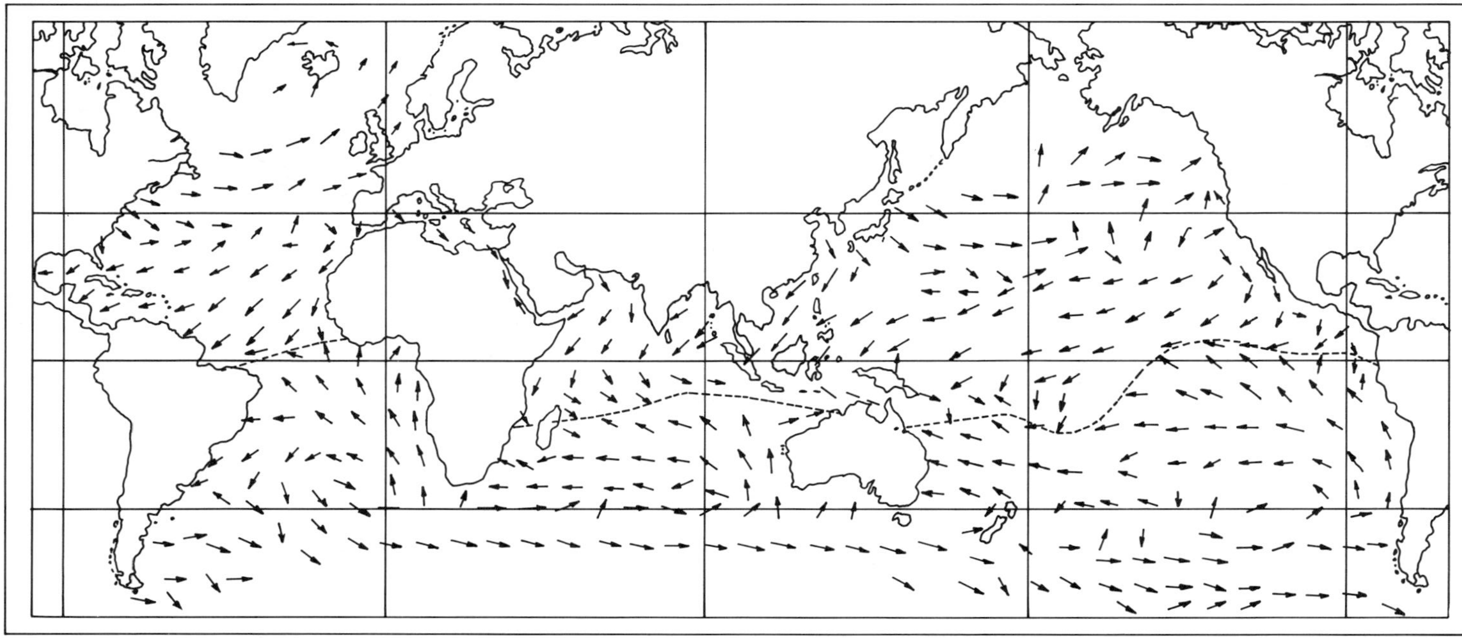

Figure 2.14. Prevailing mean winds in January (from Meteorological Office, 1986)

hemisphere. Anti-cyclones are often referred to as 'highs' and can induce wind speeds of up to 25 m/s (48.6 knots) at their edges. Such wind patterns are found largely over mid-latitudes or polar continents in the winter, and over cool ocean surfaces in the summer.

A counter-rotating weather pattern – a cyclone – can arise at the boundary region between two anti-cyclones and is induced primarily by disturbances created by temperature gradients between the two adjacent anti-cyclones. Unlike anti-cyclones, cyclones appear predominantly over the oceans in winter and over warm continents in summer. Their central regions are associated with low pressures and they characteristically induce precipitation (as rain or snow).

A large amount of qualitative information exists on the general circulation and weather induced by atmospheric winds. Accurate quantitative prediction of such phenomena is still not possible for a significant period into the future and complicates the practical requirement of weather forecasting, particularly for offshore operations. Nevertheless, recent advances in numerical modelling of the atmosphere using powerful computers, the application of satellite imaging techniques to weather observation, and a gradual improvement of fundamental knowledge of the complex interactions of solar radiation, atmospheric winds, continents and oceans, promises a steady improvement in the quality and accuracy of weather forecasts.

Atmospheric wind circulation has a direct impact on ocean water movements, both in terms of steady drift and dynamic wave action. The action of winds and other effects, described subsequently, contribute to a circulation of water currents in the oceans. An overview of the directions of these currents is presented in the map in Figure 2.15, which also names major current systems. The net movements of water due to these global currents have a profound effect on the climate of coastal areas. Classic examples are the Gulf Stream and the North Atlantic drift, which bring warm water from the tropics up to the coasts of the British Isles and maintain a climate which is relatively mild for the high latitude. The velocities of these currents generally vary between 0.25 m/s (0.49 knots) to 0.75 m/s (1.46 knots) at the ocean surface, although velocities can remain as low as 0.15 m/s (0.29 knots) in some areas, and as high as 2.5 m/s (4.9 knots) in constricted coastal regions. Current velocities vary in magnitude and direction below the sea surface, with their magnitude generally becoming smaller as depth increases and reducing to zero at the sea bed due to the boundary layer effect.

There are a number of contributory factors that drive these current systems. Surface currents are primarily induced by wind drift and by tidal action. The wind circulation pattern in the Earth's atmosphere applies a wind shear on the ocean's surface, which leads to a drift of water in the general direction of the wind vector. The relatively low resultant shear stress and the viscosity of sea water are such that ocean surface waters only drift by amounts of 1–2% of the wind speed above them. Wind induced current drift decays with depth below the sea surface and is usually undetectable at about 100 m (328 ft) depth.

Tidal action in the open oceans generally leads to negligible values of

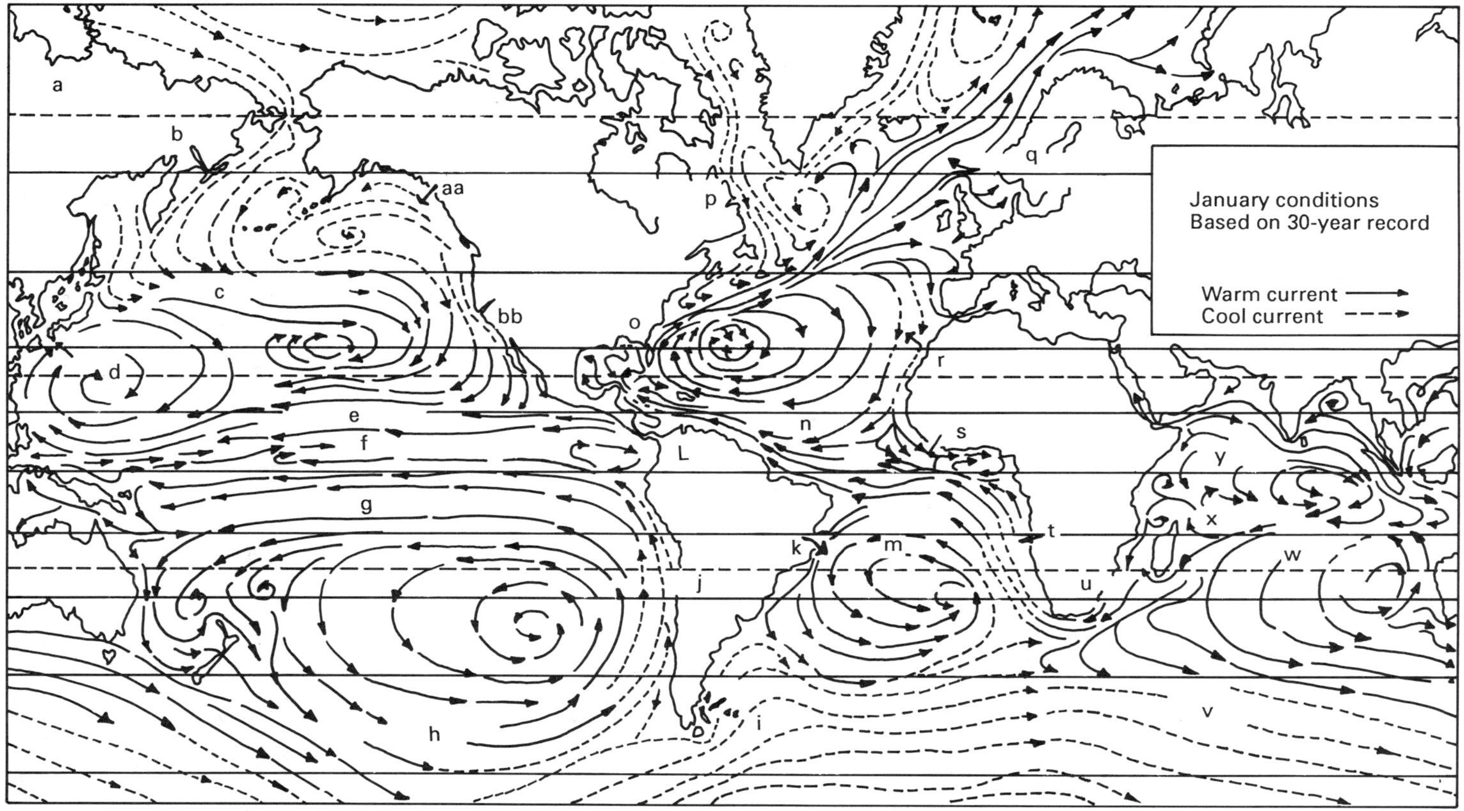

Figure 2.15. Ocean surface currents for January (from Strahler and Strahler, 1978). *Key*: a – Arctic Circle; b – Kamchatka current; c – North Pacific current; d – Kuroshio current; e – N. Equatorial current; f – Equatorial counter current; g – S. Equatorial current; h – West wind drift; i – Falkland current; j – Peru or Humboldt current; k – Brazil current; L – Equator; m – South Equatorial current; n – North Equatorial current; o – Florida current; p – Labrador current; q – North Atlantic drift; r – Canaries current; s – Guinea current; t – Benguela current; u – Agulhas current; v – West wind drift; w – South Equatorial current; x – Equatorial current; y – North Equatorial current; aa – Alaska current; bb – California current

current drift. However, the constraining effects of confining land masses and tidal action can combine to generate much larger tidal currents near land masses. However, these currents are very localized and depend on local sea bed and coast line topography.

The primary cause of both mid-depth and sea bed currents is change in sea water buoyancy (density), induced by variations in temperature and salinity. The mechanism for such global current circulation is the cooling and sinking of polar water to the sea floor and its gradual migration towards equatorial regions. As it does so, this water mixes with equatorial waters, warms up and begins to rise to the surface. This action sets up a circulation with the consequential drift of surface water from the Equator to the Poles.

The direction of current flow in the world's oceans is also governed by the action of the Coriolis force. The Coriolis force will act on any body which rotates with the Earth and which also has a radial component of velocity with respect to this axis of rotation. Therefore, the horizontal Coriolis force acting on current flow will be at a maximum near the Earth's poles and will be zero at the Equator. The force will act towards the east (the right) in the northern hemisphere and towards the west (the left) in the southern hemisphere.

Ocean waters are also subject to the oscillatory astronomical phenomenon of tidal action, caused by the gravitational attraction of the Earth to the Moon and Sun. Taking the Earth–Moon pair as an example, the gravitational attraction between these two planets is balanced by the centrifugal force induced by the lunar orbit. The gravitational attraction between the Earth and Moon is in precise balance with the centrifugal force at the centre of the Earth, but on the side of the Earth facing the Moon the net force is towards the Moon, whereas on the opposite side it is directed away from it. Thus the masses of water making up the Earth's oceans tend to bulge outwards on both sides of the Earth facing towards and away from the Moon. As the Earth turns on its axis, these bulges move relative to the Earth and in correspondence with the lunar orbit. Similar tidal forces are generated by the Sun, but although the Sun has considerably greater mass (27 million times) than the Moon, it is also 390 times further away, and therefore its tidal forces are only about 46% of those induced by the lunar orbit. The continual variation in position of the Sun and Moon in relation to the Earth produce a range of tides, with higher than average (spring) tides to lower than average (neap) tides.

At any particular location on the Earth's surface, these astronomical tidal forces are modified by a complex mix of other factors. The presence of continental masses, mid-ocean ridges and the Coriolis force induced by the Earth's rotation, combine to produce a pattern of tidal waves. Each of these wave systems is confined to its own ocean basin by the continents. Within each system the change in level of the water surface can vary from zero, at so-called nodal points, up to 15 m (49.2 ft) or more in some locally resonating basins. Typically the rise and fall of tides in the open ocean is between 1 m and 3 m (3.2–9.8 ft), with the Atlantic and Indian Oceans generally responding with twice daily tides. The Pacific Ocean also responds to both tidal periods, but in such a way that one tide is normally much larger than the other. The complex tidal patterns for a particular ocean can be quantified by using co-tidal charts, which are ocean maps with lines indicating places at which high water occurs simultaneously.

Figure 2.16 illustrates a map of the Earth marked with co-tidal lines. Each line on the map joins places that have high tide at the same instance. The number on the line denotes the time interval to high tide after Greenwich mean time, with adjacent lines indicating one hour of elapsed time between them. The resulting pattern shows how tides in the southern Atlantic basin, for example, move around a central node once every 12 h. However, the pattern of Figure 2.16 gives only an idealized picture of the global behaviour of ocean masses as they respond to the Moon's and Sun's gravitational attraction, with much of the small-scale coastal behaviour omitted.

A quantitative prediction of tidal action is obtained from a Fourier analysis of historical records of tide level with time. Such an analysis must include up to 60 frequency components in order to take account of tidal component frequencies due to astronomical movements of the Moon and Sun, ocean basin resonance effects due to surrounding land masses and sea bed topography. The frequency components revealed by such Fourier analysis will be dominated by a principal lunar semidiurnal (twice daily) component with an average period of 12 h 25 min, together with the principal solar semidiurnal component which has a period of 12 h. These two components will tend to reinforce each other at new and full Moons causing spring tides with higher than average tidal levels, and will tend to oppose each other at the first and last quarters of the Moon to yield so-called lower than average neap tides. Additional frequency components are necessary in the Fourier series to account for the fact that both the Moon and the Sun do not remain at constant distances from the Earth during the lunar orbit of the Earth and the Earth's orbit of the Sun.

The action of tides induces relatively small tidal currents in the open oceans, but these currents can be substantially magnified by shallow water at coastal locations. Tidal currents reverse in direction in line with tidal cycles and can have substantial loading effects on offshore structures operating in coastal zones. There is no reliable method of calculating the values of localized tidal currents in coastal zones. Oceanographic observations over a limited period at the location of interest are often the best guide to the magnitudes of these currents.

The regular changes in level induced by tides at a particular location increases the difficulty of relating sea bed depths or the vertical levels of an offshore structure emplaced on the sea bed to some datum water level. It is usually desirable to select a measurement datum which is such that the tide level seldom or never drops below it, so that changes of sign in the conversion of sea bed depths or offshore structure levels to another datum never occur. It is common practice to use a datum called lowest astronomical tide (LAT) for defining water level relative to the elements of a fixed offshore structure.

The surface and sub-surface layers of the oceans can also undergo a

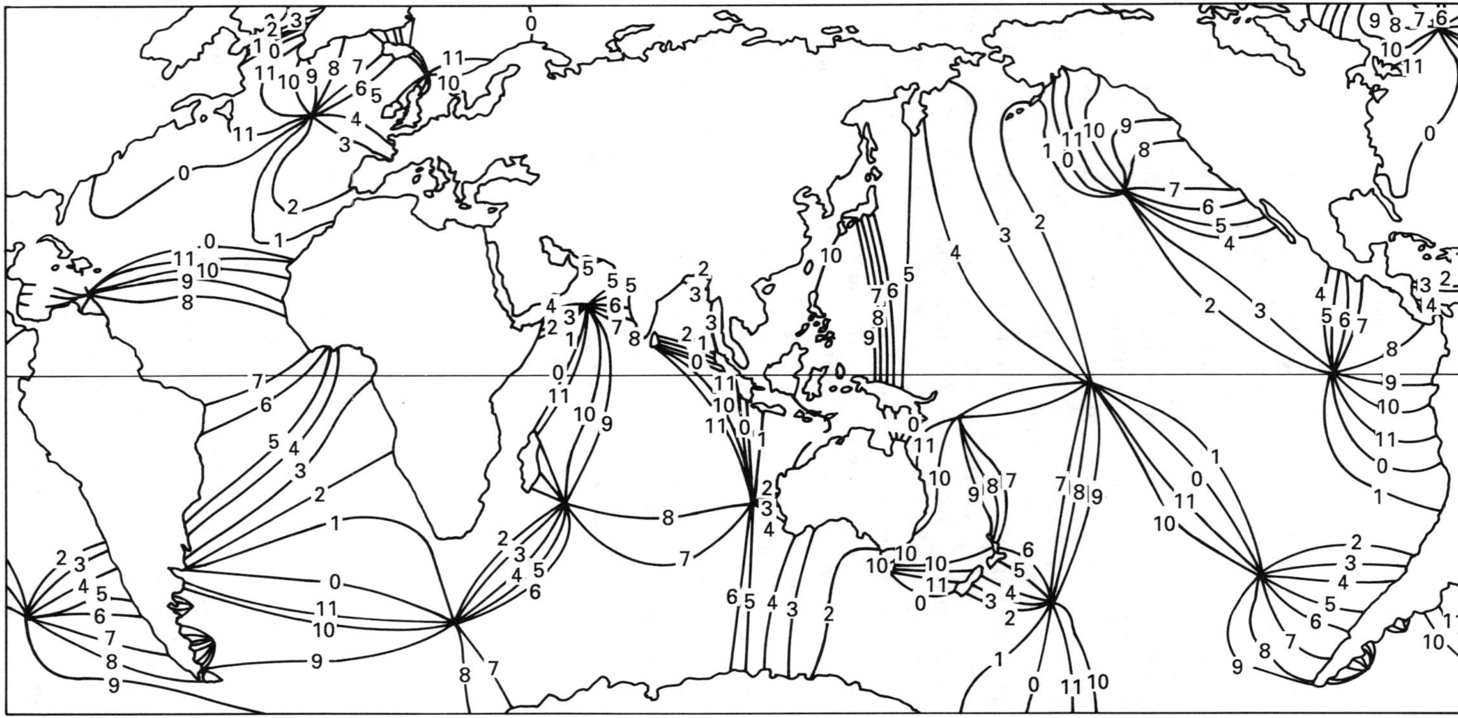

Figure 2.16. Co-tidal lines on the Earth's oceans (from Couper, 1983)

variety of oscillatory wave motions, induced most commonly by gravitational forces and by the action of wind. The tidal effects described above are an example of oscillatory changes of water level on a global scale. Large-scale oscillations in the surface of the ocean can also be induced by undersea earthquakes which can generate so-called tidal waves, or tsunamis, particularly in the earthquake belts around the Pacific. The oceans' subsurface layers also exhibit internal waves, which are wave motions of boundaries between ocean layers that are stratified due to temperature induced buoyancy changes or salinity variations. These internal waves have large periods and wave lengths, but generally have little impact on ocean operations.

At the other extreme, surface tension induced waves have lengths which are less than 1.5 cm (0.6 in) in wave length. Surface tension waves are oscillatory motions driven by the exchange of energy between surface tension and inertia forces at the water surface. However, waves of larger height and wave length induced by the oscillatory interaction of gravitational and inertia forces are of greater interest in ocean engineering and in general oceanography. These gravity waves have periods ranging from as low as 0.8 s up to 25 s, with corresponding wave lengths of from 1 m (3.3 ft) up to 1 km (3280 ft), and with wave heights as high as 33 m (108 ft) from trough to crest. The formation of gravity waves by wind action is a natural process that concentrates the rather diffused but large amount of kinetic energy in atmospheric winds into wave energy which is concentrated in a thin surface layer on the oceans. Wave energy in this layer then forms a coherent oscillatory excitation mechanism by which floating and fixed offshore structures are subjected to substantial oscillating forces throughout their working lives. The kinematics of gravity waves and the mechanics of wave loading on offshore structures thus forms a large part of the content of this book.

The action of wind blowing over an initially flat ocean surface induces distortion of this surface if the wind speed is above a critical magnitude. The mutual interaction of wind induced pressures and the deformed water surface forms a process similar to an instability, by which the size of the ocean surface distortions increase. The wave-like shape of the surface implies excesses and deficits of gravitational potential energy in the regions above and below mean sea level respectively. The gravitational potential energy in the water surface interacts with the water's inertia to induce simple harmonic vertical motion in the water surface. The interaction between adjacent layers of fluid is such that this oscillatory motion forms a travelling wave-type disturbance which moves away from its area of generation.

The action of a particular wind speed on a water surface will continually transfer energy from the wind to surface waves. This process does, however, reach a saturated or steady state for which the rate of energy transfer to surface waves equals the rate at which energy is lost from waves – primarily by wave breaking. At this point, the surface waves can be considered to be a full developed sea. Gravity waves have extremely low attenuation and can, therefore, travel vast distances without appreciable losses. A gravity wave contains a large amount of coherent, that is, systematically varying kinetic and potential energy

Table 2.6 Wind and sea scale for fully arisen sea (from Myers *et al.*, 1969)

		Wind			
Sea State	*Description*	*Beaufort wind force*	*Description*	*Range (knots)*	*wind velocity (knots)*
	Sea like a mirror	0	Calm	<1	0
0	Ripples with the appearance of scales are formed, but without foam crests	1	Light airs	1–3	2
1	Small wavelets, still short but more pronounced; crests have a glassy appearance, but do not break	2	Light breeze	4–6	5
	Large wavelets, crests begin to break. Foam of glassy appearance. Perhaps scattered white horses	3	Gentle breeze	7–10	8.5 10
2 3	Small waves, becoming larger; fairly frequent white horses	4	Moderate breeze	11–16	12 13.5 14 16
4	Moderate waves, taking a more pronounced long form; many white horses are formed (chance of some spray)	5	Fresh breeze	17–21	18 19 20
5	Large waves begin to form; the white foam crests are more extensive everywhere (probably some spray)	6	Strong breeze	22–27	22 24 24.5 26
6 7	Sea heaps up and white foam from breaking waves begins to be blown in streaks along the direction of the wind (spindrift begins to be seen)	7	Moderate gale	28–33	28 30 30.5 32

Table 2.6 (*cont.*)

Sea State	*Description*	*Wind*			
		Beaufort wind force	*Description*	*Range (knots)*	*wind velocity (knots)*
	Moderately high waves of greater length; edges of crests break into spindrift. The foam is blown in well marked streaks along the direction of the wind. Spray affects visibility	8	Fresh gale	34–40	34, 36, <u>37</u>, <u>38</u>, 40
8	High waves. Dense streaks of foam along the direction of the wind. Sea begins to roll. Visibility is affected	9	Strong gale	41–47	42, <u>44</u>, <u>46</u>
	Very high waves with long over-hanging crests. The resulting foam is in great patches and is blown in dense white streaks along the direction of the wind. On the whole the surface of the sea takes a white appearance. The rolling of the sea becomes heavy and shock-like. Visibility is affected	10	whole gale*	48–55	48, 50, <u>51.5</u>, <u>52</u>, 54
9	Exceptionally high waves (small and medium sized ship might for a long time be lost to view behind the waves) the sea is completely covered with long white patches of foam lying along the direction of the wind. Everywhere the edges of the wave crests are blown into froth. Visibility is affected	11	storm*	56–63	56, <u>59.5</u>
	Air filled with foam and spray. Sea completely white with driving spray; visibility is very seriously affected	12	hurricane*	64–71	>64

* Seas are not fully developed since required wind durations and fetches are rarely attained.
Underlined values indicate centre of Beaufort range.

Table 2.6 Wind and sea scale for fully arisen sea (from Myers *et al.*, 1969)

BWF	Sea State	Sea: wave height (m) avrg.	sig.	avrg. 1/10 hgst	Sig. range of periods	I_{max} (s)	$\bar{T}$ (s)	$\bar{I}$ average wavelength (m)	fetch (nmi)	duration (h)
0	0	0	0	0						
1		0.015	0.02	0.03	up to 1.2	0.7	0.5	0.25	5.0	18 min
2	1	0.055	0.09	0.11	0.4–2.8	2.0	1.4	2.0	8.0	39 min
3		0.08	0.30	0.37	0.8–5.0	3.4	2.4	6.1	9.8	1.7
		0.27	0.43	0.55	1.0–6.0	4.0	2.9	8.2	10.0	2.4
	2	0.43	0.67	0.85	1.0–7.0	4.8	3.4	12.2	18.0	3.8
4		0.55	0.88	1.1	1.4–7.6	5.4	3.9	15.8	24.0	4.8
	3	0.61	1.01	1.3	1.5–7.8	5.6	4.0	18.0	28.0	5.2
		0.88	1.40	1.8	2.0–8.8	6.5	4.6	22.0	40.0	6.6
5	4	1.2	1.9	2.4	2.5–10.0	7.2	5.1	27.0	55.0	8.3
		1.3	2.1	2.7	2.8–10.6	7.7	5.4	30.0	65.0	9.2
		1.5	2.4	3.0	3.0–11.1	8.1	5.7	34.0	75.0	10.0
	5	2.0	3.0	4.0	3.4–12.2	8.9	6.3	41.0	100.0	12.0
6		2.4	3.7	4.9	3.7–13.5	9.7	6.8	49.0	130.0	14.0
		2.5	4.0	5.2	3.8–13.6	9.9	7.0	50.0	140.0	15.0
	6	2.9	4.6	6.1	4.0–14.5	10.5	7.4	57.0	180.0	17.0
		3.4	5.5	7.0	4.5–15.5	11.3	7.9	65.0	230.0	20.0
		4.3	6.7	8.5	4.7–16.7	12.1	8.6	76.0	280.0	23.0

Table 2.6 (*cont.*)

BWF	*Sea State*	*Sea*: *wave height (m)* *avrg.*	*sig.*	*avrg. 1/10 hgst*	*Sig. range of periods*	I_{max} *(s)*	$\bar{T}$ *(s)*	$\bar{I}$ *average wavelength (m)*	*fetch (nmi)*	*duration (h)*
7		4.3	7.0	8.8	4.8–17.0	12.4	8.7	79.0	290.0	24.0
	7	4.9	7.9	10.1	5.0–17.5	12.9	9.1	87.0	340.0	27.0
		5.8	9.1	11.6	5.5–18.5	13.6	9.7	98.0	420.0	30.0
8		6.4	10.7	13.4	5.8–19.7	14.3	10.3	111.0	500.0	34.0
		7.0	11.3	14.2	6.0–20.5	14.9	10.5	115.0	530.0	37.0
		25.0	40.0	50.0	6.2–20.8	15.4	10.7	119.0	600.0	38.0
		28.0	45.0	58.0	6.5–21.7	16.1	11.4	135.0	710.0	42.0
9	8	31.0	50.0	64.0	7.0–23.0	17.0	12.0	150.0	830.0	47.0
		36.0	58.0	73.0	7.0–24.2	17.7	12.5	163.0	960.0	52.0
		40.0	64.0	81.0	7.0–25.0	18.6	13.1	180.0	1110.0	57.0
		44.0	71.0	90.0	7.5–26.0	19.4	13.8	198.0	1250.0	63.0
		49.0	78.0	99.0	7.5–27.0	20.2	14.3	213.0	1420.0	69.0
		52.0	83.0	106.0	8.0–28.2	20.8	14.7	224.0	1560.0	73.0
10	9	54.0	87.0	110.0	8.0–28.5	21.0	14.8	229.0	1610.0	75.0
		59.0	95.0	121.0	8.0–29.5	21.8	15.4	247.0	1800.0	81.0
11		64.0	103.0	130.0	8.5–31.0	22.6	16.3	277.0	2100.0	88.0
		73.0	116.0	148.0	10.0–32.0	24	17.0	300.0	2500.0	101.0
12		>80.0	>128.0	>164.0	10.0–(35)	(26)	(18)	—	—	—

Notes:
I_{max} = period of maximum energy of spectrum.
$\bar{T}$ = average period.
BWF = Beaufort wind force.

within its structure. The majority of this energy only dissipates when waves interact with shallow water and coast lines or, of course, when waves participate in applying loads to an offshore structure.

The kinematics of gravity waves and the various theories available to predict them quantitatively are described in Chapter 4. At this stage, however, it is useful to characterize qualitative numerical scales used to describe observed wind and wave conditions. Table 2.6 presents a generally accepted definition for these wind and sea scales. The combined action of wind and waves is described by a sea state number which can vary from 0 to 9. The corresponding wind scale is described by a Beaufort number which can vary from 0 to 12. Table 2.6 also describes the environmental features that may be observed in each of these sea states, from a very calm condition where the sea can be likened to a mirror, to an extreme hurricane condition where the sea is completely white with driving spray and visibility is very seriously affected. The wind is described by a wind speed range, whereas wave conditions are described in much more detail. The generally random variations of wave elevation are quantitifed by listing the average wave height, the significant wave height (which is the average of the highest one-third of observed wave heights), and also the average of the highest one-tenth of observed wave heights. The corresponding range of wave periods is characterized by listing the significant range of periods. Each storm condition can correspond to particular wave periods, at which the wave energy attains a maximum. This is listed in the column entitled 'I_{max}'. The average wave period is denoted by $\bar{T}$ and $\bar{I}$ is the average wave length. The last two columns denote the minimum fetch, that is, the distance over which the wind has to blow in order to develop the sea state indicated and the minimum duration for which this wind must blow. Table 2.6 offers a surprisingly accurate means of using qualitative observed indicators of wind and sea state to quantify wind speeds and gravity wave characteristics.

In recent years, the development of the offshore industry has stimulated substantial amounts of wave data collection in most of the continental shelf areas of the world, but particularly in the Gulf of Mexico and the North Sea. This wave data has tended to improve understanding of the characteristics of gravity waves and of refining aspects of their frequency content and their statistical distribution. The measurements have, in particular, been used to extrapolate the frequency of occurrence and magnitude of extreme wave conditions and to estimate the maximum wave height that a structure with a given design life ought to withstand. Some of the conclusions arising from this data collection are described further in Chapter 4.

Offshore development on the continental shelf off the northern coasts of North America has also stimulated an interest in carrying out offshore operations in ice infested waters. The wide continental shelf of the Arctic continent and the large number of islands on this shelf have been the site for numerous exploration wells and limited oil production. In these areas, both fixed and floating offshore structures have to operate in sea ice which can range from patchy floes 1.5 m (4.9 ft) thick with open areas of clear water between them to solid masses which can be more than ten times as thick. Extreme winter storms and the presence of very massive icebergs demand challenging engineering solutions for offshore exploration and production. These activities have led to more accurate definitions of the mechanical properties of both fresh water and sea water ice, in order that the interaction between free ocean water, smaller ice masses and offshore structures can be evaluated with more scientific rigour. Table 2.7 presents a summary of the properties of both fresh water and various kinds of sea water ice. As shown in the table, the density of ice is lower than that of water because of the strong hydrogen bonds that form in the molecular lattice in ice. An interesting feature of these properties is that ice has a significant mechanical strength and relatively low thermal conductivity, both of which influence engineering solutions for offshore structures in an Arctic environment. These properties of ice have already been utilized in Arctic operations in the construction of ice islands by freezing bulk water, and in the creation of large ice floes attached to land masses for transporation of offshore equipment by road.

Table 2.7 Physical properties of sea ice as a function of temperature and salinity

Property	*Tempr.*	*Salinity in ‰*					
	(°C)	*0*	*2*	*4*	*8*	*10*	*15*
Density of sea ice (kg.m^3)	—	918	919	921	924	925	929
Tensile strength (MN/m^2)	—			0.1 to 0.3			
Compressive strength (MN/m^2)	—			Approximately 12			
Youngs modulus (GN/m^2)	−20	9.75	9.27	8.78	7.81	7.33	6.12
Poisson's ratio				0.33			
Latent heat of fusion	−1	334.9	301.4	263.8	192.6	154.9	67.0
(kJ/kg)	−2	339.1	322.4	301.4	263.8	247.0	200.1
Coefficient of thermal	−2	—	−22.10	−45.89	−93.46	−117.25	−176.72
expansion (× 10^4/°C)	−4	—	−4.12	−9.92	−21.53	−27.34	−41.85
	−8	—	0.16	−1.37	−4.43	−5.95	−9.79
	−10	—	0.83	−0.02	−1.73	−2.59	−4.73
	−20	—	1.83	1.07	0.45	0.14	−0.63
Specific heat capacity	0	2.116					
(kJ/kg K)	−10	2.038					
	−20	1.960					
	−30	1.882					
	−40	1.804					
Thermal conductivity	0	2.26					
(W/m K)	−20	2.40					
	−40	2.56					
	−60	2.76					
	−80	3.00					
	−100	3.29					
	−120	3.66					

References

Camp, L. W. (1970), *Underwater Acoustics – Theory and Experiments in Underwater Sound*, McGraw Hill.

Couper Alastair (ed) (1983), *The Times Atlas of the Oceans*, Times Book, 16 Golden Square, London.

Defant, A. (1961), *Physical Oceanography*, Pergamon Press, USA.

Dietrich, G. (1963), *General Oceanography*, Interscience Publishers, Inc, USA.

Engineering Sciences Data Unit (1979), *Thermal Conductivity, Viscosity, Heat Capacity, Density and Prandtl Number of Sea Water and its Concentrates*, Item 77024 with Amendment B, December, ESDU, 251–9 Regent Street, London.

Fuchs, V. Sir (1985), *Oxford Illustrated Encyclopaedia, The Physical World*, Volume 1, Guild Publishing, London.

Hill, M. N. (ed) (1962), *The Sea, Volume I*, Physical Oceanography, Interscience Publishers Inc., USA.

Howatson, A. M., Lund, P. G. and Todd, J. D. (1972), *Engineering Tables and Data*, Chapman and Hall, London.

Meteorological Office, (1986), *Meteorology for Mariners*, 3rd edn, HMSO ISBN 0 11 400 311 4.

Myers, J. J., Holm, C. H. and McAllister, R. F. (1969), *Handbook of Ocean and Underwater Engineering*, McGraw Hill Book Co., Berks.

Neumann, G. and Pierson, W. J. (1966), *Principles of Physical Oceanography*, Prentice Hall Inc., Englewood Cliffs, NJ, USA.

Strahler, A. N. and Strahler, A. H. (1978). *Modern Physical Geography,* John Wiley and Sons, USA.

Sverdrup, H. V., Johnson, M. W. & Fleming, R. H. (1942), *The Oceans – Their Physics, Chemistry and General Biology*, Prentice Hall Inc., Englewood Cliffs, NJ, USA.

Tucker, D. G. and Gazey, B. K. (1966), *Applied Underwater Acoustics*, Pergamon Press, Oxford.

Warren, W. F. (1961), *Seawater Density in the Ocean as a Function of Depth and a Method of Utilising this Information in Design of Pressure Vessels Which Will Remain in Constant Depth Range Between Surface and Bottom*, Naval Ordinance Laboratory, NOLTR 61–179,, AD 273634.

Chapter 3

Basic fluid mechanics

An understanding of fluid mechanics is an essential step in the development and use of methods to calculate the forces exerted on offshore structures by the action of winds, currents and gravity waves. The theory of fluid mechanics is based on two fundamental properties of the structure and behaviour of a fluid. Firstly, the length scales of the vast majority of flow fields of interest are many orders of magnitude larger than the inter-molecular distances within the fluid. It is, therefore, permissible to treat the fluid as a continuous medium. Furthermore, the behaviour of fluids is governed by their property of shearing in response to applied shear forces. The fluid will then exhibit internal shear stresses which will be directly proportional to the velocity gradients within its motions. For the large majority of fluids encountered in the physical world (including air and water), it can be shown experimentally that the shear stress within the fluid is directly proportional to its velocity gradient. Thus the shear stress τ_{yz} parallel to the Oyz plane in a fluid with velocity v in the x direction only is given by

$$\tau_{yz} = \mu \frac{\partial v}{\partial x} \qquad (3.1)$$

where the constant of proportionality μ is defined as the dynamic density of the fluid with units of Ns/m^2. It follows from this property that a fluid at rest cannot exhibit any shear stress and, therefore, can only exhibit direct stresses or hydrostatic pressure within itself.

Fluids that exhibit direct proportionality between shear stress and velocity gradient are called Newtonian. Non-Newtonian fluids are defined as those where the relationship between shear stress and velocity gradient does not exhibit a direct proportionality. Fluids with long molecular chains or with high levels of suspended particles as well as emulsions have non-Newtonian characteristics, with shear stress being a function of both the shear strain history and the shear strain. Since the vast majority of fluids of interest in offshore mechanics are Newtonian, the theory of non-Newtonian fluid motions is not considered further here.

For Newtonian fluids, the ratio of dynamic viscosity μ and density ρ frequently appears in the governing equations. The ratio can, therefore, conveniently be defined as the kinematic viscosity ν, where

$$\nu = \frac{\mu}{\rho} \tag{3.2}$$

and has units of m^2/s.

Fluids also exhibit compressibility which is defined in terms of a bulk modulus of elasicity K to be

$$K = -V\frac{dp}{dV} \tag{3.3}$$

where p and V are the fluid pressure and volume respectively. In the case of a gas, K can be obtained in terms of density ρ and absolute temperature T_0 to be

$$K = \rho R T_0 \tag{3.4}$$

by using the perfect gas law with R being the universal gas constant. The bulk modulus of liquids such as water (1.993 GPa at 0 °C) is high, making them virtually incompressible, whereas gases have a low bulk modulus in comparison. For example, air at standard pressure and temperature (101 325 Pa and 15 °C) has a bulk modulus of 0.1013 MPa.

3.1 Frames of reference

Since a fluid is treated as a continuous medium, a difficulty arises in identifying a means or a frame of reference which can be used to describe the motion of all the fluid within a flow field. Just as a valid set of reference axes is required to describe the state of motion of solid bodies, the definition of a frame of reference is a pre-requisite for describing fluid motion in mathematical terms.

There are two frames of reference that may be used. The first is the Lagrangian or convective frame of reference which describes the behaviour of individual fluid particles during their motion within a flow field as a function of their initial position and time. Thus the Cartesian co-ordinates (x, y, z) describing the position of a fluid particle can be written as functions:

$$\left.\begin{aligned} x &= F_1(a, b, c, t) \\ y &= F_2(a, b, c, t) \\ z &= F_3(a, b, c, t) \end{aligned}\right\} \tag{3.5}$$

where (a, b, c) are the co-ordinates of the particle's initial position and t is time. In physical terms, the Lagrangian frame of reference corresponds to identifying a fluid particle and following its motion in space and with time. The velocities and acceleration of the particle can readily be obtained by differentiating Equations (3.5) with respect to time.

On the other hand, an Eulerian or spatial frame of reference defines a flow field by identifying fluid velocity components at discrete points within a volume of fluid and observing the characteristic variation of the fluid as it flows past that point. Thus the velocity components u, v and w are defined as

$$\left.\begin{aligned} u &= f_1(x, y, z, t) \\ v &= f_2(x, y, z, t) \\ w &= f_3(x, y, z, t) \end{aligned}\right\} \quad (3.6)$$

where (x, y, z) are the co-ordinates of a point fixed in space and t is time. In contrast to the Lagrangian frame of reference, the Eulerian scheme identifies fixed points in space and describes the velocity components of fluid particles as they pass the points. The velocity components are thus considered as 'point functions' of position in space.

There are merits and drawbacks to both these frames of reference. The Eulerian scheme is of more value for calculating forces on offshore structures since it is capable of identifying the fluid flow in a fixed volume which can also contain any structural elements that are immersed in the flow. There are occasions, however, where the Lagrangian flow description is of great value for providing physical understanding of a flow field.

In order to gain insight into the application of the Eulerian method, the fluid flow along a streamline can be considered. A streamline is defined as a line within a flow field that is everywhere tangential to the instantaneous velocity vector. Path lines, however, are lines traced out by elements of fluid moving in a flow field and can be regarded as the lines joining all the points through which a fluid particle has passed. The fluid can be given bulk by defining a stream tube as being formed by taking any closed circuit in the flow and drawing all the stream lines through the circuit. Thus no fluid crosses the walls of the resulting closed stream tube.

3.2 Rotational and irrotational flow

Since a fluid readily shears under the action of an applied shear stress it is important to examine the precise nature of this shear action and to distinguish it from other fluid motion such as rotation. This can be illustrated by the small elements of fluid shown at time $t = 0$ and $t = t_1$ in Figure 3.1. Despite the apparent similarity of the two flow fields, it can be seen that the fluid motion of Figure 3.1(a) exhibits distortion of the fluid element without rotation whereas Figure 3.1(b) demonstrates solid body rotation of the element without distortion. The flow in Figure 3.1(a) is defined as irrotational flow whereas that in Figure 3.1(b) is rotational flow.

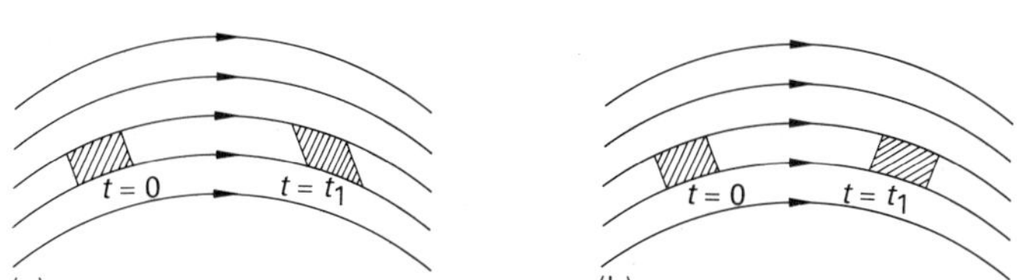

Figure 3.1. Irrotational and rotational flows

This concept of rotational and irrotational flow can also be expressed mathematically. Consider the rectangular fluid element $ABCD$ shown in Figure 3.2. Under the action of velocity components u and v parallel to the x and y axes, the element may take the position defined by $AB'C'D'$ after time δt.

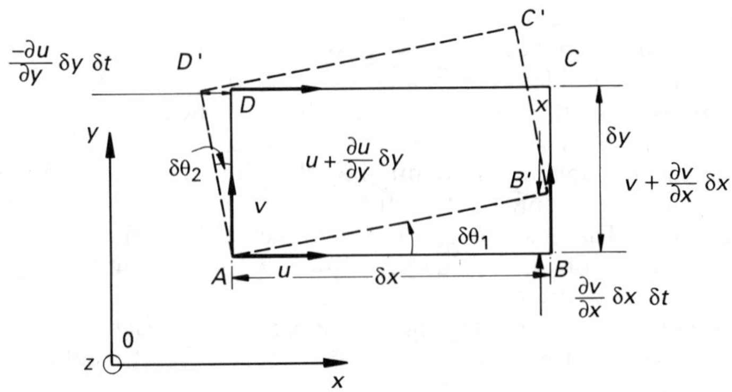

Figure 3.2. Rectangular fluid element

The rotation of this rectangular element about the Oz axis can be defined as the average of the angular velocities of two mutually perpendicular lines within the element. If the two lines are chosen to be AB and AD, then the angular velocity of AB about Oz is

$$\lim_{\delta t \Rightarrow 0} \frac{\delta\theta_1}{\delta t} = \lim_{\delta t \Rightarrow 0} \frac{(\partial v\,\delta x\,\delta t)/(\partial x\,\delta x)}{\delta t} = \frac{\partial v}{\partial x}$$

and similarly, the angular velocity of AD about Oz is

$$\lim_{\delta t \Rightarrow 0} \frac{\delta\theta_2}{\delta t} = -\frac{\partial u}{\partial y}$$

The average of the angular velocities of these two line elements is the rotation ω given by

$$\omega = \frac{1}{2}\left(\frac{\partial v}{\partial x} - \frac{\partial u}{\partial y}\right) \tag{3.7}$$

When $\omega \neq 0$, the flow is rotational and a measure of this rotationality is defined by a parameter called vorticity ζ, which is written as

$$\zeta = 2\omega = \frac{\partial v}{\partial x} - \frac{\partial u}{\partial y} \tag{3.8}$$

The definition of vorticity is conveniently selected to absorb the 2 that arises in Equation (3.7).

By repeating this analysis for the Oyz and Ozx planes and incorporating the above result, the three components of vorticity ξ, η and ζ about axes parallel to Ox, Oy and Oz can be written as

$$\left.\begin{aligned}\xi &= \frac{\partial w}{\partial y} - \frac{\partial v}{\partial z}\\ \eta &= \frac{\partial u}{\partial z} - \frac{\partial w}{\partial x}\\ \zeta &= \frac{\partial v}{\partial x} - \frac{\partial u}{\partial y}\end{aligned}\right\} \quad (3.9)$$

For irrotational flow the vorticity components become zero, and the resultant condition for irrotationality can be written as

$$\left.\begin{aligned}\frac{\partial w}{\partial y} - \frac{\partial v}{\partial z} &= 0\\ \frac{\partial u}{\partial z} - \frac{\partial w}{\partial x} &= 0\\ \frac{\partial v}{\partial x} - \frac{\partial u}{\partial y} &= 0\end{aligned}\right\} \quad (3.10)$$

These conditions are very useful for application in flows such as gravity waves where the overall flow may be taken to be irrotational since fluid particle displacements are small compared to the length scales of the bulk of fluid.

In irrotational flow with Equation (3.10) satisfied, the fluid element of Figure 3.2 may undergo pure straining deformations only where two initially perpendicular lines within the fluid rotate relative to each other. The pure shearing motion within the fluid can be defined by an equation analogous to Equation (3.7) which quantifies the rotation of line *AD* relative to *AB* about the *Oz* axis. This shear strain can be defined in terms of strain ε_{xy}, as below, together with similar equations for the other components of strain rate. Thus

$$\left.\begin{aligned}\varepsilon_{xy} &= \frac{1}{2}\left(\frac{\partial v}{\partial x} + \frac{\partial u}{\partial y}\right)\\ \varepsilon_{yz} &= \frac{1}{2}\left(\frac{\partial w}{\partial y} + \frac{\partial v}{\partial z}\right)\\ \varepsilon_{zx} &= \frac{1}{2}\left(\frac{\partial u}{\partial z} + \frac{\partial w}{\partial x}\right)\end{aligned}\right\} \quad (3.11)$$

These expressions are used subsequently in the derivation of the Navier–Stokes equations.

3.3 Stream function and velocity potential

It is cumbersome to describe the motion of a fluid as a set of displacement co-ordinates or velocity components within a flow field of large volume. A flow field can, however, be conveniently expressed in terms of mathematical functions in two or three dimensions. The stream function and velocity potential are two such mathematical functions.

The stream function ψ can describe a steady or unsteady flow pattern (or streaming) within a two-dimensional plane. Consider the two-dimensional flow field shown in Figure 3.3

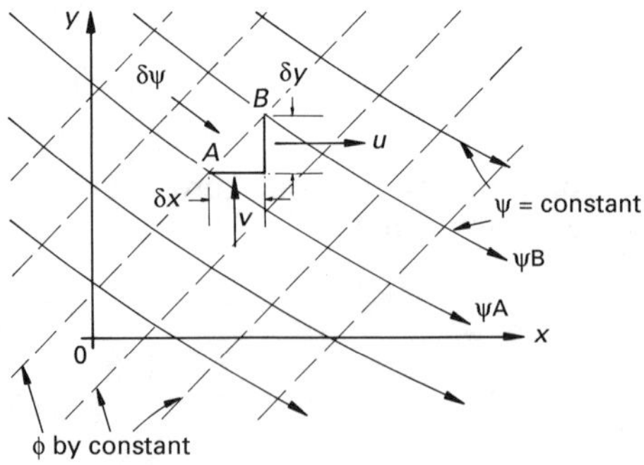

Figure 3.3. Constant ψ and ϕ lines

The lines shown on the diagram are stream lines, with the stream function ψ being a constant along them. The fluid in the plane flows between the stream lines with no fluid crossing the lines. Consider two points A and B on stream lines ψ_A and ψ_B in Figure 3.3. If the stream function ψ is regarded as the volume rate of flow or flux between the stream lines then

$$\delta\psi = \psi_B - \psi_A = u\,\delta y - v\,\delta x \tag{3.12}$$

But an increment in $\delta\psi$ can also be written as

$$\delta\psi = \frac{\partial\psi}{\partial x}\delta x + \frac{\partial\psi}{\partial y}\delta y \tag{3.13}$$

By comparing coefficients between Equations (3.12) and (3.13), a definition of the stream function can be written as

$$u = \frac{\partial\psi}{\partial y}, \quad v = -\frac{\partial\psi}{\partial x} \tag{3.14}$$

Now, taking $\delta\psi = 0$ in Equation (3.12) leads to the condition

$$\frac{\partial y}{\partial x} = +\frac{v}{u} \text{ and } \psi(x, y) = C \tag{3.15}$$

where C is a constant and the function $\psi(x, y)$ denotes a family of lines that are tangential to the local flow field. The value of ψ for any stream line is thus constant and this value of ψ relative to some datum stream line may be used as a label for the stream line. By convention, the datum stream line through the origin is taken as having a zero value for ψ.

Equations (3.14) form a definition of the stream function which has units of m^2/s. Lines of constant ψ, as shown in Figure 3.3, are stream lines along which the fluid is flowing. It can be deduced from the above definition that the distance between constant ψ lines is inversely proportional to the flow velocity between the lines. The definition of the stream function is such that it is restricted to two-dimensional flow, although it can be extended to unsteady flow and it remains valid for both rotational and irrotational flow. The stream line patterns (and thus velocity components) for any pair of two-dimensional flows can be directly added to yield stream lines and velocity components for the combined flow.

An alternative mathematical function, the velocity potential, can also be used to define a flow field. The velocity potential in three dimensions is defined as $\phi(x, y, z)$ such that

$$u = \frac{\partial \phi}{\partial x}, \quad v = \frac{\partial \phi}{\partial y} \text{ and } w = \frac{\partial \phi}{\partial z} \tag{3.16}$$

again with units of m^2/s. Lines of constant ϕ are called equipotentials and are orthogonal to the constant ψ lines illustrated in Figure 3.3. The velocity potential is also applicable to unsteady flows but its validity is restricted to irrotational flows only.

3.4 Continuity and the Laplace equation

The principle of conservation of mass states that the rate of mass accumulation within a control volume of fluid must equal the net mass flux into the volume boundaries. This principle is used to derive the equation of continuity.

Consider the infinitesimal control volume within a fluid of density ρ shown in Figure 3.4. The diagram shows the rate of mass flow across the six bounding surfaces of the control volume. Applying the principle of conservation of mass gives the equation

$$\frac{\partial(\rho\, \delta x\, \delta y\, \delta z)}{\partial t} = -\left(\frac{\partial(\rho u)}{\partial x} + \frac{\partial(\rho v)}{\partial y} + \frac{\partial(\rho w)}{\partial z}\right) \delta x\, \delta y\, \delta z \tag{3.17}$$

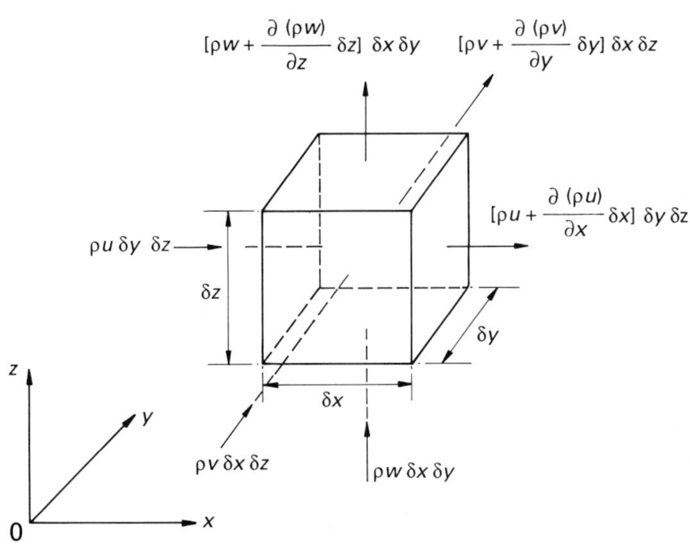

Figure 3.4. Control volume for equation of continuity

Dividing by $\delta x\,\delta y\,\delta z$ yields the general equation of continuity for compressible flow:

$$\frac{\partial \rho}{\partial t} + \frac{\partial(\rho u)}{\partial x} + \frac{\partial(\rho v)}{\partial y} + \frac{\partial(\rho w)}{\partial z} = 0 \qquad (3.18)$$

For incompressible flow, the density ρ is constant and this equation becomes

$$\frac{\partial u}{\partial x} + \frac{\partial v}{\partial y} + \frac{\partial w}{\partial z} = 0 \qquad (3.19)$$

and substituting for the velocity potential (Equation (3.16)) gives the Laplace equation

$$\frac{\partial^2 \phi}{\partial x^2} + \frac{\partial^2 \phi}{\partial y^2} + \frac{\partial^2 \phi}{\partial z^2} = 0 \qquad (3.20)$$

3.5 The Navier–Stokes equations

The dynamic behaviour of fluids is governed by the Navier–Stokes equations which extend Newton's second law of motion to a continuous fluid medium. An element in the fluid medium will be subjected to forces due to a pressure field and to body forces (such as gravity) together with direct and shear stresses induced by the motion of the fluid. These forces are balanced by the product of the fluid density and the direct as well as convective accelerations of the fluid. All of these physical effects are incorporated in the Navier–Stokes equations which are derived below for incompressible flow.

A fluid velocity component in a flow field can be regarded as a field function in that its magnitude may vary with position co-ordinates (x, y, z) as well as with time t. For example, a velocity component parallel to the x direction can be taken to have a value u at position (x, y, z) and time t as well as a value $(u + \delta u)$ at position $(x + \delta x, y + \delta y, z + \delta z)$ and time $(t + \delta t)$. Then to first order

$$u + \delta u = u + \frac{\partial u}{\partial x}\delta x + \frac{\partial u}{\partial y}\delta y + \frac{\partial u}{\partial z}\delta z + \frac{\partial u}{\partial t}\delta t$$

applies. In the limit,

$$\lim_{\delta t \Rightarrow 0}\left(\frac{\delta u}{\delta t}\right) = \frac{\partial u}{\partial t} + \frac{\partial u}{\partial x}\cdot\frac{\partial x}{\partial t} + \frac{\partial u}{\partial y}\cdot\frac{\partial y}{\partial t} + \frac{\partial u}{\partial z}\cdot\frac{\partial z}{\partial t}$$

Therefore $\qquad (3.21)$

$$\frac{\mathrm{D}u}{\mathrm{D}t} = \frac{\partial u}{\partial t} + u\frac{\partial u}{\partial x} + v\frac{\partial u}{\partial y} + w\frac{\partial u}{\partial z}$$

where the capital D denotes a substantive rate of change which includes convection induced acceleration terms.

The inertia force due to fluid density and the substantive acceleration derived above will be balanced by direct and shear stresses within the fluid. Consider an infinitesimal cuboid volume $\mathrm{d}V$ of a fluid, as shown in Figure 3.5. The shear and direct stresses acting on each surface of the volume element are represented by force vectors, with only four of these being shown in Figure 3.5 for clarity. By resolving forces in the x, y and z directions, the net force per unit volume in each of these directions can be written as

$$\mathbf{P} = \frac{\partial \mathbf{p}_x}{\partial x} + \frac{\partial \mathbf{p}_y}{\partial y} + \frac{\partial \mathbf{p}_z}{\partial z} \tag{3.22}$$

where

$$\left.\begin{aligned} \mathbf{p}_x &= \mathbf{i}\tau_{xx} + \mathbf{j}\tau_{xy} + \mathbf{k}\tau_{xz} \\ \mathbf{p}_y &= \mathbf{i}\tau_{yx} + \mathbf{j}\tau_{yy} + \mathbf{k}\tau_{yz} \\ &\text{and} \\ \mathbf{p}_z &= \mathbf{i}\tau_{zx} + \mathbf{j}\tau_{zy} + \mathbf{k}\tau_{zz} \end{aligned}\right\} \tag{3.23}$$

using **i**, **j** and **k** to denote unit vectors in the x, y and z directions respectively.

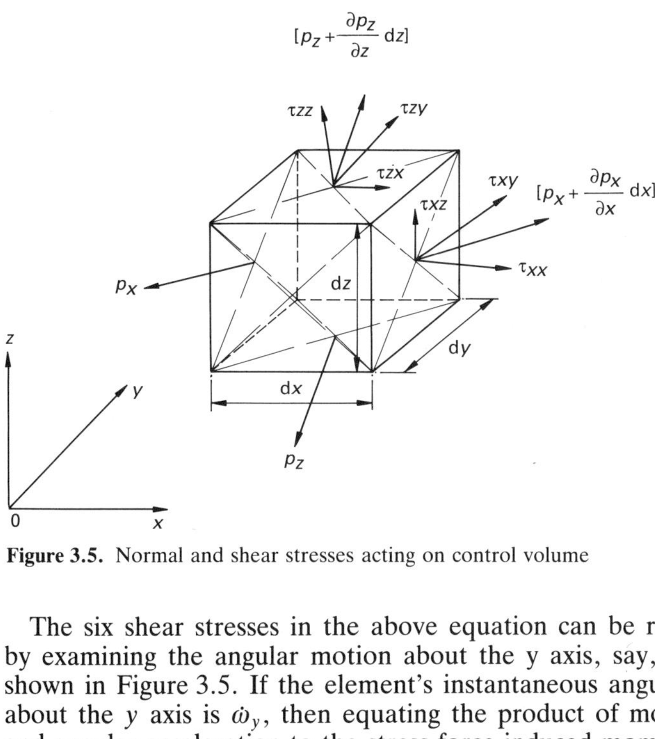

Figure 3.5. Normal and shear stresses acting on control volume

The six shear stresses in the above equation can be reduced to three by examining the angular motion about the y axis, say, of the element shown in Figure 3.5. If the element's instantaneous angular acceleration about the y axis is $\dot{\omega}_y$, then equating the product of moment of inertia and angular acceleration to the stress force induced moment yields

$$\begin{aligned} \dot{\omega}_y \, \mathrm{d}I_y &= (\tau_{xz} \, \mathrm{d}y \, \mathrm{d}z) \, \mathrm{d}x - (\tau_{zx} \, \mathrm{d}x \, \mathrm{d}y) \, \mathrm{d}z \\ &= (\tau_{xz} - \tau_{zx}) \, \mathrm{d}V \end{aligned} \tag{3.24}$$

where $\mathrm{d}I_y$ is the element's moment of inertia about the y axis. Now, as the volume of the element tends to zero, $\mathrm{d}I_y$ will reduce rapidly as the fifth power of the cuboid's dimensions (being the product of mass times distance squared), while $\mathrm{d}V$ will reduce only as the third power. Thus, in the absence of any local moments from within the fluid volume, the equation remains true for any finite $\dot{\omega}_y$ if and only if

$$\tau_{xz} = \tau_{zx} \tag{3.25}$$

Similarly, for rotation about the x and z axis it can be shown that

$$\tau_{yz} = \tau_{zy} \text{ and } \tau_{xy} = \tau_{yx} \tag{3.26}$$

reducing Equations (3.23) to

$$\left.\begin{aligned} \mathbf{p}_x &= \mathbf{i}\tau_{xx} + \mathbf{j}\tau_{xy} + \mathbf{k}\tau_{zx} \\ \mathbf{p}_y &= \mathbf{i}\tau_{xy} + \mathbf{j}\tau_{yy} + \mathbf{k}\tau_{yz} \\ \mathbf{p}_z &= \mathbf{i}\tau_{zx} + \mathbf{j}\tau_{yz} + \mathbf{k}\tau_{zz} \end{aligned}\right\} \tag{3.27}$$

with only six different stress components.

Now, applying Newton's second law along the x, y and z axis to the cuboid element of Figure 3.5, and dividing by the cuboid volume yields the equations

$$\left.\begin{aligned} \rho\frac{\mathrm{D}u}{\mathrm{D}t} &= X + \left(\frac{\partial\tau_{xx}}{\partial x} + \frac{\partial\tau_{xy}}{\partial y} + \frac{\partial\tau_{zx}}{\partial z}\right) \\ \rho\frac{\mathrm{D}v}{\mathrm{D}t} &= \mathrm{Y} + \left(\frac{\partial\tau_{xy}}{\partial x} + \frac{\partial\tau_{yy}}{\partial y} + \frac{\partial\tau_{yz}}{\partial z}\right) \\ \rho\frac{\mathrm{D}w}{\mathrm{D}t} &= \mathrm{Z} + \left(\frac{\partial\tau_{zx}}{\partial x} + \frac{\partial\tau_{yz}}{\partial y} + \frac{\partial\tau_{zz}}{\partial z}\right) \end{aligned}\right\} \tag{3.28}$$

The substantive acceleration terms on the left-hand side of the equation are derived in the same manner as for Equation (3.21), X, Y and Z denote body forces proportional to unit volume in the x, y and z directions, respectively, and the last terms of the equation are obtained from Equations (3.22) and (3.27).

Now in a Newtonian fluid, viscous stresses are postulated to be proportional to the rates of strain. The normal strains can be written as

$$\varepsilon_{xx} = \frac{\partial u}{\partial x},\ \varepsilon_{yy} = \frac{\partial v}{\partial y},\ \varepsilon_{zz} = \frac{\partial w}{\partial z} \tag{3.29}$$

whereas the three shear strains are defined in Section 3.2, Equations (3.11).

The shear stresses are taken to be directly proportional to the rates of strain with constant of proportionality 2μ to yield the equations

$$\left.\begin{aligned} \tau_{xy} &= \tau_{yx} = \mu\left(\frac{\partial v}{\partial x} + \frac{\partial u}{\partial y}\right) \\ \tau_{yz} &= \tau_{zy} = \mu\left(\frac{\partial w}{\partial y} + \frac{\partial v}{\partial z}\right) \\ \tau_{zx} &= \tau_{xz} = \mu\left(\frac{\partial u}{\partial z} + \frac{\partial w}{\partial x}\right) \end{aligned}\right\} \tag{3.30}$$

However, direct stresses within the fluid require more careful treatment.

The total direct stress within a fluid in motion will be made up of both the hydrostatic pressure p and a direct stress due to the motion of the fluid. This latter stress will have a component that is due to a change in shape of the fluid element and a further component that is due to a change in volume (that is, a change in density) of the fluid element. Thus the three components of direct stress can be written as

$$\left.\begin{aligned}\tau_{xx} &= -p + 2\mu\frac{\partial u}{\partial x} + \lambda\left(\frac{\partial u}{\partial x} + \frac{\partial v}{\partial y} + \frac{\partial w}{\partial z}\right)\\ \tau_{yy} &= -p + 2\mu\frac{\partial v}{\partial y} + \lambda\left(\frac{\partial u}{\partial x} + \frac{\partial v}{\partial y} + \frac{\partial w}{\partial z}\right)\\ \tau_{zz} &= -p + 2\mu\frac{\partial w}{\partial z} + \lambda\left(\frac{\partial u}{\partial x} + \frac{\partial v}{\partial y} + \frac{\partial w}{\partial z}\right)\end{aligned}\right\} \qquad (3.31)$$

where the hydrostatic pressure acting on the fluid element is taken as a negative stress. The constants μ and λ denote the proportionality of the direct stresses to change in shape and change in volume respectively. The definition of pressure within a fluid in motion poses a difficulty since, in general, a moving fluid will give rise to direct stresses on an element surface within it which will be a function of the surface orientation. Unlike a fluid at rest, therefore, the pressure in a moving fluid does not act equally in all directions.

A useful definition of pressure within a moving fluid is provided by taking the pressure as the negative of the average of the three normal stresses. Thus, the pressure p at a point in a moving fluid is defined as

$$p = -\tfrac{1}{3}(\tau_{xx} + \tau_{yy} + \tau_{zz}) \qquad (3.32)$$

in terms of the three normal stresses acting along any orthogonal axes at the point. It can readily be shown that the above equation is independent of the orientation of these orthogonal axes. This definition of pressure also has some physical significance in that it reflects the idea of pressure as a generalized 'squeezing' at a point due to the presence of the surrounding fluid.

Substitution of Equation (3.32) into Equation (3.31) leads to a simple relationship between μ and λ of

$$3\lambda + 2\mu = 0 \qquad (3.33)$$

Batchelor (1973) gives a rigorous proof of the above equation. For an incompressible fluid, the last terms of Equation (3.31) may be taken as zero. Equation (3.31) and (3.32) can then be written in tensor notation as

$$\tau_{ij} = -\mathrm{p}\,\delta_{ij} + 2\,\mu\varepsilon_{ij} \qquad (3.34)$$

where τ_{ij} denotes the shear stresses τ_{ij} for $i \neq j$ and the direct stress τ_{ij} for $i = j$. Here

$$\varepsilon_{ij} = \frac{\delta u_i}{\delta x_j}$$

is written with the repeated subscript summation described further in Appendix B. Substituting Equation (3.30) and (3.31) or (3.34) into

Equation (3.28) and using the equation of continuity, Equation (3.19) yields the Navier–Stokes equations for incompressible flow as

$$\left.\begin{aligned}
\frac{\partial u}{\partial t} + u\frac{\partial u}{\partial x} + v\frac{\partial u}{\partial y} + w\frac{\partial u}{\partial z} &= \frac{X}{\rho} - \frac{1}{\rho}\frac{\partial p}{\partial x} + v\left(\frac{\partial^2 u}{\partial x^2} + \frac{\partial^2 u}{\partial y^2} + \frac{\partial^2 u}{\partial z^2}\right)\\
\frac{\partial v}{\partial t} + u\frac{\partial v}{\partial x} + v\frac{\partial v}{\partial y} + w\frac{\partial v}{\partial z} &= \frac{Y}{\rho} - \frac{1}{\rho}\frac{\partial p}{\partial y} + v\left(\frac{\partial^2 v}{\partial x^2} + \frac{\partial^2 v}{\partial y^2} + \frac{\partial^2 v}{\partial z^2}\right)\\
\frac{\partial w}{\partial t} + u\frac{\partial w}{\partial x} + v\frac{\partial w}{\partial y} + w\frac{\partial w}{\partial z} &= \frac{Z}{\rho} - \frac{1}{\rho}\frac{\partial p}{\partial z} + v\left(\frac{\partial^2 w}{\partial x^2} + \frac{\partial^2 w}{\partial y^2} + \frac{\partial^2 w}{\partial z^2}\right)
\end{aligned}\right\} \tag{3.35}$$

For an inviscid flow, that is, with $v = 0$, the Navier–Stokes equations reduce to the Euler equations, which can be written as

$$\left.\begin{aligned}
\frac{\partial u}{\partial t} + u\frac{\partial u}{\partial x} + v\frac{\partial u}{\partial y} + w\frac{\partial u}{\partial z} &= \frac{X}{\rho} - \frac{1}{\rho}\frac{\partial p}{\partial x}\\
\frac{\partial v}{\partial t} + u\frac{\partial v}{\partial x} + v\frac{\partial v}{\partial y} + w\frac{\partial v}{\partial z} &= \frac{Y}{\rho} - \frac{1}{\rho}\frac{\partial p}{\partial y}\\
\frac{\partial w}{\partial t} + u\frac{\partial w}{\partial x} + v\frac{\partial w}{\partial y} + w\frac{\partial w}{\partial z} &= \frac{Z}{\rho} - \frac{1}{\rho}\frac{\partial p}{\partial z}
\end{aligned}\right\} \tag{3.36}$$

The solution of the Navier–Stokes or Euler equations of fluid motion is generally not possible for realistic flow fields because of the non-linear convective acceleration terms on the left-hand side of the equations. Even numerical solutions are only available for simple shapes and are at present impossible for flows with extensive turbulence levels.

3.6 The Bernoulli equation

The Navier–Stokes equations can be reduced to a simpler form of momentum equation if the conditions of inviscid, incompressible and irrotational flow are satisfied. Thus, taking $v = 0$, and substituting the irrotationality conditions of Equations (3.10) into the Navier–Stokes equations for incompressible flow Equation (3.35) gives

$$\left.\begin{aligned}
\frac{\partial^2 \phi}{\partial x\,\partial t} + u\frac{\partial u}{\partial x} + v\frac{\partial v}{\partial x} + w\frac{\partial w}{\partial x} &= -\frac{1}{\rho}\frac{\partial \Omega}{\partial x} - \frac{1}{\rho}\frac{\partial p}{\partial x}\\
\frac{\partial^2 \phi}{\partial y\,\partial t} + u\frac{\partial u}{\partial y} + v\frac{\partial v}{\partial y} + w\frac{\partial w}{\partial y} &= -\frac{1}{\rho}\frac{\partial \Omega}{\partial y} - \frac{1}{\rho}\frac{\partial p}{\partial y}\\
\frac{\partial^2 \phi}{\partial z\,\partial t} + u\frac{\partial u}{\partial z} + v\frac{\partial v}{\partial z} + w\frac{\partial w}{\partial z} &= -\frac{1}{\rho}\frac{\partial \Omega}{\partial z} - \frac{1}{\rho}\frac{\partial p}{\partial z}
\end{aligned}\right\} \tag{3.37}$$

where the left hand velocity term is written in term of velocity potential and the body force is expressed as a potential Ω such that

$$X = -\frac{\partial \Omega}{\partial x}, \; Y = -\frac{\partial \Omega}{\partial y} \text{ and } Z = -\frac{\partial \Omega}{\partial z} \tag{3.38}$$

Integrating Equation (3.37) with respect to x, y and z respectively yields

$$\left.\begin{aligned} \frac{\partial\phi}{\partial t} + \frac{1}{2}(u^2 + v^2 + w^2) + \frac{\Omega}{\rho} + \frac{p}{\rho} &= G_1(y, z, t) \\ \frac{\partial\phi}{\partial t} + \frac{1}{2}(u^2 + v^2 + w^2) + \frac{\Omega}{\rho} + \frac{p}{\rho} &= G_2(z, x, t) \\ \frac{\partial\phi}{\partial t} + \frac{1}{2}(u^2 + v^2 + w^2) + \frac{\Omega}{\rho} + \frac{p}{\rho} &= G_3(x, y, t) \end{aligned}\right\} \tag{3.39}$$

where G_1, G_2 and G_3 are functions that appear as constants in each partial integration. Since the left-hand sides of the above equations are identical, the right hand-side can only be a function of time to yield the equation

$$\frac{\partial\phi}{\partial t} + \frac{1}{2}q^2 + \frac{\Omega}{\rho} + \frac{p}{\rho} = G(t) \tag{3.40}$$

where $q^2 = u^2 + v^2 + w^2$. For the z axis pointing vertically upwards, this equation can also be written as

$$\frac{\partial\phi}{\partial t} + \frac{1}{2}q^2 + gz + \frac{p}{\rho} = G(t) \tag{3.41}$$

with $\Omega = \rho g z$, where g is the acceleration due to gravity and $G(t)$ is a function arising from the partial integration. For steady flow this becomes

$$\frac{1}{2}q^2 + gz + \frac{p}{\rho} = \text{constant} \tag{3.42}$$

since the $G(t)$ function may contain a constant term. Equations (3.40)–(3.42) are different forms of the Bernoulli equation.

3.7 The complex potential

The complex potential and associated conformal transformations offer a powerful method for calculating two-dimensional patterns of irrotational flow. Conformal transformations allow a wide range of flow boundaries and immersed body shapes to be deduced from algebraic transformations of the complex potential. The calculated flow pattern becomes available as a single function of a complex variable which embodies both the stream function and the velocity potential. Some of the idealized flows obtained from this technique are also useful as the building blocks of complex flows evaluated through computer based numerical methods. A classic example in the dynamics of offshore structures is the use of oscillating source–sinks in a boundary integral wave diffraction calculation (see Chapter 5).

Appendix C presents a summary of fundamental mathematical results for complex number theory. Now, a complex number $Z = x + iy$ can denote a point (x, y) on an Argand diagram or an Oxy plane. If a further complex variable $W = \phi + i\psi$ is defined such that

$$W = f(Z) \tag{3.43}$$

then this single complex potential function can denote the relationships between ϕ, ψ, x and y as

$$\phi = f_1(x, y) \text{ and } \psi = f_2(x, y) \tag{3.44}$$

But first, the meaning of the derivative $\mathrm{d}W/\mathrm{d}Z$ needs to be examined. Consider the points $P(Z = x + \mathrm{i}y)$ and $Q(Z + \delta Z = x + \delta x, \mathrm{i}\{y + \delta y\})$ on an Argand diagram, where PQ makes an angle α to the horizontal at P; W and $W + \mathrm{d}W$ may be taken as the values of a complex potential function $W = W(Z)$ at P and Q respectively. Then, to first order

$$\delta W = \frac{\partial W}{\partial x}\delta x + \frac{\partial W}{\partial y}\delta y$$

$$= \left(\frac{\partial W}{\partial x}\cos\alpha + \frac{\partial W}{\partial y}\sin\alpha\right)\delta s \tag{3.45}$$

where $PQ = \delta s$. But

$$\delta z = \delta x + \mathrm{i}\,\delta y$$

$$= (\cos\alpha + \mathrm{i}\sin\alpha)\,\delta s \tag{3.46}$$

Thus, in the limit

$$\frac{\mathrm{d}W}{\mathrm{d}Z} = \lim_{Q\Rightarrow P}\frac{\delta W}{\delta Z} = \frac{(\partial W/\partial x)\cos\alpha + (\partial W/\partial y)\sin\alpha}{\cos\alpha + \mathrm{i}\sin\alpha}$$

$$= \frac{(\partial W/\partial x) + (\partial W/\partial y)\tan\alpha}{1 + \mathrm{i}\tan\alpha} \tag{3.47}$$

This limit will not have a unique value since it depends on angle α, that is, the direction along which point Q tends to P. The limit can be taken as unique (independent of α) if and only if

$$\frac{\partial W}{\partial x} = \frac{1}{\mathrm{i}}\frac{\partial W}{\partial y} \tag{3.48}$$

If function W satisfies this condition and is continuous, finite and single-valued in a region R, then it can be said to be analytic (regular, holomorphic or monogenic) in the region R. It needs to be pointed out that analytic functions must contain x and y in explicit combinations of $x + \mathrm{i}y$ (and not $x - \mathrm{i}y$) and are uniquely differentiable everywhere except at singularities.

Thus the condition that W be analytic becomes

$$\frac{\partial W}{\partial x} = \frac{\partial \phi}{\partial x} + \mathrm{i}\frac{\partial \psi}{\partial x} = \frac{1}{\mathrm{i}}\frac{\partial W}{\partial y} = \frac{1}{\mathrm{i}}\left(\frac{\partial \phi}{\partial y} + \mathrm{i}\frac{\partial \psi}{\partial y}\right) \tag{3.49}$$

Equating real and imaginary parts gives

$$\frac{\partial \phi}{\partial x} = \frac{\partial \psi}{\partial y} \text{ and } \frac{\partial \phi}{\partial y} = -\frac{\partial \psi}{\partial x} \tag{3.50}$$

These are called the Cauchy–Reimann conditions. By differentiating the Cauchy–Riemann equations with respect to x and y, and adding, it can be shown tht ϕ and ψ both satisfy Laplace's (that is the continuity) equation.

Velocities can be obtained from the complex potential by using the Cauchy–Riemann conditions to give

$$u = \frac{\partial \phi}{\partial x} = \frac{\partial \psi}{\partial y} \tag{3.51}$$

and

$$v = \frac{\partial \phi}{\partial y} = -\frac{\partial \psi}{\partial x} \tag{3.52}$$

Also

$$\frac{\mathrm{d}W}{\mathrm{d}Z} = \frac{\partial \phi}{\partial x} + \mathrm{i}\frac{\partial \psi}{\partial x} = -\mathrm{i}\left(\frac{\partial \phi}{\partial y} + \mathrm{i}\frac{\partial \psi}{\partial y}\right) = u - \mathrm{i}v \tag{3.53}$$

This is not the velocity vector but its complex conjugate. Thus the velocity vector is given by

$$u + \mathrm{i}v = \overline{\frac{\mathrm{d}W}{\mathrm{d}Z}} \tag{3.54}$$

where the overbar denotes the complex conjugate and the square of the total velocity q may be written as

$$q^2 = u^2 + v^2 = \left(\frac{\partial W}{\partial Z}\right)\left(\overline{\frac{\partial W}{\partial Z}}\right)$$

Now it is not usually possible to solve directly for the flow past an arbitrary boundary. It is, therefore, necessary to construct more flow patterns flows by using two techniques:

1. The superposition of very simple flows whose complex potential can be readily obtained to yield useful but complicated flow fields. This technique can be applied in conjunction with a method of images.
2. The conformal transformation of simple flows or flows obtained by superposition can also lead to flow patterns that are directly useful in practical engineering problems.

3.8 Idealized flows

The stream functions, velocity potentials and complex potentials of most of the well-known simple flows are given below:

(a) *Uniform flow*
(i) *Parallel to x axis*
In this case, the velocity components parallel to the x and y axes, u and v are given by $u = U$, $v = 0$. Thus

$$u = \frac{\partial \psi}{\partial y}, \ \psi = Uy + c$$

with $c = 0$ if $\psi = 0$ at $y = 0$. Thus the stream function, velocity potential and complex potential are given by

$$\psi = Uy, \ \phi = Ux \text{ and } W = UZ \tag{3.55}$$

(ii) *Parallel to y axis*
Here the velocity components are

$$u = 0, v = V$$

such that

$$\psi = -Vx + c$$

with $c = 0$ for $\psi = 0$ at $x = 0$. Then the stream function, velocity potential and complex potential are

$$\psi = -Vx, \phi = Vy \text{ and } W = -\mathrm{i}Vz \tag{3.56}$$

(iii) *In a direction giving velocity components U, V.*
Using the velocity components $u = U$ and $v = V$, the stream function and velocity potential can be obtained by integration as above or by addition of functions for the flows in (i) and (ii). This yields

$$\psi = Uy - Vx, \phi = Ux + Vy$$

with a complex potential

$$W = (U - \mathrm{i}V)Z \tag{3.57}$$

which can also be written in another form as

$$\left.\begin{aligned} &W = Q\mathrm{e}^{-\mathrm{i}\alpha}\cdot Z \\ &\text{where} \\ &Q = \surd(U^2 + V^2) \text{ and } \alpha = \arctan\left(\frac{V}{U}\right) \end{aligned}\right\} \tag{3.58}$$

(b) *Source or sink flows*
There are artificial flow patterns which are used as building blocks for more complex flow, particularly in numerical methods which seeks to obtain steady or oscillating flows around bodies of complicated shape.

A source flow is one where fluid appears from an infinitesimal point at a rate m and flows out along radial lines. Thus in polar co-ordinates in a two-dimensional plane, the tangential velocity of the fluid $q_\theta = 0$. The radial velocity q_r can be related to the strength m of the source by equating m to the fluid crossing the circumference of a circle of radius r centred at the source. This yields

$$q_r = \frac{m}{2\pi r} \tag{3.59}$$

Since all the stream lines pass through the origin, they are defined by taking $\theta = 0$ instead of $\psi = 0$. If $\theta = 0$ is measured anti-clockwise from the x axis, then

$$\psi = \frac{m\theta}{2\pi}$$

for a stream line at an angle θ to the x axis. Since the stream lines are radial, the condition $|\theta| < \pi$ applies. Thus the stream function and velocity potential are

$$\psi = \frac{m\theta}{2\pi} \text{ and } \phi = \frac{m}{2\pi}\ln r \tag{3.60}$$

in polar co-ordinates. The resultant velocity components are

$$u = \frac{m}{2\pi}\frac{\cos\theta}{r},\ v = \frac{m}{2\pi}\frac{\sin\theta}{r}.$$

These results can be written in Cartesian co-ordinates as,

$$\left.\begin{aligned}
\psi &= \frac{m}{2\pi}\arctan\frac{y}{x} = \frac{m}{2\pi}\cos^{-1}\frac{x}{(x^2+y^2)} \\
\phi &= \frac{m}{2\pi}\ln[\sqrt{(x^2+y^2)}] \\
u &= \frac{m}{2\pi}\left(\frac{x}{x^2+y^2}\right),\ v = \frac{m}{2\pi}\left(\frac{y}{x^2+y^2}\right)
\end{aligned}\right\} \tag{3.61}$$

Also the complex potential is

$$\left.\begin{aligned}
W &= \phi + \mathrm{i}\,\psi = \frac{m}{2\pi}(\ln r + \mathrm{i}\,\theta) = \frac{m}{2\pi}\ln r\mathrm{e}^{\mathrm{i}\theta} \\
W &= \frac{m}{2\pi}\ln Z \\
\frac{\mathrm{d}W}{\mathrm{d}Z} &= \frac{m}{2\pi Z} = u - \mathrm{i}v
\end{aligned}\right\} \tag{3.62}$$

It should be noted that $\mathrm{d}W/\mathrm{d}Z$ is singular at the source origin $Z = 0$.

A sink flow is simply defined as the reverse of source flow such that fluid flows into and vanishes at a point in a plane. All the above results apply for a sink flow with m taken as negative.

In the calculation of wave flows, it is common to use pulsating or oscillating sources which have a mean of zero but oscillate sinusoidally with a prescribed amplitude strength and phase. The use of such oscillating sources in three dimensions within a boundary integral diffraction theory is described in more detail in Chapter 5.

The three simple flows described above can be used to define additional more complex flows by superposition. This is illustrated further in the next three examples.

(i) *Source–sink pair*

Consider a source and sink of equal strength placed symmetrically on the x axis of an Oxy plane at distance $\pm d$ from the origin. The stream function, velocity potential and complex potential of the flow can be obtained by superposition. Thus

$$\left.\begin{aligned}
\psi &= \frac{m}{2\pi}\left[\arctan\left(\frac{y}{x-d}\right) - \arctan\left(\frac{y}{x+d}\right)\right] \\
\phi &= \frac{m}{2\pi}\{\ln[(x-d)^2+y^2]^{1/2} - \ln[(x+d)^2+y^2]^{1/2}\} \\
W &= \frac{m}{2\pi}[\ln(Z-d) - \ln(Z+d)]
\end{aligned}\right\} \tag{3.63}$$

Since stream lines are of constant ψ, Equation (3.63) can be reduced as follows for constant ψ:

$$\tan\left(\frac{2\pi\psi}{m}\right) = \tan\left(\arctan\frac{y}{x-d} - \arctan\frac{y}{x+d}\right)$$
$$= \frac{[y/(x-d)] - [y/(x+d)]}{1 + [y^2/(x^2-d^2)]}$$

Simplification gives

$$x^2 + \left[y - d\cot\left(\frac{2\pi\psi}{m}\right)\right]^2 = d^2\operatorname{cosec}^2\left(\frac{2\pi\psi}{m}\right) \tag{3.64}$$

which proves that the stream lines around the source–sink pair are all circles that pass through the source and sink points $(\pm d, 0)$. Furthermore, as $d \Rightarrow 0$, the source and sink coincide and cancel each other out exactly.

(ii) *Doublet*

A doublet flow is obtained by placing the source–sink pair described above at co-ordinates $(0, \pm d)$ and allowing them to converge to the origin and be separated by a very small distance which is such that the doublet strength μ is a finite constant given by

$$\mu = 2md \tag{3.65}$$

where the source–sink strengths $\pm m$ would be large. Then, for small d, the source–sink functions become

$$\left.\begin{aligned} &\psi = \frac{\mu}{2\pi}\frac{y}{x^2+y^2}, \quad \phi = -\frac{\mu}{2\pi}\frac{x}{x^2+y^2} \\ &\text{and} \\ &W = -\frac{\mu}{2\pi Z} \end{aligned}\right\} \tag{3.66}$$

The equation of the stream lines become

$$x^2 + \left(y - \frac{\mu}{4\pi\psi}\right)^2 = \left(\frac{\mu}{4\pi\psi}\right)^2 \tag{3.67}$$

These lines are circles that pass through the origin with their centres lying on $x = 0$. Figure 3.6(a) illustrates a doublet flow. Note that the y axis is tangential to all the stream lines and is known as the axis of the doublet with a positive direction from sink to source. The velocity field around a doublet is given by

$$\left.\begin{aligned} u &= \frac{\partial\psi}{\partial y} = \frac{\mu}{2\pi}\frac{x^2-y^2}{(x^2+y^2)^2} = \frac{\mu}{2\pi}\frac{\cos 2\theta}{r^2} \\ v &= -\frac{\partial\psi}{\partial x} = \frac{\mu}{2\pi}\frac{2xy}{(x^2+y^2)^2} = \frac{\mu}{2\pi}\frac{\sin 2\theta}{r^2} \end{aligned}\right\} \tag{3.68}$$

(iii) *Uniform flow combined with a source–sink pair*

A uniform flow parallel to the x axis superimposed on to a source–sink pair at $(\pm d, 0)$ with the source in the upstream part of the uniform flow

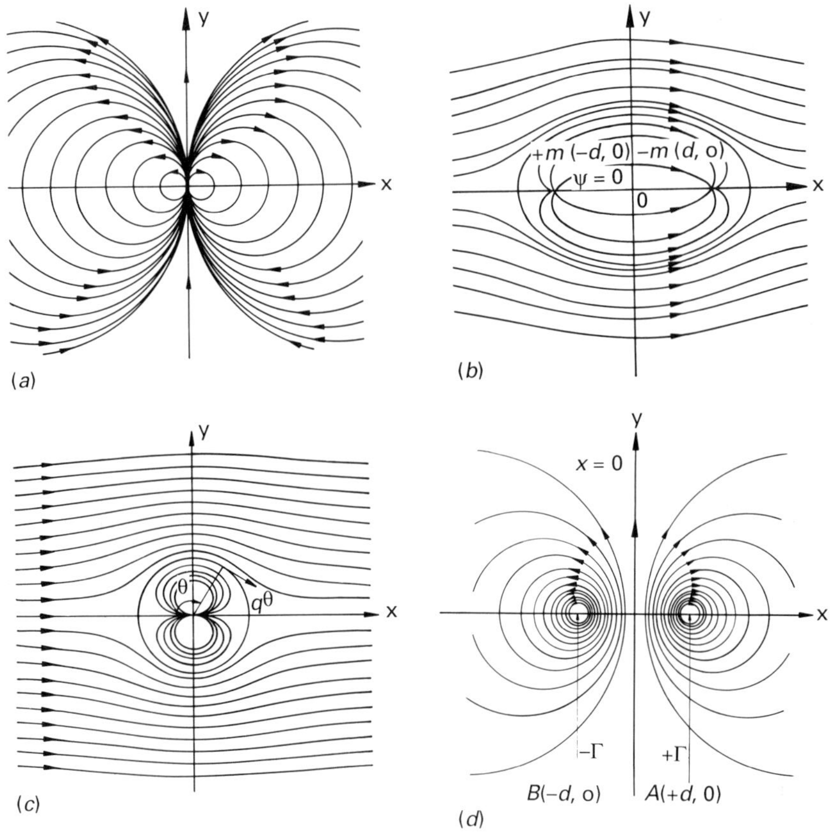

Figure 3.6. (*a*) Doublet flow. (*b*) Uniform flow with a source–sink pair. (*c*) Uniform flow with doublet. (*d*) Two-line vortices

gives a flow field of the type shown in Figure 3.6(*b*). The stream function, velocity potential and complex potential for the flows are

$$\left.\begin{aligned} \psi &= Uy + \frac{m}{2\pi}\left(\arctan\frac{y}{x+d} - \arctan\frac{y}{x-d}\right) \\ \phi &= Ux + \frac{m}{2\pi}\left\{\ln[(x-d)^2+y^2]^{1/2} - \ln[(x+d)^2+y^2]^{1/2}\right\} \\ W &= Uz + \frac{m}{2\pi}[\ln(Z-d) - \ln(Z+d)] \end{aligned}\right\} \quad (3.69)$$

It can readily be shown that the stagnation points of the flow are at $(\pm x_0, 0)$ where

$$x_0 = d\sqrt{\left(1 + \frac{m}{\pi U d}\right)} \quad (3.70)$$

and that the equation of the dividing stream line between the inner and outer surface is

$$\frac{x^2}{d^2} + \frac{y^2}{d^2} = 1 + \frac{2y}{d}\cot\left(\frac{2\pi Uy}{m}\right) \tag{3.71}$$

This dividing stream line has a shape which is known as the Rankine oval. The shape of the dividing stream line and the flow field both inside and outside this stream line are illustrated in Figure 3.6(*b*). If this dividing stream line was replaced by an oval body surface, the outer flow would still be given by Equations (3.69). This illustrates the technique for generating flow fields by superposition of simple idealized flows.

(iv) *Uniform flow combined with a doublet*
Taking the uniform flow in the positive x axis and the axis of the doublet along the negative x axis, we obtain

$$\left.\begin{aligned} \psi &= Uy - \frac{\mu}{2\pi}\frac{y}{x^2 + y^2} \\ \phi &= Ux + \frac{\mu}{2\pi}\frac{x}{x^2 + y^2} \\ W &= UZ + \frac{\mu}{2\pi Z} \end{aligned}\right\} \tag{3.72}$$

It can again be readily shown that the stagnation points of the flow are at $(\pm x_0, 0)$ where

$$x_0 = \sqrt{\left(\frac{\mu}{2\pi U}\right)} \tag{3.73}$$

and that the equation of the dividing stream lines are

$$y = 0$$

and

$$x^2 + y^2 = \frac{\mu}{2\pi U} \tag{3.74}$$

The latter equation denotes a circle of radius $\sqrt{(\mu/2\pi U)}$ centred at the origin. This flow is illustrated in Figure 3.6(*c*). The flow outside the dividing stream line is that around a circular cylinder.

Expressing Equation (3.72) in polar co-ordinates and differentiating yields the tangential velocity q_θ on the dividing stream line as being

$$q_\theta = 2U\sin\theta \tag{3.75}$$

where θ and q_θ are defined in Figure 3.6(*c*). The remaining velocity and pressure field can also be obtained from Equations (3.72).

(v) *A line vortex*
A line vortex offers an idealized flow by which the effects of vorticity or circulating irrotational flow can be incorporated into a flow field. Consider an infinitely long isolated straight vortex of circulation Γ at the origin of the Oxy plane with the vortex line being perpendicular to the plane. The flow will be parallel in all planes perpendicular to the vortex

line and will vary in the radial direction only. The radial velocity q_r around the vortex line must be zero since there is no source or sink at the vortex line, and thus the tangential velocity q_θ will be a function of radial distance r only. The integral of the tangential flow velocity with distance around a closed circuit encircling the vortex is defined as the circulation Γ. Thus

$$\Gamma = \oint q_\theta \, \mathrm{d}s \tag{3.76}$$

where q_θ is the tangential velocity and s is the distance around the closed circuit. Thus, for a circular circuit of radius r,

$$q_\theta = \frac{\Gamma}{2\pi r} \tag{3.77}$$

since q_θ is a function of radial distance r only. This equation and the earlier remarks demonstrate that the stream lines around the vortex are all circles centred at the origin but the vortex centre itself is a singularity (for $r = 0$).

In polar co-ordinates, the stream function, velocity potential and complex potential of the vortex are

$$\left.\begin{aligned} \psi &= -\frac{\Gamma}{2\pi}\ln r, \; \phi = \frac{\Gamma}{2\pi}\theta \\ W &= -\frac{\mathrm{i}\Gamma}{2\pi}\ln Z \end{aligned}\right\} \tag{3.78}$$

This can be written in Cartesian co-ordinates as

$$\psi = -\frac{\Gamma}{2\pi}\ln(x^2 + y^2)^{1/2}, \; \phi = \frac{\Gamma}{2\pi}\arctan\left(\frac{y}{x}\right) \tag{3.79}$$

and the velocity components become

$$\left.\begin{aligned} u &= -\frac{\Gamma y}{2\pi(x^2 + y^2)} \\ v &= \frac{\Gamma x}{2\pi(x^2 + y^2)} \end{aligned}\right\} \tag{3.80}$$

It should be emphasized here that the line vortex defines an idealized irrotational flow field. In real fluids, the effect of viscosity causes a central circular filament of fluid to rotate as a solid core with the tangential velocity increasing linearly with radius. This linear velocity distribution transitions into the irrotational distribution at the edge of the core.

(vi) *A vortex pair*

The flow field induced by two vortices of circulation $+\Gamma$ and $-\Gamma$ at points $(d, 0)$ and $(-d, 0)$, respectively, in the Oxy plane is given by the functions

$$\left.\begin{aligned} \psi &= -\frac{\Gamma}{4\pi}\ln\frac{x^2+y^2+d^2-2xd}{x^2+y^2+d^2+2xd} \\ \phi &= \frac{\Gamma}{2\pi}\left(\arctan\frac{y}{x-d} - \arctan\frac{y}{x+d}\right) \\ W &= -\frac{\mathrm{i}\Gamma}{2\pi}\left[\ln\left(\frac{Z-d}{Z+d}\right)\right] \end{aligned}\right\} \quad (3.81)$$

The equation of a stream line is then obtained by taking ψ as constant in Equation (3.81) and simplifying to yield

$$x^2 + y^2 - 2xd\left(\frac{1+k}{1-k}\right) + d^2 = 0 \quad (3.82)$$

where

$$k = \exp\left(-\frac{4\pi\psi}{\Gamma}\right).$$

This is a system of circles of radii varying from 0 to ∞ and centres

$$\left[\pm\frac{d(1+k)}{1-k}, 0\right]$$

Figure 3.6(d) illustrates the stream line pattern obtained.

The point vortex is commonly used in numerical calculations for representing the flow field generated by vorticity shed from a solid body in the flow field.

The flow due to a vortex pair also demonstrates another flow modelling technique called the method of images. This method is used to solve a number of problems in mathematical physics by considering optical images of physical phenomena in plane or curved walls. Thus the flow due to vortex A of circulation Γ at $(d, 0)$ in the vicinity of a plane wall $x = 0$ (see figure 3.6(d)) can be modelled by inserting an image of vortex A in the reflecting line $x = 0$. Thus an image vortex B of circulation $-\Gamma$ at $(-d, 0)$ is introduced into the flow. The effect of the 'image' and 'object' vortex flows are to make the line $x = 0$ a stream line and thus satisfy the no flow through the wall boundary condition. Thus the flow in the $x > 0$ part of the plane models the flow due to the vortex at $(d, 0)$ and the wall $x = 0$. This technique is equally valid for multiple images obtained from multiple planes or curved 'reflecting' surfaces.

3.9 Conformal transformations

The complex potential function $W = f(Z)$ offers the means of transforming simple flow patterns built from those described in the preceding section into more complicated flows which are often closer to useful applications.

In the complex potential, function Z denotes a point in the Oxy plane and W denotes a point in the $O\psi\phi$ plane. A point $P(x, y)$ on the Z plane can be transformed onto $P'(\psi, \phi)$ in the $O\psi\phi$ plane which represents the components of potential of the flow. If $W = f(Z)$ is analytic, any point P in the Z plane has a corresponding point P' on the W plane. This also holds true for regions of each plane where if P traces out the boundary of region R, then P' will trace out the boundary of a corresponding region R'.

This transformation technique can be used to distort known flow patterns to obtain further useful flow regimes. Now using W as the complex potential, Z as a point on the Oxy plane and ζ as a point on the $O\xi\eta$ plane, a function $\zeta = \zeta(Z)$ can be used to transform any point on the Z plane to one or more points on the ζ plane. There may not be one-to-one correspondence between transformed points since two or more points in the Z plane may map onto a single point on the ζ plane. Cuts are then used to obtain correspondence between regions of each plane.

It can be proved that the mapping performed by an analytic function $\zeta = \zeta(Z)$ is conformal (that is, angles within its family of curves are preserved both in magnitude and sense) at all points of the Z plane where $\zeta'(Z) \neq 0$. Elements of length are then stretched in the ratio

$$\left|\frac{d\xi}{dZ}\right|$$

Furthermore, the mapping must exhibit a one-to-one correspondence. Thus orthogonal lines of ϕ and ψ constant in the Z plane will transform into an orthogonal system in the ζ plane. A complex potential function $W = H(\zeta)$ in the plane can be transformed by an equation $\zeta = \zeta(Z)$ to form the complex potential function

$$W = H[\zeta(Z)] = G(Z) \tag{3.83}$$

in the Z plane.

The velocity components (u, v) in the Z plane are obtained from

$$\frac{dW}{dZ} = u - iv \tag{3.84}$$

and (u', v') in the ζ plane are

$$\frac{dW}{d\zeta} = u' - iv' \tag{3.85}$$

The velocities in the two planes can be related as follows

$$\frac{dW}{dZ} = \frac{dW}{d\zeta}\cdot\frac{d\zeta}{dZ} = (u' - iv')\zeta'(Z)$$

Also

$$\left|\frac{dW}{dZ}\right| = |u - iv| = |u' - iv'|\cdot|\zeta'(Z)| \tag{3.86}$$

Then, the total velocities in the Z and ζ plane are related by the equation

$$q_Z = q_\zeta \left| \frac{d\zeta}{dZ} \right| \tag{3.87}$$

A very large body of research literature exists on the multitude of transformation functions $\zeta = \zeta(Z)$ and their corresponding original and transformed flow patterns. Many of these research results have led to useful applications in aerodynamics, fluid mechanics and in fluid loading on offshore structures.

A simple example of such a transformation is presented here to illustrate this technique. A uniform flow in the Z plane is mapped onto the ζ plane by using the transformation

$$\zeta = Z^2 \tag{3.88}$$

Consider the properties of the Z to ζ plane transformation first. Using

$$\zeta = \xi + i\eta = pe^{i\lambda} = p(\cos\lambda + i\sin\lambda)$$

and

$$Z = x + iy = re^{i\theta} = r(\cos\theta + i\sin\theta)$$

with the transformation gives

$$p = r^2, \lambda = 2\theta \tag{3.89}$$

Examination of these equations shows that the positive real axis of the Z plane transforms onto the positive real axis of the ζ plane, but the positive imaginary axis of the Z plane transforms onto the negative real axis of the ζ plane. The angle between radial lines doubles during the transformation. Circles on the Z plane remain circles but change their radius according to the equation $p = r^2$. The quarter-Z plane $(x > 0, y > 0)$ becomes the positive half-ζ plane $(\eta > 0)$. Also the half-Z plane ($y > 0$ and also $y < 0$) becomes the whole ζ plane. Thus a cut can be made at the negative y axis in the Z plane and the resultant transformed surface can be used.

The uniform flow

$$W = U\zeta$$

in the ζ plane then transforms to

$$\left.\begin{aligned} W &= UZ^2 \\ \phi + i\psi &= U(x^2 - y^2) + iU \cdot 2xy \end{aligned}\right\} \tag{3.90}$$

Figure 3.7 illustrates the stream lines and velocity potentials in the ζ and Z planes with half of the ζ plane mapping on to the positive quadrant of the Z plane. The stream lines and equipotentials in the Z plane are given by

$$\left.\begin{aligned} &2xy = \text{constant} \\ &\text{and} \\ &x^2 - y^2 = \text{constant} \end{aligned}\right\} \tag{3.91}$$

respectively. The properties of a corner flow can then be deduced from the positive quadrant of the Z plane.

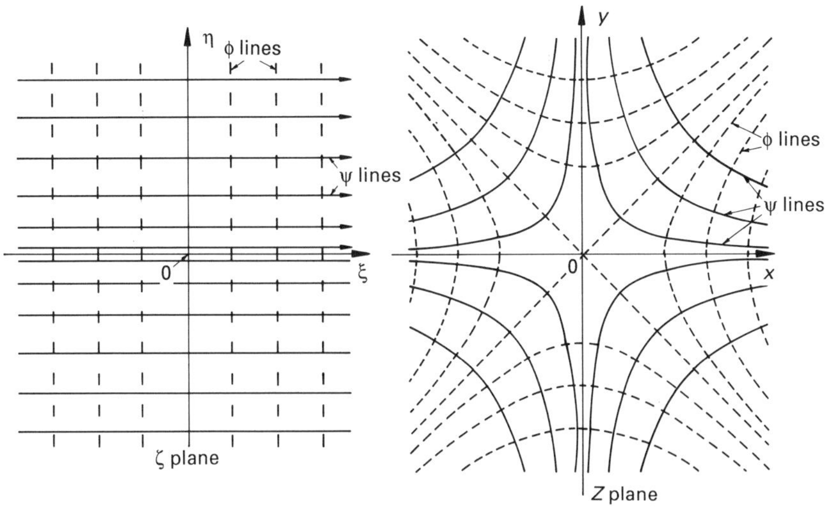

Figure 3.7. Conformal transformation planes

3.10 The added mass concept

The added mass concept arises from the tendency of a submerged body moving with an acceleration relative to the surrounding fluid to induce accelerations to the fluid. These fluid accelerations require forces which are exerted by the body through a pressure distribution of the fluid on the body. Since the submerged body, in effect, imparts an acceleration to some of the surrounding fluid, this phenomenon can be equated to the body having an added mass of fluid attached to its own physical mass. The concept is useful when performing calculations involving relative accelerations between body and fluid.

The simple case of a body moving in a straight line without rotation is first considered below. Consider a rigid body B shown in Figure 3.8 moving in a straight line (without rotation) in a perfect fluid which is unbounded and at rest, except that it will be disturbed by the motion of B.

If the instantaneous velocity of B is U then the velocity of the fluid at any point P (which occupies a fixed position relative to the body) is proportional to U. If u, v, w are the velocity components of the fluid at point P, then, the equations

$$u = Uf_u \quad v = Uf_v \quad w = Uf_w \tag{3.92}$$

apply where f_u, f_v, f_w are the values of u, v and w for unit U and are functions of the co-ordinates of P relative to the body. Thus, a rectangular element of dimensions dx, dy, dz of the fluid has kinetic energy

$$E = \tfrac{1}{2}\rho U^2(f_u^2 + f_v^2 + f_w^2)\,\mathrm{d}x\,\mathrm{d}y\,\mathrm{d}z$$

and the total kinetic energy of the fluid is

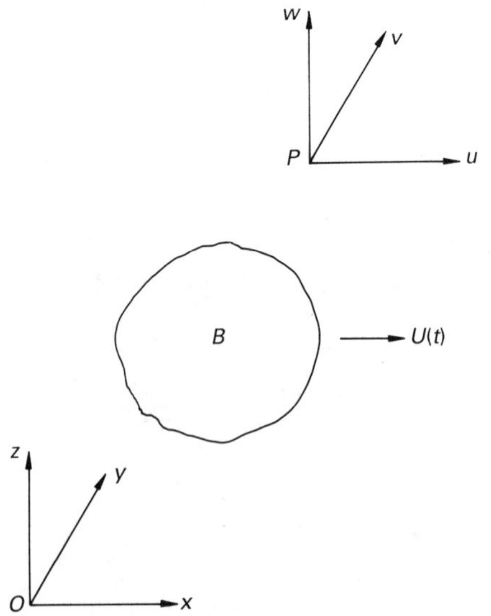

Figure 3.8. Axes system for added mass formulations

$$E = \tfrac{1}{2}\rho U^2 \iiint\limits_{\text{all fluid volume}} (f_u^2 + f_v^2 + f_w^2)\,\mathrm{d}x\,\mathrm{d}y\,\mathrm{d}z \tag{3.93}$$

where the integral covers the whole region occupied by the fluid. For convenience, Equation (3.93) can be reduced to

$$E = \tfrac{1}{2}k\rho U^2 \tag{3.94}$$

where

$$k = \iiint\limits_{\text{all fluid volume}} (f_u^2 + f_v^2 + f_w^2)\,\mathrm{d}x\,\mathrm{d}y\,\mathrm{d}z \tag{3.95}$$

is a constant for a given body and motion, its value depending on the attitude of the body to the direction of its motion.

Now if the velocity of the body varies with time, then so does the energy of the surrounding fluid and, by the principle of conservation of energy, work must be done on the fluid. Let X be the component in the direction of motion of the total force which rigid body B exerts on the fluid. Then $-X$ is the corresponding reaction which the fluid exerts on the body. The work done while B moves through a small distance $\mathrm{d}x$, is

$$X\,\mathrm{d}x = XU\,\mathrm{d}t = \mathrm{d}E$$

But

$$\mathrm{d}E = k\rho U\frac{\mathrm{d}U}{\mathrm{d}t}\cdot\mathrm{d}t$$

By equating dEs,

$$X = k\rho\frac{\mathrm{d}U}{\mathrm{d}t} \tag{3.96}$$

The body experiences a resistance to its acceleration because it is immersed in a fluid. This resistance or acceleration reaction force equals the product of the body's acceleration and $k\rho$ which can be regarded as the added mass of the body for translational motion in the direction concerned. This added mass force is zero when the acceleration is zero. Then the resistance to motion is due to skin friction and pressure drag only.

The above generalization assumes the fluid to be at rest with no obstacles or boundaries nearby. If the fluid had a fixed boundary, the values of f_u, f_v, f_w would also depend on the position of B relative to the boundary. The value of the integral contributing to k would then be modified and the resultant added mass value would change. In general, added mass calculations are performed for a perfect, inviscid fluid. It can be shown that viscosity has very small effects and results obtained from inviscid fluid theory are sufficiently accurate. Two sample added mass calculations are given below as examples, whereas Appendix G presents tables of added mass coefficients for a variety of body shapes.

(a) *Added mass of an infinite circular cylinder*

Consider a cylinder of radius a moving with velocity U in a direction Ox perpendicular to the axis of the cylinder. Since the surrounding flow would be two-dimensional for an infinite cylinder, a layer of fluid of unit depth along the axis of the cylinder is considered.

The velocity potential of the flow fluid around the cylinder is

$$\phi = -\frac{Ua^2x}{x^2+y^2} = -\frac{Ua^2\cos\theta}{r} \text{ with } x = r\cos\theta \tag{3.97}$$

in Cartesian and polar co-ordinates. Note that this flow is obtained from the velocity potential of Section 3.8(f) by subtracting the flow defined by Equation (3.72) from steady flow U. This yields the case of a cylinder moving along the positive x axis in a stationary fluid. Then

$$u = \frac{\partial\phi}{\partial x} = \frac{Ua^2\cos 2\theta}{r^2}, \quad v = \frac{\partial\phi}{\partial y} = \frac{Ua^2\sin 2\theta}{r^2}, \quad w = 0$$

and

$$f_u = \frac{u}{U} = \frac{a^2\cos 2\theta}{r^2}, \quad f_v = \frac{v}{U}\,\frac{a^2\sin 2\theta}{r^2}, \quad f_w = 0$$

Hence

$$\begin{aligned}\rho k &= \rho\int\int (f_u^2 + f_v^2)\,\mathrm{d}x\,\mathrm{d}y \\ &= \rho\int_0^{2\pi}\int_a^{\infty}\left(\frac{a^4\cos^2 2\theta}{r^4} + \frac{a^4\sin^2 2\theta}{r^4}\right) r\,\mathrm{d}\theta\,\mathrm{d}r \\ &= \rho\pi a^2 \end{aligned} \tag{3.98}$$

The added mass force is then

$$X = \rho\pi a^2\frac{\mathrm{d}U}{\mathrm{d}t} = \rho C_{\mathrm{m}}V\frac{\mathrm{d}U}{\mathrm{d}t} \tag{3.99}$$

where V is the cylinder volume, an added mass coefficient $C_{\mathrm{m}} = 1.0$

and $\rho C_m V$ can be treated as an added mass.

The added mass coefficient is used as a parameter which can account for the influence of flow around different body shapes and the effects of non-ideal features such as separating flows and turbulence which are not accounted for in potential flow calculations.

(b) *Added mass of a sphere*

The velocity potential for a sphere of radius a moving with velocity U along the Ox direction is

$$\phi = -\frac{Ua^3x}{2(x^2 + y^2 + z^2)^{3/2}} = -\frac{Ua^3x}{2r^3} \tag{3.100}$$

as given by Lamb (1975). The velocity components of the fluid are

$$u = \frac{Ua^3(3x^2 - r^2)}{2r^5}, \quad v = \frac{3Ua^3xy}{2r^5}, \quad w = \frac{3Ua^3xz}{2r^5} \tag{3.101}$$

and when substituted into Equations (3.92) and (3.93) yield the result

$$k = \tfrac{2}{3}\pi a^3$$

and the added mass force

$$X = \rho C_m\left(\frac{4}{3}\pi a^3\right)\cdot\frac{dU}{dt} \tag{3.102}$$

where the added mass coefficient $C_m = 0.5$.

3.11 Added mass coefficient – general form

The above calculation of added mass can be generalized to apply for a rigid body having six degrees of freedom, that is, linear motions along three mutually perpendicular axes and rotations about these axes. All these motions give rise to separate potential functions and the interaction between the six motions gives rise in the most general case to a (6 × 6) matrix of added mass coefficients.

Using the definition diagram of Figure 3.8, the total kinetic energy E of a fluid due to arbitrary translational and rotational motion of the body B is given by

$$E = \frac{1}{2}\rho \underset{\substack{\text{over fluid}\\ \text{volume}}}{\iiint} \left[\left(\frac{\partial\phi}{\partial x}\right)^2 + \left(\frac{\partial\phi}{\partial y}\right)^2 + \left(\frac{\partial\phi}{\partial z}\right)^2\right] dx\,dy\,dz \tag{3.103}$$

Using the second form of Green's theorem (see Appendix D or Bronshtein and Semendyayev, 1979) with $\nabla^2\phi = 0$ for the velocity potential ϕ, we get

$$E = -\frac{1}{2}\rho\int_S \phi\frac{\partial\phi}{\partial n}\,dS \tag{3.104}$$

where S is the surface enclosing the fluid volume, dS denotes a surface element and n is an outward normal to the body.

Now the velocity potential function for the general motion of an arbitrary body in three dimensions can be written as

$$\phi = \phi_1 u_1 + \phi_2 u_2 + \phi_3 u_3 + \phi_4 u_4 + \phi_5 u_5 + \phi_6 u_6 \tag{3.105}$$

where u_1 to u_6 denote the three linear and three angular velocities of the body about orthogonal axes, and ϕ_1 to ϕ_6 denote corresponding velocity potential functions for the flow field around the body.

Substituting Equation (3.105) into Equation (3.104) gives, for a body of volume V_0,

$$\left.\begin{aligned} E &= \frac{1}{2}\rho V_0 \alpha_{ij} u_i u_j \\ \text{where} & \\ \alpha_{ij} &= -\frac{1}{V_0}\int_S \phi_i \frac{\partial \phi_j}{\partial n} \cdot \mathrm{d}S \end{aligned}\right\} \tag{3.106}$$

using tensor notation for the (6×6) matrix of coefficients which are obtained by integrating the potential function over the surface of the body.

The matrix of coefficients α_{ij} can be proved to be symmetric by using the third form of Green's theorem which states that the velocity potentials (ϕ_i and ϕ_j say) satisfying the conditions

$$\nabla^2 \phi_i = 0 \quad \text{and} \quad \nabla^2 \phi_j = 0 \tag{3.107}$$

can be related by the equation

$$\int_S \left(\phi_i \frac{\partial \phi_j}{\partial n} - \phi_j \frac{\partial \phi_i}{\partial n} \right) \mathrm{d}S = 0 \tag{3.108}$$

$$\phi_i \frac{\partial \phi_j}{\partial n} = \phi_j \frac{\partial \phi_i}{\partial n}$$

and all pairs of coefficients of α_{ij} where $i \neq j$ are equal and the (6×6) matrix of coefficients is made up of 21 independent terms.

Now, using the same argument as in Section 3.10, when the generalized velocity of the body is changing with time, the rate of change of kinetic energy equals the work done by the body on the fluid through a force X_i. This force is called the acceleration reaction or added mass force. Thus for any mode of motion i

$$u_i \cdot X_i = \frac{\mathrm{d}}{\mathrm{d}t}\left(\frac{1}{2}\rho V_0 \alpha_{ij} u_i u_j \right) = \rho V_0 \alpha_{ij} u_i \, \dot{u}_j \tag{3.109}$$

and the acceleration reaction can be written as

$$X_i = \rho V_0 \alpha_{ij} \dot{u}_j \tag{3.110}$$

Since the matrix $\rho V_0 \alpha_{ij}$ acts as a mass added to the body, it is referred to as an added mass matrix with X_i referred to as the added mass force acting on the body so as to oppose its changing velocity. Thus

$$\left.\begin{aligned} X_j &= m_{ij}\dot{u}_j \\ &\text{where} \\ m_{ij} &= \rho V_0 \alpha_{ij} \end{aligned}\right\} \quad (3.111)$$

There are further simplifications that arise in the added mass matrix due to symmetries of the body relative to its reference axes.

If one plane of symmetry *Oxy* exists, then the added mass matrix is such that

$$m_{41} = m_{52} = m_{63} = m_{64} = m_{65} = 0 \quad (3.112)$$

A body with two planes of symmetry (about the *Oxy* and *Oxz* plane, say) leads to additional zeros in the added mass matrix such that only non-zero terms are given by

$$m_{11} = m_{22} = m_{33} = m_{44} = m_{55} = m_{66} = m_{26} = m_{62} \neq 0 \quad (3.113)$$

For a body with all three planes (*Oxy, Oyz* and *Ozx*) of symmetry, only the leading diagonal terms of the added mass matrix remain non-zero.

The complete determination of the added mass matrix for an arbitrary body shape is difficult to carry out. Practical results are only available for relatively simple geometries. Two of them are considered here – the circular cylinder and a non-elongated body.

(a) *Added mass matrix for a circular cylinder*

Consider an arbitrary cylinder *FG* in a system of Cartesian co-ordinates, as shown in Figure 3.9. The point *F* has co-ordinates (x_1, y_1, z_1) and *G* has co-ordinates (x_2, y_2, z_2). The cylinder has a diameter *d* with an added mass coefficient of C_m for flow normal to the cylinder axis. Thus the cylinder length and direction cosines are given by

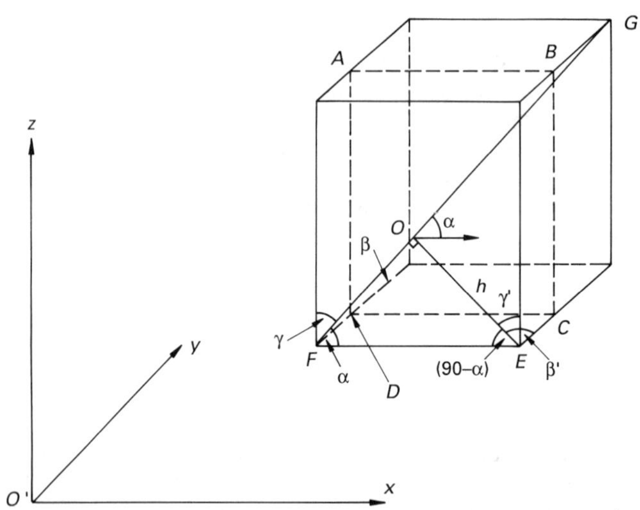

Figure 3.9. Cylinder FG at arbitrary orientations

$$\left.\begin{aligned} L &= \sqrt{[(x_2 - x_1)^2 + (y_2 - y_1)^2 + (z_2 - z_1)^2]} \\ \cos\alpha &= \frac{x_2 - x_1}{L}, \quad \cos\beta = \frac{y_2 - y_1}{L}, \quad \cos\gamma = \frac{z_2 - z_1}{L} \end{aligned}\right\} \tag{3.114}$$

Now a unit acceleration parallel to the x axis will yield acceleration components parallel and perpendicular to the cylinder axis. Only the acceleration component perpendicular to the cylinder axis will have a significant added mass force. The components of this normal flow added mass force in the three coordinate directions and the moments of this force about Ox, Oy, Oz make up the first (left-hand) column of the added mass matrix. Note that added mass equals added mass force for unit acceleration.

Thus for unit acceleration parallel to Ox, the component perpendicular to the cylinder axis is $\sin\alpha$ and the resultant added mass force in this direction is $k_1 \sin\alpha$ where $k_1 = \rho C_m \pi d^2 L/4$. Then the component force $k_1 \sin\alpha$ in the x direction is $k_1 \sin^2\alpha$ and the component of $k_1 \sin\alpha$ in the y direction is

$$(k_1 \sin\alpha)\cos\beta' = k_1 \sin\alpha(-\cot\alpha\cos\beta) = -k_1\cos\alpha\cos\beta$$

and in the z direction is

$$(k_1 \sin\alpha)\cos\gamma' = k_1 \sin\alpha(-\cot\alpha\cos\gamma) = -k_1\cos\alpha\cos\gamma \tag{3.115}$$

The angles β' and γ' are defined in Figure 3.9 and are related to angles β and γ through the equations

$$OF = h\cot\alpha$$

$$DF = CE = h\cot\alpha\cos\beta$$

$$\cos\beta' = -\cot\alpha\cos\beta$$

$$\cos\gamma' = -\cot\alpha\cos\gamma$$

Thus the added masses due to linear accelerations parallel to the x, y and z axes are

$$m_{ij}\begin{bmatrix} k_1\sin^2\alpha & -k_1\cos\alpha\cos\beta & -k_1\cos\alpha\cos\gamma \\ -k_1\cos\alpha\cos\beta & k_1\sin^2\beta & -k_1\cos\beta\cos\gamma \\ -k_1\cos\alpha\cos\gamma & -k_1\cos\beta\cos\gamma & k_1\sin^2\gamma \end{bmatrix} \tag{3.116}$$

The matrix components m_{ij} for $i = 4$ to 6 and $j = 1$ to 3 consist of added mass force induced moments about Ox, Oy, Oz due to linear accelerations along the x, y and z axes. These moments are computed by integrating the moments due to added mass forces along the cylinder length.

Unit acceleration parallel to Ox results in forces

$$\frac{m_{11}\,\mathrm{d}l}{L}, \quad \frac{m_{21}\,\mathrm{d}l}{L}, \quad \frac{m_{31}\,\mathrm{d}l}{L}$$

in the x, y and z directions, respectively, on a cylinder element of element length $\mathrm{d}l$. The total moment of the forces about the x axis is

$$m_{41} = \frac{m_{31}}{L}\int_0^L y\,dl - \frac{m_{21}}{L}\int_0^L z\,dl$$

But

$$\frac{1}{L}\int_0^L y\,dl = \frac{y_1 + y_2}{2} = y_m$$

and

$$\frac{1}{L}\int_0^L z\,dl = \frac{z_1 + z_2}{2} = z_m \qquad (3.117)$$

for

$$y = y_1 + \frac{y_2 - y_1}{L}1 \text{ and } z = z_1 + \frac{z_2 - z_1}{L}1$$

Thus

$$m_{41} = m_{31}y_m - m_{21}z_m$$

Similarly

$$m_{51} = m_{11}z_m - m_{31}x_m$$

$$m_{61} = m_{21}x_m - m_{11}y_m$$

$$m_{42} = m_{32}y_m - m_{22}z_m$$

$$m_{52} = m_{12}z_m - m_{32}x_m \qquad (3.118)$$

$$m_{62} = m_{22}x_m - m_{12}y_m$$

$$m_{43} = m_{33}y_m - m_{32}z_m$$

$$m_{53} = m_{31}z_m - m_{33}x_m$$

$$m_{63} = m_{32}x_m - m_{31}y_m$$

Now m_{ij} for $i = 1$ to 6 and $j = 4$ to 6 are added mass matrix terms due to angular accelerations about the *Ox, Oy* and *Oz* axes. In the same way as before, for unit angular accelerations about *Ox*, the linear accelerations of element *P* with co-ordinates (x, y, z) are 0, $-z$ and $+y$ along the *Ox, Oy* and *Oz* axes respectively.

Then the added mass force components on an element of length dl due to the acceleration of $-z$ are

$$-m_{12}z\,dl/L, \quad -m_{22}z\,dl/L, \quad -m_{32}z\,dl/L$$

parallel to the *Ox, Oy* and *Oz* direction and the force components due to acceleration y are

$$m_{13}y\,dl/L, \quad m_{23}y\,dl/L, \quad m_{33}y\,dl/L.$$

The integrated sum of these forces along the *Ox, Oy* and *Oz* axis will give terms m_{14}, m_{24}, m_{34} which are identical to the m_{41} terms given by Equations (3.118) above – thus confirming the matrix symmetry proved earlier. Also the summed moment of these forces about the x axis gives

$$m_{44} = -\int_0^L -\left(\frac{m_{22}z^2}{L}\right)\mathrm{d}l - \int_0^L \frac{m_{23}yz}{L}\mathrm{d}l + \int_0^L -\left(\frac{m_{32}zy}{L}\right)\mathrm{d}l + \int_0^L \frac{m_{33}y^2}{L}\mathrm{d}l$$

$$= m_{22}\int_0^L \frac{z^2}{L}\mathrm{d}l - 2m_{23}\int_0^L \frac{yz}{L}\mathrm{d}l + m_{33}\int_0^L \frac{y^2}{L}\mathrm{d}l$$

Now

$$\frac{1}{L}\int_0^L y^2\,\mathrm{d}l = y_n = \frac{1}{3}(y_1^2 + y_1y_2 + y_2^2)$$

$$\frac{1}{L}\int_0^L z^2\,\mathrm{d}l = z_n = \frac{1}{3}(z_1^2 + z_1z_2 + z_2^2) \qquad (3.119)$$

and

$$\frac{1}{L}\int_0^L yz\,\mathrm{d}l = (yz)_n = \frac{1}{6}[2y_1z_1 + 2y_2z_2 + y_1z_2 + z_1y_2]$$

using the functions of y and z given in Equation (3.117). Then

$$m_{44} = m_{22}z_n - 2m_{23}(yz)_n + m_{33}y_n \qquad (3.120)$$

Taking moments about the Oy and Oz axes also yields

$$m_{54} = -m_{21}z_n - m_{33}(xy)_n + m_{31}(yz)_n + m_{32}(zx)_n$$

and

$$m_{64} = -m_{31}y_n + m_{32}(xy)_n + m_{21}(yz)_n - m_{22}(zx)_n \qquad (3.121)$$

where

$$(xy)_n = \tfrac{1}{6}(2x_1y_1 + 2x_2y_2 + x_1y_2 + x_2y_1)$$

and

$$(zx)_n = \tfrac{1}{6}(2z_1x_1 + 2z_2x_2 + z_1x_2 + z_2x_1) \qquad (3.122)$$

Repeating the process of imposing unit angular accelerations about the Oy and Oz axes yields the remaining results:

$$m_{55} = m_{33}x_n - 2m_{31}(zx)_n + m_{11}z_n$$

$$m_{66} = m_{11}y_n - 2m_{21}(xy)_n + m_{22}x_n \qquad (3.123)$$

$$m_{65} = -m_{32}x_n - m_{11}(yz)_n + m_{21}(zx)_n + m_{31}(xy)_n$$

where

$$x_n = \tfrac{1}{3}(x_1^2 + x_1x_2 + x_2^2) \qquad (3.124)$$

with the remaining terms known from matrix symmetry.

(b) *Added mass matrix for a non-elongated body*

A simplified form of the added mass matrix can be obtained for a non-elongated body. The derivation given below applies for an arbitrary non-elongated body that has different added mass forces for acceleration components parallel to the three reference axes' directions. For a non-elongated body in a reference system $Oxyz$, the first three diagonal terms of the added mass matrix can be written as

$$\left.\begin{aligned} m_{11} &= \rho C_{m1} V \\ m_{22} &= \rho C_{m2} V \\ m_{33} &= \rho C_{m3} V \end{aligned}\right\} \quad (3.125)$$

where ρ is fluid density, V is the body volume and C_{m1}, C_{m2} and C_{m3} are added mass coefficients for body accelerations parallel to the Ox, Oy and Oz axes respectively. For a spherical body of radius a, however,

$$V = \tfrac{4}{3}\pi a^3 \text{ and } C_{m1} = C_{m2} = C_{m3} = 0.5$$

The remaining terms of the added mass matrix are obtained by calculating the acceleration reaction forces due to unit linear and angular acceleration along the reference axis. For a body with centre of volume (x_1, y_1, z_1), this yields

$$\begin{aligned} m_{12} &= m_{13} = m_{23} = 0 \\ m_{14} &= m_{25} = m_{36} = 0 \\ m_{15} &= m_{11} z_1, \quad m_{16} = -m_{11} y_1 \\ m_{24} &= -m_{22} z_1, \quad m_{26} = m_{22} x_1 \\ m_{34} &= m_{33} y_1, \quad m_{35} = -m_{33} x_1 \\ m_{44} &= m_{22} z_1^2 + m_{33} y_1^2 \\ m_{55} &= m_{33} x_1^2 + m_{11} z_1^2 \\ m_{66} &= m_{11} y_1^2 + m_{22} x_1^2 \\ m_{45} &= -m_{33} x_1 y_1 \\ m_{46} &= -m_{22} x_1 z_1 \\ m_{56} &= -m_{11} y_1 z_1 \end{aligned} \quad (3.126)$$

3.12 Body forces due to fluid inertia

The added mass force due to relative acceleration between a body and the surrounding fluid is not the only such force that acts on the body. An additional force (called the undisturbed pressure force, or the Froude–Krylov force) proportional to the acceleration of the surrounding fluid also acts on the body.

The physical argument supporting the presence of such a force is as follows:

Consider the volume of accelerating fluid shown in Figure 3.10. For a constant fluid acceleration, there must exist a corresponding pressure induced force per unit fluid volume to induce the acceleration. This force field would induce pressures on the surface of fluid volume B

(denoted by the shaded area in Figure 3.10) so as to generate a force sufficient to accelerate the fluid in element B with the same acceleration as that of the surrounding fluid. If the fluid in element B were now replaced by a submerged body, the same force would apply through a distributed pressure on the surface of the body. This force is the undisturbed fluid pressure, or Froude–Krylov force. It is a dynamic equivalent of the buoyancy force in Archimedes' principle where the force field inducing acceleration is replaced by a gravitational force field. The magnitude of the Froude–Krylov force is given by the force required to accelerate the displaced fluid volume with the acceleration of the surrounding fluid. Thus

$$F_K = \rho V \dot{q}$$

where ρ is the fluid density, V is the displaced volume of the body and $\dot{q}$ is the acceleration of the surrounding fluid.

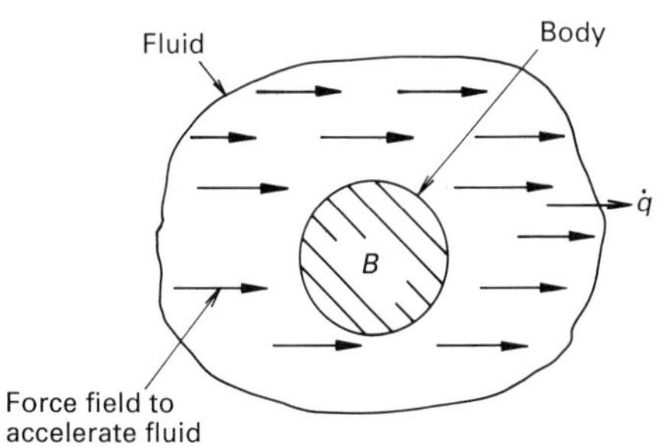

Figure 3.10. Froude–Krylov force in accelerating fluid

It is important to note that an added mass force acts whenever there is relative acceleration between a body and the surrounding fluid, whereas a Froude–Krylov force acts only in the presence of fluid acceleration. The forces listed in Table 3.1 for fluid acceleration $\dot{q}$ and body acceleration $\dot{u}$ illustrate this point. Thus the total force F_t due to fluid inertia for a body with zero acceleration in a fluid with acceleration $\dot{q}$ is given by

$$F_t = F_K + F_d = \rho V \dot{q} + \rho C_m V \dot{q} \tag{3.127}$$

Table 3.1 Added mass and Froude–Krylov forces on a submerged body. C_m is the added mass coefficient

Body acceleration	*Fluid acceleration*	*Added mass force F_d on body*	*Froude–Krylov force F_K on body*
0	0	0	0
0	$\dot{q}$	$\rho C_m V \dot{q}$	$\rho V \dot{q}$
$\dot{u}$	0	$-\rho C_m V \dot{u}$	0
$\dot{u}$	$\dot{q}$	$\rho C_m V(\dot{q} - \dot{u})$	$\rho V \dot{q}$

Then

$$F_t = (1 + C_m)\rho V\dot{q}$$
$$= C_{m0}\rho V\dot{q} \qquad (3.128)$$

where C_m is the added mass coefficient and C_{m0} $(= 1 + C_m)$ is conventionally referred to as the added inertia coefficient. Thus for a cylinder $C_{m0} = 2$ and for a sphere $C_{m0} = 1.5$.

Now consider the case when the body is accelerating with respect to fixed axes at the rate $\dot{u}$ in fluid which itself has an acceleration $\dot{q}$ with respect to the same fixed axes. If the body has volume V, added inertia coefficient C_{m0} $(= 1 + C_m)$ and physical mass M, then equating the mass times acceleration to the sum of the external forces gives

$$M\dot{u} = F + F_k + F_d \qquad (3.129)$$

where F is an external force on the body,

$$F_K = \rho V\dot{q}$$

and

$$F_d = \rho C_m V(\dot{q} - \dot{u})$$

Then

$$\left.\begin{aligned} F &= M\dot{u} - \rho V\dot{q} - \rho C_m V(\dot{q} - \dot{u}) \\ &= \underbrace{C_{m0}\rho V(\dot{u} - \dot{q})}_{\substack{\text{inertia force}\\ \text{due to relative}\\ \text{acceleration}}} + \underbrace{(M - \rho V)\dot{u}}_{\substack{\text{modification}\\ \text{due to fluid}\\ \text{acceleration}}} \end{aligned}\right\} \qquad (3.130)$$

The Froude–Krylov force equation can be generalized for three dimensions by considering a body undergoing linear and angular accelerations $\dot{u}_j (j = 1 \text{ to } 6)$ about $Oxyz$ axes in a fluid which is itself accelerating with three translational components only, $\dot{q}_j (j = 1 \text{ to } 3)$. The fluid angular acceleration components $(j = 4 \text{ to } 6)$ will be zero for the irrotational flow fields being considered. Then the resultant force on the body will arise from an added mass force as acceleration reaction due to relative acceleration of the body with the surrounding fluid, together with a Froude–Krylov force due to the absolute acceleration of the fluid only. Thus the total force F_i on the body due to fluid inertia can be written as

$$F_i = -\rho V\alpha_{ij}(\dot{u}_j - \dot{q}_j) + \rho V\dot{q}_j$$
$$= -\rho V\alpha_{ij}\dot{u}_j + \rho V(\alpha_{ij} + 1)\dot{q}_j$$

using tensor notation with summing in the repeated j suffix and variables from Section 3.11. Note that the fluid acceleration tensor contains linear components only, whereas the body acceleration tensor includes both linear and angular components.

References

Batchelor, G. K. (1973), *An Introduction to Fluid Dynamics*, Cambridge University Press, Cambridge
Bronshtein, I. N. and Semendyayev, K. A. (1979), *Handbook of Mathematics*, Van Nostrand Reinhold Company, English translation edited by K. A. Hirsch
Lamb, H. (1975), *Hydrodynamics*, Cambridge University Press, Cambridge

Chapter 4

Gravity wave theories

Gravity waves form part of a large family of surface water waves ranging from wind generated waves in the oceans, to tidal action, water surface oscillations in harbours, flood waves in rivers, tidal bores, hydraulic jumps, waves generated by siesmic disturbances, and so on. This chapter is concerned primarily with gravity waves generated by wind action, although gravity waves induced by siesmic action are also briefly considered. The unsteady fluid velocities, accelerations and pressures induced by gravity waves are responsible for a significant proportion of the environmental forces acting on offshore structures, subsea pipes and associated submerged structures.

Gravity waves can exist at the interface between any two fluids of different densities – air and water, for example – and are driven by interaction between inertia of the fluid at the interface and its potential energy under gravitational action. The physical mechanisms underlying gravity wave formation by wind action, and decay by internal losses, are complex. Winds blowing over a water surface generate waves which develop with increasing wind duration and 'fetch'. The fetch is the length of water surface over which the wind blows. The resultant 'forced' waves are classified as 'wind waves' or 'a sea'. Gravity waves are, generally, progressive (or travelling) in nature and can thus move out of their area of formation, and when running free of the winds that generated them such waves are collectively referred to as a 'swell'. Gravity waves lose energy very gradually due to air resistance, internal friction and by friction with the sea bed in shallow water. Thus progressive gravity waves can travel substantial distances across the Earth's surface with little reduction in intensity. Gravity waves are, however, dispersive in nature – their celerity (wave or phase speed) is a function of wave period or wave length so that they spread out (disperse) along their direction of wave propagation. Both the above features are illustrated in Figure 4.1 which shows that the origins of waves measured off the coast of Cornwall can be traced back to storms off Newfoundland and off the southern tip of South America. Non-progressive, that is, standing gravity waves also occur in confined waters such as harbours, inland water ways and in wave testing tanks. These waves are of no interest to the design of offshore structures and so are not considered here.

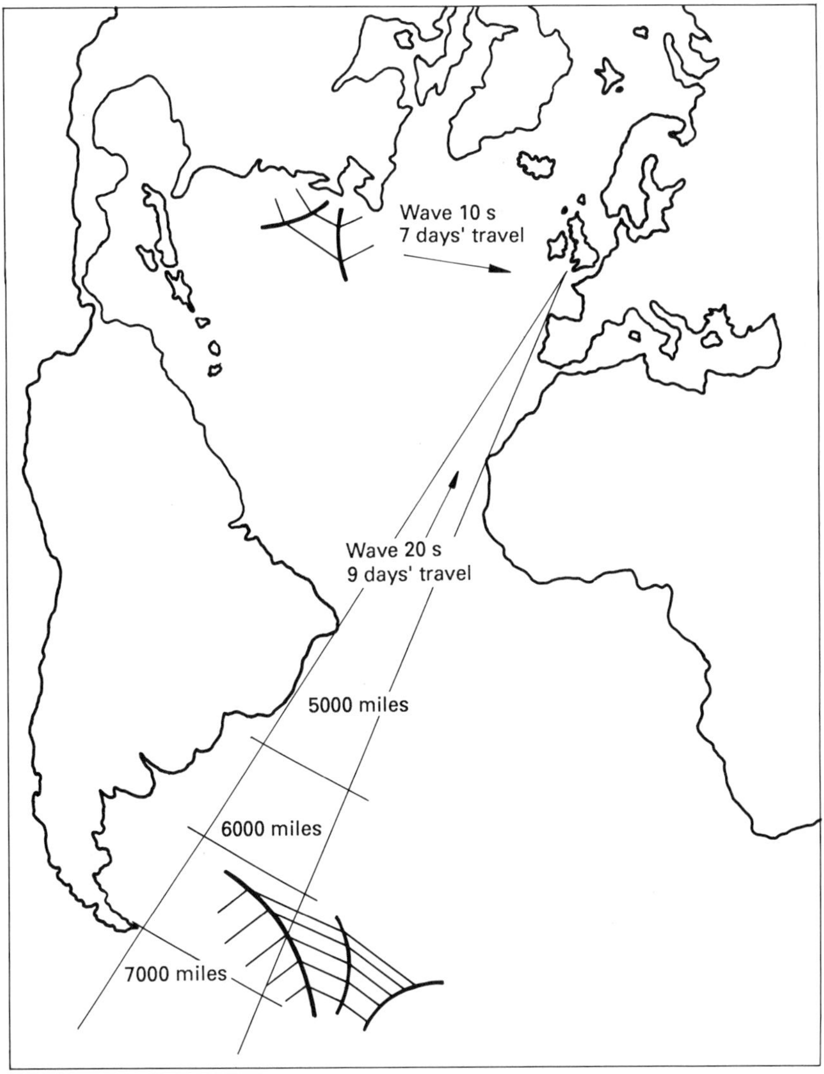

Figure 4.1. Weather map showing typical observed swell movements (from Barber and Ursell, 1948)

In nature, a large number of periodic wave components with different wave heights, periods and directions of travel occur at the same time in a given area. The superposition of all of these wave components, coupled with their dispersive behaviour, leads to a randomly varying sea surface elevation which can be treated by statistical methods. Nevertheless, the regular wave theories described in Sections 4.2–4.5 yield good mathematical models of long-crested periodic waves which are components of an irregular sea. The simplest of these theories is a linearized mathematical representation of gravity waves, known as linear wave theory. This theory permits the determination of irregular ocean waves by superposition of regular wave motions at different periods and from different directions.

The main difficulty in the study of ocean waves is the treatment of the free surface where the necessary boundary condition has to be evaluated on an unknown surface. Mathematically, this difficulty arises from the non-linear convective, inertia terms in the Navier–Stokes equations, which occur as products of velocities and their spatial gradients. However, since these velocity products are small for waves of small height compared to other terms in the equation of motion, a convenient idealization is obtained by assuming small wave height and neglecting convective inertia terms. This is the basis of linear wave theory described in Section 4.2. Other more exact mathematical solutions do not use this linearization and account for the convective terms by methods based on power series.

Initially, some basic parameters describing the characteristics of a progressive wave are defined with reference to Figure 4.2.

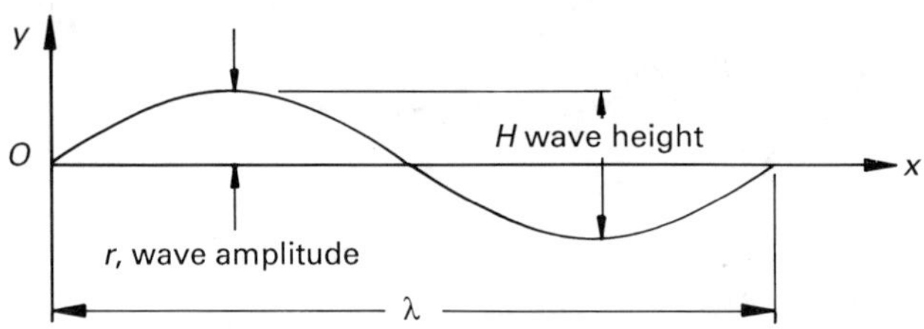

Figure 4.2. Wave elevation for a wave of amplitude r

The wave number k is defined as

$$k = \frac{2\pi}{\lambda} \tag{4.1}$$

where λ is the wave length.

Wave frequency can be defined as f in Hz, ω in radians per second or period T in seconds. These parameters are related by the equations

$$\omega = 2\pi f,\ T = \frac{2\pi}{\omega},\ T = \frac{1}{f} \tag{4.2}$$

The wave celerity of any progressive wave is related to the wave length λ and frequency f by the equation

$$c = \lambda f \tag{4.3}$$

4.1 Wave descriptions for design

The wave theories described in this chapter allow a description of a wave in terms of height, period and water depth to be translated into numerical values of fluid velocity, acceleration and pressure at any point under the wave. This pre-supposes an adequate knowledge of the wave climate at an offshore site. Such wave data has been obtained in the past by visual observations, wave recording instruments or from deductions based on meteorological data. Draper (1963) and Hogben and

Lumb (1967) give typical design data and the procedures for deriving such data from ocean wave statistics.

Wave data which is used as an 'input' to design computations can be presented in two ways.

(a) *The design wave approach*

The design wave approach is applied by defining a wave of large height H and a corresponding wave period range whose probability of occurrence is such that it represents the maximum wave that the structure will encounter in a finite but long time interval known as the return period. The low probability of occurrence is achieved by statistically defining the highest wave that is likely to be encountered in a large number (50 or 100) of years. Thus the structure can be designed to withstand this rare occurrence which is likely to be the worst encountered condition over its lifetime. Since the joint probability distribution of wind, current and wave action over such long timescales is unknown, it is conventional to apply the maximum 50- or 100-year return period values of these parameters simultaneously and acting from the same direction. This approach is realistic from the viewpoint of designing against structural failure due to large waves but does not permit fatigue damage to be considered within the design. Furthermore, for many offshore structures, the wave height of the design wave is not the worst loading case. Waves of smaller heights and periods with wave lengths that are a specific fraction of the structure dimension or with periods that excite resonant motions can induce larger structural loadings than the highest design wave.

Section 4.7.4 presents the statistical techniques required to estimate 50- or 100-year return period wave data from oceanographic observations made over a limited 1–3-year timescale. The design wave approach is used for both fixed and floating offshore structures. On fixed structures in shallow water, the design wave is applied as a quasistatic loading with the wave 'stepped' past the structure through several points in its cycle. Table 4.1 presents examples of design wave and wind conditions that have been applied for four offshore operating areas of the world.

(b) *Design to a statistical wave description*

In this approach, the occurrence of waves incident on the offshore structure is expressed in terms of a probability distribution which describes the probability of occurrence of waves with specified height, period and direction. This probabilistic data can also be expressed as the average number of waves encountered per year for each range of wave height, period and direction. The relative energy in simultaneously occurring period components can also be described by unidirectional or directional wave spectra. Such a wave description still contains extreme design waves as part of the statistical data.

Use of a statistical wave description in design does permit fatigue inducing loads to be adequately characterized. The biggest drawback of this approach, however, is that it relies on the linearity of superposition of wave components and is, therefore, prone to some error for high waves. The requirement for linear superposition is more likely to be

Table 4.1 Typical design wave and wind parameters for 100-year return period conditions

Parameter	*Offshore area*					
	Malaysia	*Persian Gulf*	*Gulf of Mexico*	*Southern North Sea*	*Central North Sea*	*Northern North Sea*
Maximum wave height (m)	10.1	10.7	11.9	15.2	22.9	30.5
Maximum one minute mean wind speed (m/s)	31.9	27.3	38.1*	51.5	51.5	51.5

* Excluding hurricane events

valid for calculating loads due to small waves on large structures where linear wave theory applies and linear inertia loading predominates over quadratic drag forces. For structures with members of small diameter, non-linear drag forces are significant and the extension of the statistical wave description to loading has to be done with caution. Current design practice for offshore structures tends to use both the above approaches and take advantage of their relative merits.

4.2 Linear wave theory

This theory, originally developed by G. B. Airy (1845) offers a powerful analytical solution for gravity waves. The following assumptions have to be invoked:

1. The water is assumed to be of uniform density with constant water depth.
2. Viscosity and surface tension effects are neglected.
3. The squares of particle velocities are considered to be negligible but vertical accelerations are not neglected. The wave heights are accordingly assumed to be very low, hence the theory is often called small amplitude wave theory.
4. The wave motion is assumed to be irrotational.

The two-dimensional linear gravity wave has to satisfy the equation of continuity. For irrotational flow, this is written in terms of velocity potential ϕ as

$$\frac{\partial^2 \phi}{\partial x^2} + \frac{\partial^2 \phi}{\partial y^2} = 0 \tag{4.4}$$

where the Oxy plane is shown in Figure 4.2. The dynamic equation of motion (or Bernoulli's equation) with irrotational flow is

$$\frac{p}{\rho} = -gy - \frac{\partial \phi}{\partial t} - \frac{1}{2}\left[\left(\frac{\partial \phi}{\partial x}\right)^2 + \left(\frac{\partial \phi}{\partial y}\right)^2\right] \tag{4.5}$$

with respect to the reference axes of Figure 4.2. Variable p denotes pressure, ρ is water density, t is time and g is the acceleration due to gravity. The Bernoulli equation incorporates the assumptions of inviscid and irrotational flow.

The boundary condition of zero flow normal to the sea bed at $y = -d$ gives

$$\left(\frac{\partial \phi}{\partial y}\right)_{y=-d} = 0 \tag{4.6}$$

The boundary condition at the free surface poses difficulties, however. The unknown water surface shape, defined by a function $\eta = f(x, y, t)$, can be regarded as a flow boundary to which the local velocity must be tangential. Thus if n_η is a local normal to the free surface, then,

$$\left(\frac{\partial \phi}{\partial n_\eta}\right)_s = 0 \tag{4.7}$$

This is referred to as a kinematic free surface boundary condition.

A further dynamic free surface boundary condition can also be specified as constant pressure on the free surface shape $\eta = f(x, y, t)$.

The non-linearity of the above system of equations arises from the fact that the free surface boundary condition is applied on an unknown surface which is obtained only after solution of the governing equations and boundary conditions.

Now the water pressure at the sea surface can be obtained from Equation (4.5):

$$\frac{p_s}{\rho} + g\eta + \left(\frac{\partial \phi}{\partial t}\right)_s + \frac{1}{2}\left[\left(\frac{\partial \phi}{\partial x}\right)^2 + \left(\frac{\partial \phi}{\partial y}\right)^2\right] = \frac{p_t}{\rho} \tag{4.8}$$

with subscript s denoting quantities defined at the water surface and p_t denoting total pressure.

Differentiating Equation (4.8) with respect to time and neglecting velocity squared terms gives

$$\frac{1}{\rho}\frac{\partial p_s}{\partial t} + g\left(\frac{D\eta}{Dt}\right) + \left(\frac{\partial^2 \phi}{\partial t^2}\right)_s = 0 \tag{4.9}$$

where the derivative of η has strictly to be the substantive derivative defined in Chapter 3, as

$$\frac{D\eta}{Dt} = \frac{\partial \eta}{\partial t} + u\frac{\partial \eta}{\partial x} + v\frac{\partial \eta}{\partial y} \tag{4.10}$$

Ignoring surface tension gives

$$p_s = p_a \text{ and } \frac{\partial p_s}{\partial t} = \frac{\partial p_a}{\partial t} = 0 \tag{4.11}$$

for constant atmospheric pressure with subscript a denoting quantities just above the water surface. Furthermore, for waves of small (mathematically infinitesimal) height,

$$\frac{D\eta}{Dt} = \frac{\partial \eta}{\partial t} = \left(\frac{\partial \phi}{\partial y}\right)_{y=0} \tag{4.12}$$

Substituting Equations (4.11) and (4.12) into Equation (4.9) yields the equation

$$g\left(\frac{\partial \phi}{\partial y}\right)_{y-0} + \left(\frac{\partial^2 \phi}{\partial t^2}\right)_{y=0} = 0 \tag{4.13}$$

This is called the linearized free surface boundary condition.

The removal of velocity squared terms limits the applicability of linear theory to infinitesimal waves in mathematical terms. Note that the linearization process not only makes the equations linear but also fixes the free surface boundary. For engineering purposes, however, the theory can be used for small to moderate wave heights.

The linear wave theory problem can thus be restated as:

Governing equation

$$\frac{\partial^2 \phi}{\partial x^2} + \frac{\partial^2 \phi}{\partial y^2} = 0 \tag{4.14}$$

$$
\left.\begin{array}{l}
\textit{Boundary conditions} \\
\textit{Sea bed} \\
\dfrac{\partial\phi}{\partial y} = 0 \text{ at } y = -d \\
\textit{Surface} \\
g\dfrac{\partial\phi}{\partial y} + \dfrac{\partial^2\phi}{\partial t^2} = 0 \text{ at } y = 0
\end{array}\right\} \quad (4.15)
$$

The above is an elliptic partial differential equation with the solution domain limited only by horizontal planes $y = 0$ and $-d$ and with no limits in the x direction. The solution for ϕ is, therefore, a simple harmonic function of x and is of the form

$$\phi = F(y)\exp[-\mathrm{i}(kx - \omega t)] \quad (4.16)$$

Substituting this into Equation (4.14) and dividing by the exponential term yields the ordinary differential equation

$$\frac{\mathrm{d}^2F}{\mathrm{d}y^2} - k^2F = 0 \quad (4.17)$$

This differential equation has the solution

$$F = A\mathrm{e}^{ky} + B\mathrm{e}^{-ky} \quad (4.18)$$

Applying the boundary condition

$$\left(\frac{\partial\phi}{\partial y}\right)_{y=-d} = \left(\frac{\mathrm{d}F}{\mathrm{d}y}\right)_{y=-d} = 0 \quad (4.19)$$

yields

$$Ak\mathrm{e}^{-kd} - Bk\mathrm{e}^{kd} = 0$$

Therefore

$$Ak\mathrm{e}^{-kd} = Bk\mathrm{e}^{kd} = \tfrac{1}{2}C,$$

say, and

$$A = \frac{C}{2k}\mathrm{e}^{kd},\ B = \frac{C}{2k}\mathrm{e}^{-kd} \quad (4.20)$$

Substituting Equations (4.18) and (4.20) into Equation (4.16) yields

$$\phi = \frac{C}{k}\cosh k(y + d)\exp[-\mathrm{i}(kx - \omega t)] \quad (4.21)$$

where the constant C still remains to be evaluated. This can be done by considering oscillatory variations due to the wave in Equation (4.8). Since the pressure in the medium above the wave surface is taken as uniform and velocity squared terms are neglected, the oscillatory wave surface elevation η is given by the equation

$$\eta = -\frac{1}{g}\left(\frac{\partial\phi}{\partial t}\right)_{y=0} \quad (4.22)$$

Substituting Equation (4.21) into the above equation for $y = 0$ gives

$$\eta = -\frac{\omega C}{gk}\cosh kd \cdot \mathrm{i}\exp[-\mathrm{i}(kx - \omega t)]$$

Combining the terms from the above equation into a constant a yields

$$a = -\frac{\omega C}{gk}\cosh kd \tag{4.23}$$

thus giving

$$\eta = a \cdot \mathrm{i}\exp[-\mathrm{i}(kx - \omega t)] \tag{4.24}$$

where a can be regarded as a wave amplitude.

Substituting Equation (4.23) into (4.21) then gives

$$\phi = -\frac{ag}{\omega}\frac{\cosh k(y + d)}{\cosh kd}\exp[-\mathrm{i}(kx - \omega t)] \tag{4.25}$$

Substituting Equations (4.21) or (4.25) into the free surface boundary condition given by (4.15) also yields the equation

$$\omega^2 = gk \tanh kd \tag{4.26}$$

Taking real parts of the complex terms in Equations (4.24) and (4.25) then gives a linear wave theory solution as

$$\phi = -\frac{ag}{\omega}\frac{\cosh k(y + d)}{\cosh kd}\cos(kx - \omega t) \tag{4.27}$$

with

$$\eta = a\sin(kx - \omega t) \tag{4.28}$$

Now using the progressive wave relationship

$$c = f\lambda = \frac{\omega}{k}, \tag{4.29}$$

and Equation (4.26), the velocity potential can also be written as

$$\phi = -\,ac\,\frac{\cosh k(y + d)}{\sinh kd}\cos(kx - \omega t) \tag{4.30}$$

Furthermore, for deep water ($d \to +\infty$), the hyperbolic terms reduce to exponentials and Equations (4.27), (4.30) and (4.26) become

$$\left.\begin{aligned} &\phi = -\frac{ag}{\omega}\,c^{ky}\cos(kx - \omega t) \\ &\text{or} \\ &\phi = -\,ace^{ky}\cos(kx - \omega t) \\ &\text{and} \\ &\omega^2 = gk \end{aligned}\right\} \tag{4.31}$$

respectively. The above equations are now used to explore some of the properties of deep water waves.

(i) *Wave period, length and celerity* Simple relationships between wave period T, wave length λ, and celerity c can readily be defined for deep water from Equations (4.26) and (4.31). These can be written as

$$\left.\begin{aligned} c &= \sqrt{\left(\frac{g\lambda}{2\pi}\right)} \\ T &= \sqrt{\left(\frac{2\pi\lambda}{g}\right)} \\ c &= \frac{gT}{2\pi} \end{aligned}\right\} \quad (4.32)$$

For example, a deep water wave of 10-s period will have a wave length of 156.1 m and a celerity of 15.61 m/s, whereas a wave of 6-s period will have a wave length of 56.2 m and a celerity of 9.37 m/s. Thus long waves travel faster than short ones so as to overtake them and produce a continually varying water surface. This 'dispersive' nature of gravity waves leads to a number of interesting properties.

(ii) *Particle motions* The co-ordinates (ξ, η) of a particle relative to its mean position are obtained using Equation (4.31) with the integrations

$$\left.\begin{aligned} \xi &= \int \left(\frac{\partial\phi}{\partial x}\right) \mathrm{d}t = a\mathrm{e}^{ky}\cos(kx - \omega t) \\ \eta &= \int \left(\frac{\partial\phi}{\partial y}\right) \mathrm{d}t = a\mathrm{e}^{ky}\sin(kx - \omega t) \end{aligned}\right\} \quad (4.33)$$

These results indicate that in deep water all particles under a gravity wave move in a circular path of radius $a\,\mathrm{e}^{ky}$ with periodic time T. The component velocities of the particles are

$$\left.\begin{aligned} u &= \frac{\partial\phi}{\partial x} = +\,a\omega\mathrm{e}^{ky}\sin(kx - \omega t) \\ v &= \frac{\partial\phi}{\partial y} = -a\omega\mathrm{e}^{ky}\cos(kx - \omega t) \end{aligned}\right\} \quad (4.34)$$

and the total velocity q is

$$q = \sqrt{[(u^2 + v^2)]} = a\omega\mathrm{e}^{ky} \quad (4.35)$$

Thus, both particle displacements and velocities reduce exponentially with depth. At one wavelength below the surface $y = -\lambda$ and the reduction ratio is $\mathrm{e}^{-2\pi}(= 0.001\,87)$. At half a wave length ($y = -\lambda/2$), the ratio is $\mathrm{e}^{-\pi}(= 0.0432)$, demonstrating that even at half a wave length below the free surface, particle motions are negligible for engineering purposes. The direction of orbital motion is such that particles in the crest of the wave move in the direction of wave propagation whereas those in the trough move in the opposite direction.

(iii) *Pressure in waves* Neglecting second-order terms in Bernoulli's Equation (4.5), writing the pressure $p = p' - p_a$ where p_a is atmospheric pressure and substituting Equation (4.31) into (4.5) yields

$$p' + \rho g y - \rho g a\mathrm{e}^{ky}\sin(kx - \omega t) = p_a \quad (4.36)$$

The distribution of pressure p' below a deep water gravity wave can

Table 4.2 Summary of water wave properties

Quantity	*Deep water*	*Shallow water* ($d/\lambda < 1/2$)
Surface profile	$\eta = a\sin(kx - \omega t)$	$\eta = a\sin(kx - \omega t)$
Velocity potential	$\phi = -ace^{ky}\cos(kx - \omega t)$	$\phi = -ac\dfrac{\cosh k(y+d)}{\sinh kd}\cos(kx - \omega t)$
Stream function	$\psi = +ace^{ky}\sin(kx - \omega t)$	$\psi = +ac\dfrac{\sinh k(y+d)}{\sinh kd}\sin(kx - \omega t)$
Horizontal water velocity	$u = \partial\phi/\partial x$ $= +kcae^{ky}\sin(kx - \omega t)$	$u = +ack\dfrac{\cosh k(y+d)}{\sinh kd}\sin(kx - \omega t)$
Vertical water velocity	$v = \partial\phi/\partial y$ $= -kcae^{ky}\cos(kx - \omega t)$	$v = -ack\dfrac{\sinh k(y+d)}{\sinh kd}\cos(kx - \omega t)$
Horizontal acceleration	$= -k^2c^2ae^{ky}\cos(kx - \omega t)$	$= -k^2c^2a\dfrac{\cosh k(y+d)}{\sinh kd}\cos(kx - \omega t)$
Vertical acceleration	$= -k^2c^2ae^{ky}\sin(kx - \omega t)$	$= -k^2c^2a\dfrac{\sinh k(y+d)}{\sinh kd}\sin(kx - \omega t)$
Wave celerity	$c = \dfrac{\lambda}{T} = \dfrac{g}{\omega} = \left(\dfrac{g\lambda}{2\pi}\right)^{1/2} = \dfrac{gT}{2\pi}$	$c^2 = \dfrac{g\lambda}{2\pi}\tanh kd^*$
Wave length	$\lambda = \dfrac{2\pi c^2}{g} = \dfrac{2\pi g}{\omega^2} = \dfrac{gT^2}{2\pi}$	
Wave number	$k = \dfrac{2\pi}{\lambda} = \dfrac{\omega^2}{g} = \dfrac{g}{c^2} = \dfrac{4\pi^2}{gT^2}$	
Pressure** $p' - p_a$	$= +\rho g a e^{ky}\sin(kx - \omega t)$.	$= +\rho g a\dfrac{\cosh k(y+d)}{\cosh kd}\sin(kx - \omega t)$
Horizontal pressure gradient** $\partial p/\partial x$	$= +\rho g a k e^{ky}\cos(kx - \omega t)$	$= +\rho g a k\dfrac{\cosh k(y+d)}{\cosh kd}\cos(kx - \omega t)$

Table 4.2 (*cont.*)

Quantity	*Deep water*	*Shallow water* ($d/\lambda < 1/2$)
Vertical pressure gradient** $\partial p/\partial y$	$= +\rho gak\mathrm{e}^{ky}\sin(kx - \omega t)$	$= \rho gak \dfrac{\sinh k(y + d)}{\cosh kd}\sin(kx - \omega t)$
Maximum wave slope	$\theta = ka = \dfrac{2\pi a}{\lambda}$	

Notes: * For very shallow water ($d \leqslant \lambda/20$), this expression becomes

$c^2 = gd$

with

$c^2 = 0.968\ gd$ at $d = \lambda/20$

For deeper water ($d > \lambda/2$), the expression becomes

$c^2 = \dfrac{g\lambda}{2\pi}$

with

$c^2 = 0.996g\lambda/(2\pi)$ for $d = \lambda/2$

** These terms exclude hydrostatic pressure.

be visualized as consisting of a series of constant pressure surfaces with the pressure value for each surface increasing with depth below the surface. The amplitude $\rho g a e^{ky}$ of subsurface pressure fluctuations has the effect of distorting these constant pressure surfaces into sinusoidal shapes with the amplitude of the sinusoids decreasing exponentially with depth. Crests and troughs of variations in the submerged constant pressure surfaces do, however, remain in the same vertical lines as the crests and troughs of the free surface elevation. Thus the phase of pressure variations remains constant with depth. Table 4.2 summarizes the velocity potential, stream function and derived properties for linear deep water gravity waves. Equivalent expressions for shallow water are also listed. The expressions for shallow water theory differ from the deep water equations in having hyperbolic variations of vertical and horizontal particle displacements, velocities and accelerations with depth. The resultant difference in wave fluid behaviour is illustrated in Figure 4.3 which shows wave particle orbits with depth for deep and shallow water. In shallow water, particle motions describe ellipses instead of circles. The ellipses become shallower with decreasing depth and at the bottom the motion is purely horizontal. In terms of numerical values, the shallow water equations tend to their deep water simplifications at water depth $d > \lambda/2$.

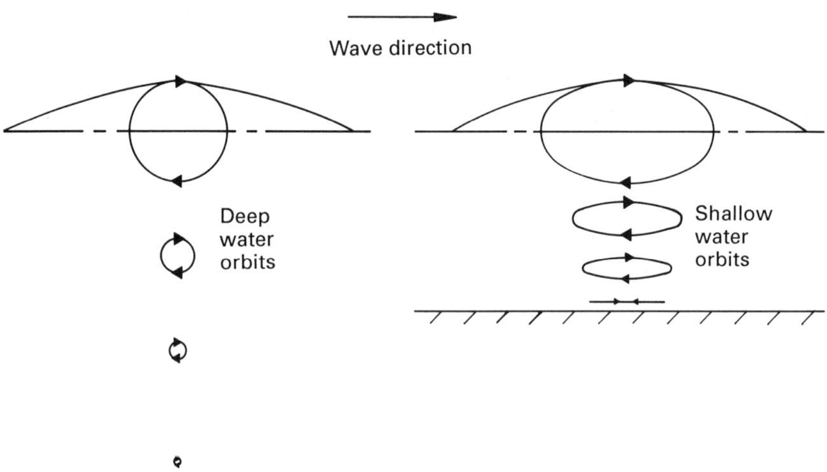

Figure 4.3. Wave particle orbits

(iv) *Total energy within a gravity wave* An expression for the potential and kinetic energy within a gravity wave is derived here using the shallow water wave equations. Consider a regular linear wave of amplitude, a (height $2a$), frequency ω and wave number $k(=2\pi/\lambda)$ in water of depth d. The energy contained within a unit thickness of water bounded by two planes, one wave length apart, is both kinetic and potential in nature. The kinetic energy arises from wave particle velocities, whereas the potential energy is due to changes in free surface elevation in the Earth's gravitational field. Thus, an excess of potential energy exists because fluid in the region of a crest is above mean sea level and fluid in the region of a trough, with a deficit of potential energy, has been removed from below mean sea level.

The total energy E for a gravity wave in water of depth d can be written in terms of the velocity potential as

$$E = \rho \iiint \left\{ \frac{1}{2} \left[\left(\frac{\partial \phi}{\partial x} \right)^2 + \left(\frac{\partial \phi}{\partial y} \right)^2 + \left(\frac{\partial \phi}{\partial z} \right)^2 \right] + gy \right\} \mathrm{d}x \, \mathrm{d}y \, \mathrm{d}z \qquad (4.37)$$

where the spatial derivatives of velocity potential represent velocities in the wave field.

Using this equation and the shallow water velocity potential of Equation (4.30), the energy of a unit thickness of fluid over one wavelength can be written as the integral

$$E_{\mathrm{w}} = \frac{\rho a^2 k^2 c^2}{\sinh^2 kd} \int_{-d}^{0} \int_{0}^{\lambda} \left[\frac{1}{2} \cosh^2 k(y + d) \sin^2(kx - \omega t) + \frac{1}{2} \sinh^2 k(y + d) \cos^2(kx - \omega t) \right] \mathrm{d}x \, \mathrm{d}y + \rho g \int_{0}^{\eta} \int_{0}^{\lambda} y \, \mathrm{d}x \, \mathrm{d}y \qquad (4.38)$$

where the potential energy in the last term of the equation is taken relative to a datum at still-water level. The $\partial\phi/\partial z$ term in Equation (4.37) vanishes due to the restriction in two dimensions. Expressing the squares of hyperbolic terms into double-angle formulae enables the outer part of the double integral to be evaluated. This gives

$$E_{\mathrm{w}} = \frac{\rho a^2 k^2 c^2}{\sinh^2 kd} \int_{0}^{\lambda} \left[\frac{\sinh 2kd}{8k} - \frac{1}{4} d \cos 2(kx - wt) \right] \mathrm{d}x + \rho \int_{0}^{\lambda} \frac{1}{2} g \eta^2 \, \mathrm{d}x \qquad (4.39)$$

Integrating again and using Equations (4.26) and (4.29) yields

$$\begin{aligned} E_{\mathrm{w}} &= \frac{\rho a^2 k^2 c^2}{\sinh^2 kd} \cdot \frac{\lambda \sinh 2kd}{8k} + \frac{1}{4} \rho g a^2 \lambda \\ &= \frac{1}{8} \rho g a^2 \lambda \, \frac{\tanh kd \sinh 2kd}{\sinh^2 kd} + \frac{1}{4} \rho g a^2 \lambda \end{aligned} \qquad (4.40)$$

and the expression for energy

$$E_{\mathrm{w}} = \tfrac{1}{2} \rho g a^2 \lambda$$

which is independent of water depth d. Therefore, the energy per unit area of water surface is

$$\tfrac{1}{2} \rho g a^2 \qquad (4.41)$$

The large amount of energy available within a typical wave can be illustrated by calculating the total energy per square metre in a wave of 5 m amplitude and 8 s period. The wave will have a wave length of 100 m and a celerity of 12.5 m/s. Using a sea water density of 1025 kg/m^3, the energy density is obtained as 125.7 kJ/m^2 or 34.9 Wh/m^2. One square kilometre of such a surface would have a total energy content of 125.7 GJ or 34.9 MW h. Since the energy content is travelling with a wave celerity of 12.5 m/s, the energy flux per unit length of wave front can be written as

$$0.1257\ \mathrm{MJ/m^2} \times 12.5\ \mathrm{m/s} = 1.57\ \mathrm{MW/m} \tag{4.42}$$

(v) *Group velocity* The dispersive nature of gravity waves, that is, their variation of celerity with wave period or wave length leads to unusual behaviour in groups of waves with slightly different wave periods travelling in the same direction. Such behaviour can be illustrated by considering a very simple wave group consisting of just two waves, each with the same wave amplitude, but slightly differing wave number and frequency. These waves are given by the expression

$$\left.\begin{aligned} \eta_{s1} &= a\cos[(k + \mathrm{d}k)x - (\omega + \mathrm{d}\omega)t] \\ &\text{and} \\ \eta_{s2} &= \mathrm{a}\cos(kx - \omega t) \end{aligned}\right\} \tag{4.43}$$

When superimposed, the resultant wave elevation is obtained as

$$\left.\begin{aligned} \eta_s &= \eta_{s1} + \eta_{s2} \\ &= A\cos\left[\left(k + \frac{\mathrm{d}k}{2}\right)x - \left(\omega + \frac{\mathrm{d}\omega}{2}\right)t\right] \\ &\text{where} \\ A &= 2a\cos\tfrac{1}{2}(\mathrm{d}k\cdot x - \mathrm{d}\omega\cdot t) \end{aligned}\right\} \tag{4.44}$$

η_s can be regarded as a wave of varying amplitude A. The joint disturbance will consist of high crests (A is positive maximum), calm water ($A = 0$), and low troughs (A is positive minimum). A is a progressive simple harmonic wave of amplitude $2a$ and very long wave length, $4\pi/\mathrm{d}k$. The behaviour of this wave pattern, illustrated schematically in Figure 4.4, is interesting because the longer (in wave length) of the two waves forming the group will have a slightly greater celerity and will, therefore, continuously overtake the accompanying wave of shorter wave length. This will mean that points of constructive and destructive interference between the two waves will be continually changing position – the net effect of this being that the wave components and the modulated shape (marked W and G, respectively, in Figure 4.4) will exhibit different wave celerities. The celerity of the wave group G, c_g, can be written as

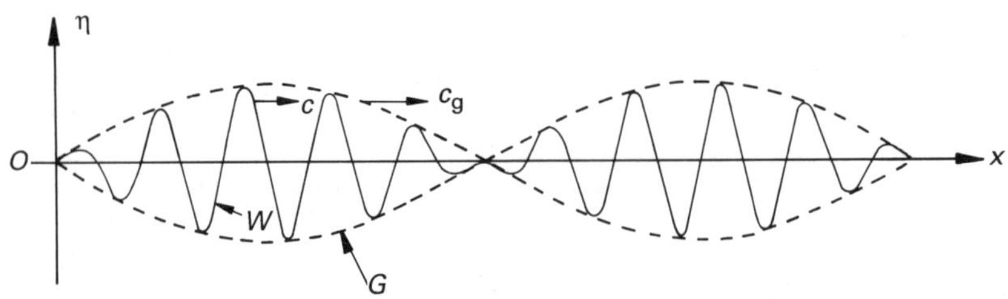

Figure 4.4. Surface elevation of wave group

$$c_g = \frac{dw}{dk} \tag{4.45}$$

For deep water waves, c_g can be related to the wave celerity c of the component waves as follows:

$$c_g = \frac{d\omega}{dk} = \frac{d(ck)}{dk} = c + k\frac{dc}{dk}$$

Since

$$k = \frac{2\pi}{\lambda}, \frac{dk}{k} = -\frac{d\lambda}{\lambda}$$

therefore

$$c_g = c - \lambda\frac{dc}{d\lambda} \tag{4.46}$$

Also since

$$c = \sqrt{\left(\frac{g\lambda}{2\pi}\right)}, \frac{dc}{c} = \frac{d\lambda}{2\lambda}$$

therefore

$$c_g = c - \tfrac{1}{2}c = \tfrac{1}{2}c$$

Thus in deep water, two wave components at periods that are close together will form wave groups with a celerity that is half the celerities for the component waves. In the wave shape of Figure 4.4, the component waves marked W will be moving at twice the celerity of the groups with the component wave profile moving into, through and beyond each wave group peak.

The analysis of Equations (4.46) is readily adapted for finite water depth and leads to the expression

$$c_g = \frac{1}{2}c\left(1 + \frac{2kd}{\sinh 2kd}\right) \tag{4.47}$$

Further aspects of linear wave theory are covered in depth by Lamb (1975) and Phillips (1966).

4.3 Stokes' third- and fifth-order wave theories

Lord Stokes (1847 and 1880) investigated the mechanics of water waves of finite height. The velocity potential was expanded using a Fourier series of the form

$$Z(W) = A_0\frac{W}{c} + B_0 + \sum_{n=1}^{\infty}\left[A_n \exp(inW/c) + B_n \exp(-inW/c)\right] \tag{4.48}$$

where $W(= \phi + i\psi)$ is the complex potential of the wave flow at a complex co-ordinate $Z(= x + iy)$, with the first two terms of the series accounting for a constant superimposed velocity c on the flow and A_0 and B_0 being constants. Complex coefficients A_n and B_n describe all the harmonic components of the wave flow.

Substitution of Equation (4.48) into the governing equations and comparing harmonic terms yields solutions of various orders. Thus, comparing terms up to $n = 3$ yields a third-order solution and comparing terms up to $n = 5$ gives a fifth-order solution.

However, the mathematical difficulty of the solution increases rapidly with the order that is used. The use of a series allows a solution which is closer to satisfying the non-linear wave surface boundary condition, but there is no closed form proof to indicate that the series will converge to the exact solution. However, Stokes' higher-order wave theories do yield predictions which are in close agreement with experimental data.

The results of Stokes' third- and fifth-order theories are presented here. No derivations are given but the results required to enable use of the theories in engineering are presented.

(i) *Stokes' third-order wave theory* Stokes' work on the series solution approach to higher order wave theories has been implemented by Skjelbreia (1959) and is also presented in usable form by Wiegel (1964). The resultant expressions for velocity potential and wave surface elevation contain higher harmonics of reducing amplitude which distort surface wave elevations. The expressions also include the primary sinusoidal oscillations arising from linear wave theory.

For Stokes' third-order wave theory, the velocity potential can be written as

$$\begin{aligned}\phi = \frac{c}{k}\Big[&F_1 \cosh k(y + d) \sin(kx - \omega t)\\ &+ \frac{1}{2} F_2 \cosh 2k(y + d) \sin 2(kx - \omega t)\\ &+ \frac{1}{3} F_3 \cosh 3k(y + d) \sin 3(kx - \omega t)\Big]\end{aligned} \tag{4.49}$$

where

$$F_1 = ka \frac{1}{\sinh kd} - (\mathrm{ka})^2 \frac{(1 + 5\cosh^2 kd)\cosh^2 kd}{8\sinh^5 kd}$$

$$F_2 = \frac{3}{4}(ka)^2 \frac{1}{\sinh^4 kd}$$

$$F_3 = \frac{3}{64}(ka)^3 \left(\frac{11 - 2\cosh 2kd}{\sinh^7 kd}\right)$$

using the axes system shown in Figure 4.1. The corresponding free surface elevation is

$$\left.\begin{aligned}\eta &= a\cos(kx - \omega t) + \frac{\pi a^2}{\lambda} f_2(kd)\cos 2(kx - \omega t)\\ &\quad + \frac{\pi^2 a^3}{\lambda^2} f_3(kd)\cos 3(kx - \omega t)\\ f_2(kd) &= \frac{(2 + \cosh 2kd)\cosh kd}{2\sinh^3 kd}\\ f_3(kd) &= \frac{3}{16}\,\frac{(1 + 8\cosh^6 kd)}{2\sinh^6 kd}\end{aligned}\right\} \tag{4.50}$$

where a is now a parameter contributing to wave height but is not equal to half the wave height. The relationship between the wave height H and a is

$$H = 2a + \frac{2\pi^2 a^3}{\lambda^2} f_3(kd) \tag{4.51}$$

Note that $f_2(kd) \rightarrow 1$ and $f_3(kd) \rightarrow \frac{3}{2}$ for $d/\lambda > \frac{1}{2}$ and in this case both Equations (4.49) and (4.50) are significantly simplified.

The wave length for a third-order wave solution can be written

$$\lambda = \underbrace{\frac{gT^2}{2\pi}}_{A} \underbrace{\tanh kd}_{B} \left[1 + \underbrace{(ka)^2 \left(\frac{14 + 4\cosh^2 2kd}{16 \sinh^4 kd} \right)}_{C} \right] \tag{4.52}$$

with the equation showing that wave length is a function of both wave period T and wave height through parameter a. It is interesting to note that in Equation (4.52) the smaller effects of shallow water and higher-order corrections can be progressively traced through terms A, B and C. The linear wave deep water result is represented by A, with B denoting the modification due to shallow water effects and C denoting the correction from the higher-order solution. Similar behaviour can be observed for the wave celerity which is

$$c^2 = \frac{g}{k} \tanh kd \left[1 + \left(\frac{ka}{2} \right)^2 \left(\frac{14 + 4\cosh^2 2kd}{16 \sinh^4 kd} \right) \right] \tag{4.53}$$

The relationship between wave celerity (c_0) and wave length (λ_0) in deep water waves for the third-order solution is

$$c_0^2 = \frac{g}{k_0} \left[1 + \left(\frac{k_0 a}{2} \right)^2 \right] \tag{4.54}$$

where the subscript 0 refers to deep water values. In order to use these equations for calculations with third-order wave theory, values of ratios a/λ and d/λ are required from wave height H, period T and water depth d. These ratios can be obtained from a solution of the equations for third-order theory as described by Skjelbreia (1959). However, for engineering purposes, it is convienent to express variations of a/λ and d/λ as carpet plots – examples of which have been given by Wiegel (1964) and are plotted in Figures 4.5 and 4.6. Here, the wave height, period and water depth are combined into non-dimensional ratios $2\pi d/gT^2$ and H/d and are used in these diagrams to extract c/λ and d/λ.

The values of water particle horizontal and vertical displacements, velocities, accelerations and pressures are obtained from the velocity potential of Equation (4.49), using the simple mathematical procedures that have been used in Section 4.2. The equations for horizontal and vertical water particle displacements for third-order wave theory reveal an interesting characteristic. Here although, the vertical component of particle displacement returns to its initial position at the end of every wave cycle, the equation for horizontal particle displacements has a term which contributes to a mean water particle drift in the direction of wave propagation. This particle drift decays with depth and the contributory term is of the form

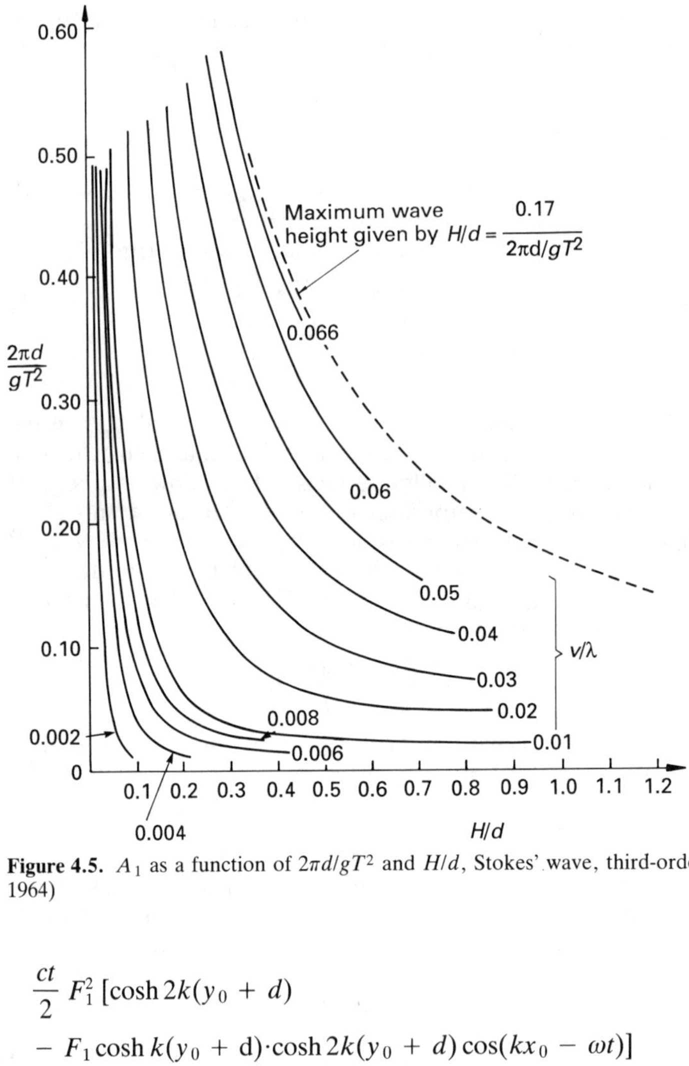

Figure 4.5. A_1 as a function of $2\pi d/gT^2$ and H/d, Stokes' wave, third-order (from Wiegel, 1964)

$$\frac{ct}{2} F_1^2 [\cosh 2k(y_0 + d)$$
$$- F_1 \cosh k(y_0 + \mathrm{d}) \cdot \cosh 2k(y_0 + d) \cos(kx_0 - \omega t)] \tag{4.55}$$

where t is time and (x_0, y_0) are the initial co-ordinates of the particle. This mean drift can be obtained in terms of an average velocity $\bar{U}_n$ to yield the expression

$$\bar{U}_n = \frac{1}{2}\left(\frac{\pi H}{T}\right)\left(\frac{\pi H}{\lambda}\right) \frac{\cosh 2k(y + d) - (1/2kd)\sinh(kd)}{\sinh^2 kd} \tag{4.56}$$

for shallow water according to Ursell (1953). For deep water this equation can be written as

$$\bar{U}_{n0} = \left(\frac{\pi H}{T}\right)\left(\frac{\pi H}{\lambda}\right)\left(e^{2kd} - \frac{1}{2kd}\right) \tag{4.57}$$

where it is assumed that the down wave transport of surface water is balanced by a uniform current in the direction opposite to that of the wave propagation.

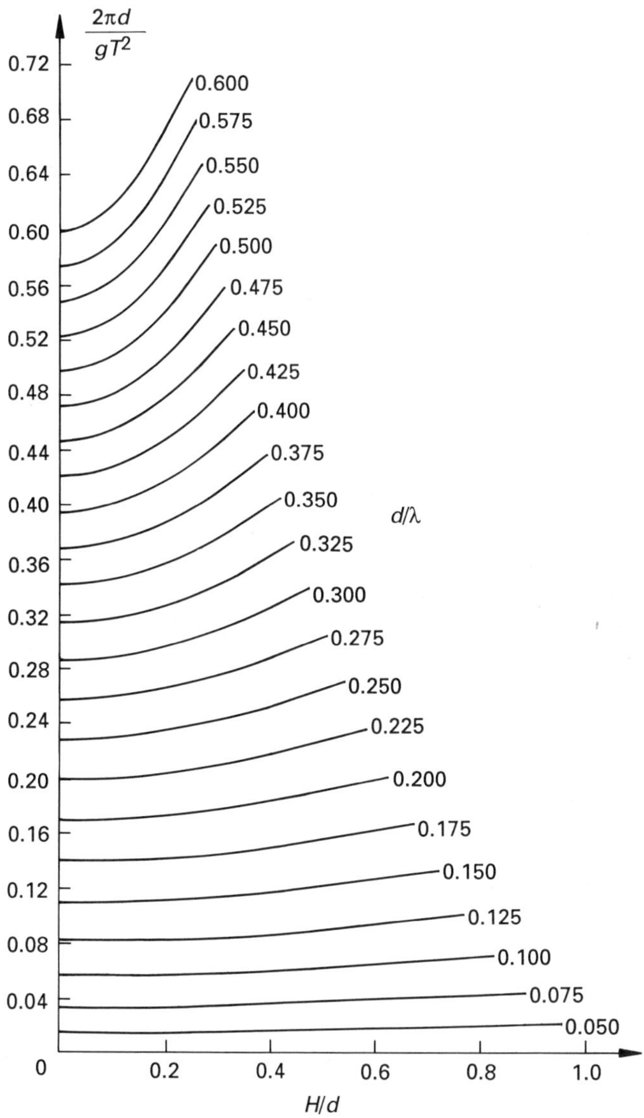

Figure 4.6. d/λ as a function of H/d and $2\pi d/gT^2$, Stokes' wave, third-order (from Wiegel, 1964)

The steady drift velocity in high waves can subject offshore structures to a substantial steady force component due to wave action. The amplitude variations in an irregular sea also induce a slowly varying drift force which has the potential to excite dynamic magnification in horizontal modes of motion for compliant offshore structures with low natural frequencies. These aspects are addressed more fully in Chapter 8. The simplifying assumptions of linear wave theory are such that this steady drift force component is not predicted. Figure 4.7 illustrates the shapes of typical water particle orbits induced by high waves, as predicted by Stokes' fifth-order theory.

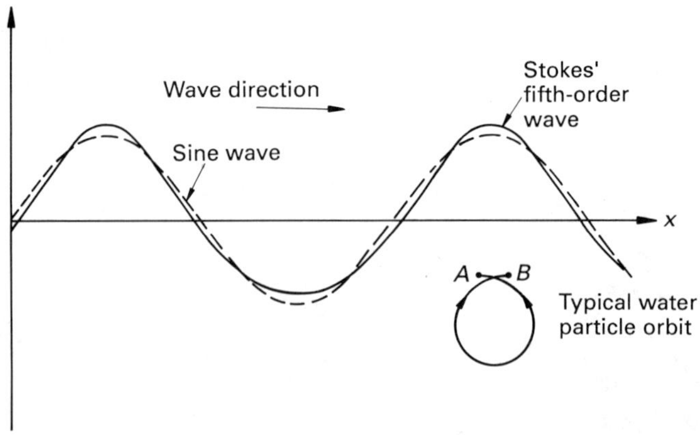

Figure 4.7. Properties of high gravity waves. A and B are at start and end of wave cycle

(ii) *Stokes' fifth-order wave theory* This implementation of Stokes' wave solution includes terms up to order five in the series solution of Equation (4.48). Skjelbreia and Hendrickson (1960) have presented a solution for fifth-order waves which has been implemented widely in computer programs used within the offshore industry. This solution is, therefore, presented here to illustrate the approach that is used. The equations for velocity potential and free surface elevation are

$$\left.\begin{aligned} \frac{k\phi}{\bar{c}} &= (\gamma A_{11} + \gamma^3 A_{13} + \gamma^5 A_{15}) \cosh ks' \sin\theta \\ &\quad + (\gamma^2 A_{22} + \gamma^4 A_{24}) \cosh 2ks' \sin 2\theta \\ &\quad + (\gamma^3 A_{33} + \gamma^5 A_{35}) \cosh 3ks' \sin 3\theta \\ &\quad + \gamma^4 A_{44} \cosh 4ks' \sin 4\theta \\ &\quad + \gamma^5 A_{55} \cosh 5ks' \sin 5\theta \end{aligned}\right\} \tag{4.58}$$

and

$$\left.\begin{aligned} k\eta &= \gamma\cos\theta + (\gamma^2 B_{22} + \gamma^4 B_{24}) \cos 2\theta \\ &\quad + (\gamma^3 B_{33} + \gamma^5 B_{35}) \cos 3\theta + \gamma^4 B_{44} \cos 4\theta \\ &\quad + \gamma^5 B_{55} \cos 5\theta \end{aligned}\right\} \tag{4.59}$$

where

$$\theta = k(x - \bar{c}t), \qquad s' = y + d$$

and

$$\left.\begin{aligned} k\bar{c}^2 &= \bar{c}_0^2(1 + \gamma^2 C_1 + \gamma^4 C_2) \\ &\text{and} \\ kK_1 &= \gamma^2 C_3 + \gamma^4 C_4 \end{aligned}\right\} \tag{4.60}$$

relates the general wave celerity $\bar{c}$ to the linear wave value $\bar{c}_0$, with K_1 being a constant. The equations use the axes system of Figure 4.1. The simple form of the above equations is deceptive. Most of their algebraic complexity is in the coefficients A, B and C, expressions for which are extracted from Skjelbreia and Hendrickson (1960) and listed in Table 4.3.

Use of the above equations requires that k and γ be found from wave height H period T and water depth d by simultaneous solution of the non-linear equations

$$\left.\begin{aligned}
&\frac{\pi H}{d} = \frac{\lambda}{d}\,[\gamma + \gamma^3 B_{33} + \gamma^5(B_{35} + B_{55})] \\
&\text{and} \\
&\frac{d}{\lambda_0} = \frac{d}{\lambda}\tanh kd\,(1 + \gamma^2 C_1 + \gamma^4 C_2)
\end{aligned}\right\} \qquad (4.61)$$

with

$$\lambda_0 = gT^2/2\pi.$$

Calculations using fifth-order wave theory are carried out routinely in the offshore industry by using a digital computer to implement the above procedure in the form of a subroutine. The procedure is supplied the input parameters H, T and d and, after execution, returns the velocities, acceleration components and pressures at the required co-ordinates. It is useful at this stage to conclude with a summary of some of the properties of waves of large height that are predicted by higher-order wave theories.

1. The observed wave profile of high waves is distorted from a sinusoidal shape by having sharper crests and flatter troughs, with the elevations of both being higher than the equivalent sinusoidal profile. This is illustrated in Figure 4.7.
2. The water particle orbits are now no longer of circular or elliptical shapes and exhibit a steady down wave horizontal drift component (see Figure 4.7). Furthermore, for high waves, the mean drift component is related to the fact that the mean water level during a wave cycle is no longer equal to still-water level.
3. Stokes' fifth-order theory is not considered reliable for $d/\lambda < \frac{1}{8}$ because the fifth- and higher-order terms of the series begin to contribute substantially to the results and reduce the numerical convergence of the series to fifth order.
4. The elevations of crests r_c above mean water level are important to the design of offshore structures in setting sufficient clearance on the underside of decks to avoid wave slamming. Typical values for crest amplitude r_c, are given approximately by the equations

$$\left.\begin{aligned}
&\frac{r_c}{H} = 0.5 + 0.67\frac{H}{T^2} \text{ for } \frac{d}{T^2} > 0.6 \text{ and } 0 < \frac{H}{T^2} < 0.268 \\
&\frac{r_c}{H} = 0.5 + 1.22\frac{H}{T^2} \text{ for } \frac{d}{T^2} = 0.3 \text{ and } 0 < \frac{H}{T^2} < 0.20
\end{aligned}\right\} \qquad (4.62)$$

where the second equation is deduced from shallow water theories of the type described in Section 4.4. Bretschneider (1960) presents r_c/H values for a wide range of values of H, T and d with Wiegel and

Table 4.3 Coefficients for fifth-order gravity wave theory

$$\bar{c}_0^2 = \frac{g}{k}(\tanh kd),$$

$$A_{11} = 1/s$$

$$A_{13} = \frac{-c^2(5c^2+1)}{8s^5}$$

$$A_{15} = \frac{-(1184c^{10} - 1440c^8 - 1992c^6 + 2641c^4 - 249c^2 + 18)}{1536s^{11}}$$

$$A_{22} = \frac{3}{8s^4}$$

$$A_{24} = \frac{(192c^8 - 424c^6 - 312c^4 + 480c^2 - 17)}{768s^{10}}$$

$$A_{33} = \frac{(13 - 4c^2)}{64s^7}$$

$$A_{35} = \frac{(512c^{12} + 4224c^{10} - 6800c^8 - 12\,808c^6 + 16\,704c^4 - 3154c^2 + 107)}{4096s^{13}(6c^2 - 1)}$$

$$A_{44} = \frac{(80c^6 - 816c^4 + 1338c^2 - 197)}{1536s^{10}(6c^2 - 1)}$$

$$A_{55} = \frac{-(2880c^{10} - 72\,480c^8 + 324\,000c^6 - 432\,000c^4 + 163\,470c^2 - 16\,245)}{61\,440s^{11}(6c^2 - 1)(8c^4 - 11c^2 + 3)}$$

$$B_{22} = \frac{(2c^2 + 1)}{4s^3}\,c$$

$$B_{24} = \frac{c(272c^8 - 504c^6 - 192c^4 + 322c^2 + 21)}{384s^9}$$

Table 4.3 (*cont.*)

$$B_{33} = \frac{3(8c^6 + 1)}{64s^6}$$

$$B_{35} = \frac{(88\,128c^{14} - 208\,244c^{12} + 70\,848c^{10} + 54\,000c^8 - 21\,816c^6)}{12\,288s^{12}(6c^2 - 1)} + \frac{(6264c^4 - 54c^2 - 81)}{12\,288s^{12}(6c^2 - 1)}$$

$$B_{44} = \frac{c(768c^{10} - 448c^8 - 48c^6 + 48c^4 + 106c^2 - 21)}{384s^9(6c^2 - 1)}$$

$$B_{55} = \frac{(192\,000c^{16} - 262\,720c^{14} + 83\,680c^{12} + 20\,160c^{10} - 7280c^8)}{12\,288s^{10}(6c^2 - 1)(8c^4 - 11c^2 + 3)} + \frac{(7160c^6 - 1800c^4 - 1050c^2 + 225)}{12\,288s^{10}(6c^2 - 1)(8c^4 - 11c^2 + 3)}$$

$$C_1 = \frac{(8c^4 - 8c^2 + 9)}{8s^4}$$

$$C_2 = \frac{(3840c^{12} - 4096c^{10} + 2592c^8 - 1008c^6 + 5944c^4 - 1830c^2 + 147)}{512s^{10}(6c^2 - 1)}$$

$$C_3 = -\frac{1}{4sc}$$

$$C_4 = \frac{(12c^8 + 36c^6 - 162c^4 + 141c^2 - 27)}{192cs^9}$$

$$s = \sinh\left(\frac{2\pi d}{\lambda}\right) \qquad c = \cosh\left(\frac{2\pi d}{\lambda}\right)$$

Beebe (1956) giving further limiting wave height values for shallow water.

5. For any water depth and wave period, there is an upper limit to the wave height which is stable before 'breaking'. This is of special importance for establishing design criteria. Generally, a minimum crest angle of 120° is attained before breaking occurs. At lower angles (for higher waves) the profile would topple over (or 'break'). This maximum or breaking limit can also be expressed as

$$\left(\frac{H}{\lambda}\right)_{max} = 0.142 \tanh\left(\frac{2\pi d}{\lambda}\right) \tag{4.63}$$

in water of depth d. Another empirical breaking condition that is often used is

$$\frac{H_{max}}{T^2} = 0.875 \tag{4.64}$$

4.4 Other wave theories

Since Stokes' third- and fourth-order wave theories break down in shallow water ($d < \lambda/8$), a number of special purpose shallow water wave theories have been developed to apply in this depth region. One of these is cnoidal wave theory – the name arising because the wave surface profile formula is expressed in terms of the Jacobian elliptic function $cn(u)$ with 'cnoidal' being analogous to sinusoidal. Laitone (1960, 1962) presents the equations and limiting conditions for cnoidal waves. The theory is not described further here because it is rarely used for offshore structures design. An alternative simpler theory of solitary waves is, however, presented since such waves do offer a useful approximation to high waves in shallow water and close to beaches.

This section also describes two numerical methods for calculating wave properties – stream function wave theory and the marker and cell method.

(i) *Solitary wave theory* A solitary wave is an unusual phenomenon in that, unlike the oscillatory motions of conventional gravity waves, water particles under a solitary wave move only in the direction of wave advance. A solitary wave is often described as a wave of translation and is generated in the laboratory by adding water impulsively to a stationary water mass or by implusively moving one end of a wave tank. In nature, solitary waves are generated by undersea earthquakes, landslides into the sea or by high waves running into shoaling waters. A train of solitary waves can be used to approximate the run-up of high open ocean waves into gradually shoaling waters. McCowan (1891) and Munk (1949) presented some of the earliest solutions to the solitary wave problem.

Since then, progessively more exact solutions have been derived by Laitone (1960) to second order, and Fenton (1972) to ninth order, with the work of many other research workers contributing to a more complete understanding of fluid motion in a solitary wave.

For engineering purposes, however, the second approximation to the solitary wave, by Laitone (1960), offers a sufficiently accurate theory, with equations that are relatively easy to apply. The main results of this theory are presented here after incorporating the further work of Laitone (1965) and Le Mehaute *et al.* (1968).

As shown in Figure 4.8, a fixed frame of reference Oxy is used with the origin at still water level in water depth d. A solitary wave of height H at time $t = 0$ is also shown in the diagram. Results from the second approximation to solitary wave theory can be expressed in compact form by using the non-dimensional physical variables

$$X = \frac{x}{d}, \; Y = \frac{y}{d}, \; \nu = \frac{\eta}{d}, \; \alpha = \frac{H}{d} \tag{4.65}$$

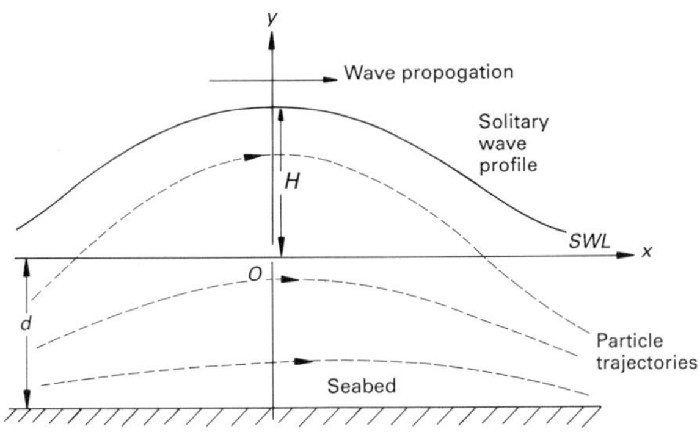

Figure 4.8. Solitary wave profile and particle trajectories

The derived variables of wave horizontal velocity u, vertical velocity, v, celerity c, and pressure p, are written as

$$U = \frac{u}{\sqrt{(gd)}}, \; V = \frac{v}{\sqrt{(gd)}}, \; C = \frac{c}{\sqrt{(gd)}}, \; P = \frac{p}{\rho g d} \tag{4.66}$$

Then the equation for wave surface elevation is

$$\mu = \alpha \operatorname{sech}^2 \gamma - \tfrac{3}{4}\alpha^2 \operatorname{sech}^2 \gamma \tanh^2 \gamma \tag{4.67}$$

where

$$\gamma = \frac{\sqrt{3}}{2}\alpha\left(1 - \frac{5}{8}\alpha\right)\left(X - \frac{ct}{d}\right) \tag{4.68}$$

The equations for horizontal and vertical velocity are

$$U = \alpha \operatorname{sech}^2 \gamma + \alpha^2 \operatorname{sech}^2 \gamma \left[\frac{1}{4} - \operatorname{sech}^2 \gamma - \frac{3}{4}(1 + Y)^2(2 - 3\operatorname{sech}^2 \gamma)\right] \tag{4.69}$$

$$V = \alpha\sqrt{(3\alpha)}(1 + Y)\operatorname{sech}^2 \gamma \tanh \gamma \left\{1 - \alpha\left[\frac{3}{8} + 2\operatorname{sech}^2 \gamma + \frac{1}{2}(1 + Y)^2(1 - 3\operatorname{sech}^2 \gamma)\right]\right\} \tag{4.70}$$

and the horizontal and vertical acceleration equations are

$$\left.\begin{aligned} \frac{1}{g}\frac{\partial u}{\partial t} &= \alpha\sqrt{(3\alpha)}\,\mathrm{sech}^2\gamma\tanh\gamma\left\{1+\alpha\left[\frac{1}{8}-2\,\mathrm{sech}^2\gamma\right.\right.\\ &\qquad \left.\left.-\frac{3}{2}(1+Y)^2(1+3\,\mathrm{sech}^2\gamma)\right]\right\}\\ \frac{1}{g}\frac{\partial v}{\partial t} &= \frac{3\alpha^2}{2}(1+Y)\,\mathrm{sech}^2\gamma(2-3\,\mathrm{sech}^2\gamma) \end{aligned}\right\}\quad (4.71)$$

Finally, the wave pressure P is

$$P = \mu - Y - \tfrac{3}{4}\alpha^2\,\mathrm{sech}^2\gamma[(1+Y)^2-1][2-3\,\mathrm{sech}^2\gamma] \quad (4.72)$$

The above expressions can be readily used in computer based calculations of solitary wave kinematics.

A simple expression, to first order, of the total volume of water above still water level in a solitary wave of crest length L can be written as

$$Q = 2L\int_0^{\infty}\eta\,\mathrm{d}x = 4Ld^2\sqrt{\left(\frac{\alpha}{3}\right)}$$

Although solitary waves are, as the name suggests, solitary, Munk (1949) showed that an effective wave length can be defined such that virtually all the water volume and the energy of a solitary wave does lie within this wave length. Thus, a train of shallow water waves can be approximated by a train of individual solitary waves by equating the actual wave length of the former to an effective wave length of the latter.

(ii) *Marker and cell method* Since digital computers play an increasing role in the design of offshore structures, parallel attempts have been made to develop numerical computation schemes for gravity wave kinematics. Two such techniques – the marker and cell method and stream function wave theory – are described here.

The marker and cell numerical scheme is based on representing the wave flow field by a rectangular mesh of cells. Numerical values of the flow's field variables (density, pressure, temperature and velocity) are associated with each cell. The position of the free surface is identified by 'markers' on each cell which indicate whether it is occupied by the wave fluid or not. The wave flow computation proceeds in time and is based on finite difference forms of equations governing the flow. Chan and Street (1970) present further details of this solution scheme and its development. The marker and cell method suffers from several serious drawbacks that have prevented its routine use for design of offshore structures. The computer code implementing the technique requires a powerful computer, the code is not readily transportable between machines and its use requires specific qualitative information on starting conditions, time step lengths and the general characteristics of the numerical scheme. Since the calculation of wave kinematics is only one of the early steps in the design process for an offshore structure, the marker and cell method has not been adopted for this purpose. Nevertheless, the technique offers an interesting and distinctive means for solving the governing equations of wave flow.

(iii) *Stream function theory* However, a stream function theory first developed by Dean (1965) and later extended by Chaplin (1980) offers a numerical method which is finding some application in design problems. The gravity wave problem for two-dimensional irrotational flow can be formulated in terms of stream function using an axes system like that of Figure 4.2 with the origin at mean water level. In order to reduce the wave problem to one of steady flow, however, the axes system is taken to be moving with constant velocity equal to the wave celerity c (parallel to the positive x axis in Figure 4.2). Then the stream function of the resultant wave flow is given by the continuity equation

$$\frac{\partial^2 \psi}{\partial x^2} + \frac{\partial^2 \psi}{\partial y^2} = 0 \tag{4.73}$$

with one boundary condition being no flow through the sea bed

$$\left(\frac{\partial \psi}{\partial x}\right)_{y=-d} = 0 \tag{4.74}$$

A kinematic free surface boundary condition imposes the free surface $[\eta = f(x, y, t)]$ as a flow boundary so that the velocity vector at this free surface must be tangential to it, giving

$$\frac{\partial \eta}{\partial x} = \frac{v}{u - c} \tag{4.75}$$

where u and v are horizontal and vertical velocity components. The dynamic free surface boundary condition introduces the constraint of constant pressure on this free surface so that from Bernoulli's equation, we get

$$\frac{(u - c)^2 + v^2}{2g} + \eta = p_0 \tag{4.76}$$

where p_0 denotes a constant total pressure head.

Equations (4.73) and (4.74) are satisfied by postulating that the stream function can be represented by a series of the form

$$\psi = -\frac{\lambda}{T} y + \sum_{n=1}^{N} X_n \sinh \frac{2\pi n(y + d)}{\lambda} \cos\left(\frac{2n\pi x}{\lambda}\right) \tag{4.77}$$

where N denotes the order of the solution. The first term in the above equations accounts for the reference axes translation, whereas the second term describes the wave disturbance. Stream function theory is applied by determining the N coefficients X_n, the wave length λ and the value of stream function ψ_η at the free surface given the wave height H and mean water depth d. The numerical method utilizes a least squares minimization of the dynamic free surface boundary condition at each of a number of sections uniformly distributed over half a wave length to obtain values for the coefficients X_n and wave length λ. The solution scheme uses assumed starting values and iteratively seeks the least squares minimum solution.

Dean (1965) used values of N of up to 19 to evaluate results for this theory, although his solution scheme requires a relatively complicated iterative computer code where the wave height is not explicitly entered into the calculation. Solutions for a particular wave height have to be obtained from tabulated data or a further iteration of complete solutions. Stream function wave theory has also been found to be in error for the limiting breaking wave height in deep water when compared with recently developed exact solutions by Cokelet (1977) for deep water waves of finite amplitude. These two factors have prompted Chaplin's (1980) further development of the theory. The stream function theory is reformulated so that wave height, period and depth are independent parameters and the wave elevations at several points along the wave profile are taken as unknowns. The solution scheme still depends on least squares minimization of the dynamic free surface boundary condition but with the advantage of more accurate starting values for the unknown surface elevations which can be provided by linear or cnoidal wave theory. Chaplin has carried out calculations using solutions of up to 45th order, which have shown good agreement for engineering purposes with the exact results of Cokelet (1977) for waves close to the breaking limit.

Recent research on waves of finite amplitude and at their limiting maximum heights has led to new theoretical developments which are likely to be utilized in the design of offshore structures. In particular, Schwartz (1974) presents analytic extensions of Stokes' series solutions to a higher order using a reformulation with a different independent parameter in the series. Cokelet (1977) presents more accurate solutions for finite amplitude gravity waves in deep water by choosing a different expansion parameter than that used by Stokes or by Schwartz, and obtaining the solution to the finite wave problem up to maximum wave height. The most significant result coming out of this work is that integral wave properties such as wave celerity, momentum and energy do not increase monotonically with wave height but attain maxima for waves that are slightly lower than the highest possible wave for any particular water depth. Thus, for waves of large height, the same wave speed, momentum or energy may be obtained by waves at two different heights. The solution presented by Cokelet (1977) is not yet suitable for direct use in engineering design but offers a numerically exact result for finite waves which serves as a bench mark for evaluating the accuracy of numerical or approximate wave theories used in engineering design.

4.5 Regions of validity

Since all wave theories that can be used for practical design work have some degree of approximation inherent in their formulation, it is necessary to determine the corresponding range of validity of each theory. The selection of a particular wave theory for use in design calculations invariably requires a compromise between the theory's accuracy and the level of computational effort required to implement it.

For example, linear wave theory is valid for small waves only and can, therefore, be used for small to moderate sea states. Linear wave theory is, of course, simple to use and has an additional advantage that, being linear, the principle of superposition applies. Complex directional irregular seas can thus be computed by summation of trains of regular linear waves.

However, if the 'design' wave approach is used to evaluate a structure, this wave will not be of small height. In this case Stokes' fifth-order theory is most commonly employed. Stokes' third-order theory tends not to be used very much, with either linear theory or fifth-order theory being selected, whichever is appropriate. Stream function theory also finds occasional use within design groups.

Both wave depth and wave height have a strong influence on the limits of validity of a wave theory. Thus linear theory becomes invalid for high waves and Stokes' fifth-order theory tends to be inaccurate for shallow water. It is, therefore, necessary to address the problem of selecting an appropriate wave theory for use and of estimating the errors that may arise from this choice.

The selection of wave theory can be made by utilizing Figure 4.9 which has been prepared by combining the results of a number of research workers (Dean, 1970, and unpublished work). The curve presents a logarithmic plot of parameters H/gT^2 against d/gT^2 for a wave of height H and period T in water depth d. The chart can be used by plotting a point on it corresponding to the wave and water depth in question. The upper left-hand area of the chart denotes a region where the corresponding wave would be unstable and would break – thus lying outside the parameter range of a physically realizable wave. The lower right-hand region of the chart, on the other hand, denotes an area in

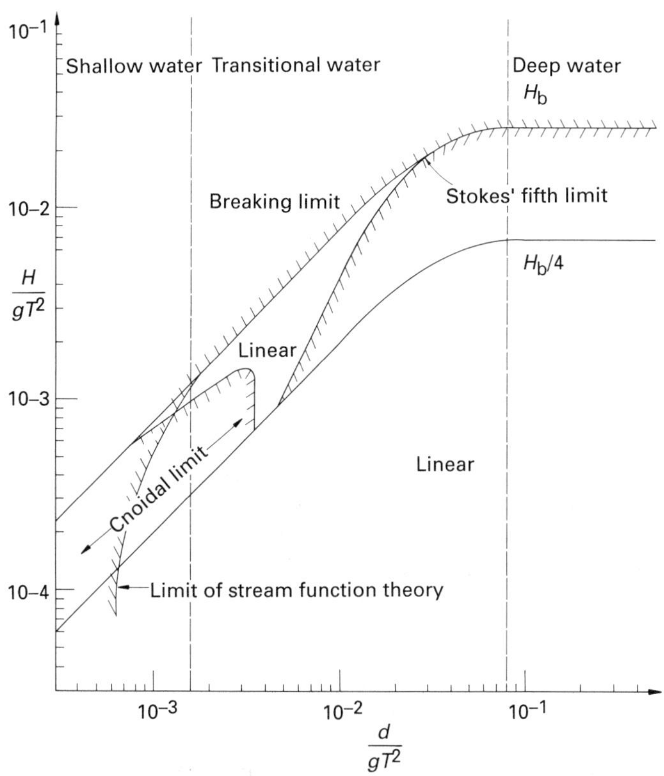

Figure 4.9. Validity of wave theories. H_b denotes the breaking wave height

which linear theory offers a reasonable representation. The banded region in between these two areas requires more careful consideration with Stokes' fifth-order or cnoidal wave theory offering the best choice for deep and shallow waters respectively. Stream function theory offers good accuracy across a wide range of water depths but it, too, breaks down for very shallow water. There is a bridging region between cnoidal and Stokes' fifth-order theory where linear theory can still provide the best analytical approach.

The percentage errors between results from each of these theories and laboratory wave data are presented in Table 4.4 and are based on work by Le Mehaute *et al.* (1968). The table presents measured water particle velocity accuracies in percentage at the sea bed, still-water level, water surface and overall. It is clear that both linear and Stokes' fifth-order theory offer sufficiently good agreement for engineering use.

Table 4.4 Percentage errors in water particle velocities from different wave theories compared with model test data

	B	*SWL*	*S*	*OA*
Airy: Weigel (1964)	1	3	4	3
Stokes III: Skjelbreia (1959)	6	8	7	8
Stokes V: Skjelbreia and Hendrickson (1960)	4	4	3	4
Cnoidal I: Laitone (1960)	10	10	7	10
Cnoidal II: Laitone (1960)	5	9	8	9
Solitary: McCowan (1891)	8	7	5	7

Notes:
B = sea bed; *SWL* = still-water level; *S* = surface; *OA* = overall.

4.6 Gravity wave phenomena

Gravity waves also exhibit behaviour similar to that observed for other types of waves (such as acoustic or electromagnetic) in the physical world. Gravity waves undergo the classical reflection, refraction and diffraction effects observed in other wave processes. They can also 'break' if a limiting stable height is exceeded. This section presents a qualitative description of these phenomena.

Waves 'break' because for any water depth and wave period there is an upper limit to the wave height, which is stable for these conditions. This 'breaking' wave height is of special interest in establishing design criteria at a given site. Knowledge of the absolute maximum possible wave height is necessary to ensure that the clearances of jacket module and semisubmersible deck undersides are sufficient to prevent wave slamming. Stokes conjectured a hypothesis that when crest particle velocities equalled the wave celerity, the crest angle equalled 120° and a limiting condition was reached beyond which the wave would begin to lose its coherent structure of water particle velocity and acceleration distributions. Recent work by Cokelet (1977) and Longuet-Higgins and Cokelet (1976) has investigated high and breaking wave behaviour, using mathematical techniques which may eventually find use in the design of offshore structures. Equations (4.63) and (4.64) also present some crude empirical rules for the breaking wave limit.

Furthermore, any gravity wave approaching shoaling water will eventually break in various ways due to the fact that the reducing water depth tends to increase the elevations and steepness of the wave surface. The form of the breaking wave depends on sea bed topography and incident wave characteristics.

Figure 4.10 presents the principal features of four possible types of wave breaking in shallow water. For prediction purposes these waves can be treated approximately as a train of solitary waves although this applies only up to the point of wave breaking.

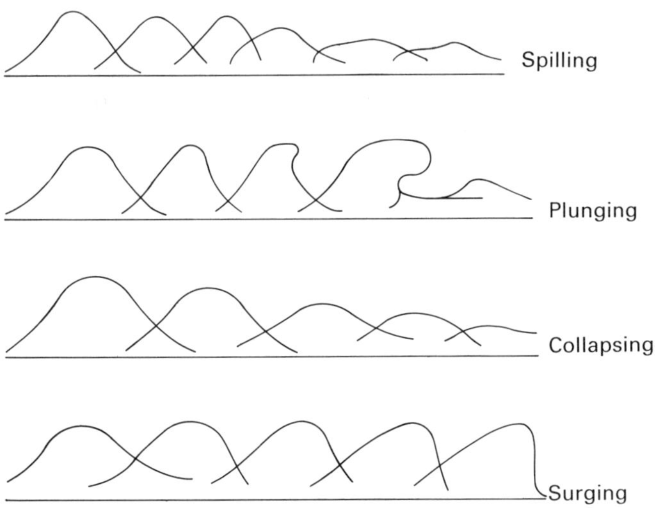

Figure 4.10. Breaking wave types

Gravity waves can also undergo diffraction, reflection and refraction effects due to interaction with submerged bodies or the effects of the sea bed. Figure 4.11(*a*) and (*b*) demonstrate these phenomena by displaying crest lines of a wave train in plan view when hitting a break water and when moving into shoaling water at an initial angle to the shore line.

Wave diffraction such as that illustrated round the break water of Figure 4.11(*a*) occurs when an incident wave train interacts with a body that is large compared to its wave length. There are two applications of this phenomena in engineering. Firstly, the design of harbours and ports has to account for wave diffraction in order to evaluate the wave climate within and outside the area. Secondly, for large offshore structures (concrete platforms and ships), wave diffraction must be used to describe the interaction between the wave and body. The modification of incident waves by the body has to be calculated in order to derive wave forces on the body.

When a travelling wave strikes a vertical flat barrier such as a sea wall, the wave is reflected. The combination of the incoming incident wave and the outgoing reflected wave results in a standing wave system of twice the incident wave amplitude in the vicinity of the reflecting wall. This larger amplitude standing wave can cause problems in waters where ships are moored by inducing severe mooring line loads. This is one of the reasons why ships are sometimes taken out to sea during storms. It also explains the importance of harbour breakwaters. The reflection process at a wall is not ideal, however, because of losses due to water viscosity as well as the elasticity and permeability of the reflecting surface.

Gravity wave reflection is analogous to light refraction in that the wave velocity in shoaling water decreases with depth in the same way as light velocity reduces in a medium such as glass. Thus the wave train approaching the beach in Figure 4.11(*b*) swings round to a direction of travel perpendicular to the shore line because of the reduced celerity of the wave front segment that encounters shallowing water first. This phenomenon may be used to concentrate waves for energy conversion by using sea bed contours as a lens or, alternatively, to control the wave environment in a shallow water area of engineering interest.

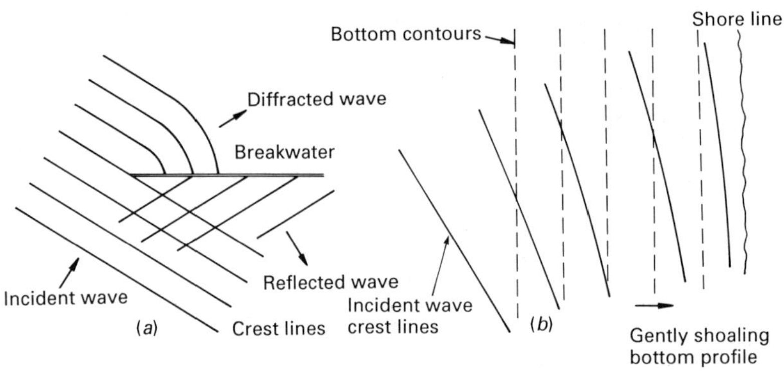

Figure 4.11. Gravity wave phenomena: (*a*) reflection and diffraction; (*b*) refraction

4.7 Ocean waves as a random process

The foregoing theories of regular, unidirectional ocean waves are applicable to simple design considerations, laboratory simulations of ocean waves and to limited types of full-scale ocean waves where a swell sea state can be approximated by long-crested regular waves.

In the general case of a confused sea state, however, the energy transferred to the sea surface spreads out along the direction of wave propagation due to dispersion and also in angular directions. The wave climate in a particular location is often due to the combined effects of local winds and swell waves propagating from other adjacent regions of wave generation. Thus the properties of realistic ocean waves are not readily definable on a deterministic basis. However, due to the very large number of physical processes that contribute to wave elevation at a particular point on the ocean surface, this wave elevation is amenable to a probabilistic description based on the well-understood theory of stationary, random processes. This theory has been extensively developed for linear systems and the application of this body of knowledge to gravity waves requires that, for design purposes at least, gravity waves are adequately described by linear wave theory. Within this restriction, however, consideration of ocean waves as a random process offers powerful methods for describing the behaviour of realistic unidirectional and spreading (long- and short-crested) ocean waves. The use of long-term probabilistic methods also makes it possible to predict the probability of extreme wave heights over such long periods of time as 50 or 100 years. Research is currently being carried out on the probabilistic behaviour of non-linear waves but this work is not sufficiently well developed for it to be applied to design.

The following section first presents an overview of the basic statistical properties of a random process. These properties are then applied to describe both long- and short-crested irregular ocean waves and to develop the methods required to predict their short- and long-term, properties. A description of the techniques adopted to measure and analyse ocean wave data is also presented.

4.7.1 Basic properties of random waves

Figure 4.12(*a*) presents a typical wave elevation, with time recorded at a point on the ocean surface. The wave elevation, $\eta(t)$ and related wave properties such as velocities, accelerations and pressures can be regarded as Gaussian random processes about a zero mean, with their probability density function $P(\eta)$, shown in Figure 4.12(*b*) for wave elevation, being Gaussian or normal. The time history of wave elevation shown in Figure 4.12(*a*) is also representative of the variation of other parameters under the wave such as pressures, particle velocities and accelerations. The random process is also assumed to be stationary and ergodic. The stationarity means that statistical properties of wave parameters do not change with absolute time. Furthermore, a stationary random process is ergodic if, in addition to all the statistical properties of any ensemble or realization being stationary with respect to changes in time scale, they are also stationary if taken at different times along the same ensemble or realization of the process. Newland (1984) gives further details.

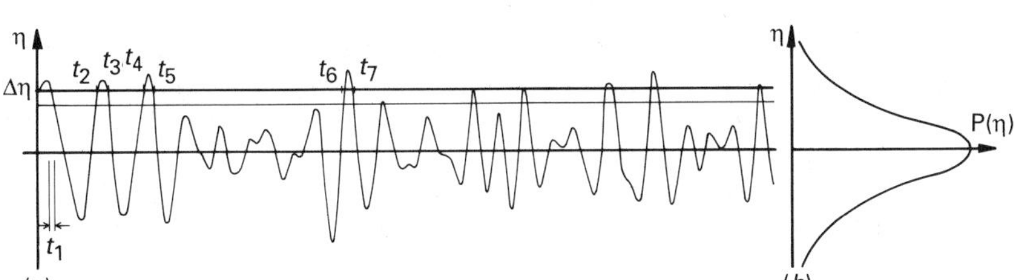

Figure 4.12. Wave elevation with time and probability density function

Now the time history $\eta(t)$ can be statistically characterized in a number of ways. At the simplest level, the mean or expected value (μ), mean square value (χ^2) and variance (σ^2) of wave elevation, $\eta(t)$, can be defined as

$$\left.\begin{aligned} \mu &= \lim_{T\to\infty} \frac{1}{T}\int_0^T \eta(t)\,\mathrm{d}t \\ \chi^2 &= \lim_{T\to\infty} \frac{1}{T}\int_0^T \eta^2(t)\,\mathrm{d}t \\ \sigma^2 &= \lim_{T\to\infty} \frac{1}{T}\int_0^T [\eta(t)-\mu]^2\,\mathrm{d}t \end{aligned}\right\} \tag{4.78}$$

In the case of linear gravity waves, the mean value μ is set to zero – thus making the mean square value and variance equal. The distribution of wave elevation $\eta(t)$ is also characterized by defining its probability density or cumulative probability distribution functions.

The probability density function (pdf) of a random variable describes the probability that the variable will have a value within a given range over a period of time. For the wave elevation time history of Figure 4.12(a), this can be expressed mathematically as

$$\begin{aligned} P(\eta) &= \lim_{\Delta\eta\to 0} \frac{1}{\Delta\eta}\{\mathrm{Prob}[\eta \leqslant \eta(t) \leqslant \eta + \Delta\eta]\} \\ &= \lim_{\Delta\eta\to 0} \frac{1}{\Delta\eta}\left(\lim_{T\to\infty} \frac{\Sigma t_i}{T}\right) \end{aligned} \tag{4.79}$$

with the t_i values as defined in Figure 4.12(a). It has been shown by observation that ocean wave elevations exhibit a zero mean Gaussian or normal probability density function given by the equation

$$P(\eta) = \frac{1}{\surd(2\pi)\sigma}\exp\left[-\frac{\eta^2}{2\sigma^2}\right] \tag{4.80}$$

The shape of this function is a characteristic bell shape shown on its side in Figure 4.12(b). Note that $P(\eta)$ is always positive for finite η.

A cumulative probability distribution function $P_c(\eta)$ is defined as the probability that $\eta(t)$ is less than or equal to a given value η. Thus

$$P_c(\eta) = \mathrm{Prob}[\eta(t) \leqslant \eta] = \int_{-\infty}^{\eta} P(\xi)\,\mathrm{d}\xi \tag{4.81}$$

gives the Gaussian cumulative probability distribution. Note that $P_c(\eta)$ is always positive within the limits of $P_c(-\infty) = 0$ and $P_c(+\infty) = 1$.

The form of both of the above functions is known to be very slightly modified for high waves, due to their attendant non-linear effects. However, these modifications are not as yet sufficiently well known for use in engineering design.

The mean value, mean square and standard deviation can also be written in terms of the probability density function as

$$\mu = \int_{-\infty}^{\infty} x\,P(x)\,\mathrm{d}x$$

$$\chi^2 = \int_{-\infty}^{\infty} x^2 P(x)\,\mathrm{d}x$$

and

$$\sigma^2 = \int_{-\infty}^{\infty} x^2 P(x)\,\mathrm{d}x - \left[\int_{-\infty}^{\infty} x\,P(x)\,\mathrm{d}x\right]^2$$

The characteristics of the probability density function can be further quantified by taking the moments of the distribution with the nth moment defined as

$$m_n = \int_{-\infty}^{\infty} \eta^n P(\eta)\,\mathrm{d}\eta \qquad (4.82)$$

The first and second moments ($n = 1$ and 2) yield the mean and mean square values respectively. Values of m_n up to $n = 5$ are used to test the fit of measured wave elevation data to the Gaussian probability density function.

The above ideas can be extended to consider the joint probability of two random variables simultaneously having values within defined ranges. The joint probability density function of two random variables $\eta_1(t)$ and $\eta_2(t)$ describes the probability that both variables will simultaneously have values within a given pair of ranges at any point in time. Thus

$$P(\eta_1, \eta_2) = \lim_{\substack{\Delta\eta_1 \to 0 \\ \Delta\eta_2 \to 0}} \left(\lim_{T\to\infty} t_{\eta 1,\eta 2}/T\right) \qquad (4.83)$$

where t_{η_1,η_2} is the total time for which $\eta_1(t)$ and $\eta_2(t)$ lie simultaneously in the range $(\eta_1, \eta_1 + \Delta\eta_1)$ and $(\eta_2, \eta_2 + \Delta\eta_2)$ respectively. The joint cumulative probability distribution function, $P_c(\eta_1, \eta_2)$, is then defined as

$$\begin{aligned} P_c(\eta_1, \eta_2) &= \mathrm{Prob}[\eta_1(t) \leqslant \eta_1, \eta_2(t) \leqslant \eta_2] \\ &= \int_{-\infty}^{\eta_1}\int_{-\infty}^{\eta_2} P(x, y)\,\mathrm{d}x\,\mathrm{d}y \qquad (4.84) \end{aligned}$$

This joint probability density concept can be extended to any number of variables with corresponding multi-parameter probability density and cumulative distribution functions.

It is also necessary to have information on the persistence of a physical process with relative time in a random time history or the frequency distribution of the time history. Such information is provided by auto- and cross-correlation functions. Thus the auto-correlation function $R(\tau)$ of variable η at time t, $\eta(t)$, and time $t + \tau$, $\eta(t + \tau)$, is defined by

$$R(\tau) = \lim_{T\to\infty} \frac{1}{T} \int_0^T \eta(t)\cdot\eta(t + \tau)\,\mathrm{d}t \qquad (4.85)$$

Figure 4.13 presents typical shapes of $R(\tau)$ functions (called auto-correlograms) for three time histories corresponding to sine wave, narrow frequency band and broad frequency band time histories. The shape of $R(\tau)$ reflects the persistence or correlation of the variable value across a separation in time of τ. It can be seen from these figures and from consideration of Equation (4.85) that $R(\tau)$ is a maximum at $\tau = 0$ and that $R(-\tau) = R(\tau)$. It can also readily be shown that

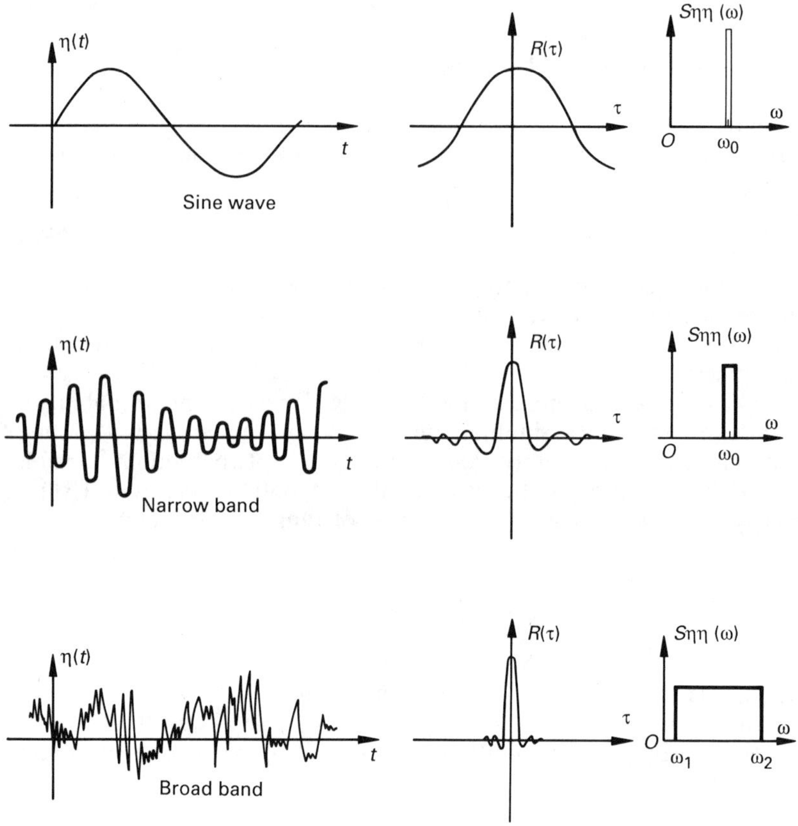

Figure 4.13. Time series autocorrelation functions and spectral densities

$$\mu = \sqrt{[R(\infty)]} \text{ and } \chi^2 = R(0) \tag{4.86}$$

Cross-correlation functions are used to describe the correlation between two sets of random variables. Thus the cross-correlation function of two time histories $\eta_1(t)$ and $\eta_2(t)$ can be defined as

$$R_{\eta_1\eta_2}(\tau) = \lim_{T\to\infty} \frac{1}{T} \int_0^T \eta_1(t)\, \eta_2(t + \tau)\, \mathrm{d}t \tag{4.87}$$

Here $R_{\eta_1\eta_2}(0)$ is not necessarily a maximum but $R_{\eta_1\eta_2}(-\tau) = R_{\eta_2\eta_1}(+\tau)$, although $R_{\eta_1\eta_2}(-\tau) \neq R_{\eta_1\eta_2}(+\tau)$; $R_{\eta_1\eta_2}(\tau) = 0$ implies that $\eta_1(t)$ and $\eta_2(t)$ are uncorrelated. Independent random variables are always uncorrelated.

Auto- and cross-correlation functions take no account of the mean value. Auto- and cross-covariance functions, however, include a correction for the mean and are defined as

$$\left.\begin{aligned} \mathrm{Cov}(\tau) &= \lim_{T\to\infty} \frac{1}{T} \int_0^T [\eta(t) - \mu][\eta(t + \tau) - \mu]\, \mathrm{d}t \\ \mathrm{Cov}_{\eta_1\eta_2}(\tau) &= \lim_{T\to\infty} \frac{1}{T} \int_0^T [\eta_1(t) - \mu_1][\eta_2(t + \tau) - \mu_2]\, \mathrm{d}t \end{aligned}\right\} \tag{4.88}$$

For wave data, since $\mu = 0$, correlation and covariance functions are identical.

The distribution of energy with frequency of a random wave elevation time history is of profound importance in calculating the effects that the wave would have on an offshore structure. The energy distribution with frequency of a time history $\eta(t)$ is expressed as a power spectral density function by taking a Fourier transform denoted by function $X(f)$ of time history $\eta(t)$ and writing

$$W_{\eta\eta}(f) = \operatorname*{Lim}_{\Delta f \to 0} \frac{1}{\Delta f} |X(f)|^2 \tag{4.89}$$

with f being the frequency in hertz. In terms of radian frequency $\omega(= 2\pi f)$ this becomes

$$W_{\eta\eta}(\omega) = \operatorname*{Lim}_{\Delta\omega \to 0} \frac{1}{\Delta\omega} |X(\omega)|^2 \tag{4.90}$$

$W_{\eta\eta}$ is also defined by taking the Fourier transform of the auto-correlation function of Equation (4.85)

$$W_{\eta\eta}(f) = \int_{-\infty}^{+\infty} R(\tau) \exp(-\mathrm{i}2\pi f\tau)\mathrm{d}\tau = 2\int_0^{\infty} R(\tau)\exp(-\mathrm{i}2\pi f\tau)\cdot\mathrm{d}\tau \tag{4.91}$$

Equation (4.91) may be re-written in terms of radian frequency, $\omega = 2\pi f$ as

$$W_{\eta\eta}(\omega) = \frac{1}{2\pi}\int_{-\infty}^{+\infty} R(\tau)\mathrm{e}^{-\mathrm{i}\omega\tau}\mathrm{d}\tau = \frac{1}{\pi}\int_0^{\infty} R(\tau)\mathrm{e}^{-\mathrm{i}\omega\tau}\cdot\mathrm{d}\tau \tag{4.92}$$

By taking a Fourier transform of the spectral density function, it can be shown that the spectral density and auto-correlation functions form a Fourier transform pair given by

$$\left.\begin{aligned} W_{\eta\eta}(\omega) &= \frac{1}{2\pi}\int_{-\infty}^{\infty} R(\tau)\mathrm{e}^{-\mathrm{i}\omega\tau}\mathrm{d}\tau \\ &\text{and} \\ R(\tau) &= \int_{-\infty}^{\infty} S(\omega)\mathrm{e}^{\mathrm{i}\omega\tau}\mathrm{d}\omega \end{aligned}\right\} \tag{4.93}$$

Figure 4.13 also presents typical spectral density functions derived using the above equations for sinusoidal, narrow band and broad band time histories. Now the mean square or variance can be directly related to the spectral density by taking.

$$\begin{aligned} \sigma^2 &= \operatorname*{Lim}_{T\to\infty} \int_{-T}^{T} \frac{1}{2T} [\eta(t)]^2 \,\mathrm{d}t \\ &= \operatorname*{Lim}_{T\to\infty} \frac{1}{2T}\frac{1}{2\pi}\int_{-T}^{T} |X(\omega)|^2 \,\mathrm{d}\omega \end{aligned} \tag{4.94}$$

using Parseval's theorem.

Then using Equation (4.90)

$$\sigma^2 = \int_{-\infty}^{+\infty} W_{\eta\eta}(\omega)\,\mathrm{d}\omega = 2\int_0^{\infty} W_{\eta\eta}(\omega)\,\mathrm{d}\omega \tag{4.95}$$

Note that the spectral power density $W_{\eta\eta}(\omega)$ is an even function and is, therefore, defined as a double-sided function with frequency values from

$-\infty$ to $+\infty$. It can also, however, be defined in single-sided form over the frequency range 0 to $+\infty$ by folding the negative part over the positive side. The definition of Equation (4.95) is of double-sided form with the corresponding single-sided form $S_{\eta\eta}$ given by

$$\sigma^2 = \int_0^{+\infty} S_{\eta\eta}(\omega)\,\mathrm{d}\omega$$

and, therefore, $S_{\eta\eta}(\omega) = 2W_{\eta\eta}(\omega)$ for positive ω. Care has to be shown in defining and maintaining consistency with the single- or double-sided definition of spectral density. The single-sided spectral density function tends to be used in offshore design practice and is, therefore, used in all subsequent equations, except when stated otherwise.

As with the probability density function, wave elevation spectra can be characterized by computing their nth moment m_n, defined as

$$m_n = \int_0^{\infty} \omega^n S_{\eta\eta}(\omega)\,\mathrm{d}\omega \tag{4.96}$$

A further characteristic for spectral density functions is their band width parameter, first defined by Cartwright and Longuet-Higgins (1956) as

$$\varepsilon^2 = \frac{m_0 m_4 - m_2^2}{m_0 m_4} \tag{4.97}$$

The parameter ε lies in the range from zero for a narrow band process to one for a broad band process. A narrow band signal has a frequency content which is confined to a small range of frequencies, whereas a broad band signal exhibits a wider frequency range. Figure 4.13 illustrates examples of narrow and broad band signals.

Analogous to the auto-spectral density, a cross-spectral density can also be defined for a pair of random processes denoted by subscripts 1 and 2. Double-sided cross-spectral densities can be defined in terms of the cross-correlation function by equations

$$W_{12}(\omega) = \frac{1}{2\pi}\int_{-\infty}^{+\infty} R_{12}(\tau)\mathrm{e}^{-\mathrm{i}\omega\tau}\mathrm{d}\tau$$

$$W_{21}(\omega) = \frac{1}{2\pi}\int_{-\infty}^{+\infty} R_{21}(\tau)\mathrm{e}^{-\mathrm{i}\omega\tau}\mathrm{d}\tau$$

The corresponding inverse relationships are

$$R_{12}(\tau) = \int_{-\infty}^{\infty} W_{12}(\omega)\mathrm{e}^{\mathrm{i}\omega\tau}\mathrm{d}\omega$$

$$R_{21}(\tau) = \int_{-\infty}^{\infty} W_{21}(\omega)\mathrm{e}^{\mathrm{i}\omega\tau}\mathrm{d}\omega$$

Cross-spectra allow the frequency content of one random process to be related to that of another. This is particularly useful for the study of dynamic systems where random signal inputs need to be related to the expected outputs.

4.7.2 Long-crested irregular waves

The statistical description of unidirectional (that is long-crested) irregular waves uses the random process theory described above by invoking four assumptions:

1. Ocean waves are taken to be a weakly stationary ergodic, Gaussian (or normal) random process with zero mean.
2. Wave spectral density functions are narrow banded.
3. Wave crest elevations (maxima) are statistically independent.
4. The statistical properties of ocean waves are homogeneous, that is they are independent of local position.

Note that the term 'stationary' applies to a random process (η(t), say) with properties σ, $R(\tau)$, and so on, which do not change if the time origin is changed. If N samples of $\eta(t)$ were considered, the stationary random process would be ergodic if the properties of the process such as σ and $R(\tau)$ calculated from one sample are equal to those calculated from any of the other $N - 1$ samples. These properties are of importance when interpreting data within a single sample function. Wave records can be considered as a realization of a stationary, ergodic random process.

A key feature of the probabilistic description of long-crested irregular waves is the spectral power density, since this provides a convenient frequency domain description of the sea state for use in design methods. A number of parametrized formulae for such power densities have been derived from observations and theoretical work. When considering such spectra, it is important to define whether these are valid for a partially or fully developed sea. A fully developed sea exists for a condition where energy input from wind into surface waves is balanced by the dissipation of wave energy in breaking. Obviously, the fully developed condition will be approached asymptotically and there are doubts as to the actual existence of such a state. However, wave spectra from the 'fully developed' condition have been proposed by various authors, the favoured being due to Pierson and Moskowitz (1964):

$$S_{\eta\eta}(\omega)\,\mathrm{d}\omega = \frac{Ag^2}{\omega^5}\exp\left[-B\left(\frac{\omega_0}{\omega}\right)^4\right]\mathrm{d}\omega \tag{4.98}$$

with $S_{\eta\eta}(\omega) = 0$ for $\omega \leqslant 0$, $A = 0.0081$, $B = 0.74$ and $\omega_0 = g/U$, where U is the wind velocity at an elevation of 19.5 m above still-water level. The shape of the spectrum is presented in Figure 4.14. The form was developed for wind speeds up to 45 knots. For U in knots the 'significant wave height' $H_{1/3}$, which is the average height of the highest one-third of waves, is given by

$$H_{1/3} = 1.82 \times 10^{-2}\mathrm{U}^2(\mathrm{ft})$$

For U in m/s

$$H_{1/3} = 2.12 \times 10^{-2}\mathrm{U}^2(\mathrm{m}) \tag{4.99}$$

An estimate of the average zero crossing wave period is

$$T_z = 0.81\,\frac{2\pi U}{g}$$

In terms of frequency f_m(Hz), the Pierson–Moskowitz spectrum, (Equation (4.98)) can be written as

$$S_{\eta\eta}(f) = \frac{Ag^2}{(2\pi)^4 f^5}\exp\left[-\frac{5}{4}\left(\frac{f}{f_m}\right)^{-4}\right] \tag{4.100}$$

where f_m is the 'peak' frequency of the spectrum, that is, the frequency at which $dS(f)/df = 0$.

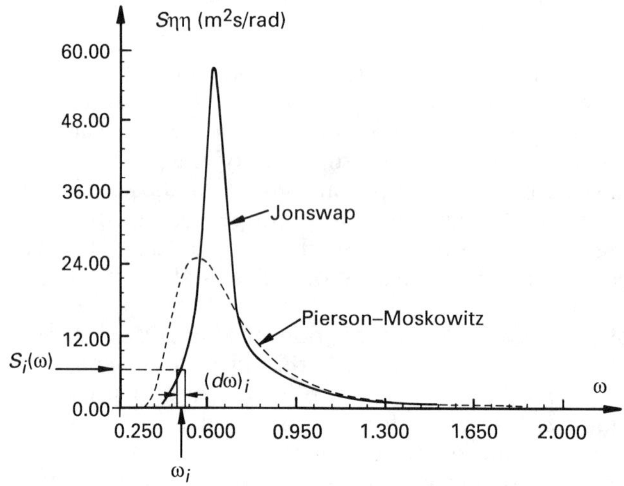

Figure 4.14. Jonswap and Pierson–Moskowitz spectra for $H_s = 5.0$ m, $T_z = 8$ s

Note that given the wind velocity U, or the significant wave height $H_{1/3}$, or the average zero crossing wave period T_z, the Pierson–Moskowitz spectrum can be calculated from Equations (4.98) and (4.99). The average zero crossing wave period is the average of the time intervals between adjacent zero up (or down) crossings in a wave elevation time history.

A modified form of Equation (4.98) has been proposed to permit the input of both $H_{1/3}$ and average zero crossing wave period T_z in evaluating the spectrum:

$$S_{\eta\eta}(\omega) = \frac{H_{1/3}^2 T_z}{8\pi^2}\left(\frac{\omega T_z}{2\pi}\right)^{-5}\exp\left[-\frac{1}{\pi}\left(\frac{\omega T_z}{2\pi}\right)^{-4}\right] \tag{4.101}$$

The majority of wave spectra used in the design of offshore structures are of the form given by Equation (4.98) which can be written in simpler form as

$$S(\omega) = \frac{P}{\omega^5}\exp\left[-\frac{Q}{\omega^4}\right) \tag{4.102}$$

The moments and statistical properties of waves with this spectral formulation are listed here as an aid to calculations. The moments are

$$\left.\begin{aligned} m_0 &= P/4Q & m_1 &= 0.306P/Q^{3/4} \\ m_2 &= \sqrt{\pi}P/(4\sqrt{Q}) & m_4 &= \infty \end{aligned}\right\} \tag{4.103}$$

The value of m_4 leads to the anomaly that the band width parameter for this spectrum is 1. Nevertheless, the narrow band assumption is employed to derive the quantities listed below:

$$\left.\begin{aligned}
&\text{Significant wave height, } H_{1/3} = \surd(4P/Q)\\
&\text{Average wave height } H_{\mathrm{m}} = \surd(\pi P/2Q)\\
&H_{1/3}/H_{\mathrm{m}} = \surd(8/\pi)\\
&\text{Modal period} = 2\pi(4Q/5)^{-1/4}\\
&\text{Average period} = 2\pi m_0/m_1 = 5.13Q^{-1/4}\\
&\text{Average zero crossing period} = 2\pi\surd(m_0/m_2) = 2\pi^{3/4}/Q^{1/4}
\end{aligned}\right\} \quad (4.104)$$

A comparison of one other family of spectra (commonly known as the JONSWAP spectrum), by Saetre (1974) and Hasselman *et al.* (1973) with the Pierson–Moskowitz spectrum is given in Figure 4.14. The JONSWAP (for Joint North Sea Wave Project) experiment suggests a peak enhancement factor which multiplies the Pierson–Moskowitz spectrum.

The JONSWAP spectrum is expressed as

$$S_{\eta\eta}(f) = \frac{ag^2}{(2\pi)^4 f^5}\exp\left[-\frac{5}{4}\left(\frac{f}{f_{\mathrm{m}}}\right)^{-4}\right]\cdot\underbrace{\gamma^{\exp\left[-\frac{(f-f_{\mathrm{m}})^2}{2\sigma^2 f_{\mathrm{m}}^2}\right]}}_{B} \quad (4.105)$$

where

$$a = 0.046\,\frac{(X)^{-0.22}}{U_{10}^2}$$

X being the fetch in metres and U_{10} is the wind velocity in m/s at 10 m elevation above still-water level, with $\gamma = 3.30$, $\alpha = 0.08$ and

$$f_{\mathrm{m}} = \frac{16.04}{(XU_{10})^{0.38}}$$

These equations were derived empirically from measured wave and wind data and represent average values. Saetre (1974) uses $\sigma = \sigma_a$ for $f < f_{\mathrm{m}}$ and $\sigma = \sigma_{\mathrm{b}}$ for $f > f_{\mathrm{m}}$ but a constant σ is a reasonable approximation in view of the scatter of data from which the spectrum is derived. The spectrum with constant σ is often referred to as the 'mean JONSWAP spectrum'. It can be seen from the equations that the mean JONSWAP spectrum is similar to the Pierson–Moskowitz spectrum with a 'peak enhancement' factor, B, and with the parameter a no longer constant. The JONSWAP spectrum has higher energies at high frequencies than the Pierson–Moskowitz spectrum and it is at these higher frequencies that the natural periods of offshore structures are often found.

The spectra defined above can be used in frequency or time domain design calculations by decomposing the random wave elevation into a summation of regular wave components. Thus the time history of a random unidirectional sea may be expressed as a linear superposition of a number of regular waves through the equation

$$\eta(t) = \sum_{i=1}^{N} a_i \sin(\omega_i t + \phi_i) \tag{4.106}$$

where the phases ϕ_i are assumed to be uniformly distributed over the range from 0 to $2\pi[P(\phi_i) = 1/(2\pi)]$. In that case, it can be shown that the probability density function of $\eta(t)$ will be Gaussian. The wave amplitude components a_i are obtained through the equation

$$a_i = \sqrt{[2S_i(\omega_i)\cdot(\mathrm{d}\omega)_i]} \tag{4.107}$$

where $S_i(\omega_i)$ is the value of spectral density at frequency ω_i and $(\mathrm{d}\omega)_i$ is the width of the frequency band being approximated by a regular wave of amplitude a_i. This approach is illustrated in Figure 4.14 through one of the bars into which the area under the spectrum is subdivided. The factor 2 above is included only if $S(\omega)$ is defined as a single-sided function from frequency 0 to ∞. For a double-sided spectral power density defined from frequency $-\infty$ to $+\infty$, the factor 2 is omitted.

4.7.3 Short-term statistical wave properties

The short-time properties of an irregular sea are defined as those within a storm duration between 20 minutes and 3 hours where the various statistical properties of wave parameters are varying very slowly or are constant. In this kind of timescale, the probability of occurrence of particular wave elevations can be derived from the Gaussian probability density function (Equation (4.80)), once the variance (σ^2) is known. However, the probability density function for the occurrence of maxima (that is, wave height divided by 2) is of much more significance for engineering design since the response of offshore structures is readily definable with respect to numerical values of individual wave height. Such a probability density function for the occurrence of maxima, P_{H}, has been derived by Longuet-Higgins (1952) and Cartwright and Longuet-Higgins (1956). It can be written as

$$P_{\mathrm{H}}(A) = \frac{1}{\sqrt{(2\pi)}}\left[\varepsilon \exp\left(-\frac{A^2}{2\varepsilon^2}\right) + A(1-\varepsilon^2)^{1/2}\exp\left(-\frac{A^2}{2}\right) \cdot \int_0^{(A/\varepsilon)(1-\varepsilon^2)^{1/2}} \exp\left(-\frac{z^2}{2}\right)\mathrm{d}z\right] \tag{4.108}$$

where $A = \eta_{\max}/\sigma$ and ε, the spectral band width parameter, is given by Equation (4.97). This expression is rather cumbersome but can be greatly simplified for the two extreme cases of $\varepsilon = 0$ and $\varepsilon = 1$ to yield the following relationships:

1. For a 'narrow band' spectrum, where $\varepsilon = 0$, the probability density of wave maxima becomes

$$P_{\mathrm{H}}(A) = A\exp\left(-\frac{A^2}{2}\right) \tag{4.109}$$

which is the 'Rayleigh' probability density function.

2. For a 'broad band' spectrum, where $\varepsilon = 1$, $P_{\mathrm{H}}(A)$ becomes

$$P_{\mathrm{H}}(A) = \frac{1}{\sqrt{(2\pi)}} \exp\left(-\frac{A^2}{2}\right) \tag{4.110}$$

which is the Gaussian probability density function.

Since the spectra of ocean waves tend to be narrow banded with ε values in the range 0.4–0.8, an assumption is made that the probability density function of wave height is given by the Rayleigh distribution of Equation (4.109). This function is plotted in Figure 4.15 and can be written in terms of wave height as

$$P_{\mathrm{H}}(H) = \frac{H}{4\sigma^2} \exp\left(-\frac{H^2}{8\sigma^2}\right) \tag{4.111}$$

Ochi (1982) shows that the approximation leads to an overestimate of wave height predictions in the region of 1–8%. Equation (4.111) may be used to predict the probability of occurrence of waves of various height in a storm of duration from 20 minutes to 3 hours as long as the sea state properties remain sufficiently stationary over this period.

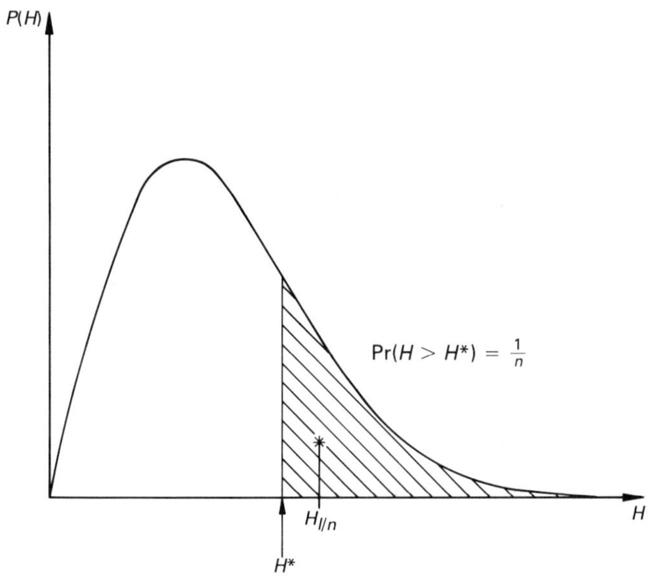

Figure 4.15. Rayleigh probability distribution: $\mathrm{Prob}\{H > H^*\} = \dfrac{1}{n}$

However, the wave height distribution arising from the Rayleigh probability density is also quantified by a parameter called the significant wave height and defined as the average of the highest one-third of wave heights in the sea state. The significant wave height is a number commonly used to characterize the severity of a sea state, with the advantage that the significant wave height value can be visually estimated with reasonable accuracy. This concept is also generalized by defining a wave height $H_{1/n}$, which is the average of the highest $1/n$th of the waves in a sea state, with n typically taking integer values from 3 to 20.

A formula for $H_{1/n}$ (including the significant wave height for $n = 3$) can be defined by finding the centroid of the area under the Rayleigh probability density function from $H = H^*$ to ∞ (see Figure 4.15). The lower limit H^* needs to be evaluated first by using the equation

$$\text{Prob}(H > H^*) = \int_{H^*}^{\infty} \frac{H}{4\sigma^2} \exp\left(-\frac{H^2}{8\sigma^2}\right) \mathrm{d}H = \frac{1}{n} \tag{4.112}$$

This yields the expression $H^* = \sqrt{(8\sigma^2 \ln n)}$ which then allows the calculation of the centroid to be completed through the equation

$$H_{1/n} = \frac{\int_{H^*}^{\infty} H\, P(H)\, \mathrm{d}H}{\int_{H^*}^{\infty} P(H)\, \mathrm{d}H} \tag{4.113}$$

Evaluation of Equation (4.113) yields the expression

$$\left.\begin{aligned} &H_{1/n} = 2\sqrt{(2)}\cdot\sigma\{\sqrt{(\ln n)} + n\sqrt{\pi}\{1 - \Phi[\sqrt{(2\ln n)}]\} \\ &\text{where} \\ &\Phi(x) = \frac{1}{\sqrt{(2\pi)}} \int_{-\infty}^{\infty} \exp\left(-\frac{u^2}{2}\right) \mathrm{d}u \end{aligned}\right\} \tag{4.114}$$

For $n = 3$, the significant wave height $H_{1/3}$ can be evaluated as

$$H_{1/3} = 4.01\sigma = 4.01\sqrt{(m_0)} \tag{4.115}$$

where m_0 is the area under the power spectral density. Other relationships for wave height are

$$\left.\begin{aligned} &H_{1/3} = 1.598\, H_{\text{mean}},\ H_{1/10} = 2.032\, H_{\text{mean}} = 1.272\, H_{1/3} \\ &H_{\text{rms}} = 1.129\, H_{\text{mean}} \end{aligned}\right\} \tag{4.116}$$

The Rayleigh probability density function for wave height can also be written

$$P_{\mathrm{H}}(H) = \frac{4H}{H_{1/3}^2} \exp\left(-\frac{2H^2}{H_{1/3}^2}\right) \tag{4.117}$$

and the cumulative probability distribution function for wave height is

$$P_{\mathrm{HC}}(H) = 1 - \exp\left(-\frac{2H^2}{H_{1/3}^2}\right) \tag{4.118}$$

The expression of Equation (4.114) can be simplified for large $n(> 50)$ when the second term is negligible to yield

$$H_{1/n} = 2\sqrt{(2\ln n)}\cdot\sqrt{(m_0)} \tag{4.119}$$

This result is useful for estimating the maximum wave height in a sea state of finite duration.

4.7.4 Long-term wave properties

One of the primary requirements in the evaluation of offshore structures is the choice of a design wave height and period. The design wave may be defined as that height which is equalled or exceeded on average once in N years, where N might be 50 or 100. An estimate of extreme wave heights is normally based on the extrapolation of instrumentally measured waves over a period of one to three years. There is, inevitably, an element of error in the extrapolation process but no alternative method of estimating the design wave height is known.

Long-term wave data is usually obtained from records of wave elevation $\eta(t)$ taken typically every three hours for a period of 15 or 20 minutes. Each sample is reduced to two representative parameters, significant wave height $H_{1/3}$ and average zero crossing period T_z, by using the methods outlined in Section 4.7.3. Data for a long time period is then represented by a bivariate histogram (or scatter diagram) of $H_{1/3}$ and T_z, as shown in Figure 4.16. This scatter diagram summarizes data for three years' measurement at the Stevenson Station in the North Sea. The number of total observations is normalized and expressed as parts per thousand. Thus each box in the scatter diagram denotes the number of occurrences of wave conditions with significant wave heights and zero crossing periods denoted by the range in the height and width of the box, respectively, and presented as parts per thousand. The diagram contains a total of 1000 recorded events recorded and so, for example, waves with significant wave height in the range 1.5–2.0 m and zero crossing period in the range 6–7 s occurred for $\frac{46}{1000}$ of the three-year period covered in the diagram.

Now the maximum wave heights that are most likely to occur over a very long time period (from 5 to 100 years) can be predicted if the probability of occurrence of $H_{1/3}$ measured over relatively short time scales (from 1 to 3 years) is plotted using distribution functions that yield straight lines. Three such distributions have been identified:

1. the logarithmic normal distribution by Jasper (1956);
2. the Weibull distribution by Weibull (1951);
3. the Gumbel distribution by Gumbel (1958) and Saetre (1974).

There is little theoretical justification for these distributions, other than that observed data generally appear as straight lines when plotted using the distribution equations. The first two of the above distributions suffer from the slight disadvantage that they give a larger emphasis to the effects of smaller waves.

Wave height data from scatter diagrams can be used in these distributions in two ways. In the first and more common approach, significant or maximum wave height data are plotted against a function of their probability of occurrence for every record in the available data set where the wave height exceeds a small threshold value. In some cases this threshold value can be set to zero. An alternative method utilizes the highest wave height in a fixed, small time interval plotted against a function of its probability of occurrence.

The detailed form of the three distributions are described below:

(i) *The logarithmic normal distribution* This distribution assumes that the natural logarithm of wave heights has a normal distribution. The probability distribution function is, therefore, described by the equation

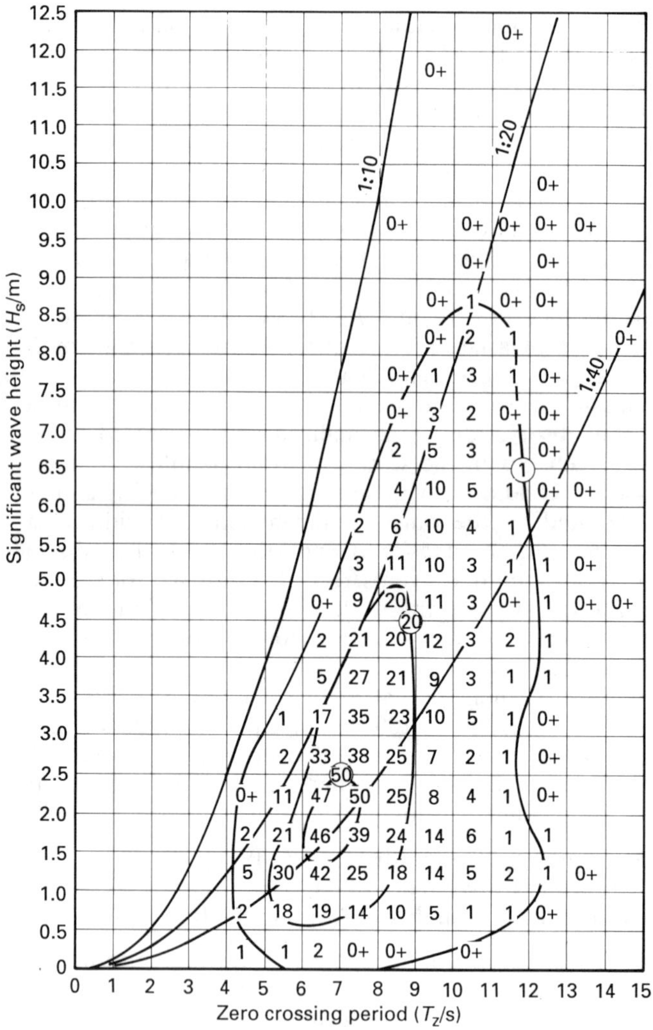

Figure 4.16. Wave scatter diagram (units indicated in parts per thousand, 0+ denotes value less than 0.5 parts per thousand (reproduced with the permission of the United Kingdom Offshore Operators' Association)

$$P_{\mathrm{L}}(H) = \frac{1}{2\pi\sigma H} \exp\left[\left(\frac{-(\ln H - \mu)^2}{2\sigma^2}\right]\right. \tag{4.120}$$

where σ and μ are constants dependent on the characteristics of the data set. Thus a plot of cumulative probability distribution against significant wave height, say, on normal probability paper should be a straight line. Measured data often deviates from a straight line, particularly at large wave heights where the linearity is necessary for accurate predictions of long-term significant wave heights.

(ii) *The Weibull distribution* The Weibull cumulative probability distribution of wave height is described by the equation

$$\mathrm{Prob}\{H > H_{\mathrm{m}}\} = P_{\mathrm{CL}}(H) = 1 - \exp\left[-\frac{(H)^{\gamma}}{\beta}\right] \qquad (4.121)$$

where $P_{\mathrm{CL}}(H)$ is the probability that any wave height chosen at random, H_{m}, is less than wave height H, and β and γ are constants.

By rearranging and taking logarithms, Equation (4.121) can be written as

$$\ln\{\ln[1 - P_{\mathrm{CL}}(H)]^{-1}\} = \gamma \ln H - \gamma \ln \beta \qquad (4.122)$$

Tests of the Weibull distribution with long-term wave data have shown that the distribution gives better fits to a straight line if the wave height H on the right-hand side of Equation (4.122) is replaced by $(H - a')$ where a' is a threshold wave height value below which wave activity is never expected to fall. In sheltered waters, a' may be taken to be zero or very small, with a large value (0.5–2.0 m) being usual for more exposed locations.

Long-term wave data can be used with the equation for the Weibull distribution by plotting $\ln\{\ln[1 - P_{\mathrm{CL}}(H)]^{-1}\}$ against $\ln(H - a')$ and using a straight line fit on the plotted points to extrapolate probabilities of occurrence of larger wave heights. This graph may also be plotted directly on to specially prepared Weibull probability paper.

(iii) *The Gumbel distribution* This is defined as

$$\left.\begin{aligned} &\mathrm{Prob}(H \leqslant h) = P_{\mathrm{CL}}(H) = \exp\{-\exp[-a(H - \beta)]\} \\ &\text{or} \\ &-\ln[-\ln P_{\mathrm{CL}}(H)] = a(H - \beta) \end{aligned}\right\} \qquad (4.123)$$

The application of these long-term distributions is illustrated by using the data of Figure 4.16 through a Weibull distribution to predict significant wave heights for 1-, 10- and 100-year return periods for wave rider buoy data recorded at the Stevenson Station in the North Sea.

Taking H as the significant wave height, the terms of the Weibull distribution equation are evaluated in Table 4.5 using a threshold value of $a' = 1.0$ m. Figure 4.17 presents plotted data from the fourth and third columns of the table which fit a straight line apart from one anamolous point due probably to insufficient occurrences recorded in the scatter diagram for high significant wave height. The 100-year return period significant wave height is predicted by taking the probability of $H < H_{100}$ to be

$$P_{100}(H) = P(H < H_{100}) = 1 - \frac{3}{100 \times 365.25 \times 24} = 0.999\,996\,58$$

Then

$$\ln\left\{\ln\left[\frac{1}{1 - P_{100}(H)}\right]\right\} = 2.533$$

gives $\ln[H_{100} - a] = 2.94$ from the graph, with H_{100}, therefore, being 19.9 m. Similarily, the 1-year and 10-year return periods' significant wave heights can be obtained as 13.7 m and 16.8 m respectively.

Table 4.5 Weibull distribution data

H/m	$P_{CL}(H)$	$\ln(H - 1.0)$	$\ln\left\{\ln\left[\frac{1}{1 - P_{CL}(H)}\right]\right\}$
2	0.370	0	−0.772
3	0.624	0.693	−0.022
4	0.783	1.099	0.424
5	0.888	1.386	0.784
6	0.940	1.609	1.034
7	0.971	1.792	1.264
8	0.981	1.946	1.377
9	0.984	2.079	1.420

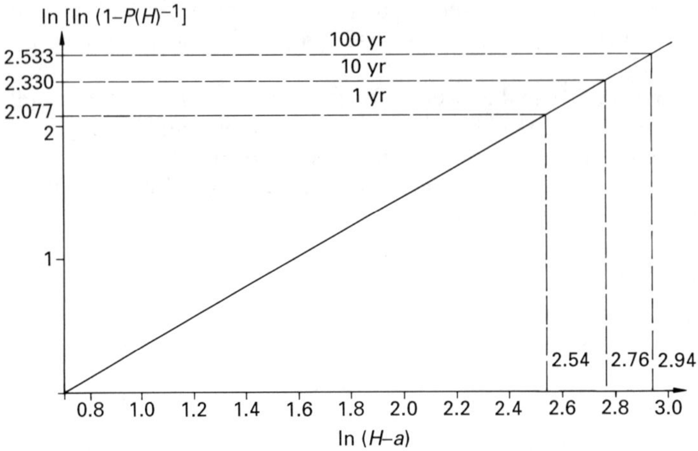

Figure 4.17. Weibull plot

A comparative evaluation of the prediction from the log-normal, Weibull and Gumbel distribution has been carried out by Saetre (1974) using three years of data. The results are listed in Table 4.6.

Table 4.6 Comparison of long-term predictions from Gumbel, Weibull and log normal distributions (as calculated by Saetre, 1974)

Wave parameter	Return period (years)	Extrapolated wave heights (m)		
		Gumbel	Weibull	Log normal
$H_{1/3}$	10	14.4	13.6	12.5
	50	16.3	15.2	14.6
H_{max} (3 h)	10	26.8	24.5	23.2
	50	30.6	27.8	27.0

The Gumbel probability distribution gives the highest extrapolated values, although Saetre concluded that the Gumbel gives the best fit to this particular set of data. The Weibull distribution offers a prediction that is midway between the three and tends to be the distribution most often used in the offshore industry. However, the recommended approach would be to use all three methods and assess the 'goodness-of-fit' to the particular data involved. The Weibull and Gumbel distributions generally show a better fit to data than the log normal distribution. Futher improvements in long-term predictions can only be achieved by continuous sets of observed data which cover periods of 10–20 years. Unfortunately, such data sets are not available at present.

4.7.5 Wave directionality

The analysis and prediction of wave properties at a point on the ocean surface needs to take account of the directionality of wave components in order to obtain a more accurate representation of the ocean wave surface. The concept of linear superposition of regular wave trains can be extended to include directionality. Consider wave elevation η, due to a single simple harmonic wave travelling at an angle θ to the x axis of the co-ordinate system, shown in Figure 4.18.

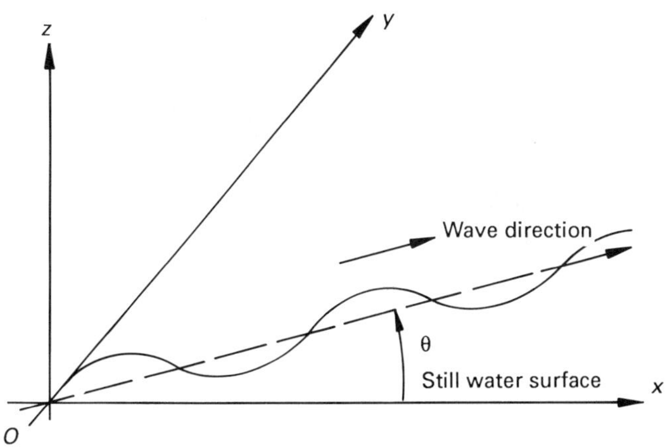

Figure 4.18. Definition diagram for directional waves

Here

$$\eta(x, y, t) = a\cos[k(x\cos\theta + y\sin\theta) - \omega t + \varepsilon']$$

with ε' being the phase of the wave at $x = y = t = 0$. The wave elevation due to a general irregular, directional sea can be considered as the summation of a large number of components, such as those given above, to yield

$$\eta(x, y, t) = \sum_{i=1}^{N} a_i \cos[k_i(x\cos\theta_i + y\sin\theta_i) - \omega_i t + \varepsilon_i'] \qquad (4.124)$$

with the random phases ε_i' taken to be uniformly distributed over the range 0–2π. The wave amplitudes a_i can be expressed in terms of the general directional single-sided spectrum $S(\omega, \theta)$ through the equation

$$\sum_{\Delta\omega}\sum_{\Delta\theta} a_i^2 = 2S(\omega, \theta)\, d\omega\, d\theta \tag{4.125}$$

In discretized form, Equations (4.124) and (4.125) can be combined to give

$$\left.\begin{aligned}\eta(x, y, t) = \sum_{j=-s}^{N}\sum_{i=0}^{r} &\cos[k_{2i+1}(x\cos\theta_{2j+1} \\ &+ y\sin\theta_{2j+1}) - \omega_{2i+1}t + \varepsilon_i'(\omega_{2i+1}, \theta_{2j+1})] \\ &\times \sqrt{[2S(\omega_{2i+1}, \theta_{2j+1})(\omega_{2i+2} - \omega_{2i})(\theta_{2j+1} - \theta_{2j})]}\end{aligned}\right\} \tag{4.126}$$

with the limits s and r chosen to cover the direction spread of $-\pi$ to $+\pi$ and the frequency range with non-zero energy content.

The general directional wave spectrum $S(\omega, \theta)$ is a two-parameter function of frequency and direction. It is, however, expressed in simple form as

$$S(\omega, \theta) = S(\omega)\cdot D(\omega, \theta) \tag{4.127}$$

where $S(\omega)$ is a point spectral density incorporating wave energy from all directions and $D(\omega, \theta)$ is an angular spreading function which is often assumed to be a function of direction only.

An understanding of typical directional spectra observed in the oceans for use in offshore design practice is only just beginning to emerge from research literature. The formulation of Equation (4.127) permits use of parametrized point spectra described in Section 4.7.2 for directional waves if the angular spreading function can be specified. Three formulations of the spreading function that may be useful in design are presented here:

(a) *Cosine square distribution*
This distribution is defined by the equation

$$\left.\begin{aligned}D(\theta) &= \frac{2}{\pi}\cos^2(\theta - \theta_0) \text{ for } (\theta - \theta_0) < \left|\frac{\pi}{2}\right| \\ &= 0 \text{ otherwise}\end{aligned}\right\} \tag{4.128}$$

where θ_0 is the mean wave direction and offers a simple formulation that is useful for design purposes. Figure 4.19 illustrates the shape of the function.

(b) *SWOP formulation*
Another directional distribution that is sufficiently amenable to practical application is presented by Cote *et al.* (1960) from analysis of data during a stereo wave observation project (SWOP). The distribution is given by

$$\left.\begin{aligned} D(\theta,\omega) &= \frac{1}{\pi}\,[1 + a\cos 2\theta + b\cos 4\theta] \text{ for } \theta < \left|\frac{\pi}{2}\right| \\ &= 0 \text{ otherwise} \\ \text{where} & \\ a &= 0.50 + 0.82\exp\left[-\frac{1}{2}\left(\frac{U\omega}{g}\right)^4\right] \\ b &= 0.32\exp\left[-\frac{1}{2}\left(\frac{U\omega}{g}\right)^4\right] \end{aligned}\right\} \quad (4.129)$$

and U is the wind speed at 19.5 m above the sea surface. Note that function $D(\omega, \theta) \neq 0$ at $\theta = \pm\pi/2$ but is truncated at these limits.

(c) *Mitsuyasu formulation*
This equation for directional distribution has been developed from work reported in Mitsuyasu *et al.* (1975) and modified by Goda and Suzuki (1975). The equation can be written as

$$D(\theta,\omega) = G_0\cos^{2s}\left(\frac{\theta-\theta_0}{2}\right)$$

where θ_0 is the mean wave direction

$$G_0 = \left[\int_{\theta_{\min}}^{\theta_{\max}} \cos^{2s}\left(\frac{\theta-\theta_0}{2}\right) d\theta\right]^{-1},$$

$$\left.\begin{aligned} S &= \begin{cases} S_{\max}\cdot(\omega/\omega_p)^5 & \text{for } \omega \leqslant \omega_p, \\ S_{\min}\cdot(\omega/\omega_p)^{-2.5} & \text{for } \omega > \omega_p, \end{cases} \\ \omega_p &= 2\pi/1.05\ T_{1/3}, \\ &\text{and } T_{1/3} \text{ is an average zero crossing period} \end{aligned}\right\} \quad (4.130)$$

Other more complex directional distributions have been proposed by Longuet-Higgins *et al.* (1961) and Hasselmann *et al.* (1976) but are not presented here because of the difficulty of using them in design calculations.

4.8 Measurement and analysis of wave data

This section examines the methods used to obtain and reduce measurements of wave properties. These measurements and their analysis underpins much of the theoretical work presented in earlier sections of this chapter. The techniques employed can readily be separated into those used for single-point wave measurements and multi-sensor arrays required for directional wave measurements.

Wave elevation is the most common single-point measurement in waves. There are a number of different methods of obtaining wave

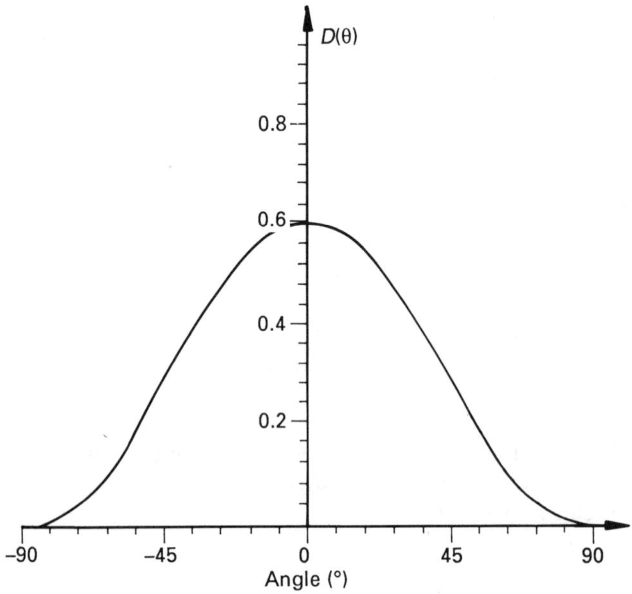

Figure 4.19. Normalized cosine squared wave directionality function

elevation measurements – the most commonly used are gauges mounted from fixed bottom mounted structures. One such type of gauge consists of two parallel stainless steel wires mounted vertically with half the wire lengths above and below still-water level. The wires are energized by a high-frequency alternating voltage and the change in resistance or capacitance is detected and converted to a direct current signal which is proportional to wave elevation. These resistance or capacitance sensing wire probes are also commonly used in laboratory wave tanks. Their calibration is dependent on the conductance (salinity) of the water, and the gauges can be in error if the water is contaminated by a surface film of pollutants. There are also other methods of transducing wave elevation by using electrical resistance or pressure sensing elements on a vertical wave staff penetrating the mean water level.

Wave elevation can also be obtained from a non-surface piercing sensor by using an underwater pressure transducer mounted one-quarter of a wave length below the free surface. Such a transducer produces reliable measurements of the wave surface elevation through the pressure induced by surface waves and is not subjected to impact loads and corrosion, as would a transducer at the free surface. Nevertheless, pressure transducers suffer from one major drawback, arising from the fact that the transfer function between sensed pressure and wave elevation is a strong function of wave frequency. In fact, waves with a length shorter than twice the depth of the transducer below the free surface are not sensed at all because of the decay of wave action with depth below the surface.

Another wave elevation measuring device is the wave rider buoy. This is a free floating, small buoy which contains vertical acceleration sensing transducers. A double integration of the measured acceleration signals yields wave elevation which is radio telemetered back to a nearby receiving station or offshore platform. Wave rider buoys are generally a reliable means of measuring wave elevation but suffer from the drawback that they are susceptible to mooring line failures in severe storms and are expensive to operate.

A modern technique for non-contact wave elevation sensing is offered by downward pointing infra-red, laser, acoustic or micro-wave transcievers which utilize the Doppler shift or time of travel between signals transmitted out, reflected and received back by the downward pointing sensor to yield a signal proportional to wave elevation. Such sensors can be mounted on fixed platforms and do not suffer from the water immersion, wave loading and corrosion problems of the 'contact' sensors described above.

A means for detecting water particle velocity components below the free surface is offered by spherical ball-type flow velocity gauges. These are spherical balls on the end of strain gauged arms – the two components of force applied by current or wave velocities on the ball are sensed by the strain gauge arm and a pre-calibration is used to deduce velocity components perpendicular to the mounting arm. The spherical balls are usually perforated to reduce the effects of vortex shedding induced force oscillations. Spherical ball sensors have proved to be a robust and reliable means of measuring flow velocity components due to current or in waves. They have been used successfully in a number of ocean wave measuring projects and are in routine use for monitoring wave particle velocities past offshore structures.

The analysis of sensor data from the single-point gauges described above follows the mathematical procedures outlined in Sections 4.7.1–4.7.4. Statistical data on probability densities and cumulative probability distributions can be readily calculated by implementing the equations outlined in these sections. Computer programs are also available for spectral analysis of sampled wave data, using either the auto-correlation or the fast Fourier transform approach, with the latter having much improved execution times. Bendat and Piersol (1971) present a detailed account of methods of analysis for frequency spectra. However, a brief summary of the principal points to note in such an analysis are presented here for completeness. There are basically three parameters involved in a spectral analysis:

1. the length of each record to be analysed (T_R);
2. the time interval Δt between successive data points in a record; and
3. the number of intervals M along the frequency axis at which the spectral value is estimated.

The conditions governing the choice of these three parameters are briefly summarized below:

1. For a record length T_R and a digitizing interval Δt, the total number of data points N equals $T_R/\Delta t$.
2. If Δt is too large, a process termed 'folding' or 'aliasing' occurs in which high-frequency components in the wave record appear at lower

frequencies, the energy being 'folded' at a 'Nyquist' frequency $\omega_N = \pi/\Delta t$. To avoid folding, a condition for Δt is

$$\Delta t \leqslant \pi/(2\omega_{max}) \tag{4.131}$$

where ω_{max} is the maximum wave frequency in the record.

3. Spectral estimates (they are, by definition, 'estimates' since in practice only a finite set of records of finite length are available with $T_R \neq \infty$) are evaluated over M frequency intervals. Thus the resolution bandwidth ω_r is given by

$$\omega_r = 2\pi/(M\cdot\Delta t) \tag{4.132}$$

4. Typically, M might be of the order of 50 to 100 depending on the anticipated shape of the spectrum. Having made the choice of T_R, Δt and M, the possible errors in spectral estimates can be defined in terms of 'confidence intervals'. Obviously, a longer record length provides a better estimate of $S(\omega)$. The standard derviation σ_s of the spectral estimate from the true value is given by

$$\sigma_s^2 = \frac{M}{N} \tag{4.133}$$

and approximately

$$\left.\begin{array}{l}(1 - \sigma_s)\ \tilde{S}(\omega) \leqslant S(\omega) \leqslant (1 + \sigma_s)\ \tilde{S}(\omega) \\ \text{for 68\% of the time and} \\ (1 - 2\sigma_s)\tilde{S}(\omega) \leqslant S(\omega) \leqslant (1 + 2\sigma_s)\tilde{S}(\omega)\end{array}\right\} \tag{4.134}$$

for 95% of the time where $\tilde{S}(\omega)$ is the estimated spectral density and $S(\omega)$ is the true spectral density.

These spectral estimates can then be used to derive significant wave height and zero crossing period values, either from the area under the spectrum and the moments of the spectrum or, alternatively, from direct statistical analysis of the wave elevation time history. Values of significant wave height and average zero crossing period over a long period of time can then be used to build up a scatter diagram describing the distribution of waves.

It is becoming increasingly necessary to take measurements of directional waves in laboratories and the open ocean. These measurements require simultaneous sensing of wave elevation, or other like property, at separate directions or at a number of points close to each other. Several suitable sensors have been developed for this purpose, the most common of these being an array of vertical wave staff gauges. These arrays are expensive to install and in addition the resolution of direction spectra available from them depends on the number of gauges used. Pitch–roll buoys are another way of measuring the simultaneous information required for directional wave resolution. A pitch–roll buoy senses the vertical acceleration for heave motions as well as two slopes of the sea surface in mutually perpendicular directions in the horizontal plane. Pitch–roll buoys offer the most reliable method of taking measurements of wave directionality in the open ocean away from a fixed structure, whereas the wave staff array is more suitable for installation from a fixed platform. A further development of the pitch–roll buoy for directional measurements is the clover leaf buoy developed originally by Mitsuyasu *et al.* (1975). This buoy records six quantities at one location such that for a wave surface elevation $\eta(x, y, t)$, the buoy records vertical acceleration $\ddot{\eta}$, two slopes in mutually perpendicular directions ($\partial\eta/\partial x$ and $\partial\eta/\partial y$) and three curvatures of the sea surface ($\partial^2\eta/\partial x^2$, $\partial^2\eta/\partial y^2$ and $\partial^2\eta/\partial x\partial y$). The additional curvature information is sensed in order to accomplish more accurate resolution of directional wave information than is possible for the pitch–roll buoy described above.

Directional wave measurements can also be made by fully submerged instruments, such as a two-dimensional current meter, together with a wave height gauge. The two-dimensional current meter is an electromagnetic flow meter that records orbital water particle velocities in two horizontal mutually perpendicular directions. A simultaneous measurement of vertical wave elevation is made at the same vertical line as the current meter measurements. Thus three quantities are simultaneously recorded, although this instrument, too, suffers from the drawback that it has to be mounted on a fixed platform. A development of the two-dimensional current meter and wave height gauge is the three-dimensional flow meter which records three orbital velocity components – two mutually perpendicular horizontal components as well as the vertical velocity components – and gives the same information as obtained from a wave staff.

A whole collection of techniques for measuring directional waves simultaneously over a large area is emerging with the use of modern radar and satellite observations. The earliest use of such photogrammetry based techniques was carried out by Cote *et al.* (1960) for a stereo wave observation project (SWOP) in 1954. The technique was implemented by taking photographs from two aeroplanes and carrying out an analysis of the resultant images. This and other similar techniques that are currently being tested offer the major advantage that information on directional waves is obtained over the complete area of observation using just one snap shot measurement. However, the equipment and data analysis techniques necessary to implement these procedures are complicated and expensive – thereby putting a question mark on their cost effectiveness for a conventional design or structural monitoring task.

The simultaneous sensor outputs from these multiple gauge arrays are analysed for directional wave information by three methods. These are the directional Fourier analysis approach as applied to wave staff arrays and to data from a pitch–roll buoy and the maximum likelihood method (MLM). The directional Fourier analysis technique as applied to wave gauge arrays is not described here and readers are referred to Panicker and Borgman (1971) for further information.

The directional Fourier analysis method for obtaining estimates of wave directionality from a pitch–roll buoy utilizes the simultaneous wave elevation, wave slope and curvature (if available) data from a buoy. The expected directional spectrum $S(\omega, \theta)$ is expressed in terms of a Fourier series of the form

$$\left.\begin{aligned} S(\omega, \theta) &= \frac{1}{2} a_0 + \sum_{n=1}^{\infty} (a_n \cos n\theta + b_n \sin n\theta) \\ \text{where} \\ a_n + ib_n &= \frac{1}{\pi} \int_{-\pi}^{\pi} S(\omega, \theta)\, e^{in\theta}\, d\theta \end{aligned}\right\} \quad (4.135)$$

where a_0, a_n, b_n are unknown coefficients, n are integers and θ is wave direction in a horizontal plane.

Now if a simple pitch–roll buoy senses only the three components – wave elevation η and slopes $\partial\eta/\partial x$ and $\partial\eta/\partial y$, denoted by subscripts 1, 2 and 3, then the auto- and cross-spectral analyses of these component can be implemented by the equations listed below:

$$\left.\begin{aligned} C_{11}(\omega) &= \int_{-\pi}^{\pi} S(\omega, \theta)\, d\theta & Q_{11}(\omega) &= 0 \\ C_{12}(\omega) &= 0 & Q_{12}(\omega) &= \int_{-\pi}^{\pi} k \cos\theta\, S(\omega, \theta)\, d\theta \\ C_{13}(\omega) &= 0 & Q_{13}(\omega) &= \int_{-\pi}^{\pi} k \sin\theta\, S(\omega, \theta)\, d\theta \\ C_{22}(\omega) &= \int_{-\pi}^{\pi} k^2 \cos^2\theta S(\omega, \theta)\, d\theta & Q_{22}(\omega) &= 0 \\ C_{23}(\omega) &= \int_{-\pi}^{\pi} k^2 \sin\theta \cos\theta S(\omega, \theta)\, d\theta & Q_{23}(\omega) &= 0 \\ C_{33}(\omega) &= \int_{-\pi}^{\pi} k^2 \sin^2\theta\, S(\omega, \theta)\, d\theta & Q_{33}(\omega) &= 0 \end{aligned}\right\} \quad (4.136)$$

where $k = \omega^2/g$ and $C_{ij}(\omega)$ for $i \neq j$ are the co-spectra (real part of the cross-spectra) of the components i and j, $C_{ii}(\omega)$ are the auto-spectra of the components i, $Q_{ij}(\omega)$ for $i \neq j$ are the quadrature spectra (imaginary part of the cross-spectra) of the components i and j and $Q_{ii}(\omega) = 0$, with i and j being 1, 2 or 3. By comparing Equations (4.135) and (4.136) the coefficients a_0, a_1, a_2, b_1 and b_2 can be written as follows:

$$\begin{aligned} a_0 &= \frac{1}{\pi} C_{11}(\omega) \\ a_1 &= \frac{1}{\pi k} Q_{12}(\omega) \\ a_2 &= \frac{1}{\pi k^2} [C_{22}(\omega) - C_{33}(\omega)] \\ b_1 &= \frac{1}{\pi k} Q_{13}(\omega) \end{aligned} \quad (4.137)$$

and

$$b_2 = \frac{1}{2\pi k^2} C_{23}(\omega)$$

It should be noted that because of the limitation in the number of measured components, that is, heave, pitch and roll, the Fourier expansion is only done to $N = 2$. In order to let the resultant partial sum of the Fourier expansion be equivalent to the infinite sum, Equation (4.135) has to be rewritten as

$$S(\omega, \theta) = \frac{a_0}{2} + \sum_{n=1}^{N} w_n(a_n \cos n\theta + b_n \sin n\theta) \qquad (4.138)$$

with $w_1 = \frac{2}{3}$ and $w_2 = \frac{1}{6}$ for $N = 2$ from Longuet-Higgins *et al.* (1961).

Then by using Equations (4.137) and (4.138), the directional wave spectrum $S(\omega, \theta)$ can be written in terms of the auto- and cross-spectra of the three measured components η, $\partial\eta/\partial x$ and $\partial\eta/\partial y$. Further details of this technique are presented by the originators Longuet-Higgins *et al.* (1961). The Fourier analysis technique described above has some drawbacks, primarily related to the fact that the three quantities recorded by the pitch–roll buoy limit subsequent analysis to the six auto- and co-spectra of these quantities. This leads to some difficulties with resolving the general angular spread of directional spectra observed in the ocean environment. These are alleviated to some extent by utilizing wave curvature data from a clover leaf buoy to obtain Fourier coefficients up to $N = 4$. The extension of the technique is described further by Cartwright and Smith (1964).

A more general technique for estimating directional spectra is offered by the maximum likelihood method (MLM). This approach was first developed, in the processing of radar and geophysical data by Capon (1969) and Lacoss (1971) and was subsequently applied to oceanography by Oakley and Lozow (1977). The technique is simple to implement and computationally quite efficient.

Consider an array of i wave elevation sensors simultaneously yielding wave elevation time histories denoted by $\eta_i(t)$. Let $f_i(\omega)$ be the Fourier transform of the ith wave elevation time history. Then the cross-power spectral density matrix $\hat{\mathbf{C}}(\omega)$ can be written

$$\hat{\mathbf{C}}_{ij}(\omega) = \overline{\left[\frac{f_i(\omega)}{|f_i(\omega)|} \cdot \frac{f_j^*(\omega)}{|f_j^*(\omega)|}\right]} \qquad (4.139)$$

where the overbar denotes averaging over sufficient separate estimates for a statistically reliable value, and the asterisk denotes a complex conjugate. The Fourier transforms are each normalized by their respective moduli to avoid errors caused by unequal sensor calibrations. Now a general directional sea state can be considered to be made up from the sum of N plane waves of power $\mathrm{S}(\omega, \theta_i)$ from directions θ_i, where integer i varies from 1 to N. The true cross-power spectral density of such a sea state in the frequency band around ω should be

$$\mathbf{C}(\omega) = \sum_{i=1}^{N} \mathbf{x}(\omega, \theta_i)\mathbf{x}^{*\mathrm{T}}(\omega, \theta_i)\mathbf{S}(\omega, \theta_i) \qquad (4.140)$$

where the mth component of vector $\mathbf{x}(\omega, \theta_i)$ is the complex phase lag between the mth sensor and the origin for a wave of frequency ω approaching from direction θ_i. We can thus write

$$\mathbf{x}_m(\omega, \theta_i) = \exp(-\mathrm{i}\,\mathbf{k}_i \cdot \mathbf{r}_m) \qquad (4.141)$$

where the vector wave number $\mathbf{k}_i$ is given by

$$\mathbf{k}_i = \frac{2\pi}{\lambda}(\cos\theta_i + i\sin\theta_i) \tag{4.142}$$

Note also that the wave length at frequency ω is denoted by λ and $\mathbf{r}_m$ is the position vector of the mth sensor. In order to derive the directional spectrum, Equation (4.140) must be inverted given that an estimate of the cross-power spectral density is available from Equation (4.139). For a direction given by angle θ_d, this inversion is

$$\left.\begin{aligned}\hat{\mathbf{S}}(\omega, \theta_d) &= \mathbf{A}_d{}^{*\mathrm{T}}\mathbf{C}(\omega)\mathbf{A}_d = \sum_{i=1}^{N} |\mathbf{A}_d{}^{*\mathrm{T}}\mathbf{x}(\omega, \theta_i)|^2 \cdot \mathbf{S}(\omega, \theta_i)\\ &= |\mathbf{A}_d{}^{*\mathrm{T}}\mathbf{x}(\omega, \theta_d)|^2\mathbf{S}(\omega, \theta_d)\\ &\quad + \sum_{i\neq d} |\mathbf{A}_d{}^{*\mathrm{T}}\mathbf{x}(\omega, \theta_i)|^2\mathbf{S}(\omega, \theta_i)\end{aligned}\right\} \tag{4.143}$$

Here, the complex vector $\mathbf{A}_d$ must be chosen so as to obtain an undistorted power component from direction θ_d whilst minimizing leakage (corruption) from all other components. In order to achieve this, the spectral estimate $\mathbf{S}(\omega, \theta_d)$ has to be minimized with respect to $\mathbf{A}_d$ under the constraint

$$|\mathbf{A}_d{}^{*\mathrm{T}}\mathbf{x}(\omega, \theta_d)| = 1 \tag{4.144}$$

This can be done by using standard Lagrangian theory to minimize

$$L(\mathbf{A}_d, \mu) = \mathbf{A}_d{}^{*\mathrm{T}}\mathbf{C}(\omega)\mathbf{A}_d + \mu(\mathbf{A}_d{}^{*\mathrm{T}}\mathbf{x}(\omega, \theta_d) - z) \tag{4.145}$$

with respect to μ, the Lagrangian multiplier and $\mathbf{A}_d$ for complex z with $|z| = 1$. This leads directly to an estimate of the power in a plane wave at frequency ω from direction θ_d of

$$\hat{\mathbf{S}}(\omega, \Theta_d) = [\mathbf{x}^{*\mathrm{T}}(\omega, \theta_d)\mathbf{C}^{-1}(\omega)\mathbf{x}(\omega, \theta_d)]^{-1} \tag{4.146}$$

It is conventional practice to scale estimates given by Equation (4.146) so that the integrated area under the directional distribution is equal to the power in the point wave spectrum. This makes the directional distribution consistent with the definition of directional wave spectra defined in Equation (4.127).

The computer time required for implementation of the MLM technique is dominated by the time needed to carry out a Fourier transform of the wave probe signals. The technique works best with sensor array dimensions which are not significantly greater than half a wave length at the highest wave frequency of interest and also should not be significantly less than the wave length at the lowest frequency of interest. The MLM also assumes that all the wave probes are measuring wave elevations which are statistically random. This condition is not satisifed particularly well in wave tanks where the wave generating mechanism or reflections from boundaries can create 'phase locked' waves which, therefore, do not satisfy the requirement for statistical similarity at all the probes. The above derivation is based on work by Jefferys *et al.* (1982). The authors also present calculations based on the MLM from measurements made in a directional wave tank. Their work shows that phase locked waves substantially degrade the resolution of the MLM and that this feature has to be considered with caution when applying the technique in directional wave tanks.

The application of directional wave measurements and calculations to the design of offshore structures is gradually increasing as research results lead to a better understanding of the underlying physics and to improved analysis techniques. This branch of offshore structures technology is poised to make substantial advances in contributing to improved design methods and more efficient structures.

References

Airy, Sir G. B. (1845), Tides and waves, *Encyc. Metrop.*, Art 192, pp. 241–396.

Barber, N. F. and Ursell, F. (1948), The generation and propagation of ocean waves and swell, I Wave periods and velocities, *Transactions of the Royal Society*, **240** (A824) 527–560.

Bendat, J. S. and Piersol, A. G. (1971), *Random Data: Analysis and Measurement Procedures*, Wiley Interscience, USA.

Bretschneider, C. L. (1960), Selection of design waves for offshore structures, *Transactions of the American Society of Civil Engineers*, Paper No, 3026.

Capon, J. (1969), High resolution frequency wave number spectrum analysis, *Proceedings of the Institution of Electrical and Electronic Engineers*, **57** (8), August, 1408–1419.

Cartwright, D. E. and Longuet-Higgins, M. S. (1956), The statistical distribution of the maxima of a random function, *Proceedings of the Royal Society, London*, **A237**, 212–232.

Cartwright, D. E. and Smith, N. D. (1964), Buoy techniques for obtaining directional wave spectra, *Buoy Technology, Marine Technical Society*, pp. 173–182.

Chan, R. K. and Street, R. L. (1970), *A Numerical Model for Water Waves*, Department of Civil Engineering, University of Stanford, TR No. 138.

Chaplin, J. R. (1980), Developments of stream function wave theory, *Coastal Engineering*, **3**, 179–205.

Cote, J. J., Davis, J. O., Marks, W., Mcghough, R. J., Mehr, E., Pierson, W. J., Ropek, J. F., Stephenson, G. and Vetter, R. C. (1960), The directional spectrum of a wind generated sea as determined from data obtained by the stereo wave observation project, *Meteorological Papers*, **2** (6) 1–88, New York University, College of Engineering, USA.

Cokelet, E. D. (1977), Steep gravity waves in water of arbitrary depth, *Philosophical Transactions, Royal Society of London*, **A286** 183–230.

Dean, R. G. (1965), Stream function representation of nonlinear ocean waves, *Journal of Geophysical Research*, **70** (18), 4561–4572.

Dean, R. G. (1970), Relative validities of water wave theories, *Proceedings of the American Society of Civil Engineers,* **96**, WW1, 105–119.

Draper, L. (1963), Derivation of a design wave from instrumental records of sea waves, *Proceedings of the Institution of Civil Engineerings London*, October, **26**, 291.

Fenton, J. (1972), A ninth order solution for the solitary wave, *Journal of Fluid Mechanics*, **53** 257–271.

Goda, Y. and Suzuki, Y. (1975), *Computation of Refraction and Diffraction of Sea Waves with Mitsuyasu's Directional Spectrum* Port and Harbour Research Institute, Japan, Technical Note No. 155, 33 pp.

Gumbel, E. J. (1958), *Statistics of extremes*, Columbia University Press, USA, 371 pp.

Hasselman, K. *et al.* (1973), Measurements of wind wave growth and swell decay during the joint North Sea wave Project (JONSWAP), *Deutsche Hydro. Zeitschr, Reihe*, **A8** (12).

Hasselman, K., Ross, D. B., Muller, P. and Sell, W. (1976), A parametric wave prediction model, *Journal of Physical Oceanography,* **6**, 200–228.

Hogben, N. and Lumb, F. E. (1967), *Ocean Wave Statistics*, HMSO, London.

Jasper, N. R. (1956), Statistical distribution patterns of ocean waves and of wave induced ship stresses and motions with engineering applications, *Transactions of the Society of Naval Architects and Marine Engineers*, **64**, 375–432.

Jefferys, E. R., Wareham, G. T., Ramsden, N. A. and Platts, M. J. (1982), Measuring directional spectra with the MLM, *Proceedings of the Conference on Directional Wave Spectra,* R L Wiegel (ed.), 14–18 September 1981, University of California, Berkeley Publications, American Society of Civil Engineers, USA.

Lacoss, (1971), Data adaptive spectral analysis methods, *Journal of Geophysics*, **36**, (4), August, 661–675.

Laitone, E. V. (1960), The second approximation to cnoidal and solitary waves, *Journal of Fluid Mechanics*, **1**, 430–444.

Laitone, E. V. (1962), Limiting conditions for Cnoidal and Stokes' waves, *Journal of Geophysical Research*, **67**, 4 April, 1555–1564.

Laitone, E. V. (1965), Series solutions for shallow water waves, *Journal of Geophysical Research*, **70**, 995–998.

Lamb, H. (1975), *Hydrodynamics*, Cambridge University Press, Cambridge.

Le Mehaute, B., Divoky, D. and Lin, A. (1968), Shallow water waves: a comparion of theories and experiment, *Proceedings of 11th Conference on Coastal Engineering*, Volume 1, 86–107.

Longuet-Higgins, M. S. (1952), On the statistical distribution of the heights of sea waves, *Journal of Marine Research*, **11**, (3), 245–266.

Longuet-Higgins, M. S., Cartwright, D. E. and Smith, N. D. (1961), Observations of the directional spectrum of sea waves using the motions of a floating buoy, *Proceedings of the Conference on Ocean Wave Spectra*, Prentice-Hall, Englewood Cliffs, NJ, USA, pp. 111–132.

Longuet-Higgins, M. S. and Cokelet, E. D. (1976), The deformation of steep surface waves on water I, A numerical method of computation, *Proceedings of the Royal Society, London*, **A350** 1–26.

McCowan, J. (1891), On the solitary wave, *Philosophical Magazine, London*, **32**, 45–58.

Mitsuyasu, H. *et al.* (1975), Observation of the directional spectrum of ocean waves using a clover leaf buoy, *Journal of Physical Oceanography,* **5**, (4), 750–760.

Munk, W. H. (1949), The solitary wave and its application to surf problems, Ocean Surface Waves, *Annals of the New York Academy of Sciences*, **51**, May, Art. 3, pp. 376–423.

Newland, D. E. (1984), *An Introduction to Random Vibrations and Spectral Analysis*, Longman, London.

Oakley, O. H. and Lozow, J. B. (1977), Directional spectra measurement by small arrays, *Proceedings of the Offshore Technology Conference*, Paper No. OTC 2745.

Ochi, M. K. (1982), Stochastic analysis and probabilistic prediction of random seas, V. T. Chow (ed.) *Advances in Hydroscience*, **13**, pp. 217–375.

Panicker, N. N. and Borgman, L. E. (1971), Directional spectra from wave gauge arrays, *Proceedings of the 12th Coastal Engineering Conference*, **1**, 117–136.

Phillips, O. M. (1966), *Dynamics of the upper ocean*, Cambridge University Press, Cambridge.

Pierson, W. J. and Moskowitz, L. (1964), A proposed spectral form for fully developed wind seas based on the similarity theory of S. A. Kitaigorodskii. *Journal of Geophysical Research*, **69**, (24).

Saetre, H. J. (1974), *On High Wave Conditions in the Northern North Sea,* Institute of Oceanographical Sciences, NERC Wormley, Report No 3.

Schwartz, L. W. (1974), Computer extension and analytic continuation of Stokes' expansion for gravity waves, *Journal of Fluid Mechanics*, **62**, 553–578.

Skjelbreia, L. (1959), *Gravity Waves, Stokes' Third Order Approximation; Tables of Functions*, The Engineering Foundation Council on Wave Research, Berkeley, California, USA.

Skjelbreia, L. and Hendrickson, H. (1960), Fifth order gravity wave theory, *Proceedings of the 7th Conference on Coastal Engineering*, The Hague, pp. 184–196.

Stokes, G. G. (1847), On the theory of oscillatory waves, *Transactions of the Cambridge Philosophical Society*, **8**, 441–455.

Stokes, G. G. (1880), On the theory of oscillatory waves, *Mathematical and Physical Papers*, Volume 1, pp. 314–326, Cambridge University Press, Cambridge.

Ursell F, (1953), Mass transport in gravity waves, *Proceedings of the Cambridge Philosophical Society*, **49**, Pt 1, January, pp. 685–694.

Weibull, W. (1951), A statistical distribution of wide applicability, *Journal of Applied Mechanics*, **18**, 293–297.
Wiegel, R. L. and Beebe, K. E. (1956), The design wave in shallow water, *Journal of Waterways and Harbours Division, American Society of Civil Engineers*, 82 WW1.
Wiegel, R. L. (1964), *Oceanographical Engineering*, Prentice-Hall, Englewood Cliffs, NJ, USA.

Chapter 5

Fluid loading on offshore structures

5.1 Introduction

Offshore structures are subjected to both steady and time dependent forces due to the action of winds, currents and waves. Winds exert predominantly steady forces on the exposed parts of offshore structures, although there are significant gust or turbulence components in winds which induce high, unsteady, local forces on structural components as well as a low frequency total force on the whole structure. Ocean currents also exert largely steady forces on submerged structures, although the localized effects of vortex shedding induce unsteady force components on structural members. However, gravity waves are by far the largest force on most offshore structures. The applied force is periodic in nature, although non-linear wave properties give rise to mean and low-frequency drift forces. Non-linearities in the wave loading mechanism can also induce superharmonic force components. Both these secondary forces can be significant if they excite resonance in a compliant structure.

In general, an air or water flow incident on an offshore structure will exert forces that arise from two primary mechanisms. A steady or unsteady flow will directly exert a corresponding steady or unsteady force with a line of action that is parallel to the incident flow direction. Such forces are sometimes called 'in-line' forces. However, the localized interaction of steady or unsteady flow with a structural member will also cause vortices to be shed in the flow and will induce unsteady 'transverse' forces with lines of action that are perpendicular to the incident flow direction. Both these force mechanisms are considered in this chapter.

The design of offshore structures requires calculation methods to translate a definition of environmental conditions into the resultant steady and time dependent forces exerted on the structure. This chapter is concerned with describing and developing methods of calculating these forces for fixed offshore structures. A general treatment of fluid flow around streamlined and bluff bodies is followed by a description of specific methods to calculate resultant body forces. However, offshore structures for deeper waters tend to be more flexible with lower natural frequencies of response. In such cases, it is necessary to calculate the loading resulting from wave action on members which may themselves have motions of significant magnitude compared to water particle motions. This would also apply to compliant or fully floating offshore platforms and is considered further in Chapters 6 and 8.

5.2 Flow around streamlined and bluff bodies

Consider an arbitrary body that is stationary in a uniform, steady fluid flow. The presence of the body will modify the velocity distribution of the fluid as it passes the body, and the resultant pressure distribution can be integrated over the body to yield a total force. The component of this force perpendicular to the undisturbed flow direction is defined as the lift or transverse force, while the component in the undisturbed flow direction is defined as the drag or in-line force. Figure 5.1(*a*) and (*b*) shows two typical bodies – a streamlined aerofoil section and a circular cylindrical bluff body. For the aerofoil section, the lift to drag ratio can be as high as 40, whereas for the bluff body the drag force is high and the mean lift force is zero, although vortex shedding can induce significant periodic lift forces.

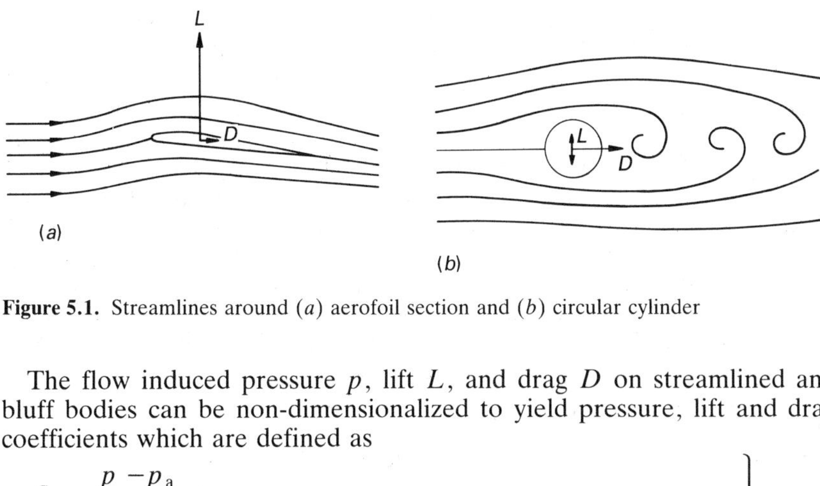

Figure 5.1. Streamlines around (*a*) aerofoil section and (*b*) circular cylinder

The flow induced pressure p, lift L, and drag D on streamlined and bluff bodies can be non-dimensionalized to yield pressure, lift and drag coefficients which are defined as

$$C_p = \frac{p - p_a}{\frac{1}{2}\rho U^2}$$

$$C_L = \frac{L}{\frac{1}{2}\rho U^2 S} \tag{5.1}$$

and

$$C_D = \frac{D}{\frac{1}{2}\rho U^2 S}$$

respectively, where ρ is fluid density, U is fluid velocity and S is usually the frontal area projected in the flow direction, although for thin streamlined bodies, such as aerofoils, S is defined as the plan area of the body. p_a is the static pressure well away from the body.

Now, the general turbulent flow around an arbitrary bluff body cannot be calculated in a mathematically rigorous manner at present because of the non-linear convective acceleration terms on the left-hand sides of the Navier–Stokes equations. However, a number of empirical techniques have been developed to enable forces due to flow around bluff bodies to be quantified for engineering design purposes. The fundamental concepts of fluid mechanics that underpin these techniques are described further in this section.

5.2.1 Non-dimensional numbers

The concept of similarity in fluid mechanics relates to the behaviour of fluid flow in systems of different physical magnitudes. For example, a question may be posed: 'How is the flow of a fluid of velocity U around a cylinder of diameter $2a$ related to the flow of a fluid of velocity $2U$ around a cylinder of diameter a?'

There are three pre-conditions that must apply before considering the similarity of two systems of fluid flow. These are: (a) that the systems are geometrically similar; (b) that the systems have similar internal constitutions, that is, both fluids must be Newtonian, although of different physical properties; and (c) that they have similar conditions at the system boundaries.

Consider a flow of velocity U incident on a body of characteristic length L where the above three conditions are satisfied. The fluid can be assumed to be homogeneous and incompressible with a constant dynamic viscosity and density ρ. The motion of the fluid around the body will be governed by forces due to accelerations (inertia forces) and also due to viscosity (viscous forces). A ratio of these two forces will yield a numerical characteristic of the relative magnitudes of these two mechanisms in the flow. Now the mass of a representative element of fluid in the flow pattern will be 'proportional' to ρL^3. Corresponding components of acceleration at these elements will be proportional to U^2/L, where U is the velocity and L is proportional to the radius of curvature of the fluid flow path at the elements. Then the inertia force, being proportional to mass times acceleration, can be written as

$$\rho L^3(U^2/L) = \rho U^2 L^2$$

For viscous forces, the velocity gradient in the fluid element will be proportional to U/L, so the shear stress will be 'proportional' to U/L and the viscous force, which is given by shear stress times area of element, will be 'proportional' to

$$\frac{\mu U}{L} \times L^2 = \mu U L$$

Thus the ratio of inertia to viscous forces becomes

$$\frac{\rho U^2 L^2}{\mu U L} = \frac{UL}{(\mu/\rho)} = \frac{UL}{\nu} \tag{5.2}$$

where $\nu = \mu/\rho$. This is defined as the Reynolds number. Two flow systems with the same Reynolds number have similar flows and thus a condition of similarity in flows where only inertia and viscous forces play a part. This answers the question posed earlier in that the two flow systems are similar (or equivalent) since the ratios of internal inertia to viscous forces are equal. Thus, given a solution for one system, the solution for the other system is simply obtained by a suitable scaling of parameters. For most flows of practical interest, viscous forces are small, giving values of Reynolds numbers of up to 5×10^6.

The Reynolds numbers and the definition of force coefficients in Equations (5.1) can also be derived using dimensional analysis. Now the drag force D on an arbitrary body in a fluid flow will be expected to be a function of fluid density ρ, velocity U, dynamic viscosity μ, and the characteristic length of the body L. Then

$$D = f(\rho, U, L, \mu) \tag{5.3}$$

and the function can be expected to have a form

$$D = k\rho^a U^b L^c \mu^d \tag{5.4}$$

where a, b, c, d and k are constants to be determined. Dimensionally, this equation can be written in terms of mass, length and time (M', L' and T') as

$$\frac{M'L'}{T'^2} \equiv \left(\frac{M'}{L'^3}\right)^a \left(\frac{L'}{T'}\right)^b L'^c \left(\frac{M'}{L'T'}\right)^d \tag{5.5}$$

Equating powers of M', L' and T' yields

$$\left.\begin{aligned} 1 &= a + d \\ 1 &= -3a + b + c - d \\ -2 &= -b - d \\ &\text{which can be simplified to} \\ a &= 1 - d,\ b = 2 - d,\ c = 2 - d \end{aligned}\right\} \tag{5.6}$$

Equation (5.4), therefore, becomes

$$\frac{D}{\rho U^2 L^2} = \text{constant}\cdot\left(\frac{UL}{\nu}\right)^{-d}$$

which can then be written as

$$C_{\mathrm{D}} = \frac{D}{\frac{1}{2}\rho U^2 S} = f\left(\frac{UL}{\nu}\right) \tag{5.7}$$

where the characteristic length squared is conveniently written as an area S, and a factor of one-half is introduced. The drag coefficient is thus a function of Reynolds number only. The most important implication of flow similarity is that the force coefficients on bodies of similar shape in uniform motion in an infinite incompressible fluid can be simply represented as a function of the Reynolds number for all such bodies. The form of the function depends solely on the shape of the body and the presence of an unconfined fluid. The concept of Reynolds number also applies to unidirectional phases within an oscillating flow.

For a viscous incompressible uniform fluid with a free surface in a uniform gravitational (body force) field, there are two numerical conditions for the similarity of two systems of fluid flow. The first is equality of Reynolds number UL/v. The second is a ratio of inertia forces to body (gravitational) forces acting on the fluid. This can be written as the non-dimensional number

$$\frac{\text{inertia force}}{\text{gravitational force}} = \frac{\rho L^3 \times U^2/L}{\rho L^3 \times g} = \frac{U^2}{gL} \tag{5.8}$$

the square root of which is called the Froude number. It refers essentially to the wave making properties of a body at the fluid surface; waves being an oscillation obtained by an interplay of fluid inertia and restoring gravity forces.

For completeness, a third ratio is introduced. The Mach number $\bar{\mathrm{M}}$ is a ratio of characteristic fluid velocity U to the local speed of sound a^*

$$\bar{\mathrm{M}}' = \frac{U}{a^*} = U\sqrt{\left(\frac{\rho}{\gamma p}\right)} = U\sqrt{\left(\frac{\rho C_v}{C_p p}\right)} \tag{5.9}$$

where C_p and C_v are the specific heats at constant pressure and volume respectively, ρ is fluid density and p is pressure; $\bar{\mathrm{M}}$ is only significant for flows where U is comparable in magnitude to a^*.

Thus for any fluid flow, the pressure, lift and drag coefficients are functions of the non-dimensional Reynolds, Froude and Mach numbers. The Reynolds number is always significant, whereas the Froude number is of interest for flows with free surfaces and the Mach number is applicable to near-sonic gas flows. Similarity of non-dimensional numbers is also a key issue in representing prototype situations in model form for wind tunnel and wave tank testing – Chapter 9 gives further details on both.

5.2.2 The boundary layer concept

The relative velocity between a flow field and a solid body is governed by the boundary condition that the fluid layer immediately adjacent to the body does not move relative to the body. This is often called the 'no-slip' boundary condition. For flows around streamlined bodies or upstream segments of flow around bluff bodies, the no-slip boundary condition gives rise to a thin layer of fluid adjacent to the surface where the flow velocity relative to the surface increases rapidly from zero at the surface to the local stream velocity at the outer edge of the layer. Such a thin sheared layer is appropriately called the boundary layer.

For many practical situations (flow around ships or aircraft, for example), the boundary layer thickness is very small compared with the distance from the forward end of the vehicle over which the flow has developed. Hence, the velocity gradients within a boundary layer in a direction perpendicular to the surface are very large in comparison to velocity gradients parallel to the surface. The former velocity gradients induce large shear stresses from the action of viscosity within the boundary layer fluid, whereas outside this layer, the velocity gradients and, therefore, the shear stresses are sufficiently small for the fluid to be regarded as inviscid. Thus, many problems concerned with flow of a fluid of small viscosity around a streamlined body can be solved by considering the fluid as inviscid everywhere except within a thin boundary layer region adjacent to the body and in the wake that forms behind it. Within the boundary layer and wake, however, the rates of shear strain are high so that the effects of viscosity and the associated shear stresses must be accounted for. The value of this shear stress at the body surface contributes to a frictional drag force which is of considerable engineering significance for streamlined bodies.

Now, as long as the boundary layer is thin, the bulk of the flow around the body is very close to that predicted by inviscid flow theory with the body surface as a boundary. Therefore, matching the far field flow to the boundary layer flow does not present a serious problem. There are, however, cases where the boundary layer is not sufficiently thin or it separates from the surface to form a thick wake. In such situations, the boundary layer has a substantial effect on the bulk of the flow around the body and on the forces experienced by the body. Boundary layer separation and the formation of a thick wake are a characteristic feature of flow around bluff bodies typically used as members of offshore structures.

Figure 5.2(*a*) present typical development of a boundary layer flow around a streamlined shape. The boundary layer grows downstream from the leading edge of the shape. During early development of the boundary layer, the flow within it is smooth, with streamlines approximately parallel to the surface. This is defined as a laminar boundary layer and has a profile characterized by the shape given in Figure 5.2(*b*). As the laminar boundary layer develops along the surface, at some position along the body, transition occurs to a boundary layer flow, characterized by significant random fluctuations in direction and velocity superimposed on a mean flow. This mean flow still remains parallel to the surface but the turbulent fluctuations in the flow can be over 10% of the magnitude of the mean velocity. Such a sheared flow is called a turbulent boundary layer. It has a fuller mean velocity profile, as shown in Figure 5.2(*b*).

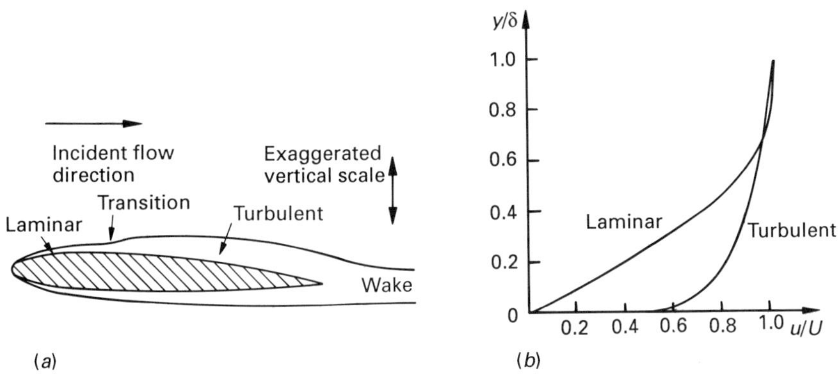

Figure 5.2. Boundary layer development: (*a*) boundary layer structure; (*b*) typical laminar and turbulent boundary layer velocity profiles

The laminar boundary layer is very sensitive to disturbances in the presence of a positive (that is, opposing) pressure gradient and will readily separate or become turbulent. Its sensitivity increases with increasing Reynolds number. For Reynolds numbers less than about 10^3, the laminar boundary layer tends to separate more readily from the surface than to become turbulent, and complete breakaway of flow may result, but in some cases the boundary layer may reattach itself in the turbulent state. The precise sequence of events depends on the pressure distribution over the surface, the nature of the surface (that is, whether rough or wavey), and turbulence in the incident flow. Laminar separation followed by reattachment can also occur at higher Reynolds numbers, with an example of offshore interest being flow around a circular cylinder, as shown later.

The phenomenon of boundary layer separation plays a very important part in flows around typical members of offshore structures. Consider the boundary layer of Figure 5.3 moving into a region of rising pressure (as over the rear of a circular cylinder or aircraft wing). As the flow progresses towards the right, it tends to lose velocity, as it would in an inviscid flow, but because of shear stresses associated with viscosity there is some interchange of momentum between adjacent fluid layers, with momentum passing from the faster to the slower layers. Consequently, the velocity decrease in the slower moving streamlines nearer the surface is not as great for a given increase of pressure as Bernoulli's equation would predict. Nevertheless, the flow slows down and will eventually reverse in direction near the surface. The boundary layer is then said to separate. At and after this separation point, the boundary layer appears to move away from the surface, with a large eddy forming between it and the surface. Such eddies are unstable and tend to move downstream from the surface with new eddies forming to replace them. The wake behind the body is then filled with a stream of eddies, some of which can be as large as the body itself (see Figure 5.1(b)).

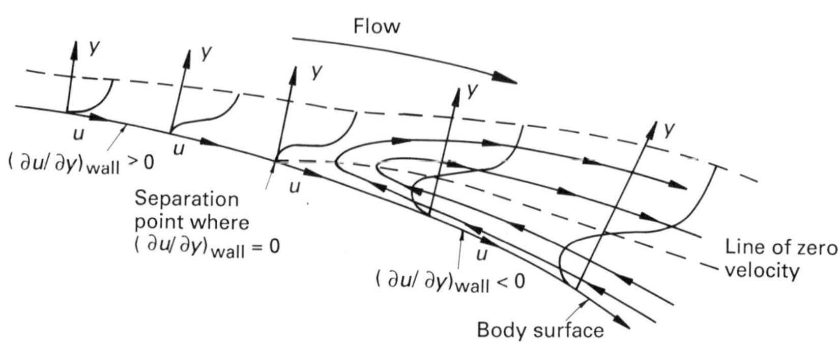

Figure 5.3. Separating boundary layer in steady flow

These eddies and wakes involve large mechanical energy losses and result in a substantial drag force on the body, mainly manifested by high suctions over the rear of the body. In contrast, the streamlined shape of Figure 5.1(a) has a small wake and a small drag force if the angle of incidence is such that no boundary layer separation occurs. In this case, the drag force is largely due to tangential viscous stresses.

The main results of boundary layer theory are presented here to illustrate the developments that have been made in this subject for the design of aircraft, ships and other streamlined structures. The Navier–Stokes equations given in Chapter 3 can be simplified to apply to boundary layers by assuming that the gradients of velocities parallel to the boundary layer (body surface) are small compared to gradients perpendicular to it and that the pressure along the thickness of the layer remains constant. The boundary layer equation can then be written in two dimensions as (see Duncan *et al.* 1962)

$$\left.\begin{aligned} &\frac{\partial u}{\partial t} + u\frac{\partial u}{\partial x} + v\frac{\partial u}{\partial y} = \frac{\partial U}{\partial t} + U\frac{\partial U}{\partial x} + \nu\frac{\partial^2 u}{\partial y^2} \\ &\frac{\partial u}{\partial x} + \frac{\partial v}{\partial y} = 0 \end{aligned}\right\} \tag{5.10}$$

where the body surface is taken to be in the Oxy plane, $U(x, y, t)$ denotes the free stream velocity and the remaining notation is from that for Equation (3.35) in Chapter 3. The above boundary layer equations can be integrated over the boundary layer thickness to yield the momentum integral equation

$$\frac{\partial \theta}{\partial x} + \frac{1}{U}\frac{\partial U}{\partial x}(\delta^* + 2\theta) + \frac{1}{U}\frac{\partial \delta^*}{\partial t} + \frac{\delta^*}{U^2}\frac{\partial U}{\partial t} = \frac{\tau_w}{\rho U^2} \tag{5.11}$$

with the displacement and momentum thicknesses (δ^* and θ) defined as

$$\delta^* = \int_0^{\delta}\left(1 - \frac{u}{U}\right)\mathrm{d}y \text{ and } \theta = \int_0^{\delta}\frac{u}{U}\left(1 - \frac{u}{U}\right)\mathrm{d}y \tag{5.12}$$

where δ is the boundary layer thickness and τ_w is the wall shear stress; δ^* and θ are useful measures of the total volume and momentum deficits due to the slower flow within the boundary layer. The momentum integral equation requires assumed boundary layer profiles in order to obtain a solution – see Duncan *et al.* (1962) for further details.

For a flat plate laminar boundary layer, that is, one with zero pressure gradient and U constant, the various quantities of interest were derived through a numerical solution by Blasius (1963). There are

$$\left.\begin{aligned} &\text{Boundary layer thickness for } u/U_0 = 0.999 \\ &\qquad \delta = 6.00x/\sqrt{(Ux/\nu)} \\ &\text{Skin friction coefficient } C_f = \frac{\tau_w}{\frac{1}{2}\rho U^2} = 0.664/\sqrt{(Ux/\nu)} \\ &\delta^* = 1.72x/\sqrt{(Ux/\nu)} \quad \theta = 0.664x/\sqrt{(Ux/\nu)} \end{aligned}\right\} \tag{5.13}$$

where x is distance from the plate leading edge and ν is kinematic viscosity. The velocity distribution across the boundary layer thickness was also derived by Blasius and is of the form shown in Figure 5.2(*b*).

Solution of equations (5.10) for laminar boundary layers with pressure gradient or unsteady free stream flow requires computer based numerical methods. Smith and Clutter (1963) and Cebeci *et al.* (1980) present details of typical modern numerical methods for the calculation of laminar boundary layers.

The equations for laminar flow can be extended to turbulent flow by separating instantaneous velocities into the sum of mean velocities and turbulent fluctuations, substituting into Equation (5.10) and taking a time mean over an interval that is sufficiently short as to average the turbulent fluctuations only. This yields the equations

$$\left.\begin{aligned} &\frac{\partial u}{\partial t} + u\frac{\partial u}{\partial x} + v\frac{\partial u}{\partial y} = \frac{\partial U}{\partial t} + U\frac{\partial U}{\partial x} + \frac{1}{\rho}\frac{\partial \tau'}{\partial y} \\ &\frac{\partial u}{\partial x} + \frac{\partial v}{\partial y} = 0 \end{aligned}\right\} \qquad (5.14)$$

where the viscous term is replaced by a turbulent shear stress term. The additional unknown introduced by this term is solved by use of an equivalent eddy viscosity or by solution of the above equation with an additional turbulent energy equation. This requirement to provide additional closure relations for solution of the equation arises due to the difficulty of quantifying the physics of turbulence. Calculations of the turbulent boundary layer thus have to rely on a number of approximations which are sufficiently well characterized for flow around streamlined bodies, but do not permit calculations of engineering use for the coupled inviscid and viscous flows around bluff bodies such as circular cylinders. Note that the momentum integral equation of Equation (5.11) applies to turbulent boundary layers as well, but requires an assumed velocity profile for calculations.

Approximate results for turbulent flow on a flat plate can be obtained by assuming a velocity profile given by a $\frac{1}{7}$th power law to yield:

$$\begin{aligned} &\frac{u}{U} = \left(\frac{y}{\delta}\right)^{1/7} \\ &\delta = \frac{0.37x}{\left(\dfrac{Ux}{\nu}\right)^{1/5}} \\ &C_{\mathrm{f}} = \frac{\tau_{\mathrm{w}}}{\frac{1}{2}\rho U^2} = \frac{0.057}{\left(\dfrac{Ux}{\nu}\right)^{1/5}} \qquad (5.15) \\ &\delta^* = \frac{0.0463x}{\left(\dfrac{Ux}{\nu}\right)^{1/5}} \quad \theta = \frac{0.0360x}{\left(\dfrac{Ux}{\nu}\right)^{1/5}} \end{aligned}$$

where the notation is the same as used before for laminar boundary layers. Typical closure relations used in numerical computations of turbulent boundary layers together with the methods used for arbitrary pressure gradients are described further by Bradshaw *et al.* (1967) and Cebeci *et al.* (1980).

The precise nature of the transition process from a laminar to a turbulent boundary layer is also not sufficiently well understood to be of quantitative use in design calculations. Nevertheless, some of the qualitative effects that influence the transition process are described here. It is known that roughness below a certain critical height has no effect on the stability of a laminar boundary layer and that the ratio of this critical height to boundary layer development length decreases with increasing Reynolds number. For any bluff body in a viscous fluid flow, there is also a Reynolds number below which the flow closes up smoothly behind the body, whilst above that Reynolds number flow breaks away from the body with a large eddying wake. This Reynolds number separating the two regimes depends on the shape of the body but is usually of the order of 50. It has been suggested that the critical height of roughness is determined by the Reynolds number of the roughness being of this order. Above it, the roughness sheds eddies into the layer and upsets its stability; below this critical height the smooth laminar nature of the flow is unaffected. Other factors which induce transition are free stream turbulence and, most important of all, a positive pressure gradient which modifies the velocity profile and makes the boundary layer go unstable and quickly become turbulent.

It is clear from the above brief description of the state of knowledge in boundary layer theory, that calculation of the behaviour of such layers on aircraft and ship hulls can be made with sufficient accuracy for engineering purposes. Such calculations are, however, restricted to streamlined bodies, whereas the members of offshore structures are invariably bluff bodied to the extent that flow about these bodies is in a complex coupled inviscid and viscous regime. Nevertheless, a qualitative understanding of such bluff body flows is provided by considering the flow around a circular cylinder over a range of Reynolds numbers. Figure 5.4(*a–f*) illustrates these flow patterns around a circular cylinder, whereas Figure 5.5 presents the resultant variation of drag coefficient with Reynolds number.

The flow regime corresponding to the Reynolds number (Re) of less than 5, is a 'Stokean' creeping motion which is of little practical significance since it corresponds to very small cylinder diameters or very slow flow velocities. An analytic solution for this flow does exist and is given by Equation (3.72). Over the Reynolds number range 5–40, separation of the laminar boundary layer occurs at the point marked L in Figure 5.4 and two stable closed eddies are formed at the rear of the cylinder with the downstream wake remaining laminar. In the Reynolds number range 40–150, disturbances downstream of the cylinder increase in intensity and well-developed vortices appear. The eddies downstream of the cylinder begin to be shed and reappear periodically to form a series of vortices known as the Karman vortex street. Boundary layer separation still remains laminar. As Re increases in the range 150–10^5, the flow pattern remains as above but the wake becomes turbulent. The non-symmetric vortex formation for Re > 40 gives rise to oscillating lift forces with a frequency defined by a Strouhal number (see Section 5.5).

For the Reynolds number range 10^5–3×10^6, the flow in the wake is entirely turbulent. On the cylinder, a laminar separation is followed by a re-attachment and turbulent separation. The turbulent boundary layer, because of its fuller velocity profile and higher turbulence, has a higher

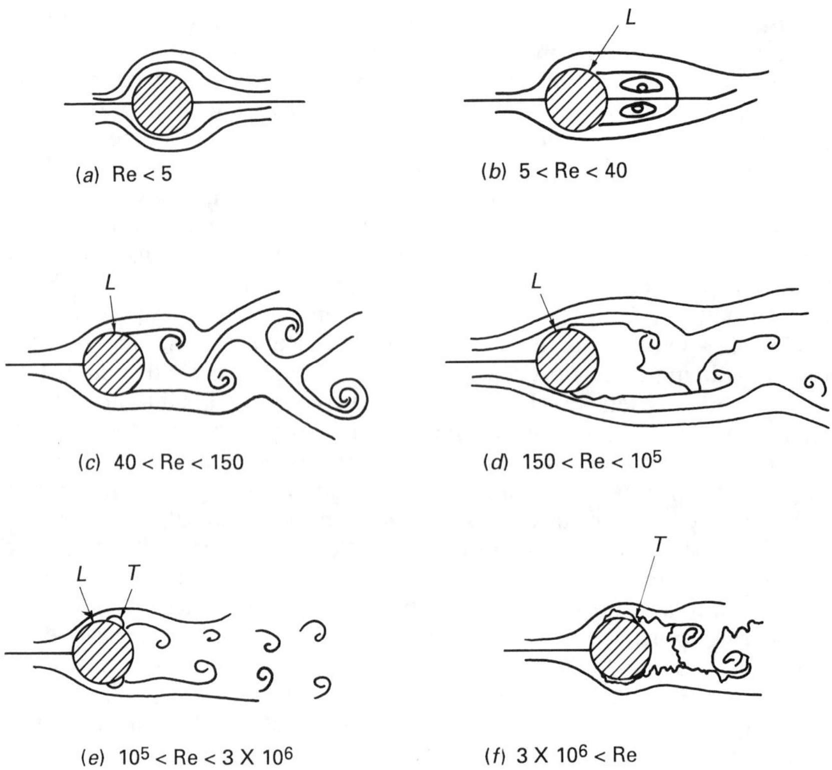

Figure 5.4. Flow around a circular cylinder for a range of Reynolds numbers: *L* and *T* denote laminar and turbulent boundary layer separation (from ESDU, 1986)

kinetic energy which can sustain a larger pressure rise as the flow decelerates over the downstream half of the cylinder. The boundary layer, therefore, remains attached for longer distances round the cylinder before separating. This causes the width of the wake to decrease and thus the pressure drag decreases giving the net effect of a sharp drop in the drag coefficient. For Reynolds numbers greater than 3×10^6, the flow, including the cylinder surface boundary layer is fully turbulent with vortex shedding occurring at a frequency defined by the Strouhal number.

5.3 Wind and current loads

The fluid loading mechanisms of wind and current on the exposed or submerged members of offshore structures are identical apart from the different natures of the incident flow fields and the physical properties of the media.

Wind forces on offshore structures account for approximately 15% of the total forces from wind, current and waves acting on the structure. In deep water, wind induced forces can have a larger relative effect on the

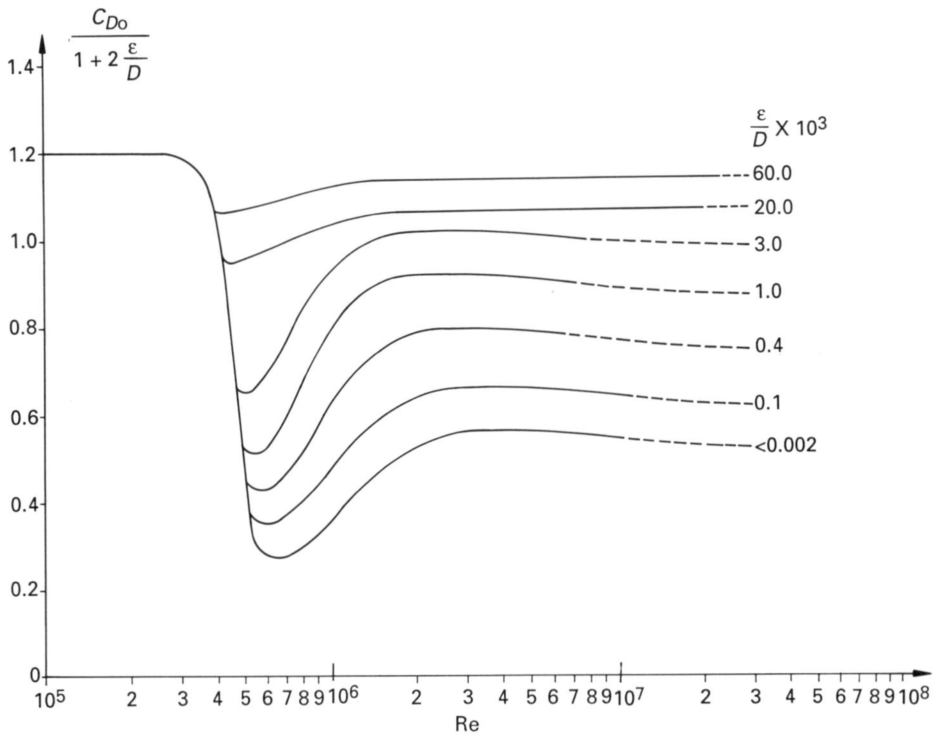

Figure 5.5. Variation of $C_{D0}/(1 + 2E/D)$ with Re. ε is the roughness height and D is cylinder diameter (from ESDU, 1986)

base overturning moment of bottom emplaced structures because of the large moment arms involved. Since the presence of high winds is always accompanied by high seas, the loadings from wind and waves are always applied simultaneously in the design process. The general equation for the wind force on a slender member of an open space framed offshore structure exposed to a wind is

$$F = \tfrac{1}{2}\rho_a U_w^2 S C_D \qquad (5.16)$$

where U_w is the flow velocity incident on the member due to the wind, ρ_a is the air density, S is the frontal area (facing the wind) and C_D is a drag coefficient obtained from experiment and experience. A value of C_D from 1.1 to 1.3 is applied for most bluff bodies although a C_D valye of 1.5 is often used for design purposes. Appendix G gives typical values of C_D for a variety of body shapes.

Two kinds of wind speeds are considered in the design of offshore structures – these being the sustained and gust wind speeds. The sustained wind speed is defined as the average wind speed over a time interval of 1-min measured at an elevation of 10 m above still-water level. For design purposes, it is usual to define a sustained wind speed value that is statistically likely to occur once in 50 or 100 years (the return period). The wind velocity above still-water level (SWL) varies significantly with height due to the boundary layer induced by viscosity.

The speed at a height Z above SWL can be related to the speed at 10 m height by the relationship

$$U_Z = U_{10}\left(\frac{Z}{10}\right)^{0.113} \tag{5.17}$$

where U_{10} is the 1-min mean sustained wind speed 10 m above still-water level.

The gust wind speed is defined as the average wind speed over a time interval of 3 s measured at an elevation of 10 m above still-water level. In the same sense as above, the 50- or 100-year return period gust wind speed can be stipulated for design purposes. Adjustments for elevation above still water level are given by the equation

$$(U_Z)_{\mathrm{GUST}} = (U_{10})_{\mathrm{GUST}}\left(\frac{Z}{10}\right)^{0.100} \tag{5.18}$$

These power law equations are semiempirical and are continually being updated by new data. Values of sustained and gust wind speeds are obtained from meteorological records or from recommendations made by certifying authorities in the offshore area of interest. Thom (1973) presents some analysis techniques for obtaining 50- and 100-year return period design wind speeds from meteorological data for a shorter interval. Thom also presents typical distributions of extreme winds over open oceans in the presence of storms. In general, design will be more accurate if the 100-year return period wind speed criteria are based on local meteorological data sources.

As explained before, actual values of drag coefficient C_D, used in Equation (5.16), are around 1.1 to 1.3, but higher values can be used in design to incorporate a safety factor. Generally, for long, slender members with length-to-diameter ratios greater than 5, C_{D} for sharp edged sections is taken in the range from 1.5 to 2.0 whereas for cylindrical sections of diameters less than 0.3 m, $C_{\mathrm{D}} = 1.2$ and for diameters greater than 0.3 m, $C_{\mathrm{D}} = 0.7$. Also, for shorter members,

$$C_{\mathrm{DS}} = C_{\mathrm{D}}\left(0.5 + \frac{0.1L}{D}\right) \tag{5.19}$$

where C_{D}, C_{DS} are drag coefficients for long and short members respectively, L is member length and D is member diameter.

If members are located behind each other in the direction of the wind, a shielding effect can be taken into account by reducing the effective area of members down wind according to the distance between the members – provided that this distance is less than seven times the up wind member diameter. More detailed guidance for circular cylinders is provided by ESDU (1986) and also in the summary of hydrodynamic coefficients in Appendix G.

Trusses or columns which are partly open can have their appropriate drag coefficients reduced. For complex structures and in cases where wind loads are vitally important (e.g., for the stability of towed structures or semisubmersibles), wind tunnel experiments are often necessary to predict design loads. Miller and Davies (1982) and Hoerner (1965) give a comprehensive survey of fluid dynamic drag coefficients for offshore structures and for a wide variety of other shapes.

Finally, for some streamlined structures, typically a hangar, substantial lift loads can be induced. The design value for such forces can be given by the equation

$$L = C_L \tfrac{1}{2}\rho_a U^2 S' \tag{5.20}$$

where S' is the plan area of the structure and C_L is an appropriate lift coefficient obtained from wind tunnel tests or aerodynamic calculations.

Although the sustained and gust wind speeds offer some quantification of the statistically random nature of wind turbulence, this feature can be defined further by writing a power spectral density of the in line (parallel to the mean wind direction) component of wind velocity fluctuations. This in-line component is generally much larger than the two components perpendicular to the wind direction and its spectral density is given by Davenport (1961) to be of the form

$$S_{vv}(f) = \frac{4k\bar{f}}{(1+\bar{f}^2)^{4/3}} \cdot \frac{\bar{U}_{10}^2}{f} \tag{5.21}$$

where f is frequency in Hz, $\bar{f}$ is a normalized frequency ($\bar{f} = fL/\bar{U}_{10}$), k is the sea surface drag coefficient (taken to be 0.005 for open sea) and $\bar{U}_{10}$ is the hourly mean wind velocity at 10 m height, L denotes a representative length scale taken here to be 1200 m. Figure 5.6 presents a typical in line gust velocity spectrum for $\bar{U}_{10} = 30$ m/s and $L = 1200$ m.

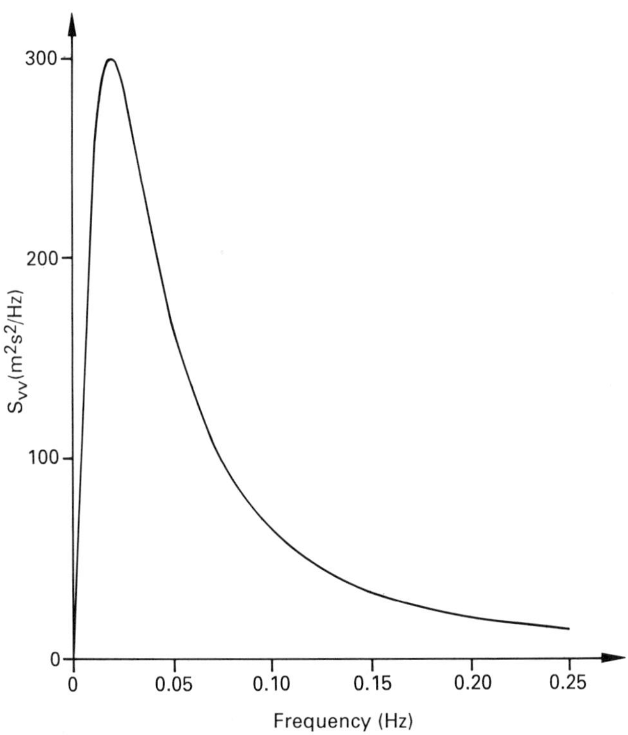

Figure 5.6. Typical wind gust spectrum with $L = 1200$ m, $U_{10} = 30$ m/s and $k = 0.005$

This in line wind gust velocity spectrum can be readily applied to a member of an open space frame structure through Equation (5.16), since the member diameter will be small compared to turbulent fluctuation length scales. The transformation from velocity spectrum to wind force will require a linearization of the velocity dependency in Equation (5.16) or the conversion of the frequency domain spectrum into a velocity time history. For very large or monolithic structures, however, calculation of gust forces must take account of the fact that a gust fluctuation length scale which is smaller or of the same dimension as the structure will tend to apply very much smaller loads. A calculation approach using aerodynamic admittance functions will then become necessary. This technique is described by Davenport (1967) but is only really necessary for calculating wind forces on very large bulky structures.

Current forces on offshore structures are calculated using the same methods as for wind forces alone apart from the fact that currents do not exhibit similar fluctuations to those of wind gusts. The current velocities used are much smaller (< 2.5 m/s) than those for wind force calculations, although, of course, the water density is higher. The magnitude of lift and drag forces due to current may be increased substantially if currents can also affect gravity waves by refraction which results in a change in the 'apparent' wavelength and direction of incident waves.

Drag force coefficients for currents are obtained in the same way as for winds. In critical applications, however, the coefficients may need to be confirmed by model tests. Furthermore, for most large structures, a substantial current can induce frictional and pressure drag as well as a residual drag which is due, primarily, to wave making at the surface. This residual drag must also be accounted for in calculations.

5.4 Flow regimes governing wave loads

The procedure for calculating wave forces on offshore structures can be split up into fundamentally different approaches depending on the size of the structural member and the height and wavelength of incident waves. These parameters can be written in the form of two ratios: structural member diameter to wavelength (D/λ) and wave height to structural member diameter (H/D). Table 5.1 uses approximate values for the ratio K (which is a measure of H/D) to separate the wave regimes into its constituent parts.

For small structural members where $D/\lambda < 0.2$, Morison's equation, which is described later in this chapter is used to estimate forces due to wave action with the implicit assumption that the diameter of the member is small enough in relation to the wave length so as not to alter incident wave characteristics to any significant extent. Morison's equation, first put forward by Morison *et al.* (1950), is based on the assumption that wave forces can be expressed as the sum of a drag force due to wave fluid velocity and an inertia force due to wave acceleration.

Table 5.1 Wave loading regimes

↓K →	*D/λ < 0.20*	*D/λ > 0.20*
K > 25	Drag dominated flow regime Morison's equation with C_M and C_D values required for computing wave forces. Drag coefficients is a function of the Reynolds number For Re > 1.5. × 10^6, C_M = 1.8, C_D = 0.62 For 10^5 < Re < 1.5 × 10^6, C_M = 1.8, C_D varies from 1.0 to 0.6	
5 < K < 25	Intermediate regime between drag and inertia domination. Morison's equation applicable but published C_M and C_D values exhibit wide scatter. Flow behaviour and consequent loading complex and uncertain. For Re > 1.5 × 10^6, C_M = 1.8, C_D = 0.62	
K < 5	Inertia dominated regime. Morison's equation or diffraction theory for computing wave forces C_M = 2.0 Effect of C_D is negligible	Morison's equation unsuitable for computing wave forces Diffraction theory required

Notes:
C_M = inertia coefficient, C_D = drag coefficient,

$$\text{Re} = \frac{U_m D}{v}, \quad \text{K} = \frac{U_m T}{D},$$

U_m = peak velocity, T = wave period, v = kinematic viscosity, Re = Reynolds number, K = Keulegan–Carpenter number

On the other hand, for larger structural members (with $D/\lambda > 0.2$), diffraction theories are necessary to account for the reflection and radiation of waves from the structural member. These potential flow methods, however, cannot account for viscous drag forces.

The second parameter of interest is the ratio H/D. Its importance is based on the fact that drag forces on structures, say a pipe of circular cross-section, in an oscillatory wave flow are dominated by the separation of flow behind the cylinder and the formation of large vortices. For a small H/D ratio (< 1.5), the wave height (and thus the orbital diameter of fluid particle motions) does not remain unidirectional long enough for the flow to initiate separation and develop or shed vortices. In this case, drag forces are very small, acceleration dependent inertia forces dominate and potential flow diffraction theory can be used to predict wave forces with confidence. At the other extreme, for approximately $H/D > 8$, the wave flow will have been unidirectional long enough for a substantial vortex flow to develop. Drag forces will then be large and a Morison formulation, which accounts for these, must be used. An intermediate region, where $1.5 < H/D < 8$ also exists where the flow regime is highly complicated and wave forces are difficult to compute. The two flow regimes corresponding to H/D of < 1.5 and > 8 are illustrated in Figure 5.7. A non-dimensional parameter called the Keulegan–Carpenter number is a more rational measure of the ratio of water particle motion double amplitude to cylinder diameter than H/D which is only valid close to the water surface. The Keulegan–Carpenter number is defined as

$$\mathrm{K} = \frac{U_{\mathrm{m}} T}{D} \tag{5.22}$$

where U_{m} is the maximum normal velocity in the oscillatory flow of period T about the cylinder of diameter D. The approximate range of numerical values encountered are from 0 to 30 m for H and from 60 m to 620 m for λ. Thus a subsea pipe or the space frame of a tower jacket structure (with typical D from 0.4 to 3 m) would give large H/D and small D/λ – indicating a drag dominated wave loading with use of Morison's equation being appropriate for computation of the loading. Alternatively, a large monolithic gravity structure (with typical $D = 80$ m) would yield small H/D and large D/λ, thus pointing to an inertia dominated loading regime and requiring potential flow diffraction theory for computing wave loads. The combination of both H/D and D/λ being small can occur but the other extreme of both H/D and D/λ being large cannot because of the height limitation for stable waves which constrains the ratio H/λ to a maximum value of approximately $\frac{1}{7}$.

Table 5.1 illustrates the wave loading regimes described above. The Keulegan–Carpenter number (K) is used as a measure of ratio H/D.

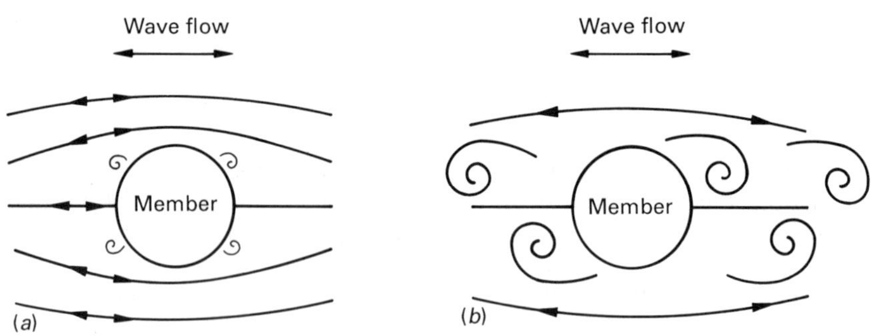

Figure 5.7. Inertia and drag dominated flows: (*a*) no vortex shedding, inertia dominated $H/D < 1.5$; (*b*) substantial vortex shedding, drag dominated $H/D > 8$

5.5 The Morison equation

The mechanics of wave loading on slender tubular subsea pipe lines or members of offshore jacket structures is described here. Such structures (with $D/\lambda < 0.2$) will be subjected to so-called Morison wave loading which is based on the assumption that wave properties are not affected by the presence of the structure and that the total wave force can be expressed as the sum of inertia forces (due to wave fluid acceleration) and drag forces (due to wave fluid velocity).

An equation to represent this loading can be written as

$$\underbrace{dF = C_M\rho\, dV\dot{U}_n}_{\text{inertia force}} + \underbrace{C_D\tfrac{1}{2}\rho\, dS|U_n|U_n}_{\text{drag force}} \tag{5.23}$$

where dF is the total wave force on a pipe line element of volume dV and frontal area, dS, U_n, $\dot{U}_n$, are instantaneous wave fluid velocities and accelerations normal to the pipe axis, ρ is the fluid density and C_M and C_D are inertia and drag coefficients. The inertia coefficient is conventionally taken as the added mass coefficient plus one to account for the Froude Krylov or undisturbed pressure force due to wave fluid acceleration (see Section 3.12). The drag force term includes a modulus sign to ensure that the drag force remains in the same direction as the wave velocity.

Although, the Morison equation was originally developed for vertical piles in shallow water, for engineering purposes it is assumed to be valid for cylinders of arbitrary orientation in deep and shallow water with the proviso that the coefficients C_M and C_D are chosen experimentally for the appropriate condition. Application of the Morison equation can be illustrated by developing expressions for the wave force on vertical and horizontal pipe segments.

Consider an element dy of a vertical circular cylinder of radius r, as shown in Figure 5.8. The total force dF acting on this element in the direction of wave propagation is the sum of inertia and drag force components. This force can be integrated to yield the total force F and moment about the sea bed as

$$F = C_M\rho\pi r^2\int_{-d}^{0}\dot{u}\, dy + C_D\rho r\int_{-d}^{0}|u|u\, dy$$

and

$$M = C_M\rho\pi r^2\int_{-d}^{0}(d + y)\dot{u}\, dy + C_D\rho r\int_{-d}^{0}(d + y)|u|u\, dy \tag{5.24}$$

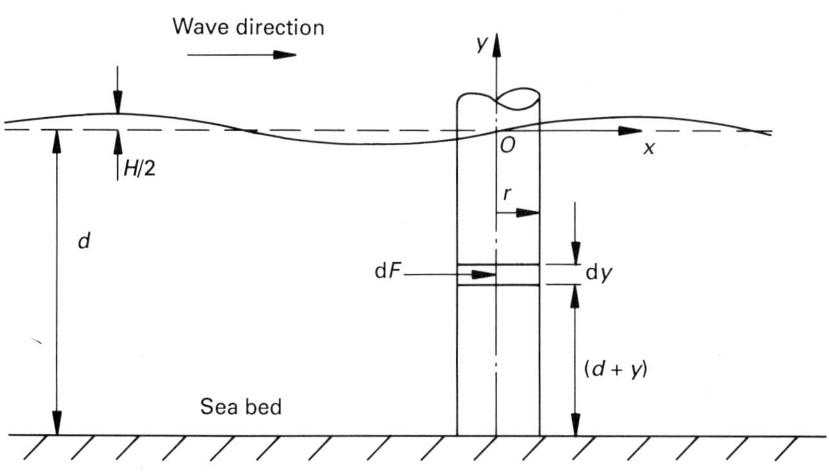

Figure 5.8. Wave force on vertical cylinder

where u and $\dot{u}$ are horizontal components of wave velocity and acceleration. Using expressions for u and $\dot{u}$ for linear wave theory from Table 4.2, with $x = 0$ and $\theta = \omega t$, Equations (5.24) can be readily integrated to yield

$$F = -\frac{2\pi\rho r H^2\lambda}{T^2}(K_2 C_M \cos\theta + K_1 C_D|\sin\theta|\sin\theta)$$

and

$$M = -\frac{2\rho r H^2\lambda^2}{T^2}(K_4 C_M \cos\theta + K_3 C_D|\sin\theta|\sin\theta)$$

where

$$K_1 = \frac{1}{16\sinh^2 kd}(2kd + \sinh 2kd) \tag{5.25}$$

$$K_2 = \frac{\pi r}{2H}$$

$$K_3 = -\frac{1}{64\sinh^2 kd}(2k^2d^2 + 2kd\sinh 2kd + 1 - \cosh 2kd)$$

$$K_4 = \frac{\pi r}{4H\sinh kd}(1 + kd\sinh kd - \cosh kd)$$

Consider also wave forces on horizontal submerged cylinders, as shown in Figure 5.9. The horizontal and vertical wave force components on the cylinder of Figure 5.9(a), with its axis normal to the direction of wave propagation, are simply written as

$$\left.\begin{aligned} F_H &= C_M\rho\pi r^2 L\dot{u} + C_D\rho r L|u|u \\ &\text{and} \\ F_V &= C_M\rho\pi r^2 L\dot{v} + C_D\rho r L|v|v \end{aligned}\right\} \tag{5.26}$$

where L is the cylinder length and the expressions for u, v, $\dot{u}$, $\dot{v}$ are obtained from Table 4.2.

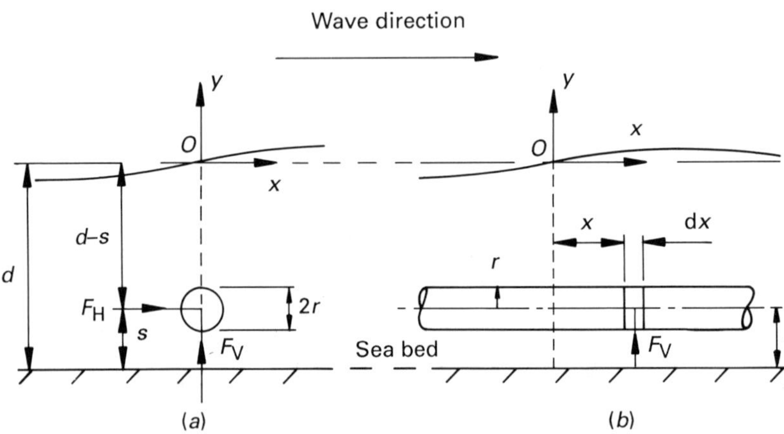

Figure 5.9. Wave force on horizontal cylinders

If, on the other hand, the horizontal cylinder is lying with its axis in the direction of wave propagation, as in Figure 5.9(*b*), then the above expression for horizontal force is no longer applicable. The expression for vertical force per unit length can still be used but the calculation does not have the benefit of the simplification due to $x = 0$. The equation for vertical force on the cylinder can then be written as

$$F_V = \int_{x_1}^{x_2} (\rho C_D r|v|v\,dx + \rho C_M \pi r^2 \dot{v}\,dx) \tag{5.27}$$

for a cylinder segment from $x = x_1$ to $x = x_2$. Evaluation of this equation requires that expressions for v and $\dot{v}$ be integrated with respect to x. Integration of the modulus sign is carried out by splitting up the integration limits into positive and negative integrand regions and evaluating each integrand separately.

The Morison equation can also be applied to arbitrarily inclined circular cylinders although the original formulation by Morison *et al.* (1950) was only intended for vertical piles. The extension of the Morison equation to inclined cylinders can be defined ambiguously depending on whether wave velocity components incident on the cylinder are first resolved into a total normal velocity component before computing forces (called the normal component approach) or the wave velocity components are combined with cylinder projected areas to compute force components in the three orthogonal directions (called the projected area method). Morison's equation may be written in vector form using the normal component approach with the wave force component normal to the cylinder axis on an element of length ds as

$$\mathrm{d}\mathbf{F} = C_M \rho \pi r^2 \dot{\mathbf{q}}_n\,\mathrm{d}s + C_D \rho r|\mathbf{q}_n|\mathbf{q}_n\,\mathrm{d}s \tag{5.28}$$

using the definition diagram of Figure 5.10 with a wave velocity vector $\mathbf{q}_n$ normal to the cylinder axis at the element of length ds. An equivalent equation for the projected area method is

$$\mathrm{d}\mathbf{F} = C_M \rho \pi r^2 \dot{\mathbf{q}}_n\,\mathrm{d}s + \tfrac{1}{2} C_D \rho \begin{bmatrix} |u|u\,\mathrm{d}A_x \\ |v|v\,\mathrm{d}A_y \\ |w|w\,\mathrm{d}A_z \end{bmatrix} \tag{5.29}$$

where $\mathbf{q} = u\mathbf{i} + v\mathbf{j} + w\mathbf{k}$ is the instantaneous wave velocity vector and $\mathrm{d}A_x$, $\mathrm{d}A_y$ and $\mathrm{d}A_z$ are the projections of the element surface area in the *x, y* and *z* directions respectively. Note that the inertia term of Equation (5.29) is still evaluated using the normal component approach and that only drag terms from the two equations will yield different results. The wave forces obtained will, in any case, depend on the added inertia and drag coefficients selected and it may be argued that the coefficients that should be selected for Equation (5.28) must also be derived by decomposing experimental data through the formulation of Equation (5.28). The same may be said for Equation (5.29). However, in terms of the physics of wave loading and the ease of applicability of the resultant equation, use of the normal component approach and Equation (5.28) is recommended and is nearly always used in design calculations. However, many fundamental questions on the mechanics of wave loading on inclined cylinders still remain open and are the subject of continuing research.

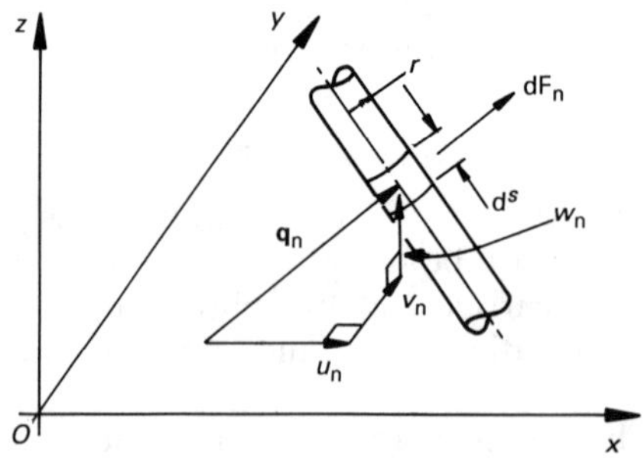

Figure 5.10. Wave force on inclined cylinder: q_n is the component of velocity normal to the cylinder axis; u_n, v_n and w_n are the components of $\mathbf{q}_n$ along the x, y and z axes respectively

A further point to note is that the formulations of Equations (5.28) and (5.29) do not give the total force on an isolated cylindrical member. There are additional axial forces due to the wave pressure added mass and drag from the flow at the cylinder ends. In particular, the axial wave pressure force at each end is obtained as the product of the local instantaneous wave pressure and cylinder cross-sectional area.

Figure 5.11 shows a typical variation of inertia force, drag force and wave elevation within a wave cycle for a vertical cylinder. The non-linear behaviour of the drag force is shown up by the shape of the curve close to the horizontal axis. The expression of Equation (5.25) for wave

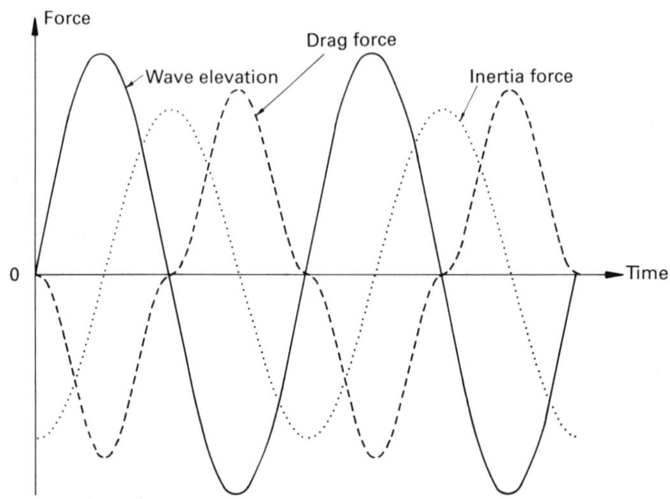

Figure 5.11. Typical wave force time history

force can be re-written as

$$F = A|\sin\theta|\sin\theta + B\cos\theta$$

where $\theta = \omega t$ and A, B are appropriate constants. The maximum value of F is given by

$$\left.\begin{aligned} &\frac{dF}{d\theta} = 2A\sin\theta\cos\theta - B\sin\theta = 0 \\ &\text{for } 0 < \theta < \pi \text{ with turning points given by} \\ &\sin\theta = 0 \quad \cos\theta = \frac{B}{2A} \end{aligned}\right\} \tag{5.30}$$

These points give the location within the wave cycle and the magnitude of the instantaneous maximum values of wave force.

5.5.1 Drag and inertia coefficients

The value of drag coefficients C_D to be inserted in Morison's equation can only be obtained experimentally. In theory, the value of the inertia coefficient C_M can be calculated (it is, for example, 2.0 for a smooth cylinder in an ideal fluid). However, measured values are used in practice, particularly when drag is the dominant force. For irregular shapes, the inertia coefficient can be calculated using diffraction theory.

Many experiments have been designed to measure values of drag and inertia coefficients in steady or planar oscillatory flows but their results can only be used with caution for wave force prediction. The most useful experimental measurements have been made in circumstances which model full scale conditions within the ocean environment. Measurements made offshore on large test structures in real sea ways are especially valuable.

One problem facing the user of Morison's equation is the large scatter in values of the inertia and drag coefficients. However, there is a useful degree of correlation between the coefficients and two flow parameters – Keulegan–Carpenter number K and Reynolds number Re. Nevertheless, the scatter, and hence some uncertainty, remains. Table 5.1 gives a summary of generally accepted added inertia and drag coefficients for flow around circular cylinders in unconfined flow remote from solid boundaries such as the sea bed. The effects of sea bed proximity on coefficient values are examined in more detail later.

Sarpkaya (1976) has carried out many systematic studies of the variations of inertia and drag forces for circular cylinders in planar ocillatory flow. Figures 5.12 and 5.13 present plots of C_M and C_D against the Reynolds and Keulegan–Carpenter numbers for smooth cylinders from data prepared by Sarpkaya and Isaacson (1981). Figures 5.14 and 5.15 also present this data plotted as a function of Keulegan–Carpenter number with parameter β being the ratio of the Reynolds number to the Keulegan–Carpenter number. It can be seen that the C_M and C_D data show consistent and physically meaningful correlations with the Keulegan–Carpenter number.

Marine growth in the ocean environment causes the surfaces of

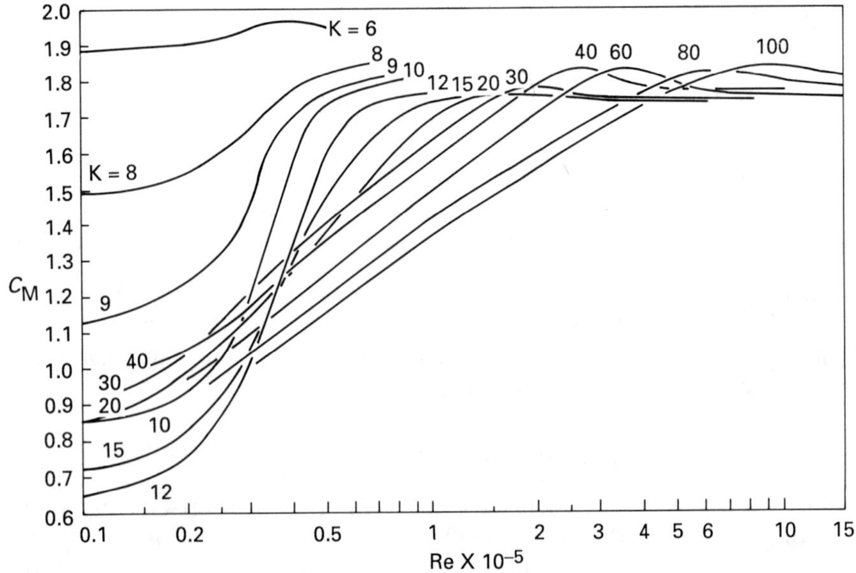

Figure 5.12. Added inertia coefficient against Reynolds number for circular cylinders at various values of Keulegan–Carpenter number (from Sarpkaya and Isaacson, 1981)

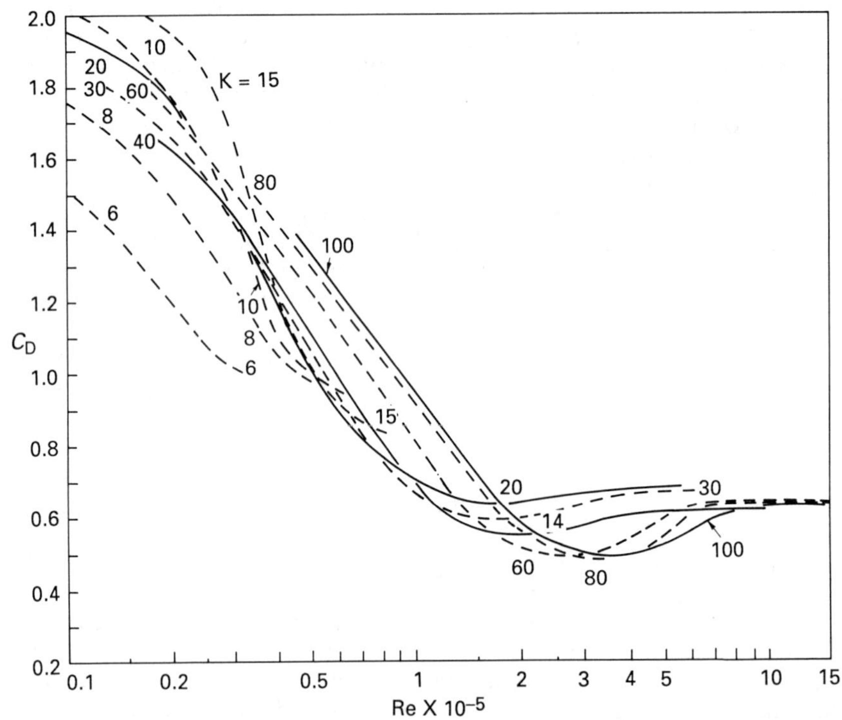

Figure 5.13. Drag coefficient against Reynolds number for circular cylinders at various values of Keulegan–Carpenter number (from Sarpkaya and Isaacson, 1981)

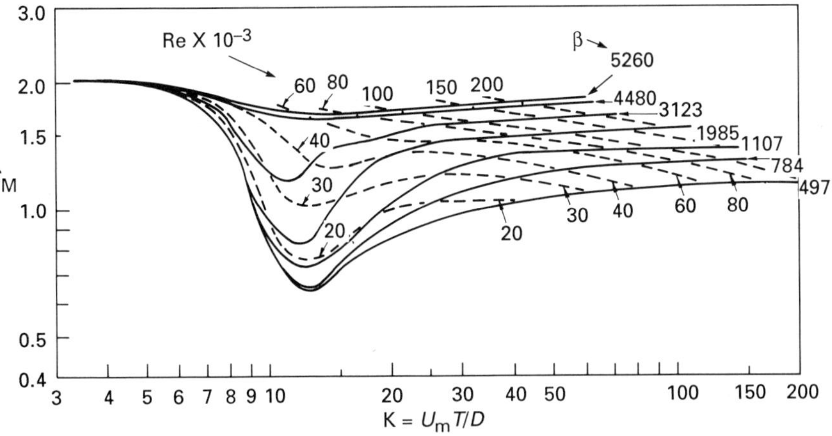

Figure 5.14. Added inertia coefficient plotted against Keulegan–Carpenter number for circular cylinders at various values of Reynolds Number Re and frequency parameter β (from Sarpkaya and Isaacson, 1981)

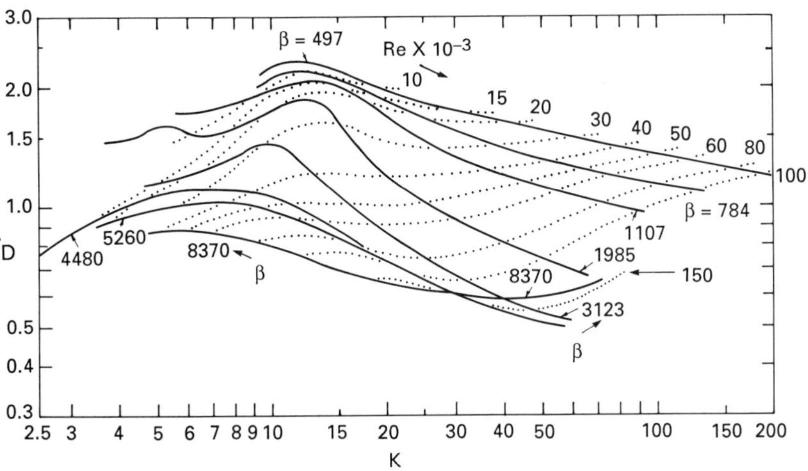

Figure 5.15. Drag coefficient against Keulegan–Carpenter number for circular cylinders at various values of Reynolds number Re and frequency parameter β (from Sarpkaya and Isaacson, 1981)

offshore structures, particularly near the splash zone, to become highly roughened. This has the effect of increasing the effective diameter of individual members and to modify the inertia and drag coefficients due to the changed interaction between the cylinder boundary layer and the roughness elements.

Roughness due to marine growth will normally increase the skin friction due to a turbulent boundary layer, but there exists a critical height below which the roughness has no effect on friction. This critical roughness height is also determined by the height for which the Reynolds number of the roughness corresponds to the development of an eddying wake. As the roughness height increase, it causes a progressive increase of drag due of the energy loss in the eddying flow behind the roughness protuberances. In a sense, the extra drag due to the roughness is predominantly a form drag, derived from the pressure distributions round the roughness elements, rather than a frictional drag, although over a surface embracing a number of roughnesses the net effect is that of an increase in skin friction. When the height of the roughnesses is large enough, the extra drag due to them becomes very much greater than the initial drag of the smooth surface. It is characteristic of the drag coefficient of a bluff body, due largely to the shedding of eddies, that it is practically independent of Reynolds number over a wide range. This also applies for large roughnesses where the drag coefficient becomes independent of Reynolds number. Fully developed roughness flow is then said to be established.

Figure 5.5 shows the effects of roughness on the steady flow drag coefficient. For planar oscillatory flow, Sarpkaya and Isaacson (1981) present recommended data for inertia and drag coefficients as a function of roughness, Reynolds number and for Keulegan–Carpenter range 20–100.

There are three other physical features that influence wave loading on cylindrical members of offshore structures. These are summarized as follows. There is very little design information available at present on the effects of flow interference in waves between groups of closely spaced cylinders, although the physics of such phenomena are being researched at present. Recent work has shown that planar oscillatory wave flow about circular cylinders induces motion of vortex pairs which can induce substantial velocity magnifications in the flow around the cylinder. The presence of other smaller cylinders around a larger diameter cylinder can interact with these velocity magnifications to give rise to significant interference effects. Grass *et al.* (1987) presents recent research work in this area.

The presence of a current superimposed on a wave seems to influence C_D but not C_M – probably because the current sweeps away the vortices shed in the flow by the obstruction. With a current running parallel to the wave direction, the current velocity should first be added vectorially to wave particle velocity before the calculation of wave forces using the Morison equation. It is generally felt that the use of coefficient values for zero current leads to conservative results for design.

Although the above inertia and drag coefficients data applies to circular cylinders, Appendix G presents a summary of these coefficients for a variety of shapes that may also be used for the members of offshore structures.

5.5.2 Vortex shedding

Vortex shedding from a circular cylinder assumes importance here because of the to and fro motion of fluid particles during wave motion. This motion initially produces vortices in the lee of the cylinder which are then forced back on to the body when the direction of flow reverses. The effect of this motion can produce very large variations in an oscillating lift force as well as an increase in the mean drag force.

The formation of vortices from a circular cylinder in uniform steady flow has been studied extensively. An empirical expression for f_s, the frequency of vortex shedding, can be written as

$$f_s = S\left(1 + \frac{19{\cdot}7}{Re}\right)\frac{U}{D} \tag{5.31}$$

where S is the Strouhal number, U is the steady velocity, D is the diameter of the structural member and the equation is approximately valid over a Reynolds number Re range of 60–2 × 10^5. Except for very small values of Re, the Reynolds number based variation can be neglected. Then

$$S = \frac{f_s D}{U} \tag{5.32}$$

For circular cylinders, $S = 0.2$ for $10^3 < Re < 2 \times 10^5$. Above $Re = 2 \times 10^5$, there is considerable variation in S. Strouhal numbers for vortex shedding from members of various cross-sectional shapes are presented in Appendix G.

Consider the horizontal component of wave particle velocity at the water surface. For deep water, the equation can be written as

$$u = \frac{\pi H}{T}\cos\theta \tag{5.33}$$

If U is taken as the average value of u above, we have

$$U = \frac{\pi H}{2T} \tag{5.34}$$

Since unidirectional flow persists for a time $T/2$ in a wave cycle of period T and the time interval between successive vortices being shed is $1/(2f_s)$ (there are two vortices per cycle of shedding), then the condition for at least one vortex to have time to form is

$$\frac{T}{2} > \frac{1}{2f_s}$$

But

$$\frac{1}{f_s} = \frac{D}{US} = \frac{2TD}{\pi HS}$$

Taking $S = 0.2$ (5.35)

$$\frac{T}{2} > \frac{1}{2}\frac{10}{\pi}\frac{TD}{H}$$

$$\frac{H}{D} > \frac{10}{\pi}$$

H/D can also be written as $U_m T/D$ where U_m is the maximum velocity in the cycle to yield the Keulegan–Carpenter number of Equation (5.22). This parameter was first introduced by Keulegan and Carpenter (1958) for correlating measured inertia and drag coefficients. The role of the Keulegan–Carpenter number in quantifying the number of vortices shed per cycle of wave oscillation leads to a physically more meaningful correlation between inertia and drag coefficients with the Keulegan–Carpenter number than is apparent with the Reynolds number. These arguments also suggest that it is beneficial to keep the Keulegan–Carpenter number constant between full scale and model scale testing conditions.

The formation of vortices as described above leads to an oscillating transverse lift force on a segment of circular cylinder in a flow. This lift force $\mathrm{d}L$ for a segment of length $\mathrm{d}s$ on a cylinder of radius r can be written as

$$\mathrm{d}L = C_{\mathrm{L}}\tfrac{1}{2}\rho(2r)U_{\mathrm{m}}^{2}\,\mathrm{d}s \tag{5.36}$$

where U_{m} is the peak flow velocity in a cycle ρ is fluid density and C_{L} is the lift coefficient. Once again, experimental data in planar oscillatory flow provides the necessary information for values of C_{L}. Figure 5.16 presents values of C_{L} against Reynolds number at a range of Keulegan–Carpenter numbers for smooth cylinders from Sarpkaya and Isaacson (1981). The force oscillation can be taken to occur at the Strouhal frequency given by taking $S = 0.2$ in Equation (5.32). For smooth cylinders with Reynolds numbers greater than 1.5×10^6, a value of $C_{\mathrm{L}} = 0.20$ may be used. Figure 5.17 (also from Sarpkaya and Isaacson, 1981) gives equivalent data for roughened cylinders with C_{L} taken to be 0.25 for large K and Re together with a Strouhal number of 0.22.

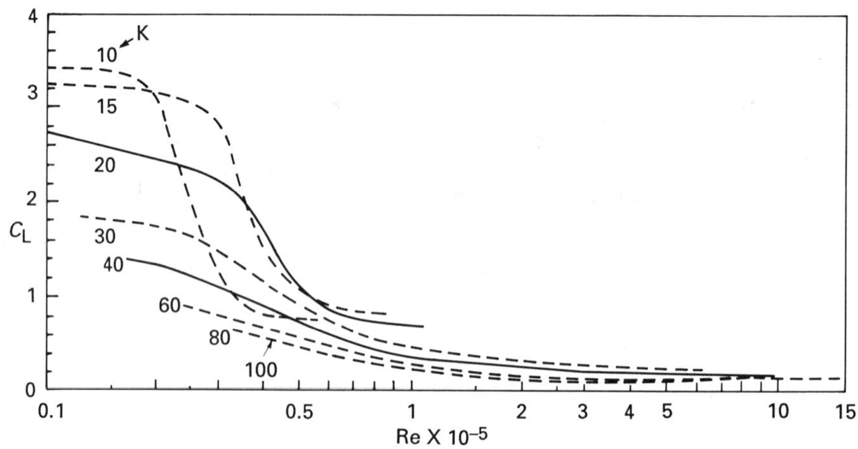

Figure 5.16. Lift coefficient against Reynolds number for circular cylinders at various values of Keulegan–Carpenter number K (from Sarpkaya and Isaacson, 1981)

The oscillatory nature of the lift force can induce dynamic excitation of the loaded member. It is known that as the vortex shedding period of the cylinder approaches its natural period, this lift force increases considerably due to cylinder motion and a dynamic instability can result. In many practical situations, structural failure has been known to occur in this fashion. Blevins (1977) gives further details.

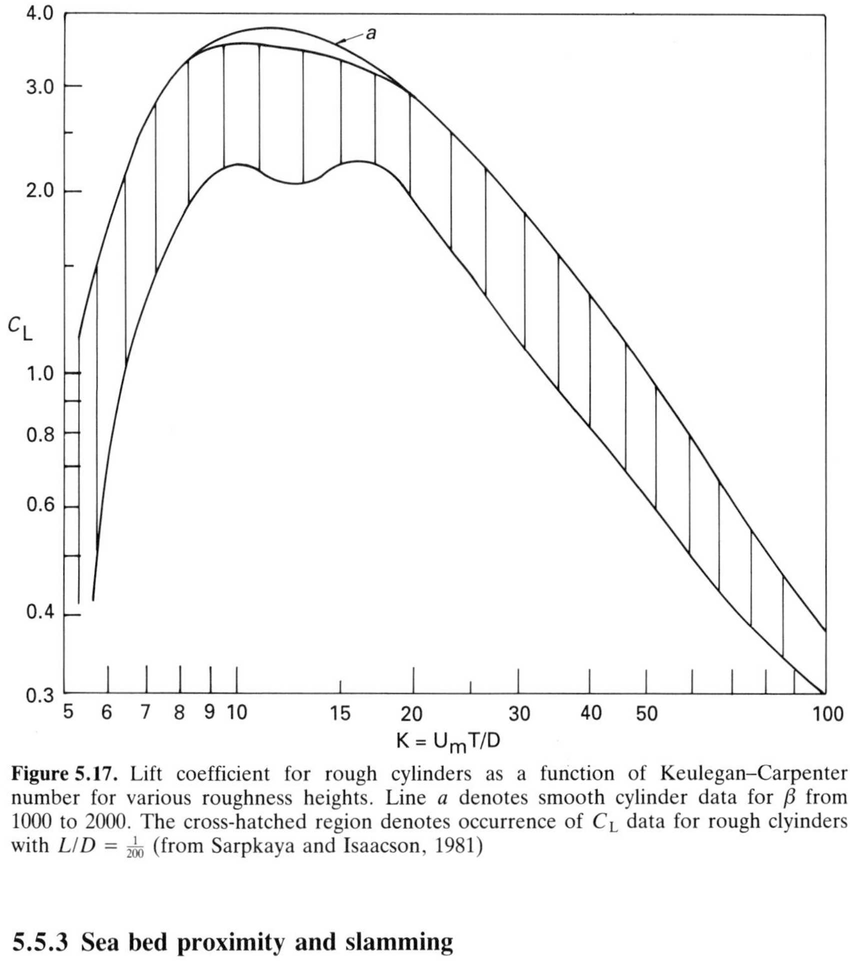

Figure 5.17. Lift coefficient for rough cylinders as a function of Keulegan–Carpenter number for various roughness heights. Line *a* denotes smooth cylinder data for β from 1000 to 2000. The cross-hatched region denotes occurrence of C_L data for rough clyinders with $L/D = \frac{1}{200}$ (from Sarpkaya and Isaacson, 1981)

5.5.3 Sea bed proximity and slamming

For pipelines laying on the sea bed or spanning some small distance above the sea bed, wave and current induced flow around the line is substantially distorted by the presence of the sea bed boundary. Experimental work for such flows shows that the Morison equation can still be used together with Equation (5.36) for lift forces provided that the added inertia, drag and lift coefficients are recognized as being strong functions of pipe distance from the sea bed. The above comments also apply for pipes that are partially buried or embedded in the sea bed. The presence of large flow velocities close to the sea bed due to a partially suspended pipe can cause sea bed scour and change local sea bed contours.

Applicable added inertia, drag and lift coefficients for pipes in close proximity to the sea bed are generally obtained from experimental data, much of which is proprietary to companies within the industry. However, the inertia coefficient for a pipe resting on the sea bed can be derived from diffraction theory to be 3.29 – see Yamamoto *et al.* (1974).

This work shows that $C_M = 3.29$ for a cylinder touching a plane wall but that this reduces rapidly to the potential flow value of 2.00 for a gap length equal to one cylinder diameter.

The best practical guide, however, is supplied by Det Norske Veritas (1982). Figures 5.18–5.21 present variations of added inertia, drag and lift coefficients with sea bed proximity. The graphs illustrate the significant increase in inertia, drag and lift forces due to the sea bed proximity. The increases occur because the presence of the sea bed restricts fluid flow largely to one side of the body and, therefore, presents a greater restriction to the flow. These increased forces must be accounted for in calculations for pipe line stability.

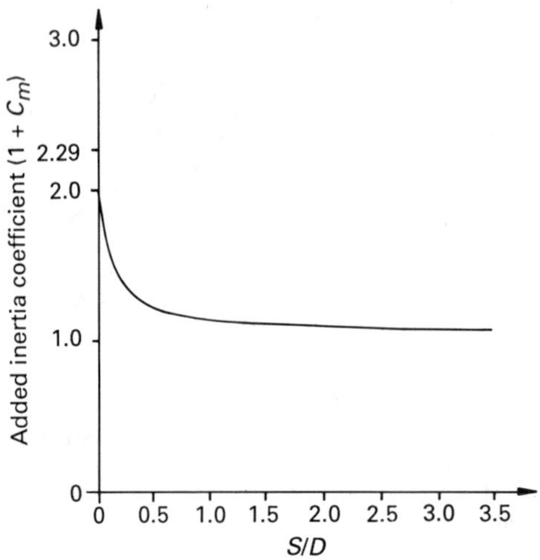

Figure 5.18. Variation of added inertia coefficient with wall proximity for a circular cylinder: *S* is distance of cylinder bottom above seabed and *D* is cylinder diameter (from Det Norske Veritas, 1982)

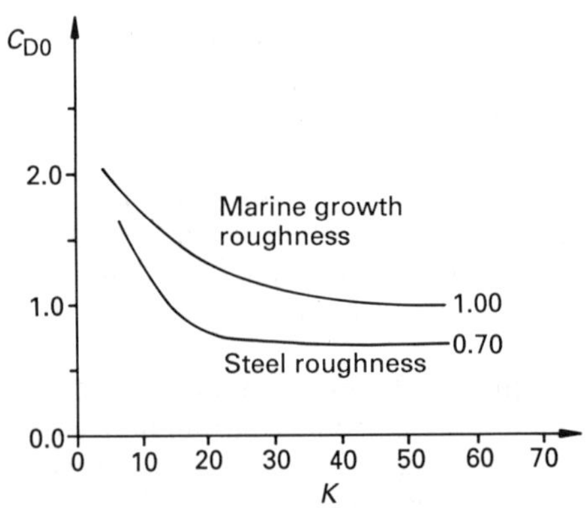

Figure 5.19. Variations of drag coefficient C_{DO} with Keulegan–Carpenter number K for a circular cylinder in unconfined flow (from Det Norske Veritas, 1982)

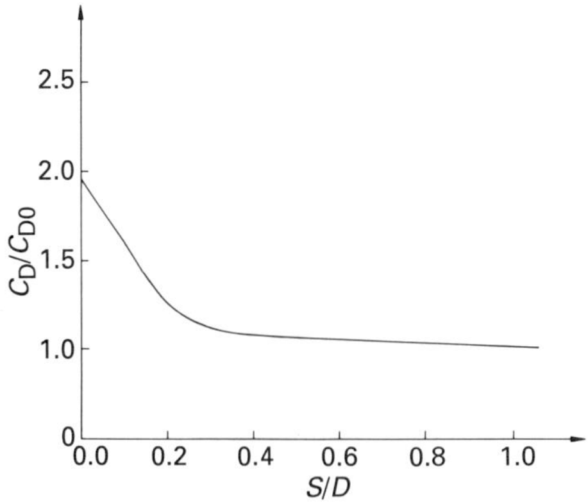

Figure 5.20. Variation of drag coefficient C_D with wall proximity ratio S/D for a circular cylinder in oscillatory supercritical flow with $K > 20$ and Reynolds number range of from 10^5 to 2×10^6. Notation as in Figure 5.18 (from Det Norske Veritas, 1982)

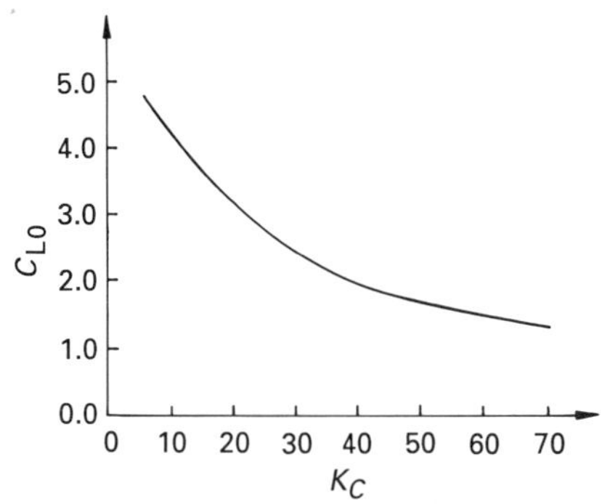

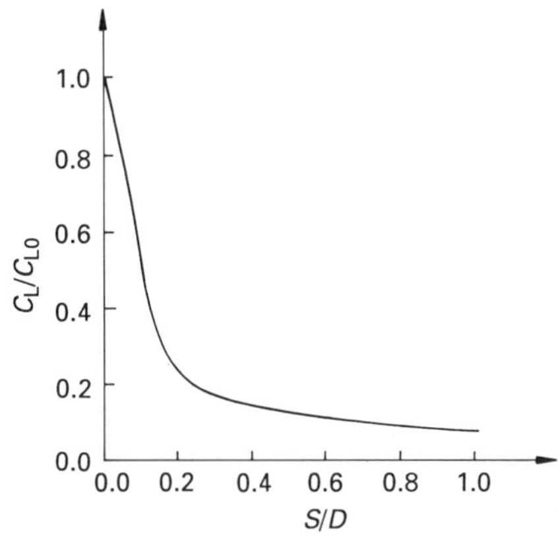

Figure 5.21. Variation of lift coefficients for a circular cylinder with Keulegan–Carpenter number K and wall proximity: C_{L0} is lift coefficient for unconfined flow and C_L is lift coefficient for flow close to a wall. Remaining notation as in Figure 5.18 (from Det Norske Veritas, 1982)

Just as structural members close to the sea bed require special attention, cylindrical members close to the sea surface need to be considered further. In particular, horizontal structural members in the splash zone are susceptible to vertical forces due to buoyancy changes and wave slam. These loads are impulsive in nature and can generate appreciable dynamic magnification.

Slamming loads occur when a moving body strikes and enters a water surface, or, conversely, when a moving water surface strikes and submerges a body. If a structural member presents a large face parallel to, or nearly parallel to, the wave surface and if the relative velocity of impact is high, slamming forces are both impulsive and large – usually much larger than the other forces (due to buoyancy, drag and inertia) experienced in fully immersed flow of the same velocity.

Peak slamming forces on the member occur due to the change of fluid momentum caused by the impact. This change is proportional to the square of the velocity of impact, and the slamming force F_s can be written as

$$F_s = \tfrac{1}{2}C_s\rho(2r)U^2 \tag{5.37}$$

per unit length for a cylinder of radius r. Here C_s is the slamming coefficient and U is the velocity of impact between the water surface and structural member. There is a large amount of scatter in the values of C_s obtained from experimental and theoretical work on this problem. Von Karman and Wattendorf (1929) and Kaplan and Silbert (1976) describe two theoretical approaches to the problem of predicting C_s which yield a value of π. Experimental determination of C_s values is particularly difficult because of the problem of separating the dynamic response of the test rig from the rigid structure values calculated by the theory and also because the dynamic response of the test rig strongly influences the measured peak load. The measured value of C_s is also very sensitive to even very small angles between the wave crest and cylinder axis. As a result, model test results for slamming coefficients show large variations in a range 2–5.2. However, work by Miller (1980) supports the use of an average C_s value of 3.5 for design.

5.6 Wave loads on large bodies – diffraction theory

Calculation of wave forces on bodies with dimensions that span a significant proportion of the wave length ($> 0.2\lambda$) must account for the fact that the wave flow will be influenced by the presence of the body. The solution of the governing Laplace equation with the usual sea bed and free surface boundary conditions, together with the additional condition of no flow through the body surface, will give rise to additional component waves in the solution which are called scattered or diffracted waves. The wave force on the body is then due to the incident wave as well as the scattered wave field arising from the presence of the body. Furthermore, the Froude–Krylov force in such a flow field with a large body can no longer be written as $\rho V\dot{U}$ (as in Section 3.12) because the wave induced pressure gradient around the body is no longer constant. At the same time, the conditions that require diffraction theory to be used ($D/\lambda > 0.2$) also imply that the ratio of wave height to body dimension H/D is less than unity since $H/\lambda < \frac{1}{7}$. For such small values of H/D, drag forces will be small since flow amplitudes will be less than the body dimension and flow separation will not occur. These fortuitous circumstances mean that a potential flow solution of the wave diffraction problem will represent a physically realistic situation.

A statement of the full wave diffraction problem in three dimensions can be written as follows using an axes system with Oxy in the still-water surface and Oz pointing vertically upwards. The governing equations are

$$\frac{\partial^2\phi}{\partial x^2} + \frac{\partial^2\phi}{\partial y^2} + \frac{\partial^2\phi}{\partial z^2} = 0 \tag{5.38}$$

with the linearized free surface boundary condition at $z = 0$ of

$$\frac{\partial^2 \phi}{\partial t^2} + g\frac{\partial \phi}{\partial z} = 0 \tag{5.39}$$

and the sea bed boundary condition at $z = -d$ of

$$\frac{\partial \phi}{\partial z} = 0 \tag{5.40}$$

together with the no-flow through the immersed body boundary condition of

$$\frac{\partial \phi}{\partial n} = 0 \tag{5.41}$$

at the body surface where n denotes a direction normal to this surface. All the assumptions inherent in the governing equations and the linearized free surface boundary condition thus apply to this linear diffraction problem, that is, irrotational and inviscid flow of small wave amplitude.

The solution to this wave diffraction problem is obtained by writing the velocity potential as the sum of incident and scattered potentials ϕ_i and ϕ_s respectively. Thus

$$\phi = \phi_i + \phi_s \tag{5.42}$$

Furthermore, the scattered waves generated by the presence of the body boundary have to be restricted in the mathematical formulation to outgoing waves only. This requires that the scattered wave potential satisfies the condition

$$\frac{\partial \phi_s}{\partial r} + \frac{1}{c}\frac{\partial \phi_s}{\partial t} = 0 \tag{5.43}$$

where r is radial distance from a point on the body surface and c is wave celerity. This can be written (see Sommerfield, 1949 and Stoker, 1957) as

$$\lim_{r \to \infty} r^{1/2}\left(\frac{\partial \phi_s}{\partial r} - ik\phi_s\right) = 0 \tag{5.44}$$

where the factor $r^{1/2}$ takes account of the directional spreading of waves.

Since the incident wave potential ϕ_i is known, the boundary condition at the body surface can also be written as

$$\frac{\partial \phi_s}{\partial n} = -\frac{\partial \phi_i}{\partial n} \tag{5.45}$$

Once ϕ_s is obtained as a solution to this problem, wave induced pressures can be obtained from the linearized Bernoulli equation and integrated to obtain forces and moments. Note that Equation (5.42) and (5.45) only apply to a stationary body. A moving body will induce additional potentials due to its motions and the above equations will need to be extended to account for this as is described subsequently in Chapter 8, Section 8.4.2.

The following section describes three solutions to the linear wave diffraction problem – the first of these is an analytic solution for vertical circular cylinders in arbitrary water depth, followed by two numerical computational approaches based on boundary integral and boundary elements methods.

5.6.1 Analytic solution for vertical cylinder

This solution was developed initially by Havelock (1940) for deep water and subsequently extended to arbitrary water depths by MacCamy and Fuchs (1954). The latter solution is presented here because it gives the wave forces on cylinders in a readily usable form.

Using the definition diagram of Figure 5.22, the problem can be re-expressed in (r, θ, z) polar coordinates with the Laplace equation becoming

$$\left.\begin{aligned} &\frac{\partial^2\phi}{\partial r^2} + \frac{1}{r}\frac{\partial\phi}{\partial\theta} + \frac{1}{r^2}\frac{\partial^2\phi}{\partial\theta^2} + \frac{\partial^2\phi}{\partial z^2} = 0 \\ &\text{and the body surface boundary condition being} \\ &\frac{\partial\phi}{\mathrm{d}r} = \frac{\partial\phi_\mathrm{i}}{\partial r} + \frac{\partial\phi_\mathrm{s}}{\partial r} = 0 \text{ at } r = a \end{aligned}\right\} \quad (5.46)$$

The incident wave potential is given by

$$\phi_\mathrm{i} = A\frac{\cosh k(z + d)}{\cosh kd}\left[\sum_{m=0}^{\infty}\beta_m J_m(kr)\cos(m\theta)\right]\mathrm{e}^{-\mathrm{i}\omega t} \quad (5.47)$$

where $A = -ag/\omega$, $\beta_m = 1$ for $m = 0$, $\beta_m = 2\mathrm{i}^m$ for $m \geq 1$, and $J_m(kr)$ is the Bessel function of the first kind of order m and argument kr. Equation (5.47) is derived from the velocity potential of Equation (4.25) by writing $x = r\cos\theta$ and expanding the exponential as an infinite series

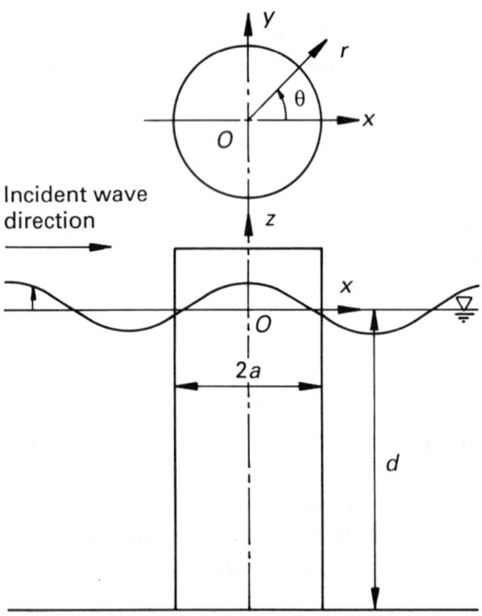

Figure 5.22. Wave past a circular cylinder

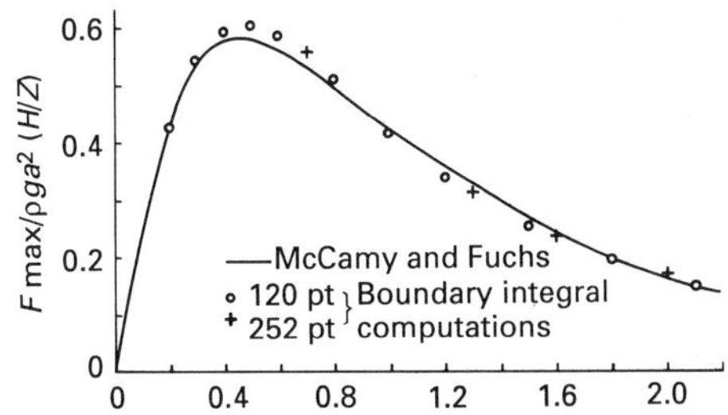

Figure 5.23. Horizontal force coefficient for circular cylinder d/a = 4.0, Full lines denote the exact solution by MacCamy and Fuchs (1954). Circles and crosses denote boundary element calculations with 120 and 252 facets respectively

of Bessel functions. An expression for the scattered velocity potential which already satisfies the Laplace equations and the sea bed and free surface boundary condition is written as

$$\phi_s = A\frac{\cosh k(z+d)}{\cosh kd}\left[\sum_{m=0}^{\infty}\beta_m B_m H_m^{(1)}(kr)\cos(m\theta)\right]e^{-i\omega t} \tag{5.48}$$

where B_m are unknown coefficients and $H_m^{(1)}(kr)$ is a Hankel function of the first kind. It can be shown that ϕ_s satisfies the far field radiation condition through the asymptotic value of $H_m^{(1)}(kr)$ for large r. Substituting Equations (5.47) and (5.48) into the boundary condition at the cylinder surface in Equation (5.46) then gives the coefficients

$$B_m = \frac{J'_m(ka)}{H_m^{(1)\prime}(ka)} \tag{5.49}$$

where $'$ denotes differentiation with respect to the argument (ka). This yields the full expression for velocity potential as

$$\phi = A\frac{\cosh k(z+d)}{\cosh kd} \times \left\{\sum_{m=0}^{\infty}\beta_m\left[J_m(kr) - \frac{J'_m(ka)}{H_m^{(1)\prime}(ka)}H_m^{(1)}(kr)\right]\cos(m\theta)\right\}e^{-i\omega t} \tag{5.50}$$

Evaluation of pressures and integration over the body surface yields forces and base overturning moment expressions of the form

$$\frac{F}{\rho g H a d} = 2\frac{G(ka)}{ka}\frac{\tanh(kd)}{kd}\cos(\omega t - \delta) \tag{5.51}$$

and

$$\frac{M}{\rho g H a d^2} = 2\frac{G(ka)}{ka}\left[\frac{kd\sinh(kd) + 1 - \cosh(kd)}{(kd)^2\cosh(kd)}\right]\cos(\omega t - \delta) \tag{5.52}$$

where

$$G(ka) = [J'^2_1(ka) + Y'^2_1(ka)]^{-1/2}$$

and

$$\delta = -\arctan\left[\frac{Y'_1(ka)}{J'_1(ka)}\right]$$

Figure 5.23 presents a curve of calculated maximum horizontal force against wave number non-dimensionalized with respect to cylinder radius (ka). The above analytic results are compared with data from a boundary integral numerical computation described in the next section.

5.6.2 Boundary integral numerical technique

For offshore structures of arbitrary geometry, it becomes necessary to develop general numerical solutions for design purposes. One such solution that is extensively used in the offshore industry is obtained by modelling the scattered or diffracted wave flow by oscillating point sources placed on facets covering the body surface. If the total incident and scattered wave potential is written as

$$\phi = \mathrm{Re}[\phi'(x, y, z)\mathrm{e}^{-\mathrm{i}\omega t}] \tag{5.53}$$

the spatial variation of potential ϕ' can be split up into its incident and diffracted components:

$$\phi' = \phi'_{\mathrm{i}} + \phi'_{\mathrm{s}}. \tag{5.54}$$

Then Lamb (1975) shows that the scattered wave potential may always be represented as a continuous distribution of point sources over the immersed body surface through a summation integral of the form

$$\phi'_{\mathrm{s}}(x, y, z) = \frac{1}{4\pi}\int_S f(\xi, \eta, \zeta)G(x, y, z; \xi, \eta, \zeta)\,\mathrm{d}S \tag{5.55}$$

where (x, y, z) denotes a point in the fluid with (ξ, η, ζ) denoting a point on the body surface, $\mathrm{f}(\xi, \eta, \zeta)$ is the source strength distribution on this surface and $\mathrm{d}S$ is a differential area on the body surface. The Green's function $G(x, y, z;\ \xi, \eta, \zeta)$ must be such that it satisfies the Laplace condition and the radiation condition as well as the sea bed and free surface boundary conditions. Such a condition is given by Wehausen and Laitone (1960) and can be written in an infinite series form as

$$G(x, y, z; \xi, \eta, \zeta) = \frac{2\pi(\nu^2 - k^2)}{k^2 d - \nu^2 d + \nu}\cosh[k(z + d)]$$

$$\cosh[k(\eta + d)][Y_0(kr) - \mathrm{i}J_0(kr)]$$

$$+ 4\sum_{k=1}^{\infty}\frac{\mu_k^2 + \nu^2}{\mu_k^2 d + \nu^2 d - \nu}\cos[\mu_k(z + d)]\cos[\mu_k(\eta + d)]\,K_0(\mu_k r)$$

where μ_k are positive roots of equation

$$\mu_k \tan \mu_k d + \nu = 0$$

$$r = \surd[(x - \xi)^2 + (y - \eta)^2] \tag{5.56}$$

$$\nu = k \tanh kd$$

and

$$k = 2\pi/\lambda$$

J_0 and Y_0 denote Bessel functions of the first and second kind, respectively, and of zero order, and K_0 denotes the modified Bessel function of the second kind of zero order. The scattered velocity potential ϕ'_s is determined by imposing a no flow boundary condition on the body surface. This condition can be implemented by imposing the condition that at an element of the immersed body surface, the flow due to a local source plus the flow due to all other sources on the body surface must be equal and opposite to the flow at the element due to the undisturbed wave. This can be written mathematically as

$$\frac{1}{2}f(x, y, z) - \frac{1}{4\pi}\int_S f(\xi, \eta, \zeta)\cdot\frac{\partial G}{\partial n}(x, y, z; \xi, \eta, \zeta)\,\mathrm{d}S$$
$$= -W(x, y, z) \tag{5.57}$$

where the undisturbed incident wave flow is given by

$$W(x, y, z) = \left[n_z\frac{\sinh k(z + d)}{\cosh kd} + \mathrm{i}n_x\frac{\cosh k(z + d)}{\cosh kd}\right]\mathrm{e}^{ikx} \tag{5.58}$$

In equation (5.57), the factor $\frac{1}{2}$ in the first term is due to the fact that half the surface source fluid will be streaming into the body's interior volume. A negative sign on the left-hand side of this equation defines fluid flow into the surface from the exterior.

Now the unknown source strengths $f(x, y, z)$ are found by satisfying Equation (5.57) at all points on the body surface. However, since the derivation of $\partial G/\partial n$ is complicated for an arbitrary body, a closed form solution is not possible and a discretized numerical method is adopted instead – see, for example, Garrison and Chow (1972). The submerged surface of the body is divided into a lattice of facets and Equation (5.57) is satisfied at nodal points on each facet to reduce the integral equation to a finite number of simultaneous linear equations. Taking f_i as the source strength at the centroid (x_i, y_i, z_i) of the ith facet of area δS_i, Equation (5.57) can be written as

$$-f_i + \alpha_{ij}f_j = 2W_i \tag{5.59}$$

where

$$\alpha_{ij} = \frac{1}{2\pi}\sum_{j=1}^{N}\frac{\partial G}{\partial n}(x_i, y_i, z_i; \xi_i, \eta_i, \zeta_i)\,\delta S_i \quad (i \neq j) \tag{5.60}$$

for a total of N facets with the summation excluded for $i = j$. The derivative of G is evaluated using the equation

$$\frac{\partial G}{\partial n} = \frac{\partial G}{\partial x}n_x + \frac{\partial G}{\partial y}n_y + \frac{\partial G}{\partial z}n_z \tag{5.61}$$

where n_x, n_y and n_z are components of the unit normal vector to the facet surface. Once the coefficient matrix α_{ij} is obtained from (5.60), Equation (5.59) is solved for all f_i by matrix inversion and the scattered potentials obtained through the equation

$$\phi'_{si} = \frac{1}{4\pi}f_j\sum_{j=1}^{N}G(x_i, y_i, z_i; \xi_i, \eta_i, \zeta_i)\,\delta S_i \tag{5.62}$$

Then taking the hydrodynamic pressure as

$$p(x, y, z, t) = -\rho\frac{\partial\phi}{\partial t}$$

$$= \mathrm{Re}\{i\rho\omega[\phi_i'(x, y, z) + \phi_s'(x, y, z)]e^{-i\omega t} \qquad (5.63)$$

the total hydrodynamic forces and moments are given by the integrals

$$\left.\begin{aligned}\mathbf{F}(t) &= -\int_S p(x, y, z, t)\mathbf{n}\,\mathrm{dS}\\ &\text{and}\\ \mathbf{M}(t) &= -\int_S p(x, y, z, t)(\mathbf{r}\times\mathbf{n})\,\mathrm{d}S\end{aligned}\right\} \qquad (5.64)$$

where **n** denotes the unit normal to the surface and **r** denotes the moment arm vector. Numerical values are computed by discretizing these equations and using the scattered wave potential from Equation (5.62).

The numerical integrations defined by Equations (5.60) and (5.62) pose some difficulties due to singularities in the Green's function form. However, the integrals are readily evaluated by using numerical integration schemes designed to cope with singularities. Furthermore, for some so-called 'irregular' wave frequencies, the matrix to be inverted for solving Equation (5.59) becomes non-positive definite, and no unique solution of the boundary integral problem is possible. This feature of the solution is not due to the numerical discretization employed but arises inherently from the source distribution representation of the scattered wave potential, see John (1950) and Murphy (1978). Irregular wave frequencies generally correspond to wavelengths which are smaller than the size of the body and are, therefore, usually at frequencies that are too high to be of concern in most wave loading calculations. Boundary integral techniques also suffer from numerical problems when modelling re-entrant structure geometries or structures with small holes or sharp corners – these problems are triggered by numerical problems due to the close proximity of adjacent surface panel sources.

Apart from these specific problem areas, however, boundary integral methods offer a robust and numerically reliable means for calculating wave forces on offshore structures. This has been recognized in the offshore industry through widespread use of diffraction analysis for calculating wave forces on both large and small bodies.

Figure 8.23 shows a typical facet geometry for a floating vessel hull. Figure 5.23 presents a comparison of wave force amplitude for an upright vertical cylinder calculated from the exact solution of MacCamy and Fuchs (1954) and also using the boundary integral method for 120 and 252 facets. The agreement between the two is seen to be reasonably good, although recent improvements in discretization techniques and numerical integration accuracy have enabled virtually exact agreement to be attained with use of fewer facets (see, for example Eatock-Taylor and Waite, 1978).

5.6.3 Boundary element numerical solution

This approach seeks to obtain a solution for the scattered potential ϕ_s' by discretizing the fluid around the submerged body into finite elements in an analogous manner to the use of this technique in elastic problems.

The problem is formulated in numerical terms by defining a functional which may be considered to be related to the total energy of the fluid domain and the work done on it across its boundaries. From the calculus of variations, it can be shown that the solution to this boundary value problem is obtained by minimizing the value of this functional which can be written as

$$\Pi = \int_V \frac{1}{2}\left[\left(\frac{\partial \phi'_s}{\partial x}\right)^2 + \left(\frac{\partial \phi'_s}{\partial y}\right)^2 + \left(\frac{\partial \phi'_s}{\partial z}\right)^2\right] \mathrm{d}V + \int_S \left(\frac{1}{2}\alpha \phi'^2_s + \beta \phi'_s\right) \mathrm{d}S \quad (5.65)$$

where $\alpha = -\omega^2/g$ on the free surface, $\alpha = -\mathrm{i}\omega/c$ on the radiation boundary and zero elsewhere, with $\beta = \partial\phi'_i/\partial n$ on the body surface and zero on all other boundaries. Here $\mathrm{d}V$ is an element of the fluid volume and $\mathrm{d}S$ is an element of the entire fluid boundary including free surface, sea bed, body surface and far fluid boundary. The finite element solution is implemented by dividing the fluid volume into smaller volumes and specifying unknown potentials ϕ_e at nodal points on these element volumes. A typical element geometry is shown in Figure 5.24. The scattered wave potential ϕ'_s within the element can then be written as

$$\phi'_s = (\mathbf{N})^{\mathrm{T}} \phi_e \quad (5.66)$$

where $\mathbf{N}$ is an interpolation function column vector. Substituting (5.66) into (5.65) and minimizing the functional Π leads to the condition

$$\frac{\partial \Pi}{\partial(\phi_e)} = \int_{V_e} \left[\left(\frac{\partial \mathbf{N}}{\partial x}\right)\left(\frac{\partial \mathbf{N}}{\partial x}\right)^{\mathrm{T}} + \left(\frac{\partial \mathbf{N}}{\partial y}\right)\left(\frac{\partial \mathbf{N}}{\partial y}\right)^{\mathrm{T}} + \left(\frac{\partial \mathbf{N}}{\partial z}\right)\left(\frac{\partial \mathbf{N}}{\partial z}\right)^{\mathrm{T}}\right] \phi'_e \, \mathrm{d}V$$
$$+ \int_{S_e} [a(\mathbf{N})(\mathbf{N})^{\mathrm{T}} \phi_e + q(\mathbf{N})] \, \mathrm{d}S = 0 \quad (5.67)$$

where subscript e refers to integration over a single element only. Equation (5.67) is then converted to a discretized form for each element

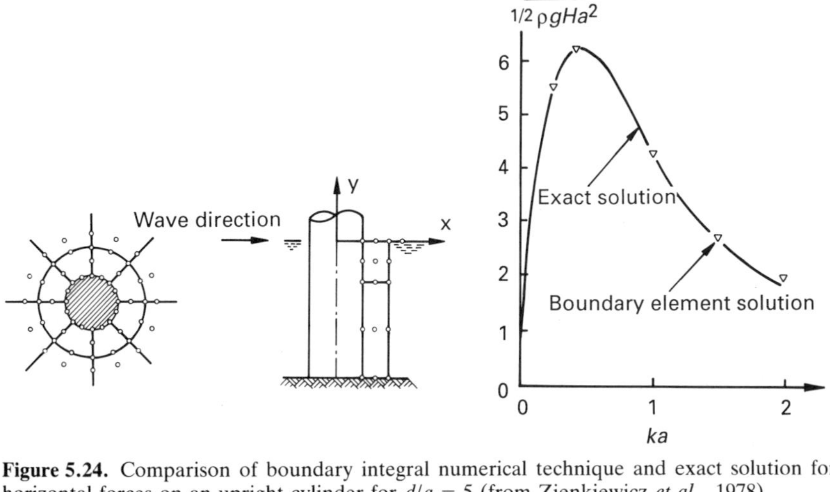

Figure 5.24. Comparison of boundary integral numerical technique and exact solution for horizontal forces on an upright cylinder for $d/a = 5$ (from Zienkiewicz *et al.*, 1978)

and all the element equations are assembled into a global matrix equation of the form

$$\mathbf{A}\Phi_{\mathrm{e}} = \mathbf{B} \tag{5.68}$$

where Φ_{e} is a discretized approximation to function ϕ_{e}. **A** and **B** can be written in integral form as

$$\left.\begin{aligned}
\mathbf{A} &= -\frac{\omega^2}{g}\int_{S_{\mathrm{s}}}(\mathbf{N})(\mathbf{N})^{\mathrm{T}}\,\mathrm{d}S - \frac{\mathrm{i}\omega}{c}\int_{S_{\mathrm{r}}}(\mathbf{N})(\mathbf{N})^{\mathrm{T}}\,\mathrm{d}S \\
&\quad + \int_V\left[\left(\frac{\partial \mathbf{N}}{\partial x}\right)\left(\frac{\partial \mathbf{N}}{\partial x}\right)^{\mathrm{T}} + \left(\frac{\partial \mathbf{N}}{\partial y}\right)\left(\frac{\partial \mathbf{N}}{\partial y}\right)^{\mathrm{T}} + \left(\frac{\partial \mathbf{N}}{\partial z}\right)\left(\frac{\partial \mathbf{N}}{\partial z}\right)^{\mathrm{T}}\right]\mathrm{d}V \\
&\text{and} \\
\mathbf{B} &= -\int_{S_{\mathrm{b}}}\frac{\partial \phi_{\mathrm{i}}'}{\partial n}(\mathbf{N})\,\mathrm{d}S
\end{aligned}\right\} \tag{5.69}$$

where subscripts b, r, and s refer to the body, the radiation surface and the free surface boundaries respectively. Equation (5.68) can be solved by matrix inversion, the scattered potentials recovered by Equation (5.66) and forces and moments calculated through the equations outlined earlier. There is some freedom in applying the technique to the discretization density and element shapes to be used, as well as the type of interpolation functions – these matters are reviewed by Zienkiewicz *et al.* (1978).

One feature of the finite element solution that has to be given special attention is concerned with satisfying the far field radiation condition in theory at an infinite distance from the body. This poses a difficulty in the numerical technique, which has been circumvented by a number of approaches, the simplest of which is to assume that the radiation condition at infinity is applied directly at a radiation boundary at a finite distance from the body. Alternatively, the flows inside and outside this radiation boundary can be matched by using an analytic series solution or a boundary integral solution for the exterior region. A third technique involves the use of infinite elements where the radiation condition can be applied directly. Zienkiewicz *et al.* (1978) reviews these methods and presents a comparison of results. Figure 5.24 presents a typical discretization and comparison of results between the boundary element, and an exact solution for a vertical circular cylinder. It can be seen that the techniques are in good general agreement.

The boundary element numerical technique for the solution of the diffraction problem offers some potential advantage for practical design problems. The technique can take account of variations in water depth around a structure by appropriate choice of boundary elements. The method can also accomodate a wider range of complicated structure geometries and has the potential to incorporate non-linear wave behaviour and coupling between structural elasticity and the dynamics of the surrounding fluid. These gains are, however, obtained at the cost of greater computational cost and ambiguities on the optimum calculation method to be used – the latter is primarily concerned with choice of the far field solution to be used and its matching with the near field boundary elements.

References

Blasius, H. (1963), Grenzschichten in Flussingkeiten mit kleiner Reibung, *Zeitschift Mathematik und Physik*, **56**, 1–37. English translation (1966), NACA TM 1256

Blevins, R. D. (1977), *Flow-induced vibrations*, Van Nostrand Reinhold, USA

Bradshaw, P., Ferris, D. H. and Attwell, N. P. (1967), Calculation of boundary layer development using the turbulent energy equation, *Journal of Fluid Mechanics*, **28**, pt 3, 593–616

Cebeci, T., Chang, K. C. and Keller, K. (1980), A general method for calculating three dimensional laminar and turbulent boundary layers on ship hulls, *Ocean Engineering*, **7**, 229–280

Davenport, A. G. (1961), The application of statistical concepts to the wind loading of structures, *Proceedings of the Institution of Civil Engineers*, Volume **19**, 449–472

Davenport, A. G. (1967), Gust loading factors, *Proceedings of the American Society of Civil Engineers (Structural Division)*, **93**, ST3, June, 11–34

Det Norske Veritas (1982), Rules for submarine pipeline systems, *Det Norske Veritas*, PO Box 300, N-1322, Hovik, Norway

Duncan, W. J., Thom, A. S. and Young, A. D. (1962), *Mechanics of Fluids*, Edward Arnold, London

Eatock-Taylor, R. and Waite, J. B. (1978), The dynamics of offshore structures evaluated by boundary integral techniques. *International Journal for Numerical methods in Engineering*, **13**, 73–92

Engineering Sciences Data Unit, (ESDU) (1986), *Mean Forces, Pressures and Flow Field Velocities for Circular Cylindrical Structures: Single Cylinder with Two-dimensional Flow*, Engineering Sciences Data Unit, Item No. 80025 with Amendment A to C, June, ESDU, 251–259 Regent Street, London

Garrison, C. J. and Chow, P. Y. (1972), Wave forces on submerged bodies, *Journal of Waterways, Harbours and Coastal Division, American Society of Civil Engineers*, **98** (WW3), 375–392

Grass, A. J., Simons, R. R. and Cavanagh, N. J. (1987), Vortex induced velocity enhancements in the wave flow field around cylinders, *Proceedings of the Offshore Mechanics and Arctic Engineering Symposium*, Volume II, American Society of Mechanical Engineers, pp. 155–164

Havelock, T. H. (1940), The pressure of water waves upon a fixed obstacle, *Proceedings of the Royal Society, London, Series A*, **963**, 175–190

Hoerner, S. F. (1965), *Fluid Dynamic Drag*, Published by the author, N York City, USA

John, F. (1950), On the motion of floating bodies II, *Communications in Pure and Applied Mathematics*, **3** 45–101

Kaplan, P. and Silbert, M. N. (1976), Impact forces on platform horizontal members in the splash zone *Proceedings of the Offshore Technology Conference*, Paper No. OTC 2498

Keulegan, G. H. and Carpenter, L. H. (1958), Forces on cylinders and plates in an oscillating fluid, *Journal of Research of the National Bureau of Standards*, May, **60** (5)

Lamb, H. (1975), *Hydrodynamics*, 6th Edn, Cambridge University Press, Cambridge

MacCamy, R. C. and Fuchs, R. A. (1954), *Wave Forces on Piles, A Diffraction Theory*, US Army Corps of Engineers, Beach Erosion Board, Tech. Memo, No. 69

Miller, B. L. (1980), *Wave Slamming on Offshore Structures*, National Maritime Institute, Report No. NMI-R81, OT-R-8041

Miller, B. L. and Davies, M. E. (1982), *Wind Loading of Offshore Structures – A Summary of Wind Tunnel Studies*, National Maritime Insititute, Report No. NMI R136, OT-R-8225, July

Morison, J. R., O'Brien, M. P., Johnson, J. W. and Schaaf, S. A. (1950), The forces exerted by surface waves on piles, *Petroleum Transactions*, **189** (TP 2846), 149

Murphy, J. E. (1978), Integral equation failure in wave calculations, *Journal of Waterways, Port, Coastal and Ocean Division*, **104** (WW4), 330–334

Sarpkaya, T. and Isaacson, M. (1981), *Mechanics of Wave Forces on Offshore Structures*, Van Nostrand Reinhold, NY, USA

Sarpkaya, T. (1976), *In Line and Transverse Forces on Smooth and Sand Roughened Cylinders in Oscillatory Flow at High Reynolds Numbers*, Report No NPS-69SL76062, Naval Postgraduate School, Monterey, California, USA

Sommerfield, A. (1949), *Partial differential equations in physics*, Academic Press, NY, USA

Smith, A. M. O. and Clutter, D. W. (1963), Solution of the incompressible boundary layer equations, *American Institute of Aeronautics and Astronautics Journal*, **1**, 2062–2071

Stoker, J. J. (1957), *Water Waves*, Interscience, NY, USA

Thom, H. C. S. (1973), Distribution of extreme winds over the oceans, *American Society of Civil Engineers, WW1*, **99**, Feb., 1–17

von Karman, T. L. and Wattendorf, B. F. (1929), *The Impact on Seaplane Floats During Landing*, NACA TN-321

Wehausen, J. V. and Laitone, E. V. (1960), Surface waves, *Handbuch der Physik*, ed. S Flugge, Springer-Verlag, Berlin, Volume IX, pp. 446–778 (p478)

Yamamoto, T., Nath, J. H. and Slotta, L. S. (1974), Wave forces on cylinders near plane boundary, *Journal of Waterways and Harbours Division*, **100**, (WW4), 345–359

Zienkiewicz, O. C., Bettes, P. and Kelly, D. W. (1978), The finite element method of determining fluid loading on rigid structures – two and three dimensional formulations. In *Numerical Methods in Offshore Engineering* (eds. O. C. Zienkiewicz, P. Lewis and K. G. Stagg), John Wiley, Chichester, England, pp. 141–183.

Chapter 6

Structural response

This chapter describes the final stages of the design process for an offshore structure where static and dynamic environmental forces calculated from the methods outlined in earlier chapters are converted to structural stresses and stress load cycles. These are then used to ensure sufficient structural strength and to estimate fatigue lives. Static and dynamic structural analyses of offshore structures are complicated by non-linear effects due predominantly to soil structure interaction in the former case and non-linearities arising from gravity wave mechanics and drag force loading in the latter. These methods of analysis and the effects of non-linear behaviour are considered in detail in this section.

6.1 Quasistatic loading and response

6.1.1 Methods of analysis

Steel framed jackets in shallow water can be analysed for structural stresses with acceptable accuracy by using a quasistatic environmental loading coupled with an elastic structural analysis modified for non-linear behaviour in the soil foundation. In this approach, wind and current are assumed to apply static loads with wave action applying a dynamic loading which is translated into dynamic structural stresses through a quasistatic stress analysis. The technique assumes that resonant frequencies of structural vibrations are sufficiently separated from wave frequencies so that dynamic magnification has a negligible effect on calculated stresses.

In the early days (before the 1960s), structural analysis of jackets was carried out by hand calculations using simple frame theory. However, the advent of computers and the development of finite element structural analysis methods has revolutionized structural calculations in the offshore industry. Finite element analysis program packages such as NASTRAN and STRUDL are used in conjunction with wave loading and foundation analysis programs to carry out complex computer aided design calculations for three-dimensional steel framed structures as well as for monolithic concrete structures.

The basic theory underlying one such finite element analysis, called the direct stiffness method, is described here. The structure to be analysed is discretized into a number of small elements – with a framed jacket structure represented by an assembly of beam elements, whereas a concrete gravity structure could also be described by beam elements or, alternatively, by cylindrical shell elements. The displacements of nodal points on the elements are taken as unknowns and the finite element method is implemented using the following steps:

1. Initially, physical data for the structure configuration, member properties and support constraints are identified. The structure is then discretized into finite elements and the unknown nodal displacements are identified and numbered using a systematic indexing procedure.
2. This is followed by evaluation of member stiffness matrices using local member axes. The member stiffness matrices are then transformed to the global structure axes and assembled into the global stiffness matrix.
3. The loads acting on the members and nodes are also assembled into a global load vector. This requires that forces acting on each member other than at the nodes are transformed into equivalent joint loads before being added on to the known loads at the nodes such that the structure is considered as being loaded only at the nodes.
4. It is then necessary to rearrange the equation relating the nodal displacement vector and stiffness matrix to the load vector in order to substructure out degrees of freedom which are constrained (that is, their nodal displacements are known or are zero) in order that a matrix equation for the unknown displacements only can be formed.
5. The penultimate step in the procedure is to use the stiffness matrix and load vector to arrive at a solution for the displacements at the nodes.
6. In the final step, these nodal displacements are used in combination with the equivalent joint loads to compute member forces and resultant internal stresses.

The mathematical formulation of the above steps is described below for a three-dimensional (space frame) beam element idealization of a typical jacket structure shown in Figure 6.1. The jacket structure consists of up to 600 beam elements to represent each of the tubular members making up the frame. A global axes system $Oxyz$ is first defined with respect to the whole structure and all the structural nodes are numbered in a systematic manner around the structure. A typical numbering system could, as shown in Figure 6.1, start from a horizontal level at the origin with increasing node numbering in a consistent direction around the nodes at that level, followed by a similar procedure at the next level down, and so on. Once this node definition is complete, structural data on nodal coordinates relative to global axes, member lengths, areas of cross-section and second moments of area can be systematically identified in terms of the node numbering. This is followed by definition of six degrees of freedom for each node – three in translation and three in rotation – parallel to and about the global axes directions Ox, Oy and Oz. Thus, the total number of degrees of freedom will be six times the number of nodes, although a small proportion of these freedoms (at the base of the jacket of Figure 6.1,

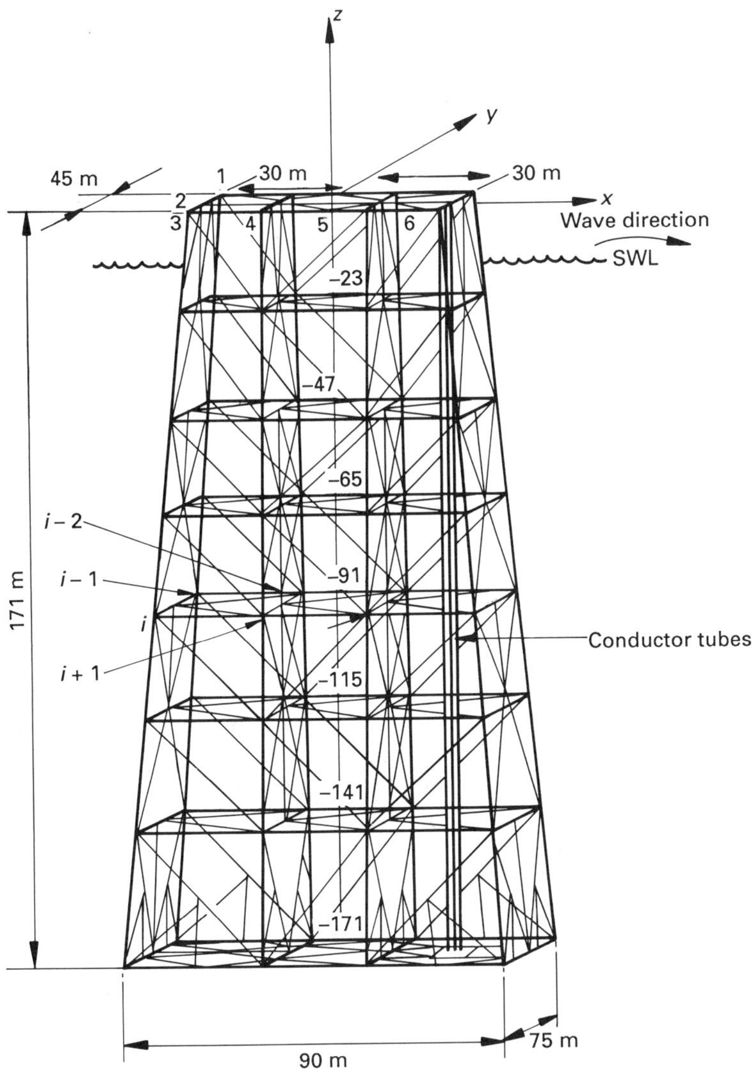

Figure 6.1. Typical idealization of a jacket structure

for example) will be constrained or given a prescribed displacement to account for the effect of the foundation.

The next step in the analysis is to determine the stiffnesses of each member in the frame work – this being done in terms of principal member axes $O_m x_m y_m z_m$ as defined in Figure 6.2. The single beam element in this diagram has 12 degrees of freedom, three translations and three rotations along and about member axes' directions at each end of the members. These are shown in Figure 6.2 with the rotational degrees of freedom denoted by double headed arrows showing the direction that the right-hand screw rule axis would point to generate the rotation. These degrees of freedom are systematically numbered by indices 1–12 with, for example, index 3 denoting displacement of the member at end j parallel to the $O_m z_m$ axis and index 6 denoting clockwise rotation at end j about the positive $O_m z_m$ axis.

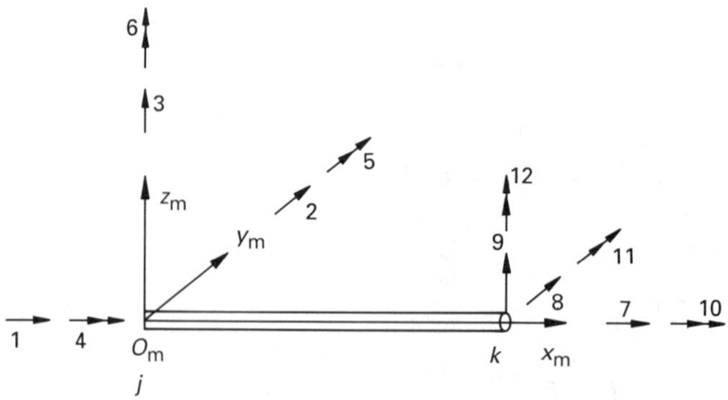

Figure 6.2. Three-dimensional beam element

Now a stiffness matrix $\mathbf{S}_{mi}$ for the ith frame member can be defined to be

$$\mathbf{F}_{mi} = \mathbf{S}_{mi}\mathbf{D}_{mi} \tag{6.1}$$

where $\mathbf{D}_{mi}$ is the vector of 12 displacements (1–12 in Figure 6.2), $\mathbf{F}_{mi}$ is a 12×1 vector of forces along these index directions and $\mathbf{S}_{mi}$ (corresponding to the 12 possible member end displacements) are obtained by defining the actions (or forces) required to obtain unit translation or rotation for each degree of freedom with all other degrees of freedom restrained to zero displacement or rotation. This is illustrated diagrammatically in Figure 6.3. Figure 6.3(a) demonstrates how a unit displacement along degree of freedom 1 requires a force of EA_x/L along 1 and will also result in a force of $-EA_x/L$ along degree of freedom 7 where E, A_x and L are the member Young's modulus, area of cross-section and length respectively. Similarly, unit rotation about the $O_m y_m$ axes only (index 5) leads to forces $-6EI_y/L^2$ and $+6EI_y/L^2$ along indices 3 and 9 as well as moments $4EI_y/L$ and $2Ei_y/L$ along indices 5 and 11, respectively, as shown in Figure 6.3(e). Similar diagrams can be obtained for unit displacements in the degrees of freedom corresponding to indices 7–12 in Figure 6.2.

These effects can be summarized into the 12×12 member stiffness matrix $\mathbf{S}_{mi}$ given in Table 6.1. The matrix rows and columns are numbered to enable a particular stiffness term to be identified with respect to the index numbering of displacements and forces. The matrix is symmetrical and can be partitioned as shown to separate the effects of each set of six displacements on themselves and on each other.

In order to combine the stiffness matrices of all frame members for a global structural analysis, the stiffness matrix for each member needs to be transformed from member to global axes. The transformation is illustrated by Figure 6.4, with generalized member axes $O_G x_G y_G z_G$ for member i with end nodes j and k and global axes $Oxyz$. It is possible that the $O_G x_G y_G z_G$ axes may not be principal axes of the member

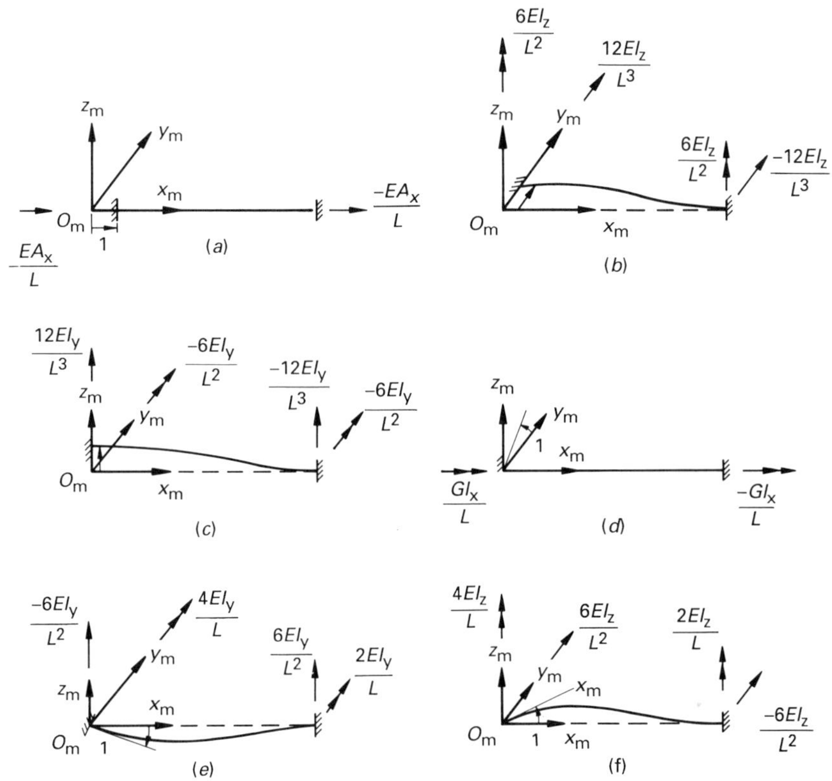

Figure 6.3. Forces and moments required for unit displacement indices 1 to 6 on beam element: L beam length, A_x area of cross-section; I_x, I_y, I_z second moments of area of cross-section about $O_m x_m$ $O_m y_m$ and $O_m z_m$ axes through centroid of cross-section; E Young's modulus of beam material

cross-section and that a rotation of angle θ (clockwise about the positive $O_G x_G$ axis) may need to be applied to transform generalized member axes to principal axes $O_m x_m y_m z_m$. Figure 6.4 also illustrates how axes $O_m x_m y_m z_m$ can be transformed so that they are parallel to axes $Oxyz$ by two rotations – through angle ϕ to make axis $O_m x_m$ parallel to the Oxy plane and then through angle ψ to make $O_m x_m$ parallel to Ox.

Now the rotation of each axis through a prescribed angle can readily be defined using vector algebra. Thus, the rotation θ to transform generalized member axes to principal axes can be written as

$$\mathbf{R}_\theta = \begin{bmatrix} 1 & 0 & 0 \\ 0 & \cos\theta & \sin\theta \\ 0 & -\sin\theta & \cos\theta \end{bmatrix} \tag{6.2}$$

and the transformations required to rotate the $O_m x_m$ axes first parallel to the global Oxy plane and then parallel to the global Ox axis can be shown to be

$$\mathbf{R}_\phi = \begin{bmatrix} C_{xz} & C_y & 0 \\ -C_y & C_{xz} & 0 \\ 0 & 0 & 1 \end{bmatrix} \tag{6.3}$$

Table 6.1 Member stiffness matrix

$$
\mathbf{S}_{Mi} = \begin{array}{c} \\ 1 \\ 2 \\ 3 \\ 4 \\ 5 \\ 6 \\ 7 \\ 8 \\ 9 \\ 10 \\ 11 \\ 12 \end{array}
\left[\begin{array}{cccccc:cccccc}
1 & 2 & 3 & 4 & 5 & 6 & 7 & 8 & 9 & 10 & 11 & 12 \\
\frac{EA_x}{L} & 0 & 0 & 0 & 0 & 0 & -\frac{EA_x}{L} & 0 & 0 & 0 & 0 & 0 \\
0 & \frac{12EI_z}{L^3} & 0 & 0 & 0 & \frac{6EI_z}{L^2} & 0 & -\frac{12EI_z}{L^3} & 0 & 0 & 0 & \frac{6EI_z}{L^2} \\
0 & 0 & \frac{12EI_y}{L^3} & 0 & -\frac{6EI_y}{L^2} & 0 & 0 & 0 & -\frac{12EI_y}{L^3} & 0 & -\frac{6EI_y}{L^2} & 0 \\
0 & 0 & 0 & \frac{GI_x}{L} & 0 & 0 & 0 & 0 & 0 & -\frac{GI_x}{L} & 0 & 0 \\
0 & 0 & -\frac{6EI_y}{L^2} & 0 & \frac{4EI_y}{L} & 0 & 0 & 0 & \frac{6EI_y}{L^2} & 0 & \frac{2EI_y}{L} & 0 \\
0 & \frac{6EI_z}{L^2} & 0 & 0 & 0 & \frac{4EI_z}{L} & 0 & -\frac{6EI_z}{L^2} & 0 & 0 & 0 & \frac{2EI_z}{L} \\
\hdashline
-\frac{EA_x}{L} & 0 & 0 & 0 & 0 & 0 & \frac{EA_x}{L} & 0 & 0 & 0 & 0 & 0 \\
0 & -\frac{12EI_z}{L^3} & 0 & 0 & 0 & -\frac{6EI_z}{L^2} & 0 & \frac{12EI_z}{L^3} & 0 & 0 & 0 & -\frac{6EI_z}{L^2} \\
0 & 0 & -\frac{12EI_y}{L^3} & 0 & \frac{6EI_y}{L^2} & 0 & 0 & 0 & \frac{12EI_y}{L^3} & 0 & \frac{6EI_y}{L^2} & 0 \\
0 & 0 & 0 & -\frac{GI_x}{L} & 0 & 0 & 0 & 0 & 0 & \frac{GI_x}{L} & 0 & 0 \\
0 & 0 & -\frac{6EI_y}{L^2} & 0 & \frac{2EI_y}{L} & 0 & 0 & 0 & \frac{6EI_y}{L^2} & 0 & \frac{4EI_y}{L} & 0 \\
0 & \frac{6EI_z}{L^2} & 0 & 0 & 0 & \frac{2EI_z}{L} & 0 & -\frac{6EI_z}{L^2} & 0 & 0 & 0 & \frac{4EI_z}{L}
\end{array}\right]
$$

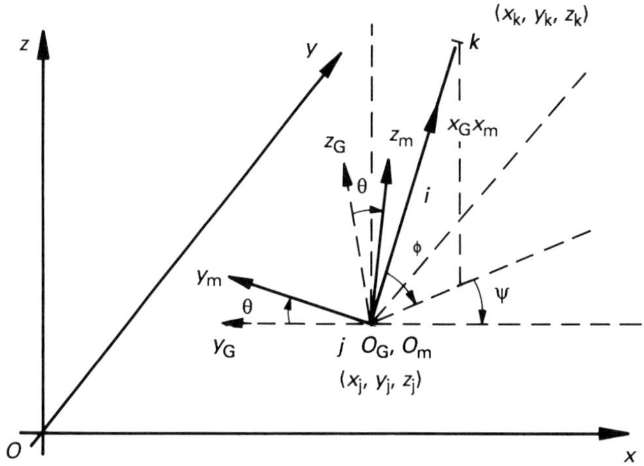

Figure 6.4. Member and global axes

and

$$\mathbf{R}_{\psi} = \begin{bmatrix} \dfrac{C_x}{C_{xz}} & 0 & \dfrac{C_z}{C_{xz}} \\ 0 & 1 & 0 \\ -\dfrac{C_z}{C_{xz}} & 0 & \dfrac{C_x}{C_{xz}} \end{bmatrix} \tag{6.4}$$

respectively, where

$$\left.\begin{aligned} &C_x = \frac{x_k - x_j}{L},\ C_y = \frac{y_k - y_j}{L},\ C_z = \frac{z_k - z_j}{L} \\ &C_{xz} = \sqrt{(C_x^2 + C_z^2)} \\ &\text{and} \\ &L = \sqrt{[(x_k - x_j)^2 + (y_k - y_j)^2 + (z_k - z_j)^2]} \end{aligned}\right\} \tag{6.5}$$

The three successive rotations required for general transformation of axes can then be written as

$$\mathbf{R} = \mathbf{R}_{\theta}\mathbf{R}_{\phi}\mathbf{R}_{\psi}$$

$$= \begin{bmatrix} C_x & C_y & C_z \\ \dfrac{-C_xC_y\cos\alpha - C_z\sin\alpha}{C_{xz}} & C_{xz}\cos\alpha & \dfrac{-C_yC_z\cos\alpha + C_x\sin\alpha}{C_{xz}} \\ \dfrac{C_xC_y\sin\alpha - C_z\cos\alpha}{C_{xz}} & -C_{xz}\sin\alpha & \dfrac{C_yC_z\sin\alpha + C_x\cos\alpha}{C_{xz}} \end{bmatrix} \tag{6.6}$$

and applied to the four sets of three degrees of freedom for each member (see Figure 6.2) by the total rotation matrix

$$\mathbf{R}_{\mathrm{T}} = \begin{bmatrix} \mathbf{R} & \mathbf{0} & \mathbf{0} & \mathbf{0} \\ \mathbf{0} & \mathbf{R} & \mathbf{0} & \mathbf{0} \\ \mathbf{0} & \mathbf{0} & \mathbf{R} & \mathbf{0} \\ \mathbf{0} & \mathbf{0} & \mathbf{0} & \mathbf{R} \end{bmatrix} \tag{6.7}$$

The member stiffness matrix $\mathbf{S}_{\mathrm{m}i}$ for member i can be transformed to the stiffness for global structure axis $\mathbf{S}_{\mathrm{m}i}$ by the matrix equation

$$\mathbf{S}_{\mathrm{ms}i} = \mathbf{R}_{\mathrm{T}i}^{\mathrm{T}}\mathbf{S}_{\mathrm{m}i}\mathbf{R}_{\mathrm{T}i} \tag{6.8}$$

It is usual during or just after this transformation to assemble the full global stiffness matrix from its constituent member matrices which are now in relation to global axes. This assembly process can be illustrated using the matrix equation

$$\begin{bmatrix} f_1 \\ f_2 \\ \hline f_i \\ \hline f_m \end{bmatrix}$$

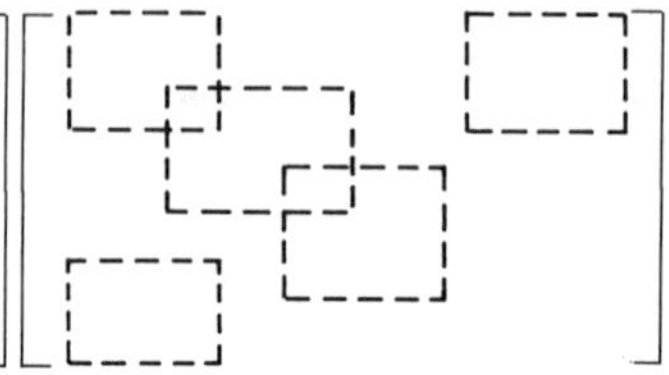

$$\begin{bmatrix} d_1 \\ d_2 \\ \hline d_i \\ \hline d_m \end{bmatrix} \tag{6.9}$$

where f_i and d_i are the vectors of forces and displacements along the m degrees of freedom of the total structure with f_{6j-5} to f_{6j} containing the forces and moments at the jth node and d_{6j-5} to d_{6j} containing the resultant displacements and rotations. The stiffness matrix of each member is then assembled within the appropriate row and column range corresponding to its degrees of freedom with overlapping terms summed to yield the assembled joint stiffness matrix. In a space frame analysis, the upper and lower 6×6 partition of each 12×12 member stiffness matrix will overlap and be added to the matrices of other adjacent members connected to the frame at that node.

In the direct stiffness method, two fundamental equations govern the structure reaction forces and member internal and applied forces. These equations can be obtained by considering the response of the structure to variations in the support and free joint dsplacements. The column vector of support reaction forces $\mathbf{A}_{\mathrm{R}}$ can be written as

$$\mathbf{A}_{\mathrm{R}} = -\mathbf{A}_{\mathrm{RC}} + \mathbf{S}_{\mathrm{RF}}\mathbf{D}_{\mathrm{F}} \tag{6.10}$$

where $\mathbf{A}_{\mathrm{RC}}$ is the column vector of actual or equivalent joint loads applied directly to the supports and $\mathbf{S}_{\mathrm{RF}}$ is the global stiffness matrix relating the forces at restrained degrees of freedom to displacements $\mathbf{D}_{\mathrm{F}}$ at the free degrees of freedom.

The column vector of internal forces for member i, $\mathbf{A}_{\mathrm{m}i}$ can also be written as

$$\mathbf{A}_{\mathrm{m}i} = \mathbf{A}_{\mathrm{mL}i} + \mathbf{S}_{\mathrm{m}i}\mathbf{D}_{\mathrm{m}i} \tag{6.11}$$

with all terms applying to member axes for each member i. Here $\mathbf{A}_{\mathrm{mL}i}$ is the column vector of fixed end actions, $\mathbf{S}_{\mathrm{m}i}$ and $\mathbf{D}_{\mathrm{m}i}$ are row i of the stiffness matrix and the column of member end displacements respectively.

It now becomes necessary to define loads acting on the structure in order to define an applied force vector which is made up of forces on the member (between its ends) and forces acting at the ends or nodes. Matrix methods of structural analysis require that the forces acting on the members be converted first to forces at the fixed ends or equivalent joint loads acting on the member ends which are used as nodes in the finite element analysis method.

These fixed end forces or equivalent joint loads are first calculated in member axes before being transformed to structure axes by the transformations described above.

A typical beam element can be subjected to concentrated loads as well as uniformly or non-uniformly distributed loads along its length. These can readily be translated into forces at the beam fixed ends by using equilibrium considerations. Table 6.2 presents fixed end forces or actions for typical loadings. The fixed end actions due to a linearly distributed force $w(x)$, given in the last row of Table 6.2, is often used in jacket structural analysis.

The fixed end actions in member axes for the ith member are written in a matrix $\mathbf{A}_{\mathrm{mL}}$ of order $12 \times m$ in which the ith column gives the vector of 12 forces and moments at the ends of the ith member – m, being the total number of degrees of freedom of the structure. The elements of matrix $\mathbf{A}_{\mathrm{mL}}$ are transformed to the equivalent joint load vector $\mathbf{A}_{\mathrm{E}}$ by first transforming $\mathbf{A}_{\mathrm{mL}}$ into global axes through the rotational transformation

$$\mathbf{A}_{\mathrm{ms}i} = \mathbf{R}_{\mathrm{T}i}^{\mathrm{T}} \mathbf{A}_{\mathrm{mL}i} \tag{6.12}$$

and then incorporating the negative of the vector $\mathbf{A}_{\mathrm{ms}i}$ into an assembly of the equivalent joint load vector $\mathbf{A}_{\mathrm{E}}$. The negative sign arises because the fixed end actions and equivalent joint loads at the member ends are in equilibrium so that the equivalent joint loads can simply be taken as the negative of the fixed end actions.

This assembly operation is similar to that shown for the stiffness matrix in Equation (6.9) with the square member stiffness matrices replaced by column vectors of member forces. Finally, the column vector $\mathbf{A}_{\mathrm{J}}$, of forces applied directly at each of the joints is then summed with $\mathbf{A}_{\mathrm{E}}$ to obtain the total combined load vector

$$\mathbf{A}_{\mathrm{C}} = \mathbf{A}_{\mathrm{E}} + \mathbf{A}_{\mathrm{J}} \tag{6.13}$$

The stiffness matrix $\mathbf{S}_{\mathrm{m}}$ and load vector $\mathbf{A}_{\mathrm{C}}$ are now known in the matrix equation

$$\mathbf{S}_{\mathrm{M}} \mathbf{D}_{\mathrm{M}} = \mathbf{A}_{\mathrm{C}} \tag{6.14}$$

Prior to solution of this equation, however, the matrix equation corresponding to the free and restrained degrees of freedom needs to be separated out. If the free and restrained degrees of freedom are denoted by suffixes F and R respectively, Equation (6.14) becomes

$$\left[\begin{array}{c|c} \mathbf{S}_{\mathrm{FF}} & \mathbf{S}_{\mathrm{FR}} \\ \hline \mathbf{A}_{\mathrm{RF}} & \mathbf{S}_{\mathrm{RR}} \end{array}\right] \begin{bmatrix} \mathbf{D}_{\mathrm{F}} \\ \mathbf{D}_{\mathrm{R}} \end{bmatrix} = \begin{bmatrix} \mathbf{A}_{\mathrm{F}} \\ \mathbf{A}_{\mathrm{R}} \end{bmatrix} \tag{6.15}$$

or

$$\mathbf{S}_{\mathrm{FF}} \mathbf{D}_{\mathrm{F}} + \mathbf{S}_{\mathrm{FR}} \mathbf{D}_{\mathrm{R}} = \mathbf{A}_{\mathrm{F}} \tag{6.16}$$

$$\mathbf{S}_{\mathrm{RF}} \mathbf{D}_{\mathrm{F}} + \mathbf{S}_{\mathrm{RR}} \mathbf{D}_{\mathrm{R}} = \mathbf{A}_{\mathrm{R}} \tag{6.17}$$

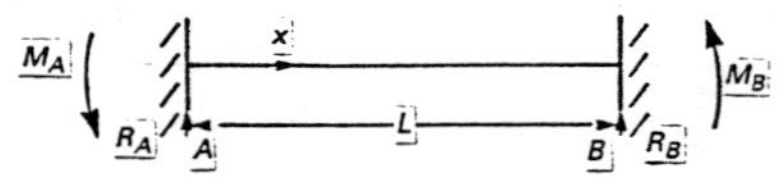

Table 6.2 Fixed end forces and moments on beams for typical loading conditions

Loading	*Forces and moments at fixed ends on beam element*			
	R_A	R_B	M_A	M_B
Concentrated lateral force P downwards at a from left-hand end and b from right-hand end	$\frac{Pb^2(3a + b)}{L^3}$	$\frac{Pa^2(a + 3b)}{L^3}$	$\frac{Pab^2}{L^2}$	$-\frac{Pa^2b}{L^2}$
Concentrated moment M with axis at right-angles to beam at a from left-hand end and b from right-hand end	$\frac{6Mab}{L^3}$	$-\frac{6Mab}{L^3}$	$\frac{Mb}{L^2}(2a - b)$	$\frac{Ma}{L^2}(2b - a)$
Uniformly distributed lateral force, w over a length a of the left-hand end	$\frac{wa}{3L^3}(2L^3 - 2a^2L + a^3)$	$\frac{wa^3}{2L^3} \cdot (2L - a)$	$\frac{wa^2}{12L^2}(6L^2 - 8aL + 3a^2)$	$-\frac{wa^3}{12L^2}(4L - 3a)$
Distributed lateral force of intensity w_0x/L	$\frac{3w_0L}{20}$	$\frac{7w_0L}{20}$	$\frac{w_0L^2}{30}$	$-\frac{w_0L^2}{20}$
Linearly distributed lateral force of intensity $w(x) = w_1 + x(w_2 - w_1)/L$	$\frac{L(7w_1 + 3w_2)}{20}$	$\frac{L(3w_1 + 7w_2)}{20}$	$\frac{L^2(3w_1 + 2w_2)}{60}$	$-\frac{L^2(2w_1 + 3w_2)}{60}$

Here only the free joint displacements $\mathbf{D}_F$ and the support reaction $\mathbf{A}_R$ are unknowns, whereas the restrained support displacements will be known or will be zero.

In computer based implementation of this technique, equations for the free joint displacements are separated out from those for the restrained displacements by rearranging the rows and columns of the stiffness matrices and the order of elements in the displacement and load column vectors in a consistent manner such that Equations (6.16) and (6.17) can be separated out. Equation (6.16) can then be solved for $\mathbf{D}_F$ using matrix inversion

$$\mathbf{D}_F = \mathbf{S}_{FF}^{-1} (\mathbf{A}_F - \mathbf{S}_{FR}\mathbf{D}_R) \tag{6.18}$$

The unknown support reactions can also be calculated by resubstituting $\mathbf{D}_F$ into the equation

$$\mathbf{A}_R = -\mathbf{A}_{RC} + \mathbf{S}_{RF}\mathbf{D}_F + \mathbf{S}_{RR}\mathbf{D}_R \tag{6.19}$$

where $\mathbf{A}_{RC}$ is the column vector of actual and equivalent joint loads applied directly to the supports, and the restrained displacements $\mathbf{D}_R$ are known and usually zero.

Internal member forces (or member end actions) $\mathbf{A}_m$ are then recovered by using the combined vector $\mathbf{D}_m$ in the equation

$$\mathbf{A}_{mi} = \mathbf{A}_{mLi} + \mathbf{S}_{mi}\mathbf{R}_{Ti}\mathbf{D}_{mi} \tag{6.20}$$

for each member i. The end forces and bending moments in $\mathbf{A}_{mi}$ are then readily converted to a distribution of stresses within the member.

It is useful at this stage to give an indication of the way in which the above equations are used to permit easier programming of the direct stiffness method for solution by a digital computer. Consider a space frame made up of n_j joints with each joint being analysed for six degrees of freedom – three translations and three rotations. If n_r degrees of freedom of the structure are restrained, the number of free degrees of freedom becomes

$$n = 6n_j - n_r \tag{6.21}$$

For the jth joint, the degrees of freedom corresponding to translation along the x, y and z directions can be systematically numbered as $6j - 5$, $6j - 4$ and $6j - 3$, respectively, with rotations about the x, y and z axis numbered as $6j - 2$, $6j - 1$ and $6j$ respectively.

Thus a member i with joint numbers j and k at its ends will have end displacements $j1$ to $j6$ and $k1$ to $k6$, as shown in Figure 6.5 and indexed as follows:

$$\left.\begin{aligned}
j1 &= 6j - 5 & k1 &= 6k - 5\\
j2 &= 6j - 4 & k2 &= 6k - 4\\
j3 &= 6j - 3 & k3 &= 6k - 3\\
j4 &= 6j - 2 & k4 &= 6k - 2\\
j5 &= 6j - 1 & k5 &= 6k - 1\\
j6 &= 6j & k6 &= 6k
\end{aligned}\right\} \tag{6.22}$$

The above indexing system is used with arrays within computer programs to simplify the extensive matrix manipulation that is required to solve for displacements. A typical example is given by the manner in which the first column of the stiffness matrix of member i in global axes contributes to the global joint stiffness matrix. This contribution is described by the equations

$$\left.\begin{aligned} (\mathbf{S}_{\mathrm{J}})j_n j_1 &= \sum \mathbf{S}_{\mathrm{ms}} + (\mathbf{S}_{\mathrm{ms}_{n,1}})i \\ (\mathbf{S}_{\mathrm{J}})k_n j_1 &= (\mathbf{S}_{m_{n+6;1i}}) \\ \text{for } n &= 1, 2, 3, 4, 5 \text{ and } 6 \end{aligned}\right\} \quad (6.23)$$

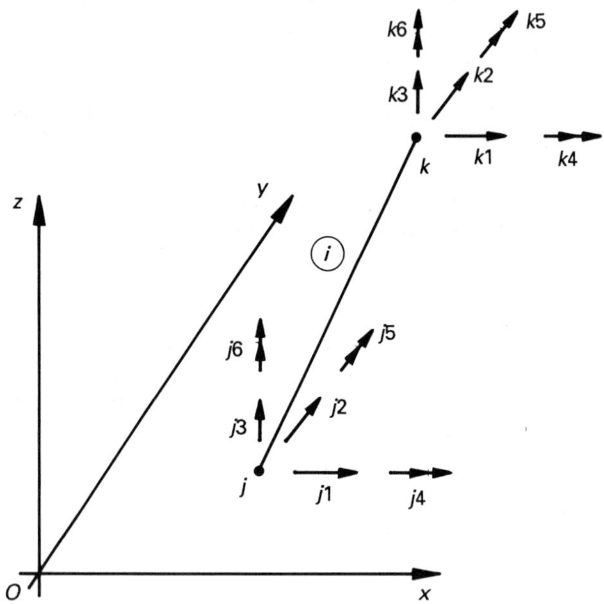

Figure 6.5. Index notation for computer implementation of the direct stiffness method: ⓘ denotes member number i

The quasi-static structural finite element analysis described above is implemented in the design through the following steps for a steel framed jacket structure:

1. The geometry of the structure is first defined with respect to a three-dimensional co-ordinate axis system with the beam elements, and their nodes and nodal displacements numbered in a consistent manner. Initial values of member diameter and wall thickness are defined, usually from experience or from a preliminary design study to examine the properties of different structural configurations. If the cross-sectional properties of a structural member vary significantly between its two ends, then it is modelled by using more than one beam element. Local joint reinforcements are normally disregarded,

although more detailed local analysis and inclusion of eccentric connection detail is worthwhile at joints between large diameter stiffened members.

2. The loads experienced by the jacket structure are then defined. These consist of the various usually vertical equipment loads from the platform top sides and lateral environmental loads from wind, current and waves. Advanced versions of finite element analysis packages for offshore applications are capable of calling subprograms which compute the loading due to current and waves on individual cylindrical members – with the latter usually done using Stoke's fifth-order wave theory. Wind force values are normally input directly. The analysis programs can automatically compute member self-weights and the buoyancy forces due to flooded or non-flooded members. Current forces are calculated by summing wave and current velocities vectorially before applying the Morison equation force formulation defined by Equations (5.28) or (5.29).
3. The soil structure interaction has a strong effect on the structural response of a piled jacket. This interaction is modelled within the finite element analysis by replacing each pile or closely spaced pile group by a substructure and using an iterative procedure coupled with *p–y* data (see Section 6.1.2) to quantify the soil reaction. The iteration is carried out by calculating jacket and substructure forces, calculating pile displacements and hence soil reactions from the *p–y* data and using the corrected jacket foundation forces to repeat the procedure until the iteration converges. A number of different iteration schemes can be used to calculate the combined structure and foundation loading. Section 6.1.2 gives more details of the pile and soil interaction analysis.
4. The finite element analysis is then carried out by stepping the design wave through the structure with small increments of wave phase (or time). Calculations are carried out for several wave periods and for simultaneous wind and current in the wave direction, as well as for different combinations of top sides loads and foundation soil properties. For each loading condition, the analysis will yield maximum shear force and overturning moment at the base of the structure, together with maximum end forces, moments, joint displacements and rotations for each member. Some programs will also give external support reactions and carry out force and moment equilibrium checks at each joint. In some cases, only the wave headings giving the highest base shear and overturning moment are analysed to reduce analysis cost.
5. The member forces are then used to compute member axial, bending, combined axial and bending, tension and shear stresses. It is common practice to use the analysis program to carry out an automatic check of these stress levels against allowable values recommended by a specified certifying authority.

The design procedure outlined above is then run repeatedly to optimize the jacket design for the number of piles and detailed member configuration. Final values for member wall thicknesses are only defined in the later stages of the detailed stress calçulations.

Application of the above technique to gravity structures differs in a number of ways. The structural analysis can be carried out through a simple beam element model using a typical idealization of the type shown in Figure 6.6 or, alternatively, more representative cylindrical shell elements can be used to model the caisson structure and vertical walls. In both cases, it is necessary to include some idealization of the foundation elasticity, either by equivalent springs as in Figure 6.6 or by continuing the finite element idealization into the foundation using solid elements as described by Penzien and Tseng (1978).

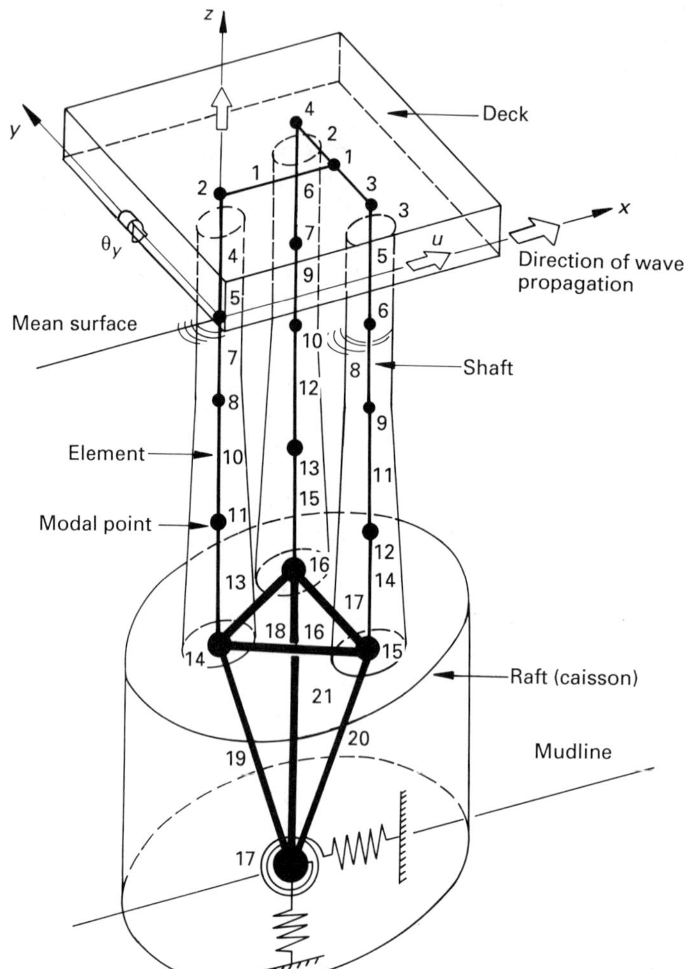

Figure 6.6. Typical simple idealization of gravity structure. Numbers on the diagram are finite element idealization node and member numbers (from Bell *et al.*, 1976)

Furthermore, wave forces are calculated using diffraction theory only or a combined Morison–diffraction approach. It is usual to calculate the forces separately and to define them as a distribution of equivalent nodal loads for input to the analysis. Since the structural deflections of these relatively rigid structures are very small compared to water particle displacements, the non-linear drag force in the Morison equation (Equation (5.28) or (5.29)) can be explicitly computed and does not present analysis problems.

6.1.2 Foundation effects

For piled jackets, the structural calculations described above have to be supplemented by a lateral pile loading analysis which quantifies the coupled response of an embedded pile and non-linear material behaviour of the surrounding soil. When a single pile is loaded by a horizontal force due to environmental loading of the jacket structure of which it is a part, the pile will deflect against the surrounding soil and take up a typical deflection and bending moment profile shown in Figure 6.7. The magnitude of deflection, rotation and bending moment at the end of the pile will depend on a balance of forces between the pile and surrounding soil. Determination of this behaviour requires data on the lateral soil reaction on the pile as a function of depth, together with a means of structural analysis of the pile itself due to a varying soil loading. Taking the former first, the soil reaction is a function of soil type, pile properties, its loading, deflection and the soil depth and may be expressed as

$$p = -E(x, y)y \qquad (6.24)$$

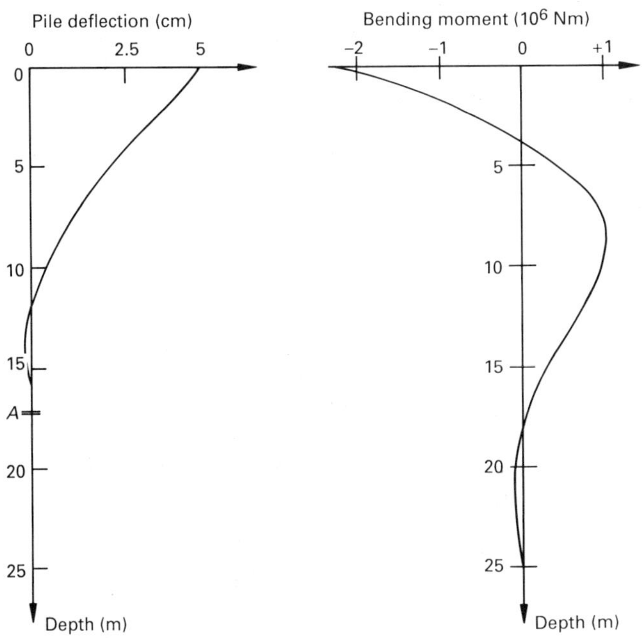

Figure 6.7. Deflection and moment of an embedded pile

where p is the distributed soil reaction force, y is the lateral pile deflection and E is the modulus of elasticity of the soil which will vary with soil deflection y, and depth x. From basic beam bending theory, the structural behaviour of the pile can be written as

$$E_p I_p \frac{d^4 y}{dx^4} = p \tag{6.25}$$

where $E_p I_p$ is the elastic modulus of the pile and p is the distributed lateral load on the pile. Combining Equations (6.24) and (6.25) yields the basic differential equation for a laterally loaded pile:

$$E_p I_p \frac{d^4 y}{dx^4} + E(x, y)y = 0 \tag{6.26}$$

One of the greatest difficulties in solving this equation arises from the fact that the soil modulus E varies both with pile deflection y and with depth x. For many soil types, E increases with depth and can be modelled by the equation

$$E = kx \tag{6.27}$$

where k is a constant. A simple but very approximate method for calculating pile behaviour is obtained by assuming that the pile is completely fixed against rotation at some point below the mud line (such as A in Figure 6.7). The position of this point is selected from experience and the forces exerted on the pile by the soil are neglected above this point. Despite the desirable simplicity of this method, it does not reflect the physical behaviour of the pile and soil and is not recommended for use.

An alternative method for computing pile–soil interaction is to solve Equation (6.26) by a finite difference numerical scheme, although an analytical solution can be obtained for the special case of constant E. In a finite difference numerical scheme, Equation (6.26) is written as

$$E_p I_p \left(\frac{y_{i-2} - 4y_{i-1} + 6y_i - 4y_{i+1} + y_{i+2}}{\delta^4} \right) + E_i y_i = 0 \tag{6.28}$$

where the pile length shown in Figure 6.8 is discretized into n elements, each of length δ. Equation (6.28) is rewritten as

$$y_{i-2} - 4y_{i-1} + \left(6 + \frac{E_i L^4}{E_p I_p n^4}\right) y_i + 4y_{i+1} + y_{i+2} = 0 \tag{6.29}$$

using $\delta = L/n$ from the discretization of Figure 6.8. The recurrence relation of Equation (6.29) represents $i - 1$ equations when applied at the discrete points 2 to n along the pile length.

Boundary conditions at the pile top and bottom ends will yield four additional equations. The shear force F and bending moment M at the top of the pile yield the equations

$$\left.\begin{aligned} &F = E_p I_p \frac{d^3 y}{dx^3} \text{ or } -y_{-2} + 2y_{-1} - 2y_2 + y_3 = \frac{FL^3}{E_p I_p n^3} \\ &\text{and} \\ &M = E_p I_p \frac{d^2 y}{dx^2} \text{ or } y_2 - 2y_1 + y_{-1} = \frac{ML^2}{E_p I_p n^2} \end{aligned}\right\} \tag{6.30}$$

At the pile bottom, a free pile tip gives zero shear force and bending moment to yield

$$\left.\begin{aligned} E_pI_p\frac{d^3y}{dx^3} &= 0 \text{ or } -y_{n-1} + 2y_n - 2y_{n+2} + y_{n+3} = 0 \\ E_pI_p\frac{d^2y}{dx^2} &= 0 \text{ or } y_n - 2y_{n+1} + y_{n+2} = 0 \end{aligned}\right\} \quad (6.31)$$

Equilibrium of horizontal shear forces and bending moments on the pile give two additional equations to bring the total number of equations to $n + 5$. Solution of these equations is most commonly carried out using a Gauss–Siedel or Jacobi iteration method and yields $(n + 5)$ values of pile lateral deflection y – of which the values at -2, -1, $n + 2$ and $n + 3$ are only used to ensure continuity of the derivatives in the formulation and do not have a physical significance. These deflections and their derivatives, together with the resultant values of F and M, provide the necessary information on pile–soil interaction.

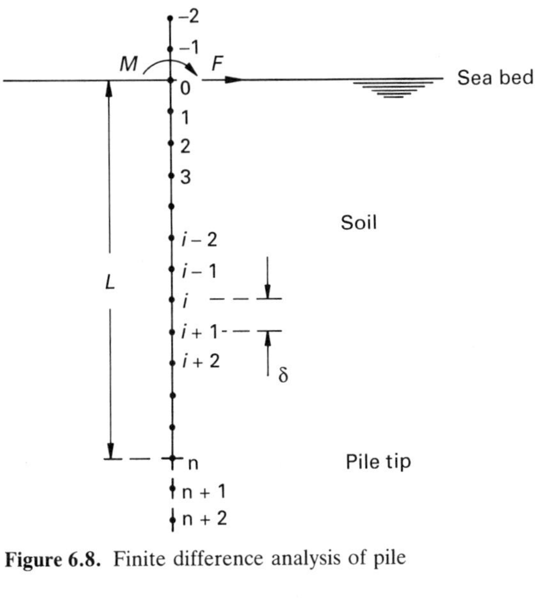

Figure 6.8. Finite difference analysis of pile

However, the analysis can only be carried out if the soil modulus of elasticity function $E(x, y)$ is known. One form of E that is commonly used is

$$E = E_L\left(\frac{x}{L}\right)^N \quad (6.32)$$

where E_L is the value of the modulus at the pile bottom and N is an empirical index equal to or greater than zero. $N = 0$ corresponds to a constant modulus whereas $N = 1$ gives a linear variation as in Equation (6.27). In general, N is taken between 0 and 0.15 for clay soils and equals 1 for granular soils. Analytic solutions for Equation (6.26) are available for $N = 0$.

For real soils, however, the relationship between soil reaction and deflection is non-linear, with the reaction reaching a limiting value when the deflection is sufficiently large. Figure 6.9 presents typical soil reaction against deflection curves for a consolidated clay soil and demonstrates the flattening out of the reaction/deflection curves at large deflections. These curves are conventionally referred to as *p–y* curves.

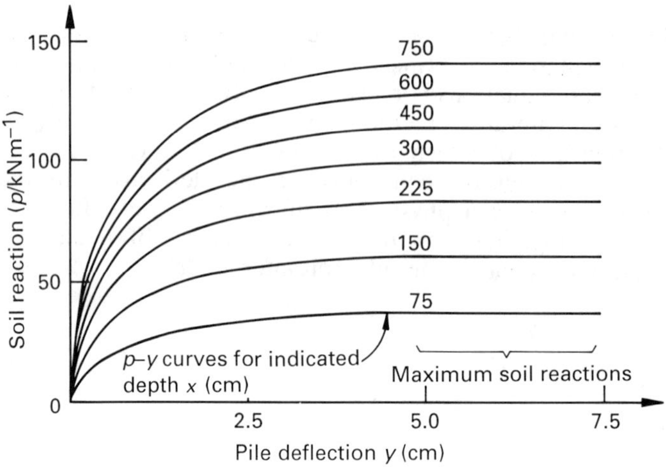

Figure 6.9. Typical *p–y* curves for a consolidated clay

The non-linear soil reaction curves are accomodated in analysis by rewriting Equation (6.26) as

$$\frac{d^2M}{dx^2} + S(x)\frac{d^2y}{dx^2} + E(x, y)y = 0 \tag{6.33}$$

where M is the pile moment at depth x and the additional effect of a variation of axial force S with depth x is accounted for (see Reese, 1977 for further details). This equation can be solved by a finite difference technique similar to that described above. The solution requires inputs of *p–y* curves at various depths (see Figure 6.9), although the method implicitly assumes that the soil behaviour at a particular depth is independent of that in adjacent layers. Equation (6.33) and the *p–y* curves are solved using an iterative finite difference technique. The resultant shear forces and bending moments at the pile surface are derived as functions of pile top displacement and rotation, respectively, and combined with a jacket finite element analysis, as in Figure 6.10, to complete the definition of the pile and structure interaction.

The *p–y* data necessary for the analysis can be obtained in three different ways – full size or model scale lateral loading tests on instrumented piles are one of the most reliable methods, although practical considerations often restrict *p–y* data sources to laboratory tests or empirical correlations based on soil survey data.

Other pile and foundation analyses also use an elastic continuous mathematical model of the surrounding soil, together with a finite beam element model of the pile to quantify pile deflections and soil behaviour – Poulos and Davis (1980) give further details. Analysis of piles for jacket platforms often need to account for pile groups and their interaction with each other and the soil – Focht and Kock (1973) and O'Neill *et al.* (1977) describe extensions of the techniques given above for pile groups.

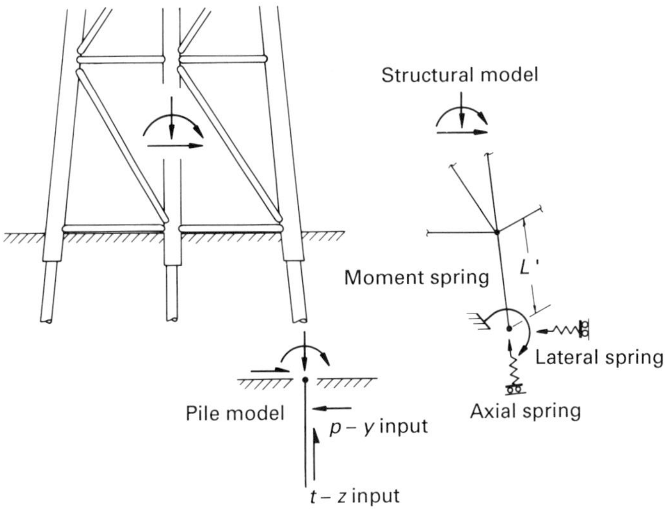

Figure 6.10. Spring model representing pile

6.2 Dynamic response

6.2.1 Governing equations

Dynamic analysis of fixed offshore structures becomes increasingly necessary as their installed water depth increases and the structural configuration becomes more slender due to the combined effects of water depth and economic pressure on costs. A dynamic analysis is also able to quantify the dynamic magnification effects of structural resonances and to provide more accurate load cycle data for fatigue calculations.

The finite element static analysis described in Section 6.1.1 can be readily extended to dynamic analysis by using Lagrange's equation to derive an equation of motion in matrix form for a multi-degree of freedom dynamic system. Lagrange's equation is written as

$$\frac{\partial}{\partial t}\left[\frac{\partial \mathbf{T}}{\partial \dot{\mathbf{q}}_i}\right] - \frac{\partial \mathbf{T}}{\partial \mathbf{q}_i} + \frac{\partial \mathbf{V}}{\partial \mathbf{q}_i} = \mathbf{Q}_i \tag{6.34}$$

where $\mathbf{T}$, $\mathbf{V}$ and $\mathbf{Q}_i$ represent kinetic energy, potential energy and generalized force, respectively, acting on a system whose motion is described in terms of generalized co-ordinates $\mathbf{q}_i$ – Appendix E gives a derivation of these equations.

Now a discretized multi-degree of freedom linear system can have its kinetic and potential energies written as

$$T' = \frac{1}{2}\sum_{j=1}^{n}\sum_{i=1}^{n} m_{ij}\dot{q}_i\dot{q}_j \tag{6.35}$$

and

$$V' = \frac{1}{2}\sum_{j=1}^{n}\sum_{i=1}^{n} k_{ij}q_iq_j \tag{6.36}$$

where n is the number of degrees of freedom of the system and the dot represents differentiation with respect to time. These energies can also be written in matrix form as

$$\mathbf{T} = \frac{1}{2}\dot{\mathbf{q}}^{\mathrm{T}}\mathbf{M}\dot{\mathbf{q}} \tag{6.37}$$

and

$$\mathbf{V} = \frac{1}{2}\mathbf{q}^{\mathrm{T}}\mathbf{K}\mathbf{q} \tag{6.38}$$

Substituting (6.37) and (6.38) into Equation (6.34) leads to the matrix equation

$$\mathbf{M}\ddot{\mathbf{q}} + \mathbf{K}\mathbf{q} = \mathbf{Q} \tag{6.39}$$

since the differential $\partial T/\partial q_i$ is zero for systems with small displacements, $\mathbf{M}$ and $\mathbf{K}$ are matrices of order $(n \times n)$ with $\mathbf{q}$ and $\ddot{\mathbf{q}}$ being column vectors of order $(n \times 1)$. The generalized force $\mathbf{Q}$ contains all non-conservative forces and external forces and with

$$\mathbf{Q} = \mathbf{F} - \mathbf{C}\dot{\mathbf{q}}, \tag{6.40}$$

the general equations of motion become

$$\mathbf{M}\ddot{\mathbf{q}} + \mathbf{C}\dot{\mathbf{q}} + \mathbf{K}\mathbf{q} = \mathbf{F} \tag{6.41}$$

where $\mathbf{C}$ is a velocity dependent $(n \times n)$ damping coefficient matrix and $\mathbf{F}$ is an $(n \times 1)$ vector of external forces.

Now all the elements of an offshore structure to be modelled by Equation (6.41) have continuous mass and stiffness properties that have to be represented by discretized matrices. A continuous infinite degree of freedom system can be reduced to a finite number of co-ordinates by using the concept of dimensionless shape functions which provide a mathematical relationship between a continuously varying function and a finite set of co-ordinates describing the function's behaviour.

Consider a beam element of an offshore structure that has an arbitrary, but continuous lateral deflection given by the function $u(x, t)$, where x is distance along the member and t is time; $u(x, t)$ can be represented in terms of finite coordinates $q_1(t)$, $q_2(t)$, . . ., $q_i(t)$, . . ., $q_n(t)$ through the equation

$$u(x, t) = q_1(t)\psi_1(x) + q_2(t)\psi_2(x) + \ldots + q_i(t)\psi_i(x) + \ldots + q_n(t)\psi_n(x) \tag{6.42}$$

where $\psi_i(x)$ are dimensionless shape functions which satisfy the geometric boundary conditions at the member ends.

The potential energy term V' in Equation (6.38) denotes the flexural strain energy which for a beam element of length L can be written as

$$V' = \frac{1}{2}\int_0^L EI(x)\left[\frac{\partial^2 v(x, t)}{\partial x^2}\right]^2 \mathrm{d}x \tag{6.43}$$

Substituting Equation (6.42) into (6.43) gives

$$V' = \frac{1}{2}\sum_{j=1}^{n}\sum_{i=1}^{n} k_{ij}q_iq_j \tag{6.44}$$

where

$$k_{ij} = \int_0^L EI(x)\,\frac{\partial^2\psi_i(x)}{\partial x^2}\cdot\frac{\partial^2\psi_j(x)}{\partial x^2}\,\mathrm{d}x \tag{6.45}$$

For a beam element of a space frame structure with 12 degrees of freedom, as shown in Figure 6.2, the shape functions ψ_1 to ψ_{12} and their first and second derivatives are presented in Table 6.3. Use of the shape functions of Table 6.3 in Equation (6.45) leads to the identical stiffness matrix as given by Table 6.1 and is an alternative 'consistent' method for deriving it.

Table 6.3 Shape functions for a three-dimensional beam element

$\psi_i(x_m)$	$\dfrac{\partial\psi_i}{\partial x_m}$	$\dfrac{\partial^2\psi_i}{\partial x_m^2}$
$\psi_1(x_m) = (x_m/L)$	$1/L$	0
$\psi_2(x) = 1 - 3\left(\frac{x_m}{L}\right)^2 + 2\left(\frac{x_m}{L}\right)^3$	$-\frac{6x_m}{L^2} + \frac{6x_m^2}{L^3}$	$-\frac{6}{L^2} + \frac{12x_m}{L^3}$
$\psi_3(x) = 1 - 3\left(\frac{x_m}{L}\right)^2 + 2\left(\frac{x_m}{L}\right)^3$	$-\frac{6x_m}{L^2} + \frac{6x_m^2}{L^3}$	$-\frac{6}{L^2} + \frac{12x_m}{L^3}$
$\psi_4(x_m) = (x_m/L)$	$1/L$	0
$\psi_5(x) = -x_m\left(1 - \frac{x_m}{L}\right)^2$	$-1 + \frac{4x_m}{L} - \frac{3x_m^2}{L^2}$	$\frac{4}{L} - \frac{6x_m}{L^2}$
$\psi_6(x) = x_m\left(1 - \frac{x_m}{L}\right)^2$	$1 - \frac{4x_m}{L} - \frac{3x_m^2}{L^2}$	$-\frac{4}{L} + \frac{6x_m}{L^2}$
$\psi_7(x) = \left(1 - \frac{x_m}{L}\right)$	$-1/L$	0
$\psi_8(x) = 3\left(\frac{x_m}{L}\right)^2 - 2\left(\frac{x_m}{L}\right)^3$	$\frac{6x_m}{L^2} - \frac{6x_m^2}{L^3}$	$\frac{6}{L^2} - \frac{12x_m}{L^3}$
$\psi_9(x) = 3\left(\frac{x_m}{L}\right)^2 - 2\left(\frac{x_m}{L}\right)^3$	$\frac{6x_m}{L^2} - \frac{6x_m^2}{L^3}$	$\frac{6}{L^2} - \frac{12x_m}{L^3}$
$\psi_{10}(x) = \left(1 - \frac{x_m}{L}\right)$	$-1/L$	0
$\psi_{11}(x) = \left(\frac{x_m^2}{L}\right)\left(1 - \frac{x_m}{L}\right)$	$\frac{2x_m}{L} - \frac{3x_m^2}{L^2}$	$\frac{2}{L} - \frac{6x_m}{L^2}$
$\psi_{12}(x) = \left(\frac{x_m^2}{L}\right)\left(\frac{x_m}{L} - 1\right)$	$-\frac{2x_m}{L} + \frac{3x_m^2}{L^2}$	$-\frac{2}{L} + \frac{6x_m}{L^2}$

A consistent mass matrix can similarly be obtained by writing the kinetic energy in terms of the shape functions.

$$T' = \frac{1}{2}\int m(x)\dot{u}(x,\, t)^2\,\mathrm{d}x \tag{6.46}$$

Substituting (6.42) into (6.46) gives

$$T' = \frac{1}{2}\sum_{j=1}^{n}\sum_{i=1}^{n} m_{ij}\dot{q}_i\dot{q}_j \tag{6.47}$$

where

$$m_{ij} = \int_0^L m(x)\psi_i(x)\psi_j(x)\mathrm{d}x \tag{6.48}$$

Use of the shape functions of Table 6.3 in Equation (6.48) yields a consistent mass matrix for a space frame beam element (see Figure 6.2) which is given in Table 6.4.

Table 6.4 Consistent mass matrix for a beam element

$\dfrac{\rho AL}{420}$ ×

1	2	3	4	5	6	7	8	9	10	11	12	
140						70						1
	156				$22L$		54				$-13L$	2
		156		$-22L$				54		$13L$		3
			$\frac{140I_p}{A}$						$\frac{70I_p}{A}$			4
		$-22L$		$4L^2$				$-13L$		$-3L^2$		5
	$22L$				$4L^2$		$13L$				$-3L^2$	6
70						140						7
	54				$13L$		156				$-22L$	8
		54		$-13L$				156		$22L$		9
			$\frac{70I_p}{A}$						$\frac{140I_p}{A}$			10
		$13L$		$-3L^2$				$22L$		$4L^2$		11
	$-13L$				$-3L^2$		$-22L$				$4L^2$	12

Notes:
ρ = density of beam material; A = area of cross-section; L = beam length; I_p = polar second moment of area.

The mass matrix can also be modelled by taking the mass of each beam element and distributing it as lumped masses at the element nodes. This lumped mass formulation gives a diagonal mass matrix where each diagonal element is made up of half or a representative proportion of the mass of all elements influenced by the acceleration of that degree of freedom. In the lumped mass formulation, a non-zero mass is only present for translational degrees of freedom, with rotational degrees of freedom having zero inertia because of the concentration of the masses at the nodal points. The lumped mass matrix has less theoretical justification than the consistent mass formulation but is easier to apply.

Whether the mass matrix for a beam element is of the lumped or consistent formulation, its terms must include the physical mass of the structure, its added mass due to lateral accelerations, the mass of fluid contained within hollow members, the masses of piles within tubulars, the mass of any attachments such as pile guides, conductor mounts, and so on. The definition of the mass matrix for elements piercing the free surface also needs special care to ensure that the added mass of the submerged element segment is correctly distributed between the nodes. Since the hydrodynamic added mass of the member due its lateral accelerations arises from the motions of the surrounding fluid distributed along the member length, it is justifiable to use the element shape function technique for computing the resultant mass matrix. The formulation of the added mass matrix using the above method implicitly assumes that only added masses due to member element motions normal to its axis are significant, with added masses due to axial translation and rotation about an axis along the member centre line being negligible. For fluids totally enclosed within a hollow member, the mass of the contained fluid must be included in the added mass due to member axial translation but not in the added inertia due to member rotation about its centre line axis. On the other hand, if the member has open ends, the fluid contained within it can move freely along its axis and the added mass contribution due to axial translation of the contained fluid should be neglected.

The damping matrix **C** is used to quantify the internal structural and foundation damping of the structure, together with the linear (due to wave radiation) or non-linear (due to drag) damping induced by vibration of the structure with fluid surrounding it.

The equation of motion (6.39) derived from Lagrange's equations can be extended to include damping and exciting force separately for a beam element by writing the beam moment due to internal energy losses as

$$M_e(x, t) = c_e I(x) \frac{\partial^3 u(x, t)}{\partial^2 x \partial t} \tag{6.49}$$

where $I(x)$ is the second moment of area of the beam section, c_e is a damping coefficient and $\partial^3 u/(\partial^2 x\, \partial t)$ is the rate of change of beam curvature with time. Therefore, the total work performed by these internal non-conservative forces is

$$W_{nc_{int}} = -\int_0^L c_e I(x) \frac{\partial^3 u(x, t)}{\partial^2 x \partial t} \cdot \delta\left(\frac{\partial^2 u(x, t)}{\partial x^2}\right) dx \tag{6.50}$$

and also by a distributed transverse non-conservative loading $f(x, t)$ is

$$W_{nc_{ext}} = \int_0^L f(x, t)\, \delta[u(x, t)]\, dx \tag{6.51}$$

where δ is an operator denoting variation of the quantity following it.

Substituting Equation (6.42) into Equations (6.50) and (6.51) and adding gives

$$\delta W_{nc_{tot}} = \sum_{i=1}^{n}\left(f_i - \sum_{j=1}^{n} C_{ij}\dot{q}_j\right)\delta q_i \tag{6.52}$$

where

$$f_i = \int_0^L w(x, t)\psi_i(x)\, dx \tag{6.53}$$

and

$$C_{ij} = \int_0^L c_e I(x) \frac{\partial^2 \psi_i(x)}{\partial x^2} \frac{\partial^2 \psi_j(x)}{\partial x^2} \, dx \tag{6.54}$$

Comparing Equation (6.52) with Equation (E5) leads to

$$Q_i = f_i - \sum_{i=1}^{n} C_{ij} \dot{q}_j \tag{6.55}$$

which can be written in matrix form as

$$\mathbf{Q} = \mathbf{F} - \mathbf{C}\dot{\mathbf{q}} \tag{6.56}$$

and leads to the full equation of motion (6.41).

The precise form of the structural damping matrix **C** is very difficult to quantify as it may not only be a function of velocities in real structures. A commonly adopted approach is to represent the structural damping as a linear combination of the mass and stiffness matrices so that

$$\mathbf{C} = \mu_1 \mathbf{M} + \mu_2 \mathbf{K} \tag{6.57}$$

with μ_1, μ_2 obtained from assuming certain small percentages of critical damping in the first two modes of vibration.

The force vector **F** of Equation (6.41) is obtained by first calculating the wave force on a small segment of a member element of the framed structure. This segment force **df** can be written as

$$\mathbf{df} = \rho \mathrm{d}V\, C_{\mathrm{M}} \dot{\mathbf{q}}_n + \frac{1}{2} \rho C_{\mathrm{D}}\, \mathrm{d}A |\mathbf{q}_{\mathrm{n}} - \dot{\mathbf{x}}_{\mathrm{n}}| (\mathbf{q}_{\mathrm{n}} - \dot{\mathbf{x}}_{\mathrm{n}}) \tag{6.58}$$

where $\mathrm{d}V$ and $\mathrm{d}A$ are the volume and frontal area of the segment, $\dot{\mathbf{x}}_n$ is the velocity of the segment normal to the member axis and $\mathbf{q}_{\mathrm{n}}$, $\dot{\mathbf{q}}_{\mathrm{n}}$ are wave fluid velocity and acceleration components normal to the member axis; C_{M} and C_{D} are added inertia and drag coefficients which may be obtained from recommended values presented in Appendix G. The wave force on the segment is written as a distributed force and the resultant fixed end actions and equivalent joint loads are calculated using numerical integration based on equations in the last row of Table 6.2. The equivalent joint loads are transformed into global axes and assembled into a force vector **F** using the techniques given in Section 6.1.

Equation (6.58) can be linearized by writing $\sigma_{\dot{\mathrm{V}}_{\mathrm{r}}}$ as the root mean square of the relative velocity $|\dot{\mathbf{q}}_{\mathrm{n}} - \mathbf{x}_{\mathrm{n}}|$ so that

$$\mathbf{df} = \rho\, \mathrm{d}V\, C_{\mathrm{M}} \dot{\mathbf{q}}_{\mathrm{n}} + \frac{1}{2} \rho\, C_{\mathrm{D}}\, \mathrm{d}A \sqrt{(8/\pi)} \cdot \sigma_{\dot{\mathrm{V}}_{\mathrm{r}}} (\dot{\mathbf{q}}_{\mathrm{n}} - \dot{\mathbf{x}}_{\mathrm{n}}) \tag{6.59}$$

see Malhotra and Penzien (1970).

6.2.2 Solution techniques

The matrix equation of motion (6.41) for dynamic response can be solved in either the frequency or time domain. For solution in the frequency domain, it is necessary to use linear wave theory to obtain wave properties and to linearize the drag force term. Once these simplifications are made, a frequency domain solution offers readily usable transfer functions of structure displacements, internal forces and stresses which can be applied to deriving statistical response results.

The frequency domain solution can be obtained in two ways. A direct substitution of the form of solution

$$\mathbf{q} = \mathbf{q}_0 e^{-i\omega t} \tag{6.60}$$

in Equation (6.41) will yield the unknown displacements through matrix inversion of the term in square brackets in the equation

$$[\mathbf{k} - \omega^2\mathbf{M} - i\omega\mathbf{C}]\mathbf{q}_0 = \mathbf{F}_0 \tag{6.61}$$

where

$$\mathbf{F} = \mathbf{F}_0 e^{-i\omega t} \tag{6.62}$$

is the linearized form of the wave exciting force. Solution of Equation (6.61) requires the inversion of a large matrix with complex number elements. This is a numerically cumbersome operation and has prompted the development of computationally more efficient solution techniques.

One such method uses modal superposition. This technique relies on the fact that despite the large number of degrees of freedom of a typical offshore structure, its dynamic response is limited to only a few modes of vibration. The modal superposition technique uses this feature to reduce the solution computation time. The method initially requires determination of the undamped natural frequencies and vibration modes of the structure. The equation for free undamped motions is

$$\mathbf{M}\ddot{\mathbf{q}} + \mathbf{K}\mathbf{q} = 0 \tag{6.63}$$

Assuming a solution of the form

$$\mathbf{q} = \mathbf{q}_0 e^{-i\omega t} \tag{6.64}$$

and substituting into (6.63) yields

$$(\mathbf{K} - \mathbf{M}\omega^2)\mathbf{q}_0 = 0 \tag{6.65}$$

Apart from the trivial solution of $\mathbf{q}_0 = 0$, Equation (6.65) has solutions given by

$$\mathrm{Det}[\mathbf{K} - \mathbf{M}\omega^2] = 0 \tag{6.66}$$

which leads to a polynomial of degree n where n is the number of degrees of freedom in the governing Equation (6.41). Solution of the determinant yields n values of ω which are the natural frequencies of the system. Each natural frequency ω_r has a corresponding set of values of $\mathbf{q}_{0r}$ given by

$$(\mathbf{K} - \mathbf{M}\omega_r^2)\mathbf{q}_{0r} = 0 \tag{6.67}$$

although the elements of vector $\mathbf{q}_{0r}$ are only defined relative to each other rather than in absolute terms.

For systems with many degrees of freedom, the determination of the eigenvalues ω_r and the eignvectors $\mathbf{q}_{0r}$ requires considerable computational effort, although modern computer based algorithms make the task simpler than it would be otherwise – see Bathe and Wilson (1973) for a survey of available technqiues.

It can also be shown that the eigenvectors for any two vibration modes satisfy an orthgonality condition. Consider vibration modes r and s. Using subscripts r and s in Equation (6.67) these can be written as

$$\mathbf{Kq}_{0r} = \mathbf{M}\omega_r^2\mathbf{q}_{0r} \tag{6.68}$$

and

$$\mathbf{Kq}_{0s} = \mathbf{M}\omega_s^2\mathbf{q}_{0s} \tag{6.69}$$

Taking the transpose of Equation (6.68) gives

$$\left.\begin{aligned} (\mathbf{Kq}_{0r})^{\mathrm{T}} &= \omega_r^2(\mathbf{Mq}_{0r})^{\mathrm{T}} \\ &\text{or} \\ \mathbf{q}_{0r}^{\mathrm{T}}\mathbf{K} &= \omega_r^2\mathbf{q}_{0r}^{\mathrm{T}}\mathbf{M} \end{aligned}\right\} \tag{6.70}$$

since $\mathbf{K}$ and $\mathbf{M}$ are symmetric and $\mathbf{K} = \mathbf{K}^{\mathrm{T}}$ and $\mathbf{M} = \mathbf{M}^{\mathrm{T}}$. If Equation (6.70) is post-multiplied by $\mathbf{q}_{0s}$ and Equation (6.69) is pre-multiplied by $\mathbf{q}_{0r}^{\mathrm{T}}$, then

$$\mathbf{q}_{0r}^{\mathrm{T}}\mathbf{Kq}_{0s} = \omega_r^2\mathbf{q}_{0r}^{\mathrm{T}}\mathbf{Mq}_{0s} \tag{6.71}$$

and

$$\mathbf{q}_{0r}\mathbf{Kq}_{0s} = \omega_s^2\mathbf{q}_{0r}^{\mathrm{T}}\mathbf{Mq}_{0s} \tag{6.72}$$

Subtraction of (6.72) from (6.71) yields

$$(\omega_r^2 - \omega_s^2)\mathbf{q}_{0r}^{\mathrm{T}}\mathbf{Mq}_{0s} = 0 \tag{6.73}$$

so that for $\omega_r \neq \omega_s$, the orthogonality condition for the modes of vibration are obtained as

$$\mathbf{q}_{0r}^{\mathrm{T}}\mathbf{Mq}_{0s} = 0 \tag{6.74}$$

and

$$\mathbf{q}_{0r}^{\mathrm{T}}\mathbf{Kq}_{0s} = 0 \tag{6.75}$$

The eigenvectors are often normalized using Equation (6.74) with $r = s$ so that for the rth normalized eigenvector $\boldsymbol{\phi}_r$ the condition

$$\boldsymbol{\phi}_r^{\mathrm{T}}\mathbf{M}\boldsymbol{\phi}_r = 1 \tag{6.76}$$

holds. This can be achieved by calculating the scale factor

$$S_r = \mathbf{q}_{0r}^{\mathrm{T}}\mathbf{Mq}_{0r} \tag{6.77}$$

and getting the normalized eigenvector as

$$\boldsymbol{\phi}_r = \frac{\mathbf{q}_{0r}}{\sqrt{(S_r)}} \tag{6.78}$$

With this normalization procedure, if all the eigenvectors $\boldsymbol{\phi}_r$ are collected in an $m \times m$ square matrix Φ, then the condition

$$\boldsymbol{\Phi}^{\mathrm{T}}\mathbf{M}\boldsymbol{\Phi} = \mathbf{I} \tag{6.79}$$

where $\mathbf{I}$ is an $m \times m$ matrix with unit diagonal terms and zero elsewhere.

The scene is now set for further development of the modal superposition technique. The eigenvectors or mode shapes can be said to be the building blocks of the dynamic response of a general N degree of freedom system to the extent that it is sufficient to describe the response in terms of these mode shapes. Thus the solution $\mathbf{q}$ of an $\mathbf{N}$ degree of freedom dynamic system described by the matrix equation

$$\mathbf{M}\ddot{\mathbf{q}} + \mathbf{C}\dot{\mathbf{q}} + \mathbf{K}\mathbf{q} = \mathbf{F}(t) \tag{6.80}$$

can be written in terms of eigenvectors (or mode shapes) through the matrix equation

$$\mathbf{q} = \boldsymbol{\Phi}\mathbf{Y} \tag{6.81}$$

where $\boldsymbol{\Phi}$ is the $N \times N$ matrix of normalized eigenvectors and $\mathbf{Y}$ is the $(N \times 1)$ column vector of modal amplitudes. Thus Equation (6.81) transforms from geometric variables $\mathbf{q}$ to modal co-ordinates $\mathbf{Y}$ which describe the amplitudes of excited modes during dynamic response. These modal amplitudes are sometimes called normal co-ordinates.

Equation (6.80) can be written in terms of the modal co-ordinates $\mathbf{Y}$ by substitution of (6.81) and pre-multiplying by the transpose of the eigenvector matrix $\boldsymbol{\Phi}^{\mathrm{T}}$, to give

$$\boldsymbol{\Phi}^{\mathrm{T}}\mathbf{M}\boldsymbol{\Phi}\ddot{\mathbf{Y}} + \boldsymbol{\Phi}^{\mathrm{T}}\mathbf{C}\boldsymbol{\Phi}\dot{\mathbf{Y}} + \boldsymbol{\Phi}^{\mathrm{T}}\mathbf{K}\boldsymbol{\Phi}\mathbf{Y} = \boldsymbol{\Phi}^{\mathrm{T}}\mathbf{F}(t) \tag{6.82}$$

now, as a consequence of the orthogonality condition, all components except for the ith column vector term vanish in the mass and stiffness terms. The orthogonality does not apply to the damping term but the modal superposition technique assumes that terms other than those in the ith mode also vanish in the terms of Equation (6.82) that are derived from the damping matrix. Thus the ith mode is entirely decoupled and satisfies the equation

$$\left.\begin{array}{l} m_{ii}y_i + c_{ii}y_i + k_{ii}y_i = F_i(t) \\ \text{where} \\ m_{ii} = \boldsymbol{\Phi}_i^{\mathrm{T}}\mathbf{M}\boldsymbol{\Phi}_i \\ k_{ii} = \boldsymbol{\Phi}_i^{\mathrm{T}}\mathbf{K}\boldsymbol{\Phi}_i \\ c_{ii} = \boldsymbol{\Phi}_i^{\mathrm{T}}\mathbf{C}\boldsymbol{\Phi}_i \\ \text{and} \\ F_i(t) = \boldsymbol{\Phi}_i^{\mathrm{T}}\mathbf{F}(t) \end{array}\right\} \tag{6.83}$$

where m_{ii}, c_{ii} and k_{ii} are the ith diagonal terms in the mass, damping and stiffness matrices **M, C** and **K** respectively.

Equation (6.83) can be solved for all the modal co-ordinates y_i and the geometric co-ordinates $\mathbf{q}$ can be recovered from Equation (6.81). The major advantage of the modal superposition technique is that the structure dynamic response is usually made up of the superposition of only a few of all the modes that are likely to be excited. The number of times Equation (6.83) needs to be evaluated can, therefore, be reduced for computational efficiency by evaluating the eigenfrequencies, eigenvectors and Equation (6.83) for significant modes only. Reduction in the number of eigenfrequencies and vectors to be evaluated contributes to most of the reduction in computation time.

It can be shown that a damping matrix of the form of Equation (6.57) will satisfy the orthogonality condition, since the mass and stiffness matrices also will each satisfy this condition. Thus, if the damping matrix is approximated by Equation (6.57), the modal superposition technique will uncouple the damping term without resorting to the approxmation of neglecting off diagonal terms in Equation (6.82).

Note that the uncoupled equations of motion in modal co-ordinates given by Equation (6.83) can be solved by any method in the frequency or time domain.

Solution of the equation of motion by frequency domain and modal superposition techniques, as described above, is only applicable to linear systems with harmonic excitation, whereas the equations of motion for drag dominated structures can have a substantial non-linear character, as illustrated by Equation (6.58) for wave forces. This non-linear nature can be preserved by solving the equations of motion through a time stepping procedure. This also permits the wave kinematics to be more accurately represented by Stoke's fifth-order theory for regular waves at least. The disadvantage of this approach is its complexity and the fact that output results are available only as time histories. Time domain solutions are also computationally expensive.

Two methods for time step integration are described here. The first, called the central difference method, is an explicit algorithm which is based on a Taylor series expansion of structural displacements, velocities and accelerations. The equations of motion in matrix form are

$$\mathbf{M}\ddot{\mathbf{x}} + \mathbf{C}\dot{\mathbf{x}} + \mathbf{K}\mathbf{x} = \mathbf{F} \tag{6.84}$$

in the usual notation; $\mathbf{x}_i$ and $\mathbf{x}_{i+1}$ are taken to be the values of vector $\mathbf{x}$ at times i and $i + 1$ with $\Delta t = t_{i+1} - t_1$. Expanding $\mathbf{x}_{i+1}$ and $\mathbf{x}_i$ as a Taylor series gives

$$\left.\begin{aligned} &\mathbf{x}_{i+1} = \mathbf{x}_i + (\Delta t)\,\dot{\mathbf{x}}_i + \frac{1}{2}(\Delta t)^2\ddot{\mathbf{x}}_i + \ldots \\ &\text{and} \\ &\mathbf{x}_{i-1} = \mathbf{x}_i - (\Delta t)\,\dot{\mathbf{x}}_i + \frac{1}{2}(\Delta t)^2\,\ddot{\mathbf{x}}_i - \ldots \end{aligned}\right\} \tag{6.85}$$

By substracting and adding Equations (6.85) and neglecting terms of power 3 and higher we get

$$\left.\begin{aligned} &\dot{\mathbf{x}}_i = (x_{i+1} - x_{i-1})/(2\Delta t) \\ &\text{and} \\ &\ddot{\mathbf{x}}_i = \{x_{i+1} - 2x_i + x_{i-1}]/(\Delta t)^2 \end{aligned}\right\} \tag{6.86}$$

Direct substitution in Equation (6.84) with $\mathbf{x} = \mathbf{x}_i$ gives

$$\left[\mathbf{M} + \frac{1}{2}(\Delta t)\,\mathbf{C}\right]\mathbf{x}_{i+1} = (\Delta t)^2\mathbf{F}_i + \left[2\mathbf{M} - (\Delta t)^2\mathbf{K}\right]\mathbf{x}_i + \left[\frac{1}{2}(\Delta t)\,\mathbf{C} - \mathbf{M}\right]\mathbf{x}_{i-1} \tag{6.87}$$

Equation (6.87) is used for successive time steps to derive the displacement at time $i + 1$ from displacements at times i and $i - 1$. The central difference method requires the condition

$$\frac{\Delta t}{T_n} \leqslant 0.318 \tag{6.88}$$

to be satisfied for numerical stability where T_n is the period of the highest vibration mode of the system.

The second time step integration method considered here is the Newmark-β technique, which assumes that displacements and velocities at the end of a time interval can be related to the displacement, velocity and acceleration at the beginning of the time interval by the equation

$$\left.\begin{aligned} \dot{\mathbf{x}}_{i+1} &= \dot{\mathbf{x}}_i + \frac{1}{2}(\Delta t)(\ddot{\mathbf{x}}_i + \ddot{\mathbf{x}}_{i+1}) \\ &\text{and} \\ \mathbf{x}_{i+1} &= \mathbf{x}_i + (\Delta t)\dot{\mathbf{x}}_i + \left(\frac{1}{2} - \beta\right)(\Delta t)^2\ddot{\mathbf{x}}_i + \beta(\Delta t)^2\ddot{\mathbf{x}}_{i+1} \end{aligned}\right\} \tag{6.89}$$

The variable β does have physical significance, in that $\beta = \frac{1}{4}$ in Equation (6.89) corresponds to a constant acceleration variation from i to $i + 1$, whereas $\beta = \frac{1}{6}$ converts the equation to apply for linearly varying acceleration.

Writing Equation (6.84) for time periods $i - 1$, i and $i + 1$ gives

$$\mathbf{M}\ddot{\mathbf{x}}_{i-1} + \mathbf{C}\dot{\mathbf{x}}_{i-1} + \mathbf{K}\mathbf{x}_{i-1} = \mathbf{F}_{i-1} \tag{6.90}$$

$$\mathbf{M}\ddot{\mathbf{x}}_i + \mathbf{C}\dot{\mathbf{x}}_i + \mathbf{K}\mathbf{x}_i = \mathbf{F}_i \tag{6.91}$$

$$\mathbf{M}\ddot{\mathbf{x}}_{i+1} + \mathbf{C}\dot{\mathbf{x}}_{i+1} + \mathbf{K}\mathbf{x}_{i+1} = \mathbf{F}_{i+1} \tag{6.92}$$

Multiplying Equations (6.90) and (6.92) by $\beta(\Delta t)^2$, Equation (6.91) by $(1 - 2\beta)(\Delta t)^2$ and adding gives

$$\begin{aligned} &(\Delta t)^2\mathbf{M}\left\{\left[\beta\ddot{\mathbf{x}}_{i+1} + \left(\frac{1}{2} - \beta\right)\ddot{\mathbf{x}}_i\right] - \left[\beta\ddot{\mathbf{x}}_i + \left(\frac{1}{2} - \beta\right)\ddot{\mathbf{x}}_{i-1}\right]\right. \\ &\left. + \left[\frac{1}{2}\ddot{\mathbf{x}}_i + \frac{1}{2}\ddot{\mathbf{x}}_{i-1}\right]\right\} + (\Delta t)^2\mathbf{C}\left\{\left(\frac{1}{2}\dot{\mathbf{x}}_i\right) + \left(\frac{1}{2}\dot{\mathbf{x}}_{i-1}\right)\right. \\ &\left. + \left[\beta\left(\dot{\mathbf{x}}_{i+1} - \dot{\mathbf{x}}_i\right)\right] + \left[\left(\frac{1}{2} - \beta\right)\left(\dot{\mathbf{x}}_i - \dot{\mathbf{x}}_{i-1}\right)\right]\right\} \\ &+ (\Delta t)^2\mathbf{K}[\beta\mathbf{x}_{i+1} + (1 - 2\beta)\mathbf{x}_i + \beta\mathbf{x}_{i-1}] \\ &= (\Delta t)^2[\beta\mathbf{F}_{i+1} + (1 - 2\beta)\mathbf{F}_i + \beta\mathbf{F}_{i-1}] \end{aligned} \tag{6.93}$$

Substituting from Equation (6.89), simplifying and rearranging gives

$$\begin{aligned} &\left[\mathbf{M} + \frac{1}{2}(\Delta t)\mathbf{C} + \beta(\Delta t)^2\mathbf{K}\right]\mathbf{x}_{i+1} \\ &= (\Delta t)^2\,[\beta\mathbf{F}_{i+1} + (1 - 2\beta)\mathbf{F}_i + \beta\mathbf{F}_{i-1}] \\ &+ [2\mathbf{M} - (\Delta t)^2(1 - 2\beta)\mathbf{K}]\mathbf{x}_i \\ &- \left[\mathbf{M} - \frac{1}{2}(\Delta t)\mathbf{C} + \beta(\Delta t)^2\mathbf{K}\right]\mathbf{x}_{i-1} \end{aligned} \tag{6.94}$$

Thus the displacements at time $i + 1$ can be obtained from the displacement at times i and $i - 1$. Displacement $\mathbf{x}_1$ at time Δt can be obtained by a special case of Equation (6.94) obtained from Equations (6.91), (6.92) and (6.89) as

$$\left[\mathbf{M} + \frac{1}{2}(\Delta t)\mathbf{C} + \beta(\Delta t)^2\mathbf{K}\right]\mathbf{x}_1$$

$$= (\Delta t)^2\,\beta\mathbf{F}_1 + (\Delta t)^2\left[\left(\frac{1}{2} - \beta\right)\mathbf{I} + \left(\frac{1}{4} - \beta\right)\Delta t\mathbf{C}\mathbf{M}^{-1}\right]\mathbf{F}_0 \qquad (6.95)$$

with $x_0 = \dot{x}_0 = 0$ at $t = 0$.

Comparison of Equations (6.87) and (6.94) shows that the central difference and Newmark-β methods are equivalent for $\beta = 0$. It can be shown from stability analysis that the Newmark-β method is unconditionally stable for $\beta > \frac{1}{4}$. For $\beta < \frac{1}{4}$, the stability conditions are given by the equations

$$\frac{\Delta t}{T_n} = \begin{cases} 0.318 \text{ for } \beta = 0 \\ 0.450 \text{ for } \beta = \dfrac{1}{8} \\ 0.551 \text{ for } \beta = \dfrac{1}{6} \end{cases} \qquad (6.96)$$

Once the condition for stability is met for either method, the accuracy of the solution must be investigated to ensure that a sufficiently small value of Δt is used for accuracy but which is not too small that computation time is prohibitive.

Felippa and Park (1978) present more information on these and other time step integration techniques whereas Godeau *et al.* (1977) presents some results for non-linear behaviour of a fixed offshore structure in irregular waves.

Despite the more exact solution of structural dynamics offered by a time step integration, research has continued to examine ways in which frequency domain solutions can be extended to predict the effects of the drag force non-linearity by using an iterative procedure based on minimization of mean square error (see Penzien and Tseng, 1978, and Taudin, 1978). Eatock-Taylor and Rajgopalan (1982) present a perturbation technique to examine the higher harmonics generated due to the drag force non-linearity.

6.3 Structural failure and fatigue life calculations

The safe structural design of a jacket structure ultimately depends on the detailed design of individual tubular joints. The analyses of Sections 6.1 and 6.2 readily ensure that members between joints do not fail as a result of buckling or over stressing in tension or bending. There are, however, several failure mechanisms that have to be designed against in the definition of a tubular joint. A tubular joint brace loading the main member in compression can initiate failure by local buckling or by punching shear of the main member wall. Tensioned braces can cause failure due to crack growth or lamellar tearing of the main member wall.

However, a large amount of structural design studies, finite element analyses and model tests has ensured that these failure mechanisms can be almost completely avoided. The majority of useful recommendations made by such work has been incorporated into codes of practice of which those issued by the American, Petroleum Institute (1987) and the British Standards Institution (1982) are good examples.

At present, the largest source of structural failures in operation is fatigue damage compounded by the effects of corrosion. A metal will fail under repeated cycles of relatively low stress levels due to the growth of internal cracks that can be initiated from local fabrication defects or at stress concentration sites. Such metal fatigue failures are quantified from experimental data by the use of *S–N* curves which are plots of stress range *S* against number of cycles to failure *N*, with both axes plotted in logarithmic form. The fatigue behaviour of a variety of steels can, therefore, be characterized by *S–N* curves of the type shown in Figure 6.11 obtained from British Standard 6235 (1982). These curves show that steel specimens in air do exhibit a low stress range level (the fatigue limit) below which the metal would have a theoretically infinite life, although this property disappears if the effects of combined stress range cycles and corrosion are considered.

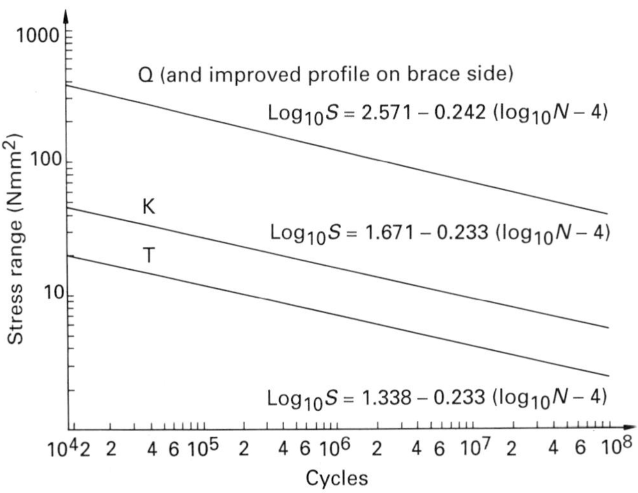

Figure 6.11. Typical *S–N* lines for welded tubular joints. The Q line is used for general fatigue life calculations with the K and T lines being used for an appropriate 'punching shear' method (from the British Standards Institution, 1982)

The Palmgren–Miner cumulative fatigue damage rule is used to quantify fatigue damage caused by a large number of cycles at low stress ranges. The rules give the resultant cumulative damage ratio *D* as

$$D = \sum_{i=1}^{N} \frac{n_i}{N_i} \leqslant 1 \qquad (6.97)$$

where N denotes the total number of stress range intervals being considered, n_i is the number of cycles experienced by the structure within the ith stress range and N_i is the number of cycles to failure given by the S–N curve for this ith stress range. Failure will occur when parameter D reaches unity. If the number of stress range cycles used in the calculation correspond to one year's operation, then the fatigue life of the structure in years can be given by the reciprocal of the cumulative damage.

In design practice, the fatigue life is calculated using a deterministic approach illustrated by the left-hand column of the flow diagram of Figure 6.12. The starting point for such an analysis is a significant wave height against average zero crossing period scatter diagram (see Figure 4.16 for an example), typically averaged for one year's weather conditions. The number of occurrences (expressed in parts per thousand) in each row of the scatter diagram are converted to the number of waves in one year by using the appropriate average zero crossing period and adding up to obtain a variation of maximum wave height (converted from the significant height in the scatter diagram) against the number of waves that exceeded this height, as in box number 2 of Figure 6.12. Each wave height range in this diagram has an associated period range that is also presented in the diagram. The wave height and period ranges are used through the wave loading and response analyses described earlier to yield an equivalent curve of stress range against number of loading cycles that will be applied at this range in box 3. It is usual practice in the calculation to select a wave period for any wave height that will produce the largest stress although the choice of an average between the range of wave periods will yield a less conservative result. A curve such as that in box 3 in Figure 6.12 is obtained for each of the fatigue prone zones on the structure. A stress concentration factor is then applied to the data of box 3 and combined with the Palmgren–Miner damage rule and a material S–N curve to yield the annual cumulative damage D and fatigue life of $1/D$ years.

A probabilistic approach can also be applied to estimation of the stress range exceedence of box 8 by using a parameterized spectral density to represent each significant wave height – zero crossing period range in the scatter diagram of box 1. A linear structural response analysis is then carried out to yield the stress spectrum (as in box 6) at all points of interest. A Rayleigh probability distribution of stress peaks (box 7) is then assumed to estimate the probability of occurrence of each stress range from which the figure in box 8 can be derived and the fatigue analysis completed. It should be noted that the average zero crossing period of the stress spectrum in box 6 will be different from that of the wave spectrum of box 5 and will need to be calculated in order to determine the numbers of stress cycles applied in a year.

The effect of sea water corrosion and the value of stress concentration factors to be used are two of the most important sources of uncertainity in fatigue analysis of an offshore structure. The American Petroleum

Institute (1987) and the British Standards Institution (1982) give some guidance on these matters.

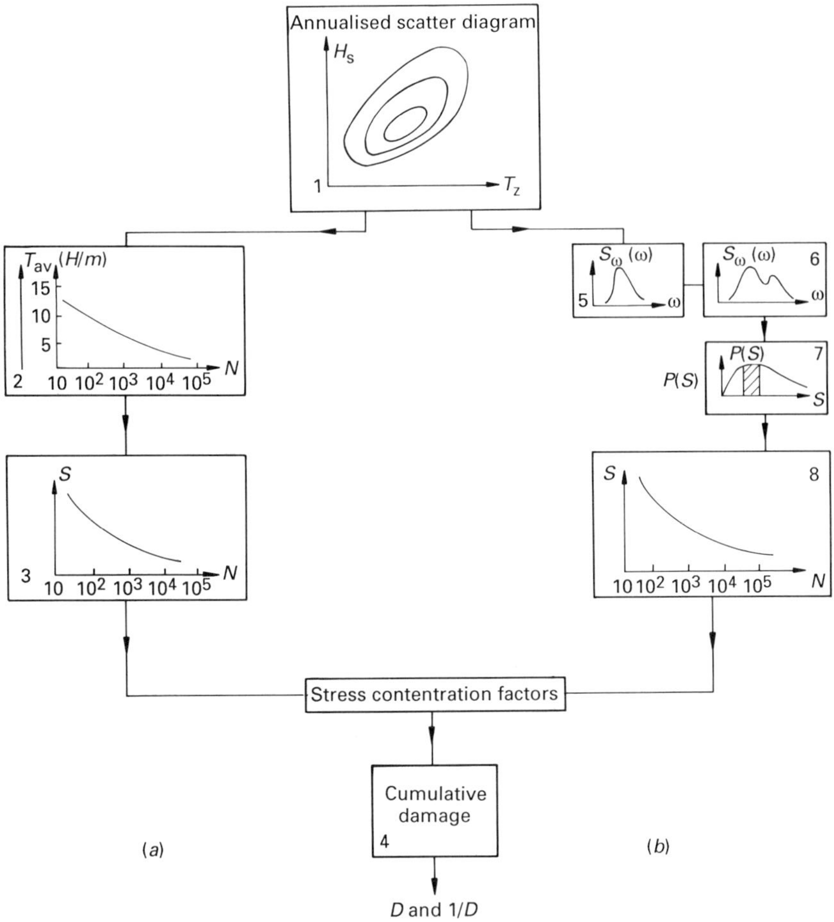

Figure 6.12. Fatigue life calculation procedures: (*a*) deterministic; (*b*) stochastic

Much recent progress has been made on quantifying fatigue crack growth in tubular joints by the application of the theory of fracture mechanics coupled with more accurate in service crack detection techniques – see, for example, the work of Dover and Connolly (1986) and Dover and Wilson (1986). Their combined use is leading to more reliable monitoring and rectification of fatigue crack growth during the life of an offshore structure.

6.4 Design and certification requirements

It is important to appreciate that the design procedures for jacket structures outlined in the last three chapters are only a small part of the total design process. In order to illustrate this point, Figure 6.13 presents a flow chart showing the design procedures that need to be followed from the initial specification through to commencing operation of a typical offshore structure. The jacket has to have sufficient strength as it is assembled during the fabrication stage and loaded out of the yard. It also has to meet the naval architectural and structural requirements of tow out, up-ending and installation, as well as surviving a 20–40 year life. Some of the supplementary design tasks not covered in this text book include determining the response of the structure to earthquakes, the provision of corrosion protection and in-service structural monitoring. The design procedure for large jackets invariably contains a model test phase for critical operations such as up-ending during installation. The documentation of the material, structural and welding details of the design during its certification, fabrication and service life pose an engineering management problem.

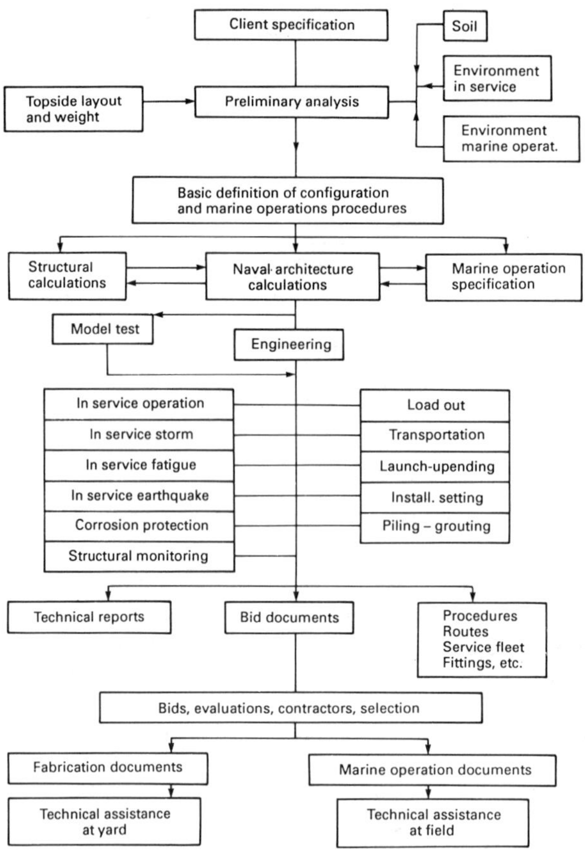

Figure 6.13. Design procedure for jacket structure

Certifying authorities play a key role in the design procedure for an offshore structure. The major certifying authorities in the United Kingdom, Norway and the United States have built up extensive codes of practice which reflect research work, in-service experience and the results of failure investigations over many years of operation. Certifying authorities also provide an independent check of many of the calculations and decisions that need to be made during a typical design. There tends to be close technical collaboration between research establishments, designers and the operators of offshore structures.

References

American Petroleum Institute, (1987), *Recommended Practice for Planning, Designing and Constructing Fixed Offshore Platforms,* Dallas, Texas, USA Report No. API-RP-2A (revised annually).

Bathe, K. J. and Wilson, E. L. (1973), Solution methods for eigen-value problems in engineering, *International Journal for Numerical Methods in Engineering*, **6**, 213–216.

Bell, K., Hansteen, O. E., Larsen, P. K. and Smith, E. K. (1976), Analysis of a wave-structure-soil system – case study of a gravity platform, *Proceedings of the Conference on Behaviour of Offshore Structures*, Volume 1, BOSS 1976, pp. 846–863.

British Standards Institution, (1982), *Code of Practice for Fixed Offshore Structures,* BS6235, BSI, 2 Park Street, London, W1A 2BS.

Dover, W. D. and Connolly, M. P. (1986), *Fatigue Fracture Mechanics Assessment of Tubular Welded Y and K Joints*, Paper No. C141/86, Institution of Mechanical Engineers, London.

Dover, W. D. and Wilson, T. J. (1986), *Corrosion Fatigue of Tubular Welded T-joints*, Paper No. C136/86, Institution of Mechanical Engineers, London.

Eatock-Taylor, R. and Rajgopalan, A. (1982), Dynamics of offshore structures, *Journal of Sound and Vibration*, **83**, 401–431.

Felippa, C. A. and Park, K. C. (1978), Computational aspects of time integration procedures in structural dynamics, *Journal of Applied Mechanics*, **45**, 596–611.

Focht, J. A. Jr and Kock, K. J. (1973), Rational analysis of the lateral performance of offshore pile groups, *Proceedings of the Offshore Technology Conference*, OTC 1896.

Godeau, A. J., Deleuil, G. and Heas, J. Y. (1977), Statistical analysis of nonlinear dynamic response of fixed structures to random waves, *Proceedings of the Offshore Technology Conference*, OTC 3030.

Malhotra, A. K. and Penzien, J. (1970), Nondeterministic analysis of offshore tower structures, *Journal of Engineering Mechanics Division, American Society of Civil Engineers*, **96**, (EM6), 985–1003.

O'Neill, M. W., Ghazzaly, O. I. and Ho Boo Ha (1977), Analysis of three-dimensional pile groups with nonlinear soil response and pile-soil-pile interaction, *Proceedings of the Offshore Technology Conference*, OTC 2838.

Penzien, J. and Tseng, W. S. (1978), Three dimensional dynamic analysis of fixed offshore platforms, *Numerical Methods in Offshore Engineering*, O. C. Zienkiewicz *et al.* (eds.) Wiley, NY, USA.

Poulos, H. G. and Davis, E. H. (1980), *Pile Foundation Analysis and Design*, John Wiley and Sons, NY, USA.

Taudin, P. (1978), Dynamic response of flexible structures to regular waves, *Proceedings of the Offshore Technology Conference*, OTC 3160.

Reese, L. C. (1977), Laterally loaded pile; program documentation, *Journal of the Geotechnical Engineering Division, American Society of Civil Engineers*, **103**, (GT4) 287–305.

Chapter 7

Hydrostatics of floating bodies

Hydrostatic pressures within a fluid at rest can exert very large forces on the submerged parts of offshore structures, particularly at large water depths. Hydrostatic pressures also impart rather more subtle properties which affect the stability of floating bodies. Both of these features have a profound influence on the design of floating and fixed offshore structures and are considered here in more detail.

7.1 Basic properties of a fluid at rest

A fluid may be defined as a substance that deforms due to the effect of a shear stress, however small. It follows from this statement that no shearing stresses can exist in a fluid at rest with the reactions between adjacent layers of fluid being confined to normal stresses only. These normal stresses are called pressures and are defined as the normal force per unit area on an infinitesimal plane surface at any orientation in the fluid. By resolving pressure induced forces on an infinitesimal tetrahedral fluid element with three mutually perpendicular faces, it can readily be shown that the pressure at a point in a fluid in equilibrium is the same in all directions. This is also true for a fluid in bulk motion in which there are no shearing stresses. The pressure at a point in a fluid undergoing generalized motion is considered further in Section 3.5

Although the above properties apply both to liquids and gases, the hydrostatics of offshore structures is principally concerned with liquids of large density (compared to gases) which induce substantial pressure gradients in the Earth's gravitational field – see, for example, Figure 2.6(b) for such gradients in the oceans and the atmosphere. A qualitative characteristic that distinguishes liquids from gases is the fact that a gas will expand in volume indefinitely to fill its containing volume fully, whereas a liquid will be of an essentially constant volume with no definite shape.

Sea water at rest in the oceans, relative to the Earth, is subjected primarily to a uniform gravitational force field acting towards the Earth's centre. Thus, there are no external forces to consider apart from the weight of sea water, and the resultant hydrostatic pressure p at a point distance z below the water surface can be proved to be

$$p = p_a + \rho g z \qquad (7.1)$$

by considering force equilibrium of a vertical column of fluid of constant cross-section from the water surface down to depth z. In Equation (7.1), p_a is atmospheric pressure, ρ is sea water density and g is the acceleration due to gravity. It demonstrates that there is a pressure increase of 1 atmosphere for every 10.08 m increase of depth for sea water of density 1025 kg/m^3.

Taking horizontal force equilibrium on any horizontal cylinder of fluid can also show that in a fluid at rest under gravity, the pressure is the same at all points in a horizontal plane.

Now the pressure field defined by Equation (7.1) will exert forces on structures immersed in the fluid. For an arbitrary plane surface, the pressure field will induce a system of parallel forces of differing magnitude acting normal to the surface. The point in the plane where the resultant of these pressure forces acts is defined as the centre of pressure. For Cartesian axes Oxy in an arbitrary plane surface, the co-ordinates (ξ, η) of the centre of pressure can be obtained by the following integrations:

$$\left.\begin{aligned} \xi &= \frac{\int\int px \,\mathrm{d}x\,\mathrm{d}y}{\int\int \mathrm{p}\,\mathrm{d}x\,\mathrm{d}y} \\ \eta &= \frac{\int\int py \,\mathrm{d}x\,\mathrm{d}y}{\int\int p\,\mathrm{d}x\,\mathrm{d}y} \end{aligned}\right\} \qquad (7.2)$$

where the function $p = f(x, y)$ defines the linear pressure gradient of Equation (7.1) in terms of axes Oxy. Also, the total pressure induced force T on the plane surface is given by

$$T = \int\int p \,\mathrm{d}x\,\mathrm{d}y \qquad (7.3)$$

The total thrust and centre of pressure on a curved surface are similarly found by integration of the pressure field of Equation (7.1) over the surface using an appropriate co-ordinate system.

Both analytic and numerical integration techniques may be utilized, depending on the complexity of the problem – Section 7.3 gives an example of both of these.

7.2 Buoyancy forces and stability

The resultant of hydrostatic pressure induced forces on the surfaces of a fully or partly submerged body immersed in a fluid at rest is called the buoyancy force. The magnitude of this force can be calculated using the summation of pressure induced forces defined above, but a simpler and more elegant approach is offered by the principle of Archimedes. This states that the buoyancy force on a wholly or partially submerged free body in a fluid at rest is equal and opposite to the weight of fluid displaced by the body and acts in a vertical line through the centre of gravity of the displaced fluid. For a fluid of constant density, this centre of gravity will be at the centre of volume of the displaced fluid.

This result can be readily proved by using arguments similar to those in Section 3.12. Consider an arbitrary boundary, n, drawn in a fluid at rest, B, in Figure 7.1(a) and (b). Since the fluid is at rest, the pressure gradient within it will be supporting the weight of fluid contained inside boundary A such that the total force F_B across the boundary will be vertical and given by

$$F_B = \rho g V \tag{7.4}$$

where V is the fluid volume contained in boundary A and ρ is the density of fluid B; F_B will also act through the centre of gravity of the fluid in volume V. Now if the fluid volume inside the boundary was replaced by a solid body of identical shape, the above pressure induced force, the buoyancy force, would still act. The point through which it acts is also called the centre of buoyancy of the submerged volume.

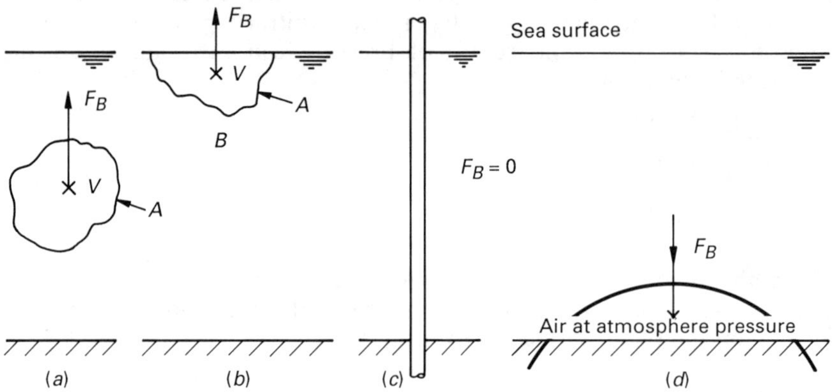

Figure 7.1. Buoyancy forces

It is important to appreciate that Archimedes' principle only applies to freely floating bodies such as are illustrated in Figure 7.1(a) and (b). There are, however, many structures used in offshore engineering that are submerged but not freely floating. Figure 7.1(c) and (d) gives two examples – the former being a vertical 'riser' pipe connection between the atmosphere and a drill hole or hydrocarbon reservoir below the sea bed, whereas the latter could be a subsea 1-atmosphere habitat. Since a perfectly vertical riser pipe is only subjected to horizontal pressure induced forces, the net vertical buoyancy force on it must be zero. On the other hand, the subsea habitat could be said to have a negative (downward acting) buoyancy force. It is clear, therefore, that the buoyancy force for bodies which are not freely floating must be calculated from first principles by integration of pressure induced forces on their surface. This exercise is often necessary in the design of offshore structures.

7.2.1 Initial stability

A freely floating body, such as that illustrated in Figure 7.2, will be in vertical equilibrium if its weight W acting through its centre of gravity G is equal to the buoyancy force $\rho g V$ acting through the centre of buoyancy (or volume) B, where V is the body's submerged volume. However, the relative positions of points G and B will govern the body's stability in roll and pitch. The body will be in stable, neutral or unstable equilibrium if the centre of buoyancy B, is above, coincides with or is below the centre of gravity G. A body in stable equilibrium will, if given a small displacement and then released, return to its original position. If the equilibrium is unstable, the body will not return to its stable position but will move further from it. In neutral equilibrium, the body will neither return to its original position nor increase its displacement following the initial disturbance, but, rather, will simply adopt a new position. The condition for stable equilibrium requires a restoring couple to return the vessel to its original position following angular perturbation.

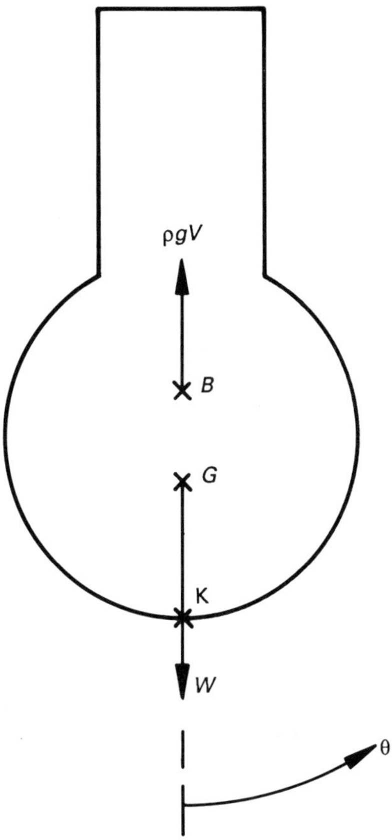

Figure 7.2. Submerged body

This restoring couple can be written as

$$M = \rho g V(KB - KG)\sin\theta = W(KB - KG)\sin\theta \tag{7.5}$$

where θ is the rotation angle of the body from its equilibrium position and K is a reference point on the vessel keel.

Now, for a fully submerged rigid body without free surfaces in internal tanks, the positions of points B and G relative to the body will remain fixed and the stability will be fully defined by the above considerations. However, for a freely floating body at the free surface, the shape of the submerged volume, and hence the position of the centre of volume (or buoyancy), will shift with inclination of the body.

This aspect complicates consideration of the stability of a floating body at the free surface and requires the problem to be examined from first principles. Consider the arbitrary body in Figure 7.3 floating at the free surface with submerged volume V. The shape of the surface boundary of this volume is described with respect to Cartesian axes $Oxyz$ that are fixed in the body with Oxy initially in the water plane. At equilibrium, the constant body weight W will be equal to the buoyancy force $\rho g V$ with the latter acting through the centre of buoyancy B. Now, if the body is given a small inclination β about the Oy axis, the condition that the submerged volume must remain constant before and after inclination is

$$V = \int_A S\,\mathrm{d}A = \int_A (S - x\beta)\,\mathrm{d}A \tag{7.6}$$

where the function $S = f(x, y)$ describes the z co-ordinate of the submerged body surface relative to axes $Oxyz$. Simplifying Equation (7.6) and carrying out a similar procedure for rotation about the x axis leads to the conditions

$$\int_A x\,\mathrm{d}A = 0, \quad \int_A y\,\mathrm{d}A = 0 \tag{7.7}$$

which prove that rotation of the body at constant submerged volume, and, therefore with equilibrium maintained, can only take place about axes through the centroid of the water plane area. This point is defined as the centre of flotation.

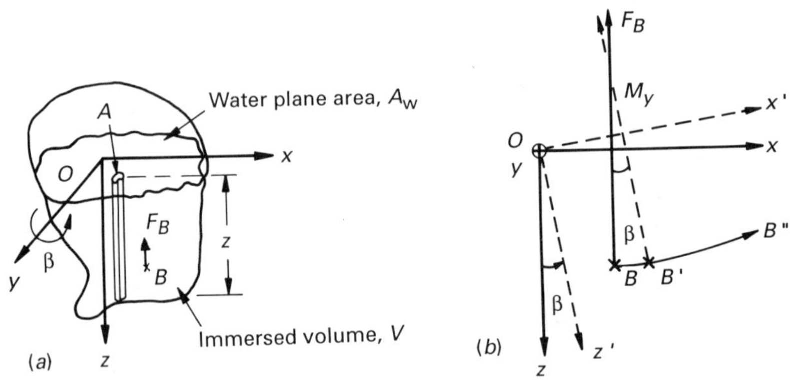

Figure 7.3. Arbitrary body floating freely at the water surface

The co-ordinates of the centre of buoyancy before inclination $(\bar{x}, \bar{y}, \bar{z})$ can be defined by the equations

$$\left.\begin{aligned}\bar{x} &= \frac{1}{V}\int_A Sx\,\mathrm{d}A \\ \bar{y} &= \frac{1}{V}\int_A Sy\,\mathrm{d}A \\ \bar{z} &= \frac{1}{V}\int_A \frac{1}{2}S^2\,\mathrm{d}A\end{aligned}\right\} \quad (7.8)$$

After an inclination of β about the Oy axis, the centre of buoyancy shifts to co-ordinates $(\bar{x}', \bar{y}', \bar{z}')$

$$\left.\begin{aligned}\bar{x}' &= \frac{1}{V}\int_A (S - x\beta)x\,\mathrm{d}A \\ \bar{y}' &= \frac{1}{V}\int_A (S - x\beta)y\,\mathrm{d}A \\ \bar{z}' &= \frac{1}{V}\int_A \frac{1}{2}(S - x\beta)(S + x\beta)\,\mathrm{d}A\end{aligned}\right\} \quad (7.9)$$

with the moments being taken about the initial co-ordinate axes. The movement of the centre of buoyancy parallel to the y axis can then be obtained as

$$\bar{y} - \bar{y}' = \frac{\beta}{V}\int_A xy\,\mathrm{d}A \quad (7.10)$$

This equation indicates that if the cross-product of area of the water plane about Ox and Oy is zero, that is Oxy are principal axes of the water plane area, then the centre of buoyancy will move parallel to the Oxz plane during inclination about the y axis. The condition of zero cross-product of water plane area is satisfied if either Ox or Oy are lines of symmetry of the water plane. Since this condition is satisfied for the majority of floating vessels used in offshore operations, $\bar{y} = \bar{y}'$ and $\bar{x} = \bar{x}'$ are taken to be valid for inclination about the Oy and Ox axes respectively. These conditions also imply that rotation about the Ox and Oy axes can be treated independently in hydrostatic analysis if Oxy are principal axes of the water plane area. On the other hand, if the cross-product of water plane area is non-zero, it is recommended that the hydrostatic analysis is carried out from first principles without the use of the relationships derived in this section.

The vertical movement of the centre of buoyancy can be written as

$$\bar{z} - \bar{z}' = \frac{1}{2V}\beta^2\int_A x^2\,\mathrm{d}A \quad (7.11)$$

with the β^2 indicating that the movement is small or of second order. Also, the horizontal movement of the centre of buoyancy is given by

$$\bar{x} - \bar{x}' = \frac{\beta}{V}\int_A x^2\,\mathrm{d}A_{\mathrm{w}} \quad (7.12)$$

The effect of this is illustrated by Figure 7.3(b) where $BB'B''$ indicates the locus of centre of buoyancy as β increases. The upward shallow curvature of the line is due to the vertical movement in B being of second order compared to the horizontal movement. Furthermore, for small β, the vertical through the centre of buoyancy (B') goes through a fixed point M_y whose position can be defined by using triangle BM_yB' in Figure 7.3(b) to obtain

$$BM_y = \frac{BB'}{\beta} = \frac{\bar{x} - \bar{x}'}{\beta} = \frac{1}{V}\int_A x^2\,\mathrm{d}A = \frac{I_y}{V} \tag{7.13}$$

with I_y being the second moment of water plane area about Oy. Using a similar derivation for rotation about the Ox axis, the relationship

$$BM_x = \frac{I_x}{V} \text{ where } I_x = \int_A y^2\,\mathrm{d}A \tag{7.14}$$

can be obtained. Therefore, for small rotations about the Ox and Oy axes, the centre of buoyancy will move along a surface which is concave upwards but with different curvatures in the Oy and Ox directions – the centres of curvatures being the metacentres M_x and M_y respectively. These M_x and M_y metacentre positions are given by Equations (7.13) and (7.14) and may be considered as the effective point of application of the buoyancy force after rotations about the Ox and Oy axes respectively. At any inclination of the body, the tangent plane to the surface of buoyancy will be parallel to the water plane. Note also that if the axes Oxy are not principal axes of the water plane (that is, $\int_A xy\,\mathrm{d}A \neq 0$), the metacentre positions cannot in general exist since the lines through B and B' in Figure 7.4(b) will not necessarily intersect.

For stable equilibrium, the metacentre positions must both lie above the centre of gravity, although the centre of buoyancy need not do so. The righting moments in the Oyz and Oxz planes, M_x and M_y can then be written as

$$\left.\begin{aligned} M_x &= \rho gV\cdot GM_x\cdot\sin\theta = \rho gV\cdot GZ_x \\ M_y &= \rho gV\cdot GM_y\cdot\sin\beta = \rho gV\cdot GZ_y \\ GZ_x &= GM_x\sin\theta, \quad GZ_y = GM_y\sin\beta \end{aligned}\right\} \tag{7.15}$$

where GM_x, GM_y are metacentric heights, θ and β are inclinations about the Ox and Oy axes respectively and GZ_x, GZ_y are moment arms of the restoring couples.

Calculations for initial stability of a freely floating vessel are carried out by identifying the position of five points within or around the vessel. The lower most point on the vessel vertical centre line, denoted by K (for keel level), is conventionally used as a reference. The positions of vessel centre of gravity G, centre of buoyancy B, and the longitudinal and transverse metacentres M_y and M_x are defined with respect to the keel reference K and using the Ox axis pointing forwards and Oz vertically downwards. The longitudinal and transverse metacentric heights are given by

$$\left.\begin{aligned} GM_y &= KB + BM_y - KG \\ &\text{and} \\ GM_x &= KB + BM_x - KG \end{aligned}\right\} \tag{7.16}$$

since BM_x and BM_y can be directly calculated from Equations (7.14) and (7.13) respectively.

Table 7.1 presents data on typical positions of these points for four floating vessels used on both non-offshore and offshore applications. The relative numerical values for GM_y and GM_x reflect the large length to beam ratios of ship shape vessels compared to semisubmersibles.

Table 7.1 Typical values of initial stability parameters

Vessel type	*Distances (m)*					
	KB	KG	KM_x	KM_y	GM_x	GM_y
Cargo vessel	4.70	8.81	11.93	298	3.12	289
Frigate	2.56	4.94	5.92	250	0.98	245
Destroyer	2.95	6.95	7.80	396	0.85	389
Semisubmersible*	6.98	14.63	20.14	21.04	5.51	6.41

*At drilling draught.

7.2.2 The wall sided formula and large-angle stability

As the angle of inclination of a freely floating vessel at the free surface increases beyond small values, the analysis of the above section no longer holds true. The shape and properties of the water plane area change sufficiently that the metacentre position is no longer fixed. Furthermore, the shape of the submerged vessel volume can change substantially as keel or deck edges emerge or submerge. Large-angle stability is, therefore, important for the design of floating offshore structures where safety or performance at larger angles of inclination need to be evaluated.

An extension of the results of the preceeding section to larger angles can be derived by the so-called 'wall sided' formula. A wall sided vessel is one in which all the vessel's water plane intersecting 'sides' are vertical for some distance above and below the waterline. Most semisubmersibles and conventional ship shape vessels can be considered wall sided to a good degree of approximation since a major proportion of such vessels' sides are perpendicular to the water plane.

For a floating vessel with all its sides perpendicular to the water plane an extension to the small-angle hydrostatic stability result can be derived by considering the cross-section of the vessel, as shown in Figure 7.4, with two water planes corresponding to angles of inclination of zero and β. Since the volumes of the immersed and submerged wedges during inclination will be equal for a wall sided vessel with symmetry about the fore and aft vertical centre plane, the volume of the vessel before and after inclination can be taken as a constant V. Due to inclination through a large angle β, the centre of buoyancy will exhibit a large horizontal shift BB' and a smaller vertical shift $B'B_\beta$. The horizontal shift is given by taking moments of volume in the transverse direction about an axis through O. Thus

$$BB' \cdot V = 2 \times (y \cdot y \tan\beta \cdot \tfrac{1}{2} \cdot L) \cdot \tfrac{2}{3} y \qquad (7.17)$$

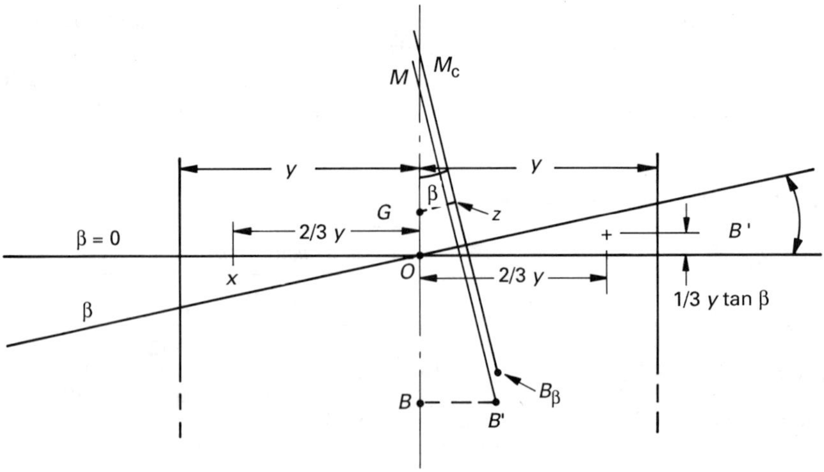

Figure 7.4. Definition diagram for wall sided formula

where L is the length of the vessel with a uniform cross-section. This yields

$$BB' = \frac{I}{V}\tan\beta \text{ for } I = \frac{2}{3}y^3L \tag{7.18}$$

where I is the second moment of area of the initial (at $\beta = 0$) water plane about the axis through O. Furthermore $B'B_\beta$ is given by taking moments of volume in the vertical direction

$$B'B_\beta \cdot V = 2\cdot\tfrac{1}{2}y^2L\tan\beta\cdot\tfrac{1}{3}y\tan\beta$$

$$B'B_\beta = \frac{1}{2}\frac{I}{V}\tan^2\beta \tag{7.19}$$

Taking M and M_c as initial and large angle metacentre positions respectively, BM_c is given by

$$BM_c = BM + MM_c$$

$$= BB'\cot\beta + B'B_\beta$$

$$BM_c = BM(1 + \tfrac{1}{2}\tan^2\beta) \tag{7.20}$$

The wall sided formular result can also be written in terms of GZ (see Equation 7.15) to yield

$$GZ = (GM + \tfrac{1}{2}BM\tan^2\beta)\sin\beta \tag{7.21}$$

The wall sided formula is valid for wall sided vessels up to angles at which deck edge submergence or keel edge submergence occurs. However, as pointed out earlier, most ship shape and semisubmersible vessels are non-wall sided for a small proportion of their water plane cutting surfaces such that the wall sided formula only offers a good approximation to their stability at small to moderate angles.

Calculations of hydrostatic stability beyond the range of validity of the wall sided formula have been carried out in conventional naval architecture by using manual or computer based numerical methods to calculate the submerged volume, position of centre of buoyancy and righting moments from the hull surface definition. Rawson and Tupper (1976) present a summary of manual computations. Some of these have been converted to computer based methods which derive hydrostatics information by more extensive numerical operations on the hull form data. All of these large-angle methods are not directly applicable to the non-ship-shape hull forms used in offshore engineering. A more exact approach based on integration of hydrostatic pressures has been developed for use on offshore vessels and is described in more detail in Section 7.3.

The behaviour of a typical vessel floating at the free surface and inclined at large angles is illustrated by the resultant movements of the centres of buoyancy and of the metacentres. The surface defined by the movement of the centre of buoyancy is called the surface of buoyancy, whereas the surface defined by the centres of curvature (the metacentre) of the surface of buoyancy is defined as the evolute of the surface of buoyancy. Figure 7.5(*a*) and (*b*) show typical shapes of these surfaces. The surface of buoyancy is like an ellipsoidal shape, whereas its evolute is made up of four downward or upward pointing cusps around the body shape.

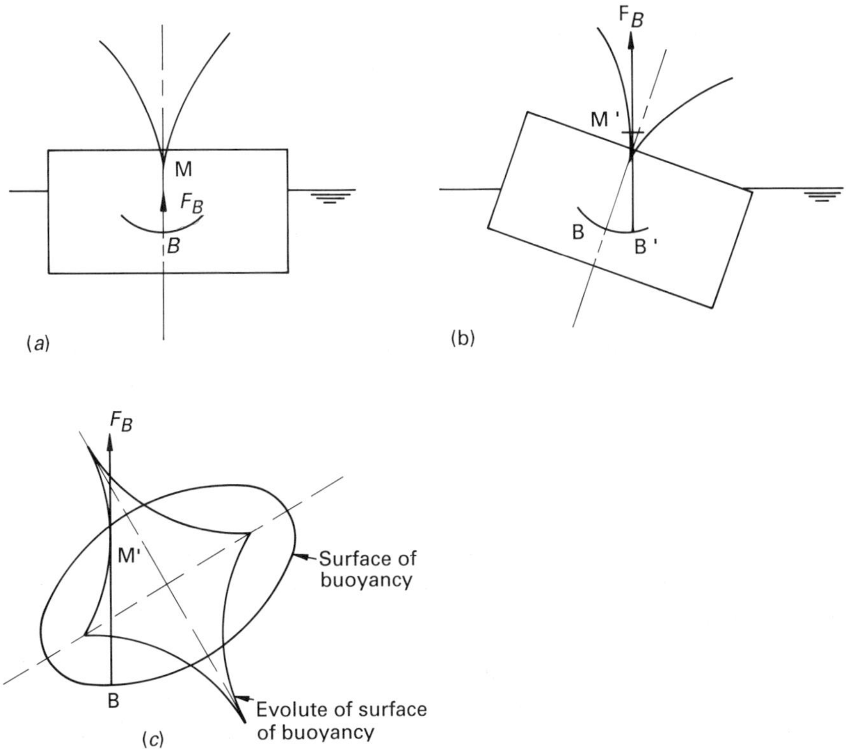

Figure 7.5. Large-angle behaviour of surface of buoyancy and evolute of surface of buoyancy

In practical terms, the large-angle stability characteristics of a hull form can be defined by tabulated or plotted data presented in various ways. Cross-curves of stability (see Figure 7.6(*a*)) present stability moment arms (GZ) against vessel displacement for a range of heel angles. In some cases, the variability of centre of gravity position, G is removed from GZ by replacing it with SZ, where S is an arbitrary fixed point or pole on the vessel vertical centre line; SZ can readily be related to GZ if the positions of S and KG are known. Statical stability curves define the variation of GZ, SZ or righting moment against angle of heel (see Figure 7.6(*b*)). Statical stability curves are often drawn on the same axes as wind heeling curves to check that the vessel has sufficient hydrostatic restoring moment to recover from a heeling moment due to wind and to estimate the angle of heel that will be induced by wind. The area under either of these curves is a measure of the work done to heel the vessel over to any particular angle. A practical means of ensuring stability under wave induced motions is to require that the area under the restoring moment curve exceeds the area under the wind heeling curve by a specified fraction at least. This feature is described in more detail in Section 7.4.

Curves of form can be used to define the values of displacement, KB, KM_y, KM_x, BM_y, BM_x as functions of vessel draught – see Figure 7.6(*c*). Safe vessel loading conditions can be defined by allowable KG curves (Figure 7.6(*d*)) which give the highest permissible centre of gravity heights above the keel as a function of vessel draught. All these curves are usually presented in tabulated and graphical form in the stability manual of a floating offshore vessel.

7.2.3 Stability losses

There are a number of physical effects which can contribute to a loss in the hydrostatic stability of a vessel. These effects arise from the fact that the mass distribution of a vessel can change with angle of inclination and the resultant centre of gravity shift is invariably such as to reduce the hydrostatic restoring moment arm GZ. The stability reductions can occur due to the effects of tanks of fluids with free surfaces inside a vessel, due to freely suspended loads from cranes and also due to loads applied by catenary moorings or riser pipe connections between the floating platform and sea bed. Calculation of these stability losses is essential during definition of the hydrostatic stability of a floating vessel.

The most common source of stability loss is the so-called free surface effect. Consider a floating vessel which has partially filled tanks with free surfaces on board. Inclination of the vessel causes these free surfaces to move such that the fluid in the tanks ‘piles’ up towards the side of tanks closest to the submerging part of the ship. This causes the centre of gravity of the tank fluid contents, and therefore of the vessel, to shift in the direction of movement of the vessel hull’s centre of buoyancy as it inclines. There is a consequent reduction in the magnitude of the restoring couple between vessel weight and the buoyancy force. An expression for this stability loss is obtained as follows.

Consider a tank A with fluid contents of weight w on board a vessel B of weight W as shown in Figure 7.7. When the vessel inclines through a small angle β, the horizontal shift in centre of gravity of the tank contents gg' induces a horizontal shift GG', in the vessel centre of gravity such that

$$W \cdot GG' = w \cdot gg' \tag{7.22}$$

But gg' is identical to the centre of buoyancy shift that would occur if the shape of the tank fluid contents were to be considered as the submerged volume of a freely floating body. Then Equation (7.22) can be written as

$$W \cdot GG' = w \frac{I_1}{V_1} \beta \tag{7.23}$$

where I_1 is the second moment of area of the fluid free surface about an axis through the centroid of the free surface parallel to the axis of rotation of the ship, and V_1 is the volume of liquid in the tank. Converting the centre of gravity shift as a loss in metacentric height, taking $w = \rho_1 g V_1$, where ρ_1 is the density of fluid in the tank, and $W = \rho g V$ where ρ is the density of sea water and V is the submerged volume of the vessel, the metacentric height loss is obtained as

$$\Delta(GM) = \frac{GG'}{\beta} = \frac{\rho_1 I_1}{\rho V} \tag{7.24}$$

For a ship with a large number of internal tanks (say N) with free surfaces, the total metacentric height loss is

$$\Delta(GM) = \frac{1}{\rho V} \sum_{i=1}^{N} \rho_i I_i \tag{7.25}$$

where ρ_i is the fluid density for the contents of the ith tank and I_i is its free surface second moment of area about an axis through the free surface's own centroid and parallel to the vessel's axis of rotation.

Note that the free surface effect is independent of the volume of fluid in the tanks and would disappear if a tank was completely full – in this case the mass of fluid would act as a rigid body during vessel rotation. Free surface effects are normally reduced in the design of a floating offshore vessel by compartmentation of tanks to reduce the value of I_i substantially for each of the tanks. The loss of metacentric height or free surface correction is then subtracted from the rigid vessel metacentric height to give a net value to be considered for vessel evaluation.

Freely suspended masses on board a vessel have a similar effect to free surfaces in that the centre of gravity is shifted in the same horizontal direction as the centre of buoyancy and, therefore, tends to reduce stability. For a vessel with N suspended masses on board, each of mass m_i and with free suspension lengths l_i, the centre of gravity shift is

$$GG' = \beta \frac{1}{W} \sum_{i=1}^{N} m_i g l_i \tag{7.26}$$

Using the same approach as before, the metacentric height loss can be

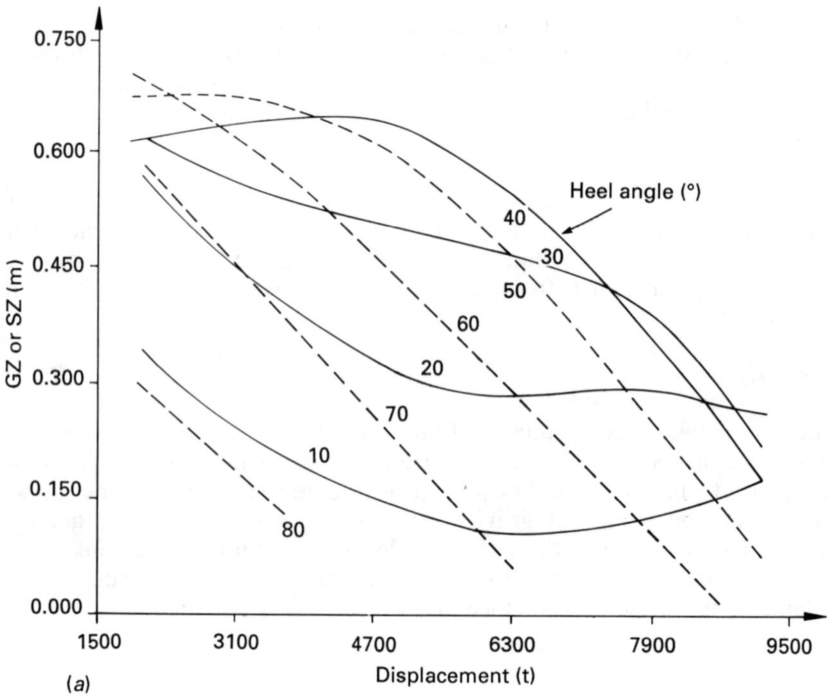
0.750
0.600
0.450
0.300
0.150
0.000
GZ or SZ (m)
Heel angle (°)
40
30
50
60
20
70
10
80
1500
3100
4700
6300
7900
9500
Displacement (t)

(*a*)

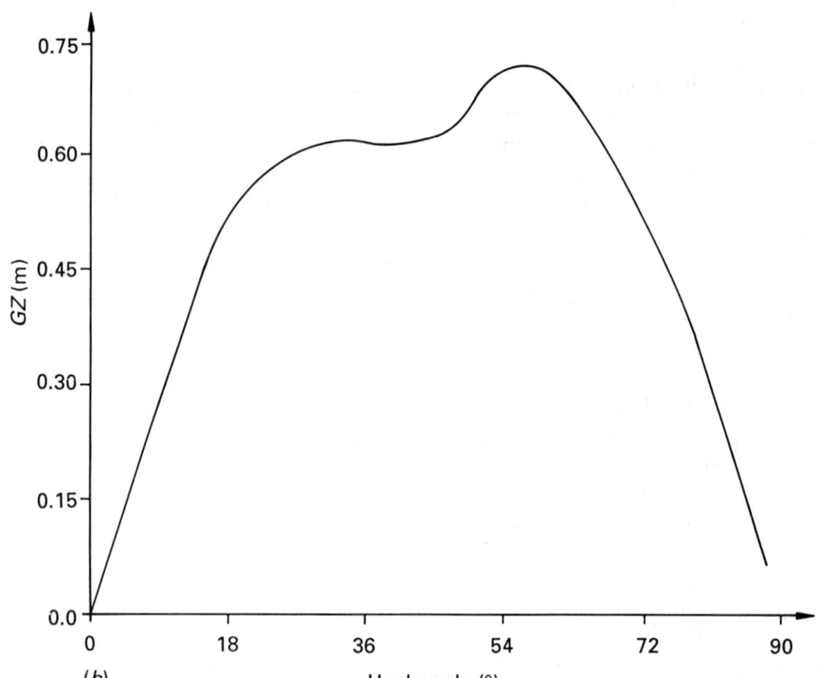
0.75
0.60
0.45
0.30
0.15
0.0
GZ (m)
0
18
36
54
72
90
Heel angle (°)

(*b*)

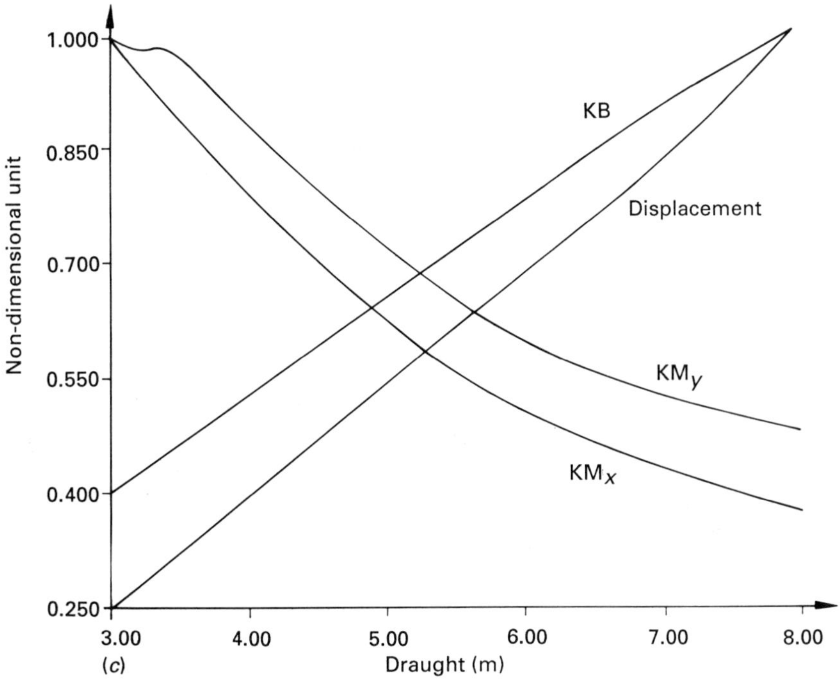

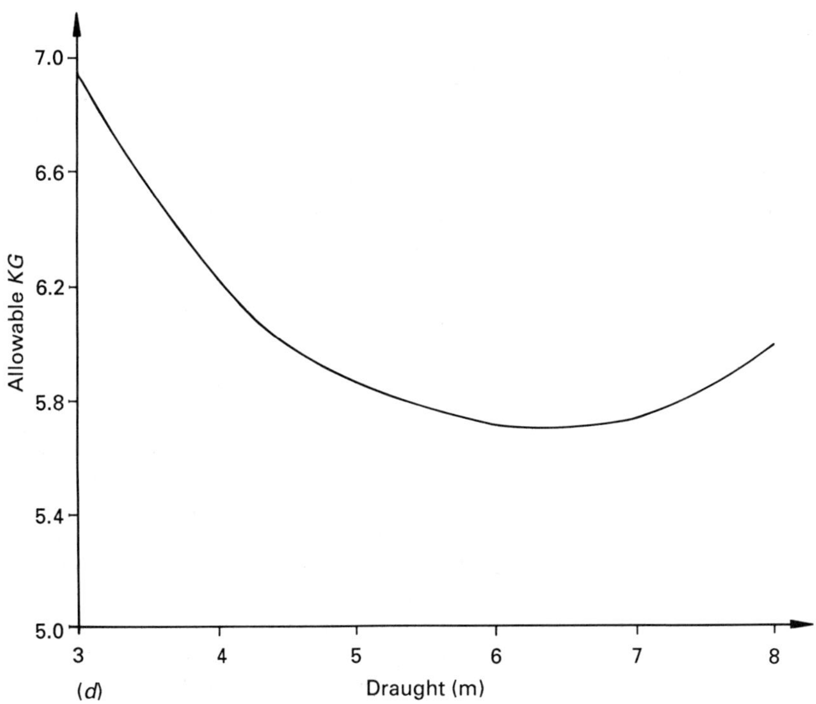

Figure 7.6. Hydrostatic stability curves: (*a*) cross-curves of stability; (*b*) static stability curve; (*c*) curve of form; (*d*) allowable *KG* curve

written as

$$\Delta(GM) = \frac{1}{\rho V}\sum_{i=1}^{N} m_i l_i \qquad (7.27)$$

This correction to metacentric height can be applied by subtracting it from the rigid vessel metacentric height. Alternatively, the centre of gravity calculations of the vessel can be carried out by placing the masses m_i at the top end of the suspension lengths l_i. This has the same effect as the correction defined above.

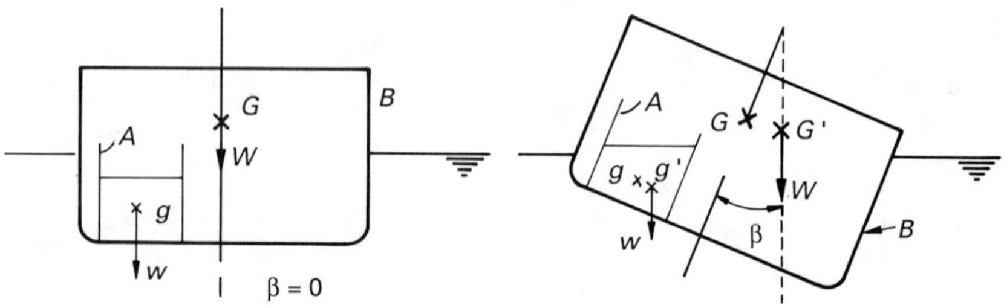

Figure 7.7. Free surface effect

The forces applied on a vessel by a catenary mooring spread can vary with angle of inclination of the vessel to contribute to either some gain or loss in stability, but such corrections are usually small and are neglected for hydrostatics calculations. The generally constant downward force applied by a marine riser to a drilling or production vessel has no effect on vessel hydrostatic stability. However, both moorings and the riser apply net downward forces on the vessel which have to be accounted for in the weight and buoyancy force balance of the vessel.

The hydrostatics of compliant vessels with rigid components that may be connected by freely articulating joints requires special treatment – Patel and Walker (1983) gives an example of one such calculation.

7.3 The pressure integration technique

The difficulty of calculating large-angle hydrostatic stability using submerged volume and water plane area properties for offshore floating vessels has prompted the development of a more fundamental approach. This technique uses an integration of pressure distribution acting over the submerged body surface to yield all the necessary hydrostatic characteristics. The method is computationally more efficient when transformed into a programmed set of instructions for an arbitrary structure at any orientation. Mathematically, the surface integral of pressure can be transformed into a volume integral such that numerical quantities used in conventional naval architecture can be derived directly from the surface integrals. The computational requirement to divide all the vessel surface into panels is analogous to a requirement for surface panels in potential flow boundary integral techniques (see Section 5.6.2).

Consider an arbitrary body shown in Figure 7.8 floating at a free surface with the fluid above the free surface being at a constant pressure. At any point $\mathbf{x}(x, y, z)$ in the fluid below the free surface, there will be an increase in pressure p above the free surface pressure given by the equation

$$p = \rho g(d - z) \tag{7.28}$$

The incremental force $\mathrm{d}\mathbf{F}$ acting on an element of the body surface due to this fluid pressure is

$$\mathrm{d}\mathbf{F} = \rho g(d - z)\mathbf{n}\,\mathrm{d}S \tag{7.29}$$

By integrating over the submerged surface, the total force $\mathbf{F}$ is given by

$$\mathbf{F} = \int_S \rho g(d - z)\mathbf{n}\,\mathrm{d}S \tag{7.30}$$

where $\mathbf{n}$ is the unit normal vector acting into the body and is a function of position $\mathbf{x}$. Similarly, the incremental moment $\mathrm{d}\mathbf{M}$ is

$$\begin{aligned}\mathrm{d}\mathbf{M} &= \mathbf{x} \wedge \mathrm{d}\mathbf{F} \\ &= \mathbf{x} \wedge \rho g(d - z)\mathbf{n}\,\mathrm{d}S\end{aligned} \tag{7.31}$$

Integrating,

$$\mathbf{M} = \int_S \mathbf{x} \wedge \rho g(d - z)\mathbf{n}\,\mathrm{d}S \tag{7.32}$$

Now these surface pressure induced forces and moments will be in equilibrium with the weight of the body acting vertically downwards. Resolving forces and moments due to body weight, therefore, gives

$$\mathbf{F} = W\mathbf{k} \quad \text{and} \quad \mathbf{M} = \mathbf{x}_c \wedge W\mathbf{k} \tag{7.33}$$

where W is the weight of the body and $\mathbf{x}_c$ is the centre of gravity.

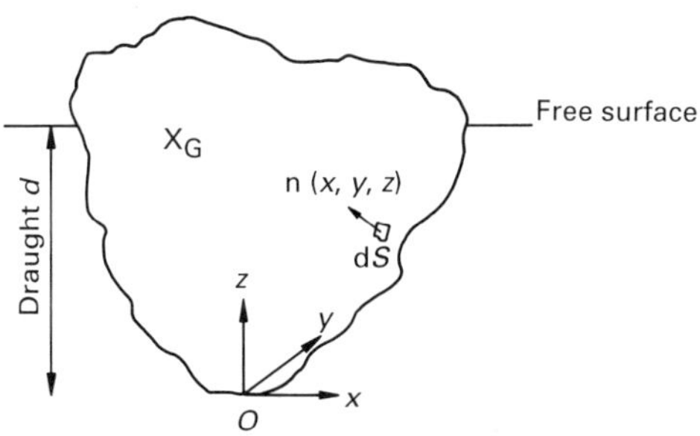

Figure 7.8. Floating body in equilibrium

For static equilibrium, Equations (7.33) must be satisfied since the fluid exerts an upward thrust on the body equal and opposite to the body's weight. This upward thrust, the buoyancy force, must lie along the same line of action as the weight.

Now the divergence theorem (see Appendix D) states that

$$-\int_S \mathbf{f}\cdot\mathbf{n}\,\mathrm{d}S = \int_V \nabla\cdot\mathbf{f}\,\mathrm{d}V \tag{7.34}$$

where $\mathbf{f}$ is a vector field and ∇ is the gradient operator. The negative sign arises because the divergence theorem is written here with the unit normal vector pointing into the surface. From Equation (7.30), the buoyancy force is

$$\mathbf{F} = \int_S \rho g(d - z)\mathbf{k}\cdot\mathbf{n}\,\mathrm{d}S \tag{7.35}$$

Using the divergence theorem, Equation (7.35) becomes

$$\begin{aligned}\mathbf{F} &= -\rho g\int_V \frac{\partial}{\partial z}(d - z)\,\mathrm{d}V\cdot\mathbf{k}\\ &= \rho g V\cdot\mathbf{k}\end{aligned} \tag{7.36}$$

This is Archimedes' principle – the buoyancy force acting on a floating body is equal to the weight of the volume of fluid displaced by the body.

The solutions of Equations (7.33) determine the magnitude of the buoyancy force and the horizontal co-ordinates of the centre of buoyancy. The vertical centre of buoyancy cannot be determined from these equations since the solution for moment only gives the line of action. The solution for the vertical centre of buoyancy is given from the fact that it lies at the centroid of the submerged volume. Thus

$$\mathbf{x}_B\cdot\mathbf{k} = d - \frac{\rho g}{|\mathbf{F}|}\int_S \frac{(d - z)^2}{2}\mathbf{k}\cdot\mathbf{n}\,\mathrm{d}S \tag{7.40}$$

Relating the right-hand side of Equation (7.37) to the right-hand side of Equation (7.34) for the divergence theorem gives

$$\nabla\cdot\mathbf{f} = d - z \tag{7.38}$$

The solution to Equation (7.38) is

$$\mathbf{f} = -\frac{(d - z)^2}{2}\mathbf{k} \tag{7.39}$$

Substituting Equation (7.39) into Equation (7.37) via Equation (7.34) results in

$$\mathbf{x}_B\cdot\mathbf{k} = d - \frac{\rho g}{|\mathbf{F}|}\int_S \frac{(d - z)^2}{2}\mathbf{k}\cdot\mathbf{n}\,\mathrm{d}S \tag{7.40}$$

All the hydrostatic properties can therefore be determined in terms of surface integrals. The advantage of these surface integrals is that they do not have to be evaluated over the water plane area since they go to zero on that surface.

The other characteristics of the water plane can also be derived from surface integrals. Taking $Ox'y'$ as principal axes of the water plane area, these integrals can be written as follows. The second moments of water plane area about the x and y axes are

$$\left.\begin{aligned} I_{x'x'} &= \int_S (\mathbf{x}'\cdot\mathbf{i})^2\mathbf{n}\cdot\mathbf{k}\,\mathrm{d}S\\ &\text{and}\\ I_{y'y'} &= \int_S (\mathbf{x}'\cdot\mathbf{j})^2\mathbf{n}\cdot\mathbf{k}\,\mathrm{d}S\end{aligned}\right\} \tag{7.41}$$

The water plane area A_{w} is given by

$$A_{\mathrm{w}} = \int_S \mathbf{n}\cdot\mathbf{k}\,\mathrm{d}S \tag{7.42}$$

The co-ordinates of the centre of flotation $(x_{\mathrm{F}}, y_{\mathrm{F}})$ are given by

$$x_{\mathrm{F}} = \frac{1}{A_{\mathrm{w}}}\int_S (\mathbf{x}\cdot\mathbf{i})\mathbf{n}\cdot\mathbf{k}\,\mathrm{d}S$$

and

$$y_{\mathrm{F}} = \frac{1}{A_{\mathrm{w}}}\int_S (\mathbf{x}\cdot\mathbf{j})\mathbf{n}\cdot\mathbf{k}\,\mathrm{d}S \tag{7.43}$$

Solutions to Equations (7.33) and (7.40) give the body's equilibrium position. However, nothing is known about the stability of the body in this position. The classical approach to stability is to apply a small linear or rotational displacement to a body and test whether it returns to its equilibrium position of its own accord or if it goes to another equilibrium position. In the theory described here, a small rotation is applied to the body. The point of rotation may be anywhere in space provided that the buoyancy force remains constant. The centre of buoyancy moves to a new position $\mathbf{x}'_B$ and the centre of gravity moves to $\mathbf{x}'_{\mathrm{G}}$. Note that the centre of gravity need not move relative to the rigid body but it would move in a fixed global frame of reference.

Taking moments,

$$\begin{aligned}\mathbf{M}_{\mathrm{R}} &= (-\mathbf{x}'_{\mathrm{G}} \wedge W\mathbf{k}) - (\mathbf{x}'_B \wedge -W\mathbf{k}) \\ &= W(\mathbf{x}'_B - \mathbf{x}'_{\mathrm{G}}) \wedge \mathbf{k}\end{aligned} \tag{7.44}$$

where $\mathbf{M}_{\mathrm{R}}$ is the restoring moment vector. If $\mathbf{M}_{\mathrm{R}}$ is greater than zero in the direction of rotation, then the body is stable. If $\mathbf{M}_{\mathrm{R}}$ is less than zero, the body is unstable and will find a new equilibrium position.

The solution by pressure integration is obtained by evaluating the surface integrals described above. The surface of the body to be analysed is fully covered with a set of m surface elements. A vessel of complex geometry can, therefore, be represented by a collection of simpler surface elements. Equations (7.30) and (7.32) then become

$$\mathbf{F} = \sum_{i=1}^{m} \mathbf{F}_i$$

and

$$\mathbf{M} = \sum_{i=1}^{m} \mathbf{M}_i \tag{7.45}$$

where $\mathbf{F}_i$ and $\mathbf{M}_i$ are the forces and moments acting in the ith surface element; $\mathbf{F}_i$ and $\mathbf{M}_i$ are given by

$$\mathbf{F}_i = \int_{S_i} \rho g(d - z)\mathbf{n}_i\,\mathrm{d}S_i$$

and

$$\mathbf{M}_i = \int_{S_i} \mathbf{x} \wedge \rho g(d - z)\mathbf{n}_i\,\mathrm{d}S_i \tag{7.46}$$

The advantage of the above is that a complex surface integral has been reduced to a summation of a set of known surface integrals. Depending on the type of surface element, Equations (7.46) may be evaluated analytically, numerically, or by a combination of analytic and numerical methods.

Two methods for the solution of these equations are presented here as examples. The first is a complete numerical summation over the element's surface. The element is treated as a region and is subdivided into a set of rectangular subelements. Equations (7.46) then become

$$\left.\begin{aligned} \mathbf{F}_i &= \sum_{p=1}^{M_p} \sum_{q=1}^{M_q} \rho g(d - \mathbf{x}_{pq} \cdot \mathbf{k}) \mathbf{n}_{pq} \Delta h \Delta v \\ &\text{and} \\ \mathbf{M}_i &= \sum_{p=1}^{M_p} \sum_{q=1}^{M_q} \mathbf{x}_{pq} \wedge \rho g(d - \mathbf{x}_{pq} \cdot \mathbf{k}) \mathbf{n}_{pq} \Delta h \Delta v \end{aligned}\right\} \tag{7.47}$$

where $\mathbf{x}_{pq}$ is the co-ordinate of the centroid of the rectangular subelement with dimensions Δh and Δv and a unit normal $\mathbf{n}_{pq}$. There are M_p of these subelements along one submerged edge of the element and M_q along the other.

The second method is a numerical summation of a series of line integrals of finite width. The line integrals are evaluated analytically and are, therefore, exact. Equations (7.46) then become

$$\left.\begin{aligned} \mathbf{F}_i &= \sum_{p=1}^{M_p} \rho g \, \Delta h \int_0^{L_p} (d - \mathbf{x} \cdot \mathbf{k}) \mathbf{n} \, \mathrm{d}l \\ \mathbf{M}_i &= \sum_{p=1}^{M_p} \rho g \, \Delta h \int_0^{L_p} \mathbf{x} \wedge (d - \mathbf{x} \cdot \mathbf{k}) \mathbf{n} \, \mathrm{d}l \end{aligned}\right\} \tag{7.48}$$

where L_p is the submerged length of the pth line element. This reduction of a complete surface integral to a sum of element surface integrals is equally applicable to all the surface integral equations. This solution method is preferred because it is more accurate and computationally faster.

Further details of the analytic solution for line integrals are given by Witz and Patel (1985).

The pressure integration technique can be implemented in a computer program to utilize a library of rectangular, triangular, circular and curved plate elements to discretize the surface of complex ship or semisubmersible hull forms. The surface integrals for each element are evaluated and then summed over the entire surface to yield hydrostatic properties. Two examples of the use of this technique are presented here. Figure 7.9 offers a comparison between the pressure integration technique and the wall sided formula for a rectangular box 10 m by 10 m in plan view with 4 m depth and 2 m draught floating in sea water. The box offers a special case where the wall sided formula is valid over the entire 180° of rotation. The pressure integration technique is in excellent agreement with the analytic result. The two techniques are also compared in Figure 7.11 for the more realistic case of a semisubmersible of the type illustrated in Figure 7.10 by a line drawing of the plate geometry used to model the vessel.

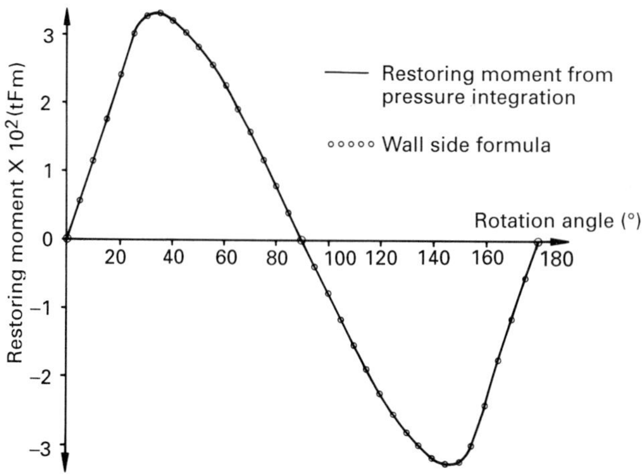

Figure 7.9. Restoring moment of a 10 m × 10 m × 4 m box of 250 tonnes displacement in sea water

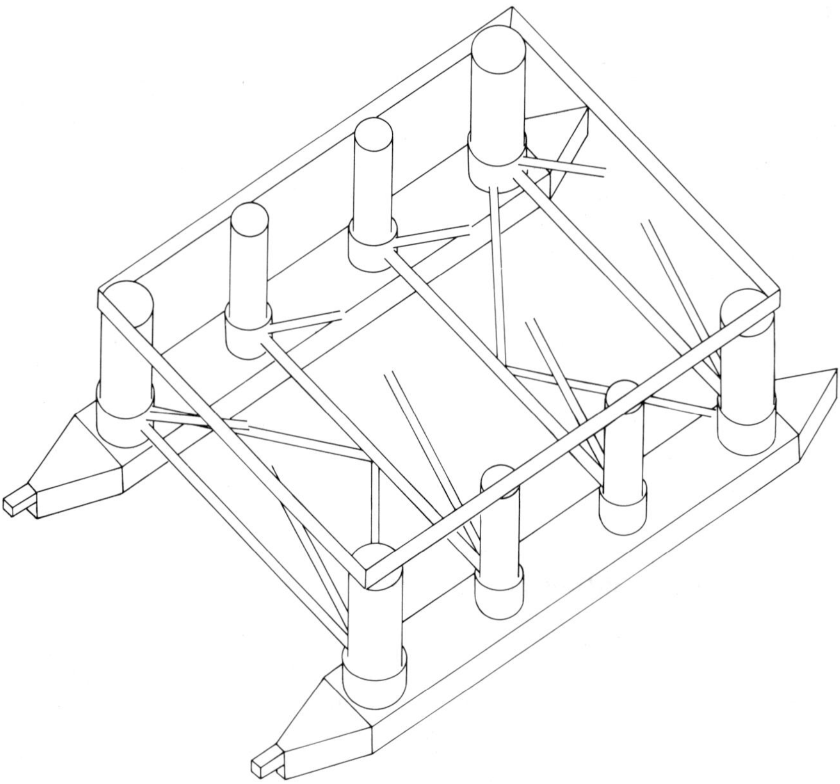

Figure 7.10. Discretization of semisubmersible for pressure integration

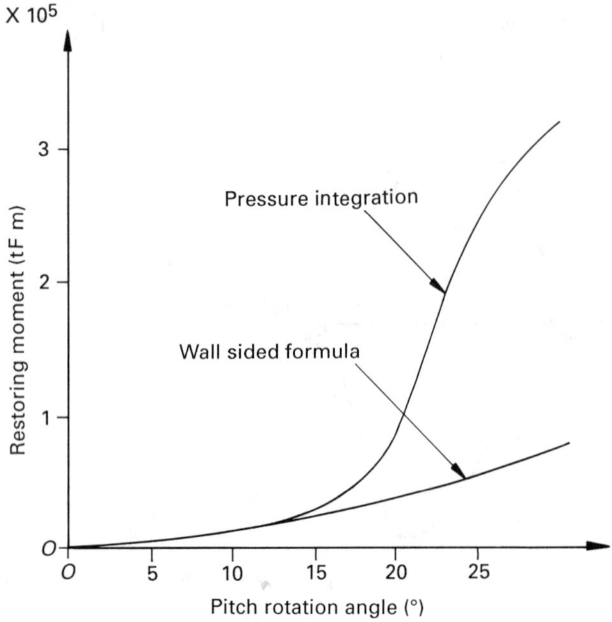

Figure 7.11. Semisubmersible restoring moments

The wall sided formula is often used as the first approximation in evaluating the large-angle restoring moments of geometries which are not strictly wall sided. It is clear here that there are substantial changes in righting moment as the deck edge submerges, with the wall sided formula only remaining valid up to approximately 12°. The semisubmersible case illustrates the advantages and disadvantages of both the classical approach and the pressure integration method to hydrostatic analysis. The classical approach for initial stability (that is, at and around zero heel and trim angles) is easy to apply but is only valid for a small range of heel angles. Extensions of the classical approach to larger heel angles are, however, difficult to apply for complicated vessel geometries. The pressure integration method allows complex geometries such as the semisubmersible of Figure 7.10 to be modelled accurately for large-angle rotations. Its disadvantage is that it is a numerical method which requires the use of a computer.

7.4 Certifying authority regulations

Major ship and offshore structure classification societies such as Lloyds Register of Shipping, Det Norske Veritas and the American Bureau of Shipping act as national certifying authorities for overseeing the design, construction, installation and operation of offshore structures. The societies set out regulations for minimum standards on all aspects of the design and operation of offshore structures. They also produce guidelines for recommended practice in design procedures. The Department of Energy (1986), Lloyds Register of Shipping (1986), Det Norske Veritas (1987) and the American Bureau of Shipping (1987), publish codes of practice listing hydrostatic stability requirements that must be met for certification of a structure.

These national statutory requirements for hydrostatic stability have tended to become more stringent since the mid-1970s. A typical set of requirements for acceptable hydrostatic stability would contain the following major components:

1. Vessel light weight and centre of gravity: this part of the regulations ensure that reliable mandatory procedures are laid down to determine the vessel light weight and centre of gravity position accurately. This is done by means of an inclining test to measure the metacentric height of the light ship vessel coupled with a careful weight audit of the vessel during operations.
2. Definition of operating conditions and hydrostatic particulars: the regulations also require the definition of several common loading and operating conditions for which full hydrostatic particulars need to be computed and made available in the operating manual. The stability data includes the type of information presented in Figure 7.6.
3. Intact stability criteria: these are designed to ensure that the righting moment against angle of heel curve for the intact (that is, undamaged) vessel is of sufficient magnitude and range to withstand static overturning moments due to wind and also to tolerate wave induced overturning moments and heel angles. The latter dynamic aspect of vessel stability is difficult to quantify for design purposes but an approximate approach has been evolved by defining minimum area ratios under the hydrostatic righting moment and wind heeling overturning moment curves. These criteria are described in more detail below.
4. Damage assumptions and stability: the rules require that the vessel remains safe in terms of sinking or capsizing following a certain level of reasonable damage and flooding. The rules lay down recommended damage assumptions and the stabiity criteria to be met after flooding. These are normally somewhat less stringent than those for the intact vessel.
5. Survival to extreme damage: recent capsizing accidents with semisubmersible vessels have also led to extreme damage rules which require that a specified level of vessel survivability is maintained after extreme damage such as loss of an entire vertical column.
6. Watertight integrity and freeboard: the rules for these aspects of a vessel's hydrostatic design are linked to meeting the requirements for 3, 4 and 5 above but are concerned with crew safety, vessel survival after an accident and wave induced damage in severe storms.

Most rules for mobile offshore units differentiate between ship shape, semisubmersible and jack-up rigs during the transportation phase in setting slightly different criteria for each.

In order to illustrate the above requirements further, a summary of typical hydrostatic stability rules for a semisubmersible unit are presented below:

(i) *Operating conditions*

1. transit condition at low draught with pontoons piercing the water surface;
2. operating conditions at normal drilling draught with various combinations of payload and riser pipe connections;
3. survival condition with a draught slightly less than that used for normal operations to improve vessel sea keeping in extreme waves.

(ii) *Intact stability criteria* These are expressed in relation to the example statical stability curve presented in Figure 7.12 where the wind heeling moment curve intercepts the hydrostatic righting moment curve at angles of heel of β_1 and β_2 with A, B, and C denoting the areas between the curve boundaries shown:

1. Wind speed in operating and transit condition = 36.0 m/s.
   Wind speed in survival condition = 51.5 m/s.
2. The initial metacentric height after correction for free surface effects shall not be less than 1.0 m at transit, survival and operating draft and shall not be less than 0.3 m at temporary draft.
3. The area ratio of righting moment curve to wind heeling moment curve to the assumed down flooding point or second intercept should not be less than 1.3, that is,

$$\frac{A + B}{B + C} \geqslant 1.3 \tag{7.49}$$

   The down flooding point refers to a condition or angle of heel at which a vessel opening will fall below still water level and cause progressive flooding to occur.
4. The angle of heel due to the wind heeling moment should not be greater then 15°, that is $\beta_1 \leqslant 15°$. This is the angle at the first crossing of the righting moment and wind heeling moment curves.
5. The second crossing of the righting moment and heeling moment curves shall occur at an angle of heel of at least 30°, that is, $\beta_2 \geqslant 30°$.

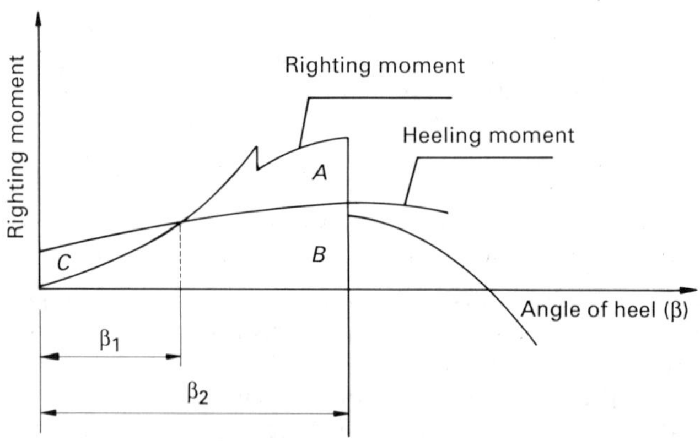

Figure 7.12. Typical righting moment and wind heeling moment against angles of heel

(iii) *Damaged stability criteria* Following damage and flooding, a typical statical stability curve will look similar to that in Figure 7.12 but the curve will be shifted to the right by a steady heel angle β_0. Then, typical criteria are:

1. Wind speed = 25.8 m/s
2. Damage stability shall be investigated at transit and operating drafts.
3. Any one compartment adjacent to the sea shall be assumed to be flooded regardless of location.
4. In additon to 3, the damage is normally assumed to occur at any water level between 5 m above and 3 m below the waterline in question and at least one watertight bulkhead should be assumed damaged. The vertical extent is assumed to be 3 m within the above zone with the horizontal extent being taken to be 3 m measured along the periphery of the vessel's external shell. The penetration of the damage zone into the vessel is assumed to be a horizontal distance of 1.5 m measured radially from the shell.
5. After damage, the angle of heel should not be greater than 15°, that is $\beta_0 \leqslant 15°$.
6. The area under the righting moment curve up to the second crossing or any lesser angle shall be equal to or greater than the area under the wind heeling moment curve to the same limiting angle. Thus

$$\frac{A + B}{B + C} \geqslant 1.0 \tag{7.50}$$

(iv) *High-energy extreme damage criteria* The semisubmersible will need to be provided with sufficient means of buoyancy in the deck structure to remain afloat and satisfy the criteria described below, upon the loss of buoyancy equivalent to the volume of the whole – or a major part – of any one column. The loss of buoyancy is assumed to occur when the platform is at maximum operating draught and with the maximum allowable KG.

After this extreme loss of buoyancy, the semisubmersible should satisfy the following criteria:

1. The angle of heel should not exceed 35° in any direction.
2. Not taking any effect of wind and waves into account, the waterline in the final condition of equilibrium should be at least 0.6 m below the lower edge of any opening through which progressive flooding may take place.
3. The righting moment GZ-curve in the damaged condition shall have a positive extent of at least 20° beyond equilibrium together with a maximum GZ-value of at least 1.0 m.

These criteria for high-energy extreme damage can only be met by ensuring that the vessel deck is a watertight structure capable of providing large righting moments after major loss of buoyancy in a vertical column.

References

American Bureau of Shipping, (1987), *Rules for Building and Classing Mobile Offshore*

Drilling Units, ABS, 45 Eisenhower Drive, PO Box 910, Paramus, NJ, USA 07653-0910

Department of Energy, (1986), *Offshore Installations, Guidance on design and construction*, Part III, Section 4.3, London, HMSO

Det Norske Veritas, (1987), *Rules for Classification of Mobile Offshore Units*, Det Norske Veritas, PO Box 300, N-1322, Hovik, Oslo, Norway

Lloyds Register of Shipping, (1986), *Rules and Regulations for the Classification of Mobile Offshore Units*, January, Part IV, Chapter 1, Sections 2, 3, 4 and 5, Lloyds Register of Shipping, 71 Fenchurch Street London, EC3 4BS

Patel, M. H. and Walker, S. (1983), On the hydrostatics of floating bodies with articulated appendages, *Transactions of the Royal Institution of Naval Architects*, **125**, July, 229–236

Rawson, K. J. and Tupper, E. C. (1976), *Basic Ship Theory*, Volume 1, Longman, London

Witz, J. A. and Patel, M. H. (1985), A pressure integration technique for hydrostatic analysis, *Transactions of the Royal Institution of Naval Architects*, **127**, 285–294

Chapter 8

Dynamic response of floating structures in waves

Floating offshore structures such as semisubmersibles, ships and loading buoys are playing an increasing role in offshore operations as the escalating cost of bottom emplaced structures becomes uneconomic for small fields or greater water depths. Compliant structures such as tensioned buoyant platforms, articulated towers and single anchor leg moored vessels are also examples of floating structures with a greater level of mooring restraint than a free or catenary moored floating vessel. The wave induced motions of these vessels and consequent structural loads are crucial to their operating effectiveness. In particular, wave induced motions contribute to down time in severe weather when excessive vessel motions prevent connection of the vessel to a drilling or production work site on the sea bed. Thus the design of a floating vessel with low wave induced motions, a high level of station keeping in response to steady environmental forces, large payload capacity and low cost offers the best combination of characteristics for improving the effectiveness of offshore operations.

Section 8.2 shows that the fluid structure interactions of a floating vessel can be reduced to the classical second-order linear differential equations of motion for a multi degree of freedom system with inertial, damping and stiffness components. It is, therefore, instructive to examine the free and forced response of such a dynamic system first.

8.1 Dynamic response analysis

The dynamic response of a single degree of freedom linear, second-order system is considered first. Such a system can be described by the equation of motion

$$m\ddot{z} + b\dot{z} + kz = F(t) \tag{8.1}$$

where z is the motion component, m is the mass, b is a linear damping coefficient, k is stiffness and $F(t)$ is the exciting force. The coefficients m, b and k are all taken to be positive.

8.1.1 Free response

A free response of the above equation can be obtained by solving the characteristic equation obtained by substituting the solution

$$z = Ae^{\lambda t} \tag{8.2}$$

into Equation (8.1) with the right-hand side set to zero. The two roots of the characteristic equation are

$$\lambda = -\frac{b}{2m} \pm \sqrt{\left[\left(\frac{b}{2m}\right)^2 - \frac{k}{m}\right]} \tag{8.3}$$

If $(b/2m)^2 \geqslant (k/m)$, then both roots are negative and real, giving no oscillatory motion. If $(b/2m)^2 < (k/m)$, then the two roots are complex conjugates and the motion takes the form of a harmonic decaying vibration described by the equation

$$\begin{aligned} z(t) = {} & A_1 \exp[-(\mathrm{b}/2\mathrm{m})\mathrm{t}] \cdot \exp\left\{\mathrm{i}\sqrt{\left[\frac{k}{m} - \left(\frac{b}{2m}\right)^2\right]} \cdot t\right. \\ & + A_2 \exp[-(\mathrm{b}/2\mathrm{m})\mathrm{t} \cdot \exp\left\{-\mathrm{i}\sqrt{\left[\frac{k}{m} - \left(\frac{\mathrm{b}}{2m}\right)^2\right]} \cdot t\right. \end{aligned} \tag{8.4}$$

This equation can be rewritten by using the undamped natural frequency ω_0,

$$\left.\begin{aligned} &\omega_0 = \sqrt{\left(\frac{k}{m}\right)} \\ &\text{the damping ratio } \beta, \\ &\beta = \frac{b}{2\sqrt{(mk)}} \\ &\text{and the damped natural frequency } \omega_{\mathrm{d}}, \\ &\omega_{\mathrm{d}} = \sqrt{\left[\frac{k}{m} - \left(\frac{b}{2m}\right)^2\right]} \\ &\quad\;= \omega_0\sqrt{[(1-\beta^2)]} \end{aligned}\right\} \tag{8.5}$$

Thus

$$z(t) = \exp(-\beta\omega_0 \mathrm{t})(A_1 \cos \omega_{\mathrm{d}} t + A_2 \sin \omega_{\mathrm{d}} t) \tag{8.6}$$

The constants A_1 and A_2 are found from boundary conditions. Taking z_0 and $\dot{z}_0$ as the initial displacement and velocity at time $t = 0$, A_1 and A_2 can be obtained and substituted into Equation (8.6) to give

$$z(t) = \exp(-\beta\omega_0 \mathrm{t})\left(\frac{\dot{z}_0 + \beta\omega_0 z_0}{\omega_{\mathrm{d}}} \sin \omega_{\mathrm{d}} t + z_0 \cos \omega_{\mathrm{d}} t\right) \tag{8.7}$$

Figure 8.1 illustrates one possible form of $z(t)$ with finite z_0 and $\dot{z}_0$ equal to 0. It is observed that a decaying oscillatory motion response is obtained at a 'natural' frequency of oscillation ω_{d}, and an amplitude decay given by the expontential term in Equation (8.7). As the damping coefficient β tends to zero, the response of Figure 8.1 becomes an oscillation at constant amplitude and undamped natural frequency ω_0. At the other extreme, however, as β tends to 1, the damped natural frequency ω_{d} tends to zero with

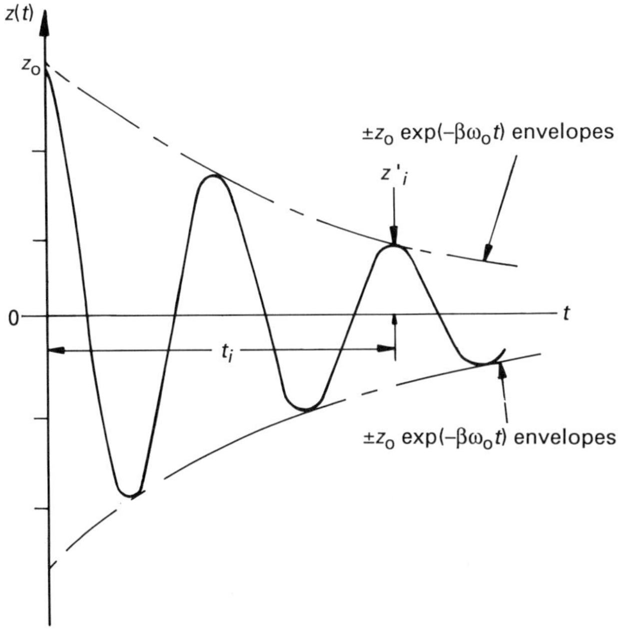

Figure 8.1. Response of single degree of freedom system with linear damping and initial displacement

$$\left(\frac{b}{2m}\right)^2 = \frac{k}{m} \text{ or } b = 2\sqrt{(mk)}$$

In this case, Equation (8.7) becomes

$$z(t) = \exp(-\omega_0 t)[\dot{z}_0 t + (1 + \omega_0 t)z_0] \qquad (8.8)$$

which describes a decaying motion that does not reverse direction. This motion, shown in Figure 8.2, is described as being critically damped. Thus, for a general decaying response, the damping ratio β also defines the non-dimensionalized damping as a proportion of critical damping.

Most floating offshore structures exhibit values of damping ratio in the range 0.05–0.25. For this range of damping ratios the undamped and damped natural frequencies are nearly equal since for $\beta = 0.25$,

$$\omega_d = \omega_0\sqrt{(1 - 0.0625)} = 0.9375\omega_0$$

The damping present in a floating structure or its scaled model can be measured by recording its free vibration in still water and noting the peak values of consecutive cycles. Then the logarithmic decrement δ is defined as

$$\delta = \ln\left(\frac{z'_i}{z'_{i+1}}\right)$$

where z'_i and z'_{i+1} are the ith and $(i + 1)$th peaks of the damped response shown in Figure 8.1. The general form of this equation must exclude the first apparent peak at $t = 0$, although for very lightly

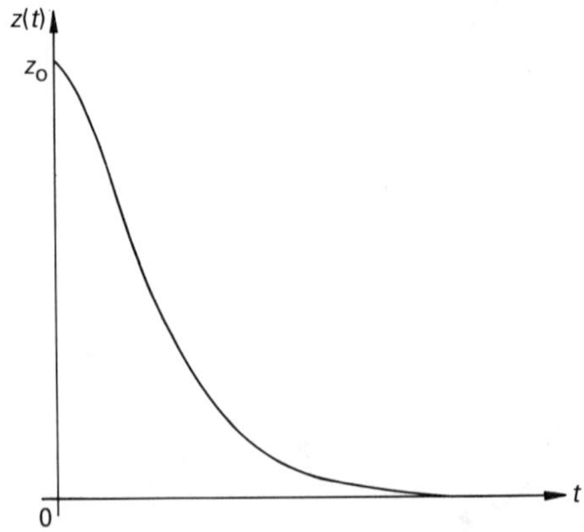

Figure 8.2. Response of single degree of freedom system with critical linear damping and initial displacement

damped systems, the value of z at $t = 0$ can be regarded as the first peak. Taking T_d as the damped natural period, the above equation can be written as

$$\delta = \ln\left\{\frac{\exp(-\beta\omega_0 t_i)}{\exp[-\beta\omega_0(t_i + T_d)]}\right\} = \ln[\exp(\beta\omega_0 T_d)]$$

$$= \beta\omega_0 T_d = \frac{2\pi\beta}{\sqrt{(1-\beta^2)}} \tag{8.9}$$

For lightly damped systems, $\sqrt{(1-\beta^2)} \simeq 1$ and β can be obtained by a least squares fit to the equation

$$\ln z_i' = -\beta\omega_0 t_i + \ln z_0 \tag{8.10}$$

where t_i and z_i' are the co-ordinate of the ith peak value of the damped response (see Figure 8.1) and z_0 is an initial displacement.

The free response of the system can also be used to develop the idea of an impulse response function. If an impulse Q is applied to the mass m when it is initially at rest at time $t = 0$, the mass will acquire an initial velocity of Q/m and from Equation (8.7) the system response is given by

$$z(t) = \frac{Q}{m\omega_d}\exp(-\beta\omega_0 t)\cdot\sin\omega_d t \text{ for } t \geqslant 0 \tag{8.11}$$

Therefore, if the impulse Q had been applied at time $t = \tau$, then the resulting motion would be

$$z(t) = \frac{Q}{m\omega_d}\exp[-\beta\omega_0(t-\tau)]\sin\omega_d(t-\tau) \text{ for } t \geqslant \tau$$

$$= Q\,h(t-\tau) \tag{8.12}$$

where

$$h(t-\tau) = \frac{1}{m\omega_d}\exp[-\beta\omega_0(t-\tau)]\sin\omega_d(t-\tau) \tag{8.13}$$

is the response function at time t for unit impulse at time τ.

The impluse response function can be extended to determine system response to a stationary random force time history $F(t)$ by considering this force to be made up of a series of impulses of magnitude $F(\tau)\cdot\delta\tau$. The response $z(t)$ is then the sum of the responses to each of these impulses. Thus

$$z(t) = \int_0^t F(\tau)h(t - \tau)\,\mathrm{d}\tau \tag{8.14}$$

This formulation for the response is considered further in the next section.

8.1.2 Forced response to sinusoidal and random inputs

A sinusoidal force may be applied to the single degree of freedom system of Equation (8.1) by substituting

$$F(t) = F_a e^{i\omega t} \tag{8.15}$$

Taking a particular integral of the form

$$z = z_a e^{i\omega t} \tag{8.16}$$

and substituting this with Equation (8.15) into (8.1) gives

$$\begin{aligned} z &= \frac{F_a(k - m\omega^2 - ib\omega)}{(k - m\omega^2)^2 + b^2\omega^2} e^{i\omega t} \\ &= H(\omega)F_a e^{i\omega t} \end{aligned} \tag{8.17}$$

where function $H(\omega)$ is the complex transfer function between force and motion amplitude. Taking the real parts of Equations (8.15) and (8.17) also yields the governing equation and solution to be of the form

$$\left.\begin{aligned} &m\ddot{z} + b\dot{z} + kz = F_a \cos\omega t \\ &z = \frac{F_a[(k - m\omega^2)\cos\omega t + b\omega\sin\omega t]}{(k - m\omega^2)^2 + b^2\omega^2} \\ &\quad = \frac{F_a}{[(k - m\omega^2)^2 + b^2\omega^2]^{1/2}}\cos(\omega t - \varepsilon) \\ &\text{where} \\ &\varepsilon = \arctan\left(\frac{b\omega}{k - m\omega^2}\right) \end{aligned}\right\} \tag{8.18}$$

The total solution for Equations (8.1) and (8.15) is made up of the sum of the complementary function representing the transient response of Equation (8.6) and the particular integral describing the steady state response of Equations (8.17) or (8.18).

If z_s is taken to be the displacement to a static force of magnitude F_a, r as the ratio of forcing frequency to natural frequency and β as the ratio of actual damping to critical damping, the magnitude of the transfer function $H(\omega)$ can be written in terms of a dynamic amplification factor as

$$\left.\begin{aligned}
&\frac{z_0}{z_s} = k|H(\omega)| = \frac{1}{[(1 - r^2)^2 + (2\beta r)^2]^{1/2}} \\
&\text{where} \\
&r = \omega/\omega_0,\ \beta = b/2\surd(mk) \text{ and } z_s = F_a/k \\
&\text{and the phase lag is} \\
&\varepsilon = \arctan\left(\frac{2\beta r}{1 - r^2}\right)
\end{aligned}\right\} \quad (8.19)$$

Figure 8.3 describes the variation of z_0/z_s and ε with frequency ratio r for a range of values of the damping ratio β. This diagram illustrates the effects of frequency on a resonant system with large dynamic amplification at frequency ratio of 1, a quasistatic response at very low frequency ratios and an asymptotic response tending to zero at large frequency ratios. The phase lags also vary from 0° to 180° with a value of 90° at resonance.

The impulse response function defined in Section 8.1.1 can be related to the transfer function $H(\omega)$ through a Fourier transform. An arbitrary function $z(t)$ can be related to its frequency composition $Z(\omega)$ by the well-known Fourier transform through the pair of equations

$$\left.\begin{aligned}
&z(t) = \int_{-\infty}^{\infty} Z(\omega)e^{i\omega t}d\omega \\
&\text{and} \\
&Z(\omega) = \frac{1}{2\pi}\int_{-\infty}^{\infty} z(t)e^{-i\omega t}dt
\end{aligned}\right\} \quad (8.20)$$

If the system described by Equation (8.1) is excited by a sinusoidal force of amplitude F_a and frequency ω, then the response $z(t)$ is

$$z(t) = H(\omega)F_a e^{i\omega t} \quad (8.21)$$

Summation by integration of many such sinusoidal components will lead to the equation

$$z(t) = \int_{-\infty}^{\infty} H(\omega)F_a(\omega)e^{i\omega t}d\omega \quad (8.22)$$

with $F_a(\omega)$ written as

$$F_a(\omega) = \int_{-\infty}^{\infty} F(t)e^{-i\omega t}dt \quad (8.23)$$

Comparing Equations (8.19) and (8.22) then yields

$$Z(\omega) = H(\omega)F_a(\omega) \quad (8.24)$$

Now, if the system is excited by the unit impulse defined by

$$\left.\begin{aligned}
&\delta(t) = 0 && \text{for } t \neq 0 \\
&\delta(t)\cdot\Delta t \to 1 && \text{for } t = 0
\end{aligned}\right\} \quad (8.25)$$

then $F_a(\omega)$ in Equation (8.23) becomes unity and the dynamic system response will be the impulse response function $h(t)$. Equation (8.22)

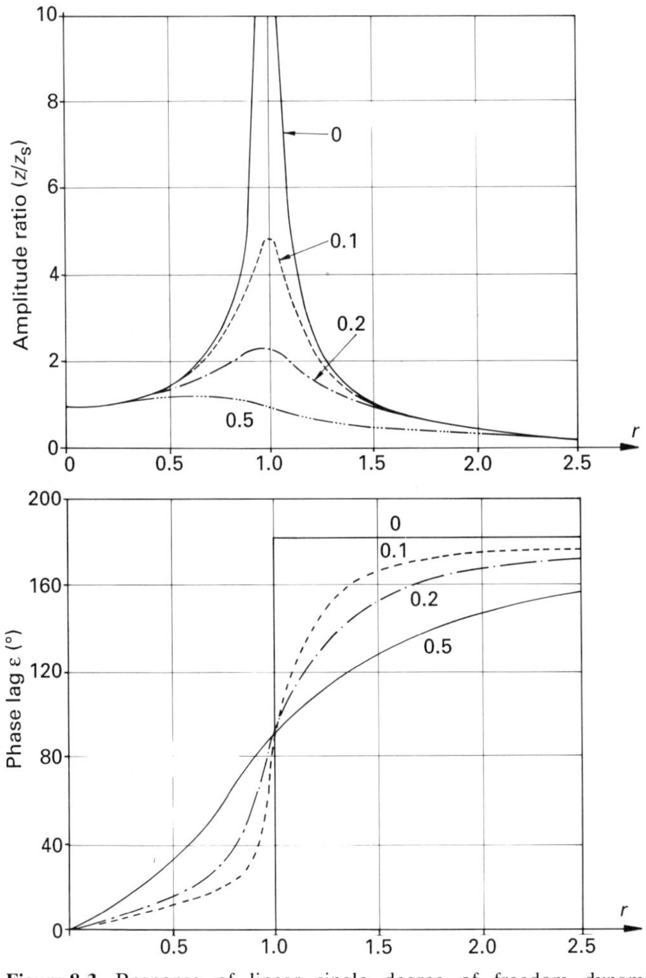

Figure 8.3. Response of linear single degree of freedom dynamic system to forced vibration. Curve labels indicate values of damping ratio β

then becomes

$$\left.\begin{aligned} h(t) &= \int_{-\infty}^{\infty} H(\omega)\mathrm{e}^{\mathrm{i}\omega t}\mathrm{d}\omega \\ &\text{Its Fourier transform is} \\ H(\omega) &= \frac{1}{2\pi}\int_{-\infty}^{\infty} h(t)\mathrm{e}^{-\mathrm{i}\omega t}\mathrm{d}t \end{aligned}\right\} \quad (8.26)$$

Thus the impulse response function and transfer function form a Fourier transform pair.

The response of a single degree of freedom dynamic system to random excitation is now obtained by extending the ideas of random processes in Section 4.7.1 to examining system response. Consider the governing equation given by Equation (8.1) and take $F(t)$ to denote a stationary random excitation force with frequency spectrum $S_f(\omega)$ and auto-correlation function $R_f(t)$. The dynamic system response z will then have some corresponding frequency spectrum $S_z(\omega)$ and auto-correlation function $R_z(t)$. The latter can be rewritten here as

$$R_z(\tau) = \operatorname*{Lim}_{T\to\infty} \frac{1}{T}\int_0^T z(t)z(t+\tau)\,\mathrm{d}t \tag{8.27}$$

using its definition from Section 4.7.1.

The response of a dynamic system to random excitation is equivalent to the system responding to a series of impulses of magnitude $F(\tau)\cdot\delta\tau$ with a response equation given by Equation (8.14). Replacing $t-\tau$ by τ_1 in this equation yields

$$z(t) = \int_0^\infty F(t-\tau_1)h(\tau_1)\mathrm{d}\tau_1 \tag{8.28}$$

and also

$$z(t+\tau) = \int_0^\infty F(t+\tau-\tau_2)h(\tau_2)\mathrm{d}\tau_2 \tag{8.29}$$

by replacing t with $t+\tau$ and τ_1 by τ_2 in Equation (8.28). Substituting Equations (8.28) and (8.29) into (8.27) yields

$$R_z(\tau) = \int_0^\infty h(\tau_1)\int_0^\infty h(\tau_2)\cdot$$

$$\operatorname{Lim}\frac{1}{T}\int_0^T F(t-\tau_1)F(t+\tau-\tau_2)\,\mathrm{d}t\,\mathrm{d}\tau_2\,\mathrm{d}\tau_1$$

$$= \int_0^\infty h(\tau_1)\int_0^\infty h(\tau_2)R_f(\tau-\tau_2+\tau_1)\,\mathrm{d}\tau_2\,\mathrm{d}\tau_1$$

But

$$S_z(\omega) = \frac{1}{\pi}\int_{-\infty}^\infty R_z(\tau)\mathrm{e}^{-\mathrm{i}\omega\tau}\mathrm{d}\tau$$

$$= \frac{1}{\pi}\int_{-\infty}^\infty \mathrm{e}^{-\mathrm{i}\omega\tau}\int_0^\infty h(\tau_1)\int_0^\infty h(\tau_2)R_f(\tau-\tau_2+\tau_1)\,\mathrm{d}\tau_2\,\mathrm{d}\tau_1\,\mathrm{d}\tau \tag{8.30}$$

Since $h(\tau) = 0$ for $t < 0$, this equation becomes

$$S_z(\omega) = \frac{1}{\pi}\int_{-\infty}^\infty h(\tau_1)\exp(\mathrm{i}\omega\tau_1)\int_{-\infty}^\infty h(\tau_2)\exp(-\mathrm{i}\omega\tau_2)$$

$$\times\int_{-\infty}^\infty R_f(\tau-\tau_2+\tau_1)\exp[-\omega(\tau+\tau_1-\tau_2)]\mathrm{d}\tau\cdot\mathrm{d}\tau_2\,\mathrm{d}\tau_1$$

$$= \int_{-\infty}^\infty h(\tau_1)\exp(\mathrm{i}\omega\tau_1)\int_{-\infty}^\infty h(\tau_2)\exp(-\mathrm{i}\omega\tau_2)$$

$$\times\frac{1}{\pi}\int_{-\infty}^\infty R_f(\tau_3)\exp(-\mathrm{i}\omega\tau_3)\mathrm{d}\tau_3\,\mathrm{d}\tau_2\,\mathrm{d}\tau_1 \tag{8.31}$$

where $\tau_3 = \tau - \tau_2 + \tau_1$. Substituting from Equations (8.26) and (8.30) into Equation (8.31) yields

$$S_z(\omega) = H(-\omega)H(\omega)S_f(\omega) \tag{8.32}$$

Inspection of the form of $H(\omega)$ in Equation (8.17) shows that $H(-\omega)$

and $H(\omega)$ are complex conjugates permitting Equation (8.32) to be written as

$$S_z(\omega) = |H(\omega)|^2 S_f(\omega) \tag{8.33}$$

Thus, if the system is excited by a stationary, random force of Gaussian probability distribution and spectral density $S_f(\omega)$, the system response will also have a Gaussian probability distribution with spectral density $S_z(\omega)$.

All the properties of random processes described in Section 4.7 can then be used to deduce the properties of the system response. Note that the motion of an offshore structure in waves can be regarded as the product of two transfer functions – the first from the wave elevation to the exciting force and the second from the exciting force to motion response. If the wave elevation spectrum is narrow banded, then the exciting force and motion response spectra will also be narrow banded, although their shapes could be substantially distorted by the transfer functions involved.

The dynamic response analysis techniques presented above can be readily extended to multiple degrees of freedom by writing the equations of motion in matrix form

$$\mathbf{M}\ddot{\mathbf{X}} + \mathbf{B}\dot{\mathbf{X}} + \mathbf{K}\mathbf{X} = \mathbf{F}(t) \tag{8.34}$$

where the $(n \times 1)$ column vector of displacements $\mathbf{X}$ holds the n degrees of freedom that characterize the response of the structure, and $\mathbf{F}(t)$ denotes the $(n \times 1)$ column vector of applied forces. All the coefficients of the n equations of motion implied in Equation (8.34) are incorporated in $(n \times n)$ matrices of mass, damping and stiffness ($\mathbf{M}$, $\mathbf{B}$ and $\mathbf{K}$). The magnitude and distribution of coefficients in these matrices reflect the dynamic characteristics of each degree of freedom and of the couplings between them.

The free and forced response analysis presented in this section can be readily extended to apply to systems with multiple degrees of freedom. A detailed description of this extension and the available methods of solution are described in Sections 6.1 and 6.2 where these techniques are applied to solve for structural deformations of fixed offshore structures in very many degrees of freedom. For floating structures which can be regarded as rigid bodies (that is, structural deformations are very small compared to wave induced motions), it is usual to use only the six degrees of freedom for the surge, sway, heave, roll, pitch and yaw motions of the structure. Section 8.2 gives more details of these solutions whereas Warburton (1976) presents further fundamental details.

8.2 Application to floating bodies

8.2.1 Heave motion of cylindrical buoys

The transformation of a problem concerned with fluid–structure interaction of a floating body into the governing equation of a second order dynamic system (Equation (8.1)) is illustrated by considering heave motion of the small buoy shown in Figure 8.4. Co-ordinate y denotes vertical motion of the water surface and z denotes the resultant buoy vertical motion. Now the total force acting on the buoy will be due to added mass, damping and hydrostatic stiffness induced forces due to differential motion between the buoy and the surrounding fluid. The net force due to these effects will serve to accelerate the buoy. Thus the equation of motion can be written as

$$m\ddot{z} = A(\ddot{y} - \ddot{z}) + B(\dot{y} - \dot{z}) + C(y - z) \tag{8.35}$$

where m is the buoy mass and A, B, C are coefficients expressing the added mass, damping and hydrostatic stiffness induced forces on the buoy. Here the damping force is taken to be linear and the Froude–Krylov force – applying through dynamic wave pressure on the buoy base – is given by the stiffness term Cy for a buoy geometry where draught h is very small compared to incident wave lengths.

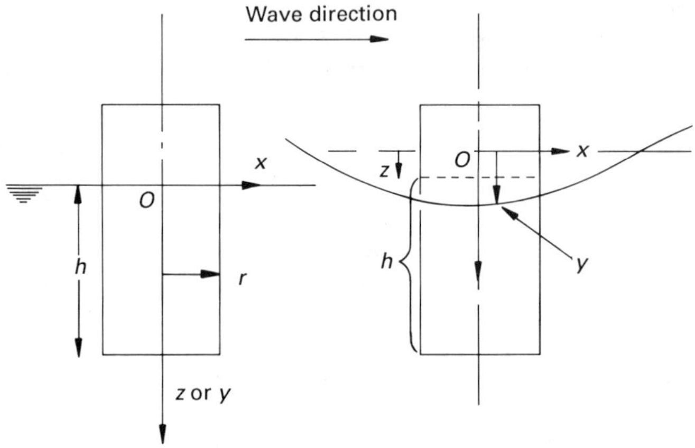

Figure 8.4. Heave motion of a buoy

Rearranging Equation (8.35) to bring unknown buoy motion terms to the left-hand side gives

$$(m + A)\ddot{z} + B\dot{z} + Cz = F(t) \tag{8.36}$$

where

$$F(t) = A\ddot{y} + B\dot{y} + Cy \tag{8.37}$$

Thus, there are two equations governing the behaviour of the buoy in waves. Equation (8.37) is concerned with the mechanism of wave elevation motion y exerting an exciting force $F(t)$ on the buoy, whereas Equation (8.36) is of a similar form to those of Equation (8.1) for dynamic response. The form of the above equations is typical of that obtained for all floating bodies in that two transfer functions (from wave elevation to force and from wave force to motion) are involved in deriving a body motion response from wave elevation.

A typical but very simplified solution procedure for the above equations is presented here to illustrate these ideas further. Taking buoy height h as very small compared to wave length, a wave of amplitude a with elevation

$$y = a \cos \omega t \tag{8.38}$$

can be used with Equation (8.37) to give

$$F(t) = a[(C - A\omega^2)\cos \omega t - B\omega \sin \omega t]. \tag{8.39}$$

Thus the wave force is a function of A, B, C frequency ω, and wave amplitude a, and can be expressed as an amplitude and phase transfer function of the form

$$\left.\begin{aligned} &F(t) = F_{\mathrm{a}} \cos(\omega t + \varepsilon) \\ &\text{where} \\ &\frac{F_{\mathrm{a}}}{a} = [(C - A\omega^2)^2 + (B\omega)^2]^{1/2} \\ &\text{and} \\ &\varepsilon = \arctan\left(\frac{-B\omega}{C - A\omega^2}\right) \end{aligned}\right\} \tag{8.40}$$

Equation (8.40) possesses all the features of a second-order dynamic system transfer function (see Equation (8.18)), with its amplitude and phase varying with frequency of excitation. Equation (8.36) then becomes

$$(m + a)\ddot{z} + B\dot{z} + Cz = F_{\mathrm{a}} \cos(\omega t + \varepsilon) \tag{8.41}$$

and has a solution of the form

$$\left.\begin{aligned} &z = z_{\mathrm{a}} \cos(\omega t + \varepsilon + \phi) \\ &\text{where} \\ &\frac{z_{\mathrm{a}}}{F_{\mathrm{a}}} = \frac{1}{\{[C - (m + A)\omega^2]^2 + (B\omega)^2\}^{1/2}} \\ &\text{and} \\ &\phi = \arctan\left[\frac{B\omega}{C - (m + A)\omega^2}\right] \end{aligned}\right\} \tag{8.42}$$

Note that the wave force leads the wave elevation by ε and the buoy motion lags the wave force by ϕ. When $m \to 0$, $\varepsilon + \phi = 0$ and suggests that a buoy of zero mass will follow the wave profile exactly without phase lag. The amplitude and phase transfer functions from wave force to buoy motion will be of the form given by Figure 8.3, but the overall wave elevation to buoy motion amplitude and phase transfer functions will be distorted by the presence of the wave elevation to wave force transfer function. It is recommended that both these transfer functions be evaluated independently during the motions analysis of a floating body to obtain a clearer physical understanding of its behaviour. Figures 8.8, 8.10, 8.12 and 8.13 present examples of such typical transfer functions derived in the following pages.

For a cylindrical buoy of radius r, and height h, floating in water of density ρ, the terms m, C and A can be given by

$$\left.\begin{aligned} m &= \rho\pi r^2 h \\ C &= \rho\pi g r^2 \\ A &= \frac{1}{3}\pi\rho r^3 \end{aligned}\right\} \quad (8.43)$$

where the added mass A for vertical motion of the cylinder end is approximated to that of a hemisphere of radius r and added mass coefficient 0.5 on the cylinder end. Other approximations for A involve treating the cylinder end as a flat disc moving along its axis of symmetry. Evaluation of a linear damping coefficient B is examined in more detail in the next section. Furthermore, Appendix G presents a summary of added mass and drag coefficients which may be used to deduce values for coefficients A and B.

8.2.2 Heave motion of semisubmersibles

Semisubmersibles are floating platforms with a geometry that is considerably different from the conventional ship shape, or monohull, form. This leads to significant differences in the hydrostatic and hydrodynamic behaviour of semisubmersibles in comparison with monohull vessels. Semisubmersibles consist of deeply submerged pontoons which are connected to the deck by several large-diameter columns, together with bracing members. A semisubmersible possesses low wave induced motions due to having a large proportion of its submerged volume at a deep draught where wave pressures have decayed rapidly with depth. The small water plane area of the vessel and the large submerged volume yield long natural periods in heave, roll and pitch. These periods are above the periods of predominant wave action, contributing further to reducing motions. However, the heave or vertical motion of a semisubmersible is still a significant factor in curtailing marine operations such as drilling or oil production. Therefore, it is desirable to refine the heave motion of a semisubmersible during design to reduce its down time to weather.

It is relatively simple to evaluate the undamped natural frequency or period of a semisubmersible vessel. An example of how this can be done is illustrated for a twin-pontooned vessel with n legs on each side, as shown in Figure 8.5. The undamped equation of vertical motion can be written as

$$(m + A)\ddot{z} + Cz = 0 \quad (8.44)$$

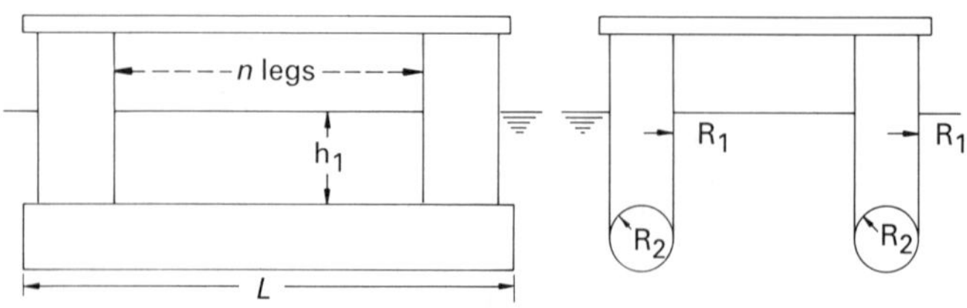

Figure 8.5. Twin-pontooned semisubmersible structure

using the notation of Section 8.2.1, and the resultant undamped natural period is

$$T_0 = 2\pi\sqrt{\left(\frac{m + A}{C}\right)}$$

Using Figure 8.5,

$$m = \rho[2\pi R_2^2 L + 2n\pi R_1^2 h_1]$$

$$A = 2\pi\rho R_2^2 L \tag{8.45}$$

and

$$C = 2\pi\rho g n R_1^2$$

using an added mass coefficient of 1.0 for vertical motion of the circular pontoons. Then

$$T_0 = 2\pi\sqrt{\left[\frac{1}{g}\left(h_1 + \frac{2R_2^2 L}{nR_1^2}\right)\right]}$$

$$= 2\pi\sqrt{\left[\frac{h_1}{g}(1 + 2\alpha)\right]} \tag{8.46}$$

if α is the ratio of pontoon volume to submerged column volume. This equation demonstrates that the natural period in heave motion is strongly influenced by pontoon submergence h_1 and, to a lesser extent, by the volume ratio α.

Also, if the semisubmersible of Figure 8.5 had pontoons of rectangular cross-section of width b_2 and height h_2, then terms m and A in the above equations change to

$$m = \rho(2b_2h_2L + 2n\pi R_1^2 h_1)$$

and

$$A = 2\rho C_m b_2 h_2 L \tag{8.47}$$

where the added mass coefficient C_m is a function of the aspect ratio (b_2/h_2) of the pontoon cross-section

Then

$$T_0 = 2\pi\sqrt{\left\{\frac{1}{g}\left[h_1 + \frac{(1 + C_m)b_2h_2L}{\pi n R_1^2}\right]\right\}}$$

$$= 2\pi\sqrt{\left\{\frac{h_1}{g}[1 + \alpha(1 + C_m)]\right\}} \tag{8.48}$$

which is of a similar form to Equation 8.46.

The natural period in heave of a single circular column terminated at its base by a circular footing of larger radius (see Figure 8.6) can also be defined through the following equations using a subscript 1 to denote a single column:

$$m_1 = \rho\pi(R_1^2 h_1 + R_2^2 h_2)$$

$$C_1 = \pi\rho g R_1^2 \tag{8.49}$$

$$A_1 = \pi\rho R_2^3 C_{m1}$$

where C_{m1} is a function of the geometry of the structure through R_1, R_2 and h_2.

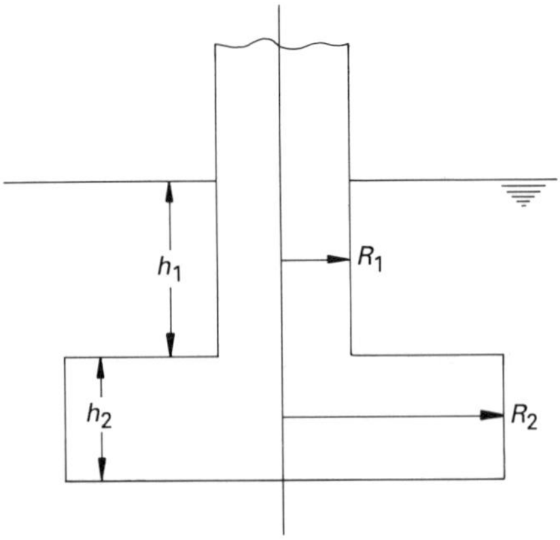

Figure 8.6. Column and pontoon footing structure

Then the natural period

$$\left.\begin{aligned} T_1 &= 2\pi\sqrt{\left[\frac{h_1}{g}\left(1 + \frac{R_2^2 h_2}{R_1^2 h_1} + \frac{R_2^3 C_{m1}}{R_1^2 h_1}\right)\right]} \\ &= 2\pi\sqrt{\left\{\frac{h_1}{g}\left[1 + \alpha\left(1 + \frac{C_{m1} R_2}{h_2}\right)\right]\right\}} \end{aligned}\right\} \quad (8.50)$$

is again of a similar form to that found before, with α being the ratio of the volume of the column footing to the column itself. If a semisubmersible is made up of n identical column and footings of the type shown in Figure 8.6, then for this vessel

$$m_n = n \cdot m_1, \; C_n = n \cdot C_1, \; A_n = n \cdot A_1 \quad (8.51)$$

and, therefore, the natural period remains the same as for the single column and footing arrangement. This result assumes that the added masses of each column and footing are unaffected by the presence of others in their vicinity.

The heave motion in waves of semisubmersibles of simple geometry can be evaluated by treating the vessel as a collection of slender members on which the loading can be calculated using the Morison equation with deep water linear wave theory. In this particular derivation, the damping and wave induced drag forces are neglected for simplicity. Thus the equation of vertical motion for a four-column, twin-pontooned semisubmersible of the configuration shown in Figure 8.7 can be written as

$$(m + A)\ddot{z} + Cz = F(t) \quad (8.52)$$

For a simple four-column, twin-pontooned semisubmersible of the type shown in Figure 8.7, the coefficients of this equation are

$$m = 4\rho(A_v d + A_h L)$$
$$A = 4\rho C_m A_h L$$
$$C = 4\rho g A_v$$

where A_v and A_h are the areas of cross-section of the vertical columns and rectangular pontoons, respectively, and C_m is the added mass coefficient for the pontoons in vertical motion.

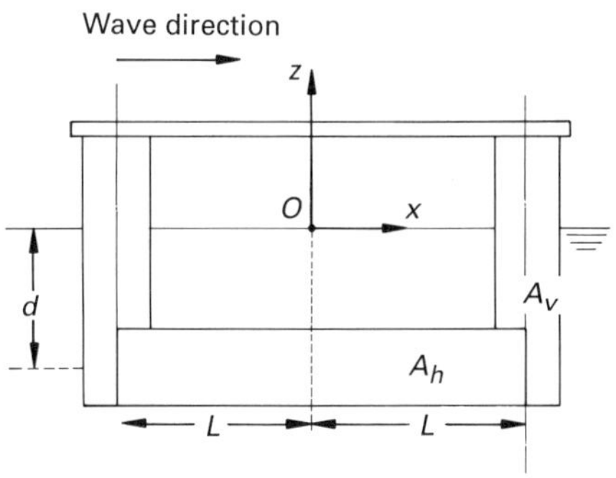

Figure 8.7. Side view of four-column semisubmersible

The vertical wave exciting force on the vessel can be taken as the sum of wave forces on the pontoons and columns to yield

$$F(t) = \int_{-L}^{L} 2\rho A_h(1 + C_m)\dot{w} \cdot \mathrm{d}x + 2A_v(p_f + p_r) \tag{8.53}$$

where $\dot{w}$ is the vertical wave particle acceleration at $z = -d$ and p_f, p_r are wave pressures at $z = -d$ and $x = +L$ and $-L$ respectively. Note that the pontoon/column intersections have been approximated by lengths d and L to simplify the above algebra.

The Froude–Krylov force normal to the pontoon axis is calculated using an inertia coefficient $(1 + C_m)$, whereas the axial Froude–Krylov force on the vertical columns is calculated as a product of dynamic wave pressure and column area of cross-section. The former approach is valid for calculating normal forces on a slender member about which wave properties can be considered constant, whereas the wave pressure formulation is necessary for axial forces on members whose length is comparable to incident wave length.

Using the Oxz axes of Figure 8.7, the wave particle vertical acceleration and pressure can be written as

$$\dot{w} = -a\omega^2 \mathrm{e}^{-kd} \sin(kx - \omega t)$$

and

$$p = +a\,\frac{\rho\omega^2}{k}\,\mathrm{e}^{-kd} \sin(kx - \omega t) \tag{8.54}$$

and substituted into Equation (8.53) to yield the wave force relationship

$$F(t) = 4a\rho g e^{-kd}[-A_v \cos kL + A_h(1 + C_m)\sin kL]\sin \omega t \tag{8.55}$$

Evaluation of Equation (8.55) followed by solution of Equation (8.52) yields the wave induced force and motion response which can be written as

$$\left.\begin{aligned} F(t) &= F_a \sin(\omega t + \varepsilon) \\ &\text{and} \\ z(t) &= z_a \sin(\omega t + \varepsilon + \phi). \end{aligned}\right\} \tag{8.56}$$

Figure 8.8 presents the variations of $|F_a|/|a|$, $|z_a|/|a|$, ε and $(\varepsilon + \phi)$ as functions of frequency for a semisubmersible with $d = 13.5$ m, $L = 36$ m, $A_h = 70\ \text{m}^2$ and $A_v = 90\ \text{m}^2$. A value of 1.2 is used for the added mass coefficient. These curves exhibit several interesting features. The wave force amplitude cancels out to zero at certain wave frequencies when instantaneous wave forces on the pontoons balance wave forces on the columns. At these cancellation frequencies, the phase ε shifts by 180° such that over some wave frequencies, the vertical force on the vessel is out of phase with wave elevation. The motion response transfer function shows the combined effects of the wave force cancellation and resonance with the total phase difference between vessel motion and wave elevation around the resonant frequency exhibiting some complex behaviour.

Now a major proportion of the damping of slender space frame floating structures arises from drag forces. Inclusion of this drag force induced damping yields an equation of motion of the form

$$(m + A)\ddot{z} + B_v|\dot{z}|\dot{z} + Cz = F(t) \tag{8.57}$$

One approach that permits solution of Equation (8.57) is to use a linearized equivalent damping coefficient B that is selected so that the non-linear and equivalent linear damping coefficients dissipate the same energy at resonance. For a linear damping term of the form given in Equation (8.36), the energy dissipated at resonance can be written as

$$W_L = 2\int_{-R}^{R} B(R\omega_0 \sin \omega_0 t)\, dz \tag{8.58}$$

where $z = R\sin(\omega_0 t - 90°)$ is the resonant motion at frequency ω_0. Writing the integral in terms of time gives

$$\begin{aligned} W_L &= 2\int_0^{\pi} B\omega_0 R^2 \sin^2 \omega_0 t\, d(\omega_0 t) \\ &= \pi B \omega_0 R^2 \end{aligned} \tag{8.59}$$

For non-linear drag induced damping, the work done becomes

$$\begin{aligned} W_D &= 2\int_{-R}^{R} B_v(R^2\omega_0^2 \sin^2 \omega_0 t)\, dz \\ &= 2\int_0^{\pi} B_v\omega_0^2 R^3 \sin^3 \omega_0 t\, d(\omega_0 t) \\ &= \frac{8}{3} B_v \omega_0^2 R^3 \end{aligned} \tag{8.60}$$

Equating the work done by linear and non-linear damping yields

$$B = \frac{8}{3\pi}(\omega_0 R)B_v$$
$$= \frac{8}{3\pi}(\dot{z}_{max})B_v \qquad (8.61)$$

Equation (8.57) can be solved by replacing $B_v|\dot{z}|$ by B from Equation (8.61). The equation has, therefore, to be solved by iteration in that a value of $\dot{z}_{max}$ has to be determined which satisfies the equation. The first solution for the iteration is usually found by assuming that the damping is linear and 10% of critical. The new value of $\dot{z}_{max}$ obtained from this initial solution is then placed in the damping term and the procedure repeated until successive values of $\dot{z}_{max}$ are within a small value of each other. This iteration converges very quickly away from resonance and within approximately ten solutions at resonance. The resultant solution for z will now also be a function of the amplitude of the applied force F_a and of wave amplitude, a.

8.2.3 Pitch motion of articulated columns

The procedures used in the two preceding sections can also be applied to an articulated column which is freely pinned at a fixed point below the sea surface, and remains upright in still water due to an excess buoyancy in the structure. Figure 8.9 shows a column that may typically be used as a loading or mooring terminal for an oil tanker. The equation of motion of the slender column can be written as

$$(I + I_a)\ddot{\theta} + B_v|\dot{\theta}|\dot{\theta} + C\theta = M(t) \qquad (8.62)$$

where I and I_a are moments of inertia of physical mass and of added mass where as B_v and C are the non-linear damping coefficient and angular stiffness respectively. All of these parameters are defined relative to pitch angle θ of the column about a horizontal axis through O; $M(t)$ is also the wave induced pitching moment about this axis.

These coefficients are obtained by integrating along the column length as follows

$$\left.\begin{aligned} I_a &= \rho C_m \pi r^2 \int_0^L z^2\,dz = \frac{1}{3}\rho C_m \pi r^2 L^3 \\ B_v &= \rho C_d r \int_0^L |z| z^2 dz = \frac{1}{4}\rho\, C_d r L^4 \\ I &= mk^2 \\ C &= \rho g \pi r^2 L L_B - F_W L_G \end{aligned}\right\} \qquad (8.63)$$

where C_m and C_d are added mass and drag coefficients respectively and F_W is the column weight. The physical inertia can be written as the product of column mass m and radius of gyration k, squared, whereas the angular stiffness C arises from the balance of moments induced by buoyancy and weight forces acting through the centres of buoyancy and gravity (B and G) of the column.

The wave induced pitching moment is obtained by summing the contributions of the wave down the stationary upright column by the equation

$$M(t) = \rho(C_{\mathrm{m}} + 1)\pi r^2 \int_0^L \dot{u} z' \mathrm{d}z' + \rho C_{\mathrm{d}} r \int_0^L |u| u \, z' \mathrm{d}z' \qquad (8.64)$$

where u, $\dot{u}$ are horizontal components of wave particle velocities and accelerations, due to a wave of amplitude a.

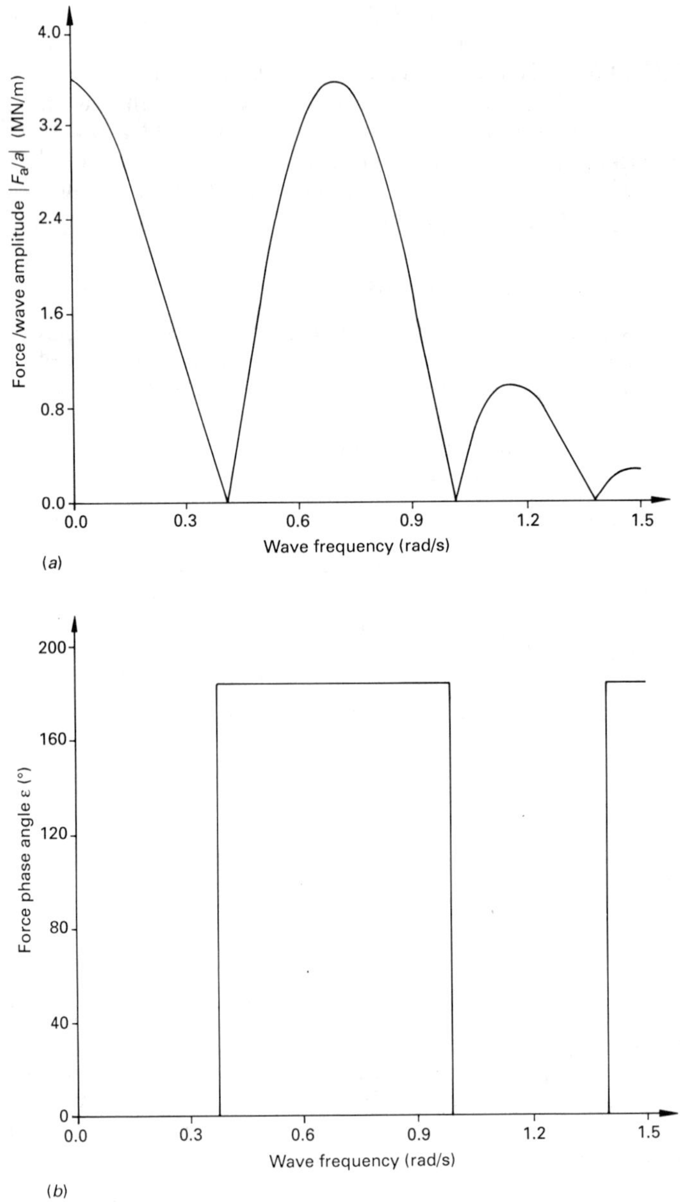

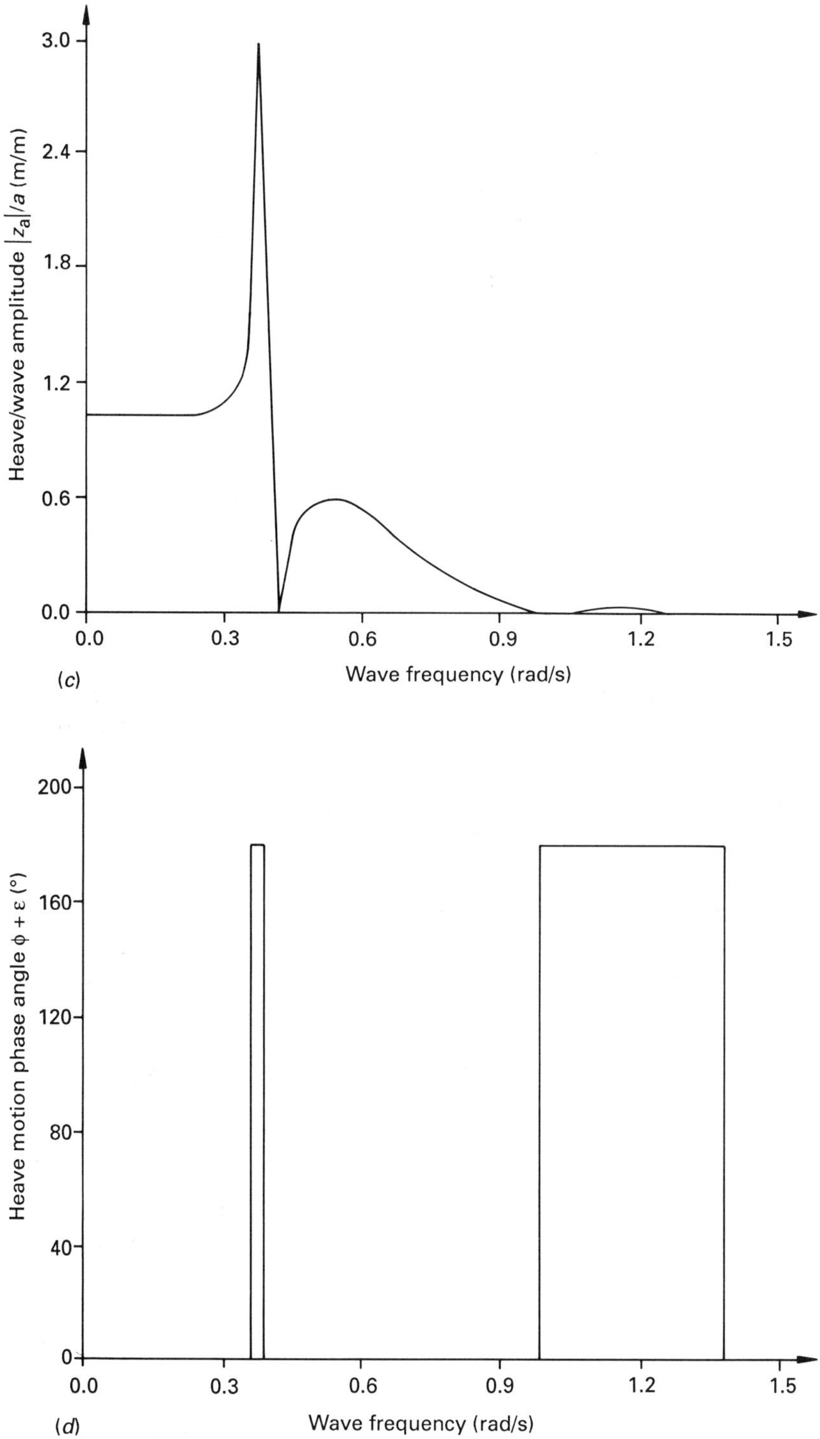

Figure 8.8. Heave force and motion response of four-column semisubmersible

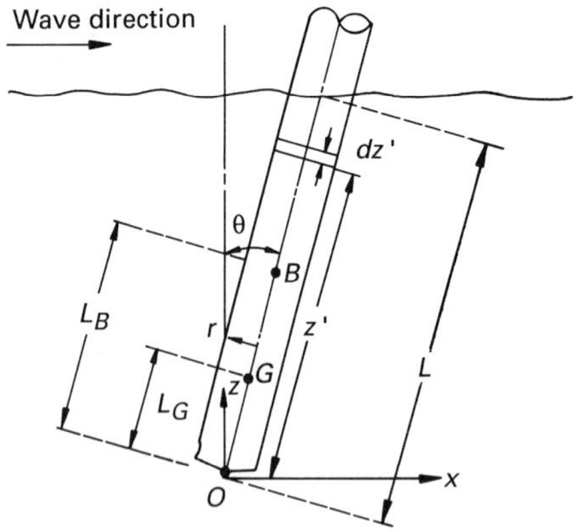

Figure 8.9. Articulated column

The horizontal wave velocities and accelerations can be taken as

$$\left.\begin{aligned} u &= +a\omega \mathrm{e}^{k(z'-L)}\sin(kx - \omega t) \\ &\text{and} \\ \dot{u} &= -a\omega^2 \mathrm{e}^{k(z'-L)}\cos(kx - \omega t) \end{aligned}\right\} \quad (8.65)$$

for deep water linear waves using an axes system of Oxz' with origin at the articulated column pivot (see Figure 8.9). Substituting Equations (8.65) into (8.64) yields

$$M(t) = \rho(C_\mathrm{m} + 1)\pi r^2 a\omega^2\left(-\frac{L}{k} + \frac{1}{k^2}(1 - \mathrm{e}^{-kL})\right)\cos\omega t + \frac{8}{3\pi}\rho C_\mathrm{d} r a^2\omega^2\left(-\frac{L}{2k} + \frac{1}{4k^2}(1 - \mathrm{e}^{-2kL})\right)\sin\omega t \quad (8.66)$$

with application of the linearization procedure outlined earlier. Note that deep water linear wave theory is used here to simplify the algebra but, therefore, also restricts the validity of the solution to wave lengths that are less than twice the column length.

The above equations are valid for small θ and can be solved using the same procedure as in Section 8.2.2 together with linearization of the damping coefficient. Figure 8.10 shows typical results for an articulated column of 812 t mass, 7 m diameter and a sea bed to water line length of 42 m with a moment of inertia of physical mass of 1.249×10^9 kg m^2 ($k = 41.88$ m) and L_G of 18 m. The added mass and drag coefficients are taken to be 1.0 and 1.2 respectively. Figure 8.10 presents the wave induced moment and motion response in terms of amplitude and phase using the equations

$$M = M_0 \sin(\omega t + \varepsilon)$$

$$\theta = \theta_0 \sin(\omega t + \varepsilon + \phi)$$

The equations of motion for an articulated column can also be formulated for large angles and solved numerically using time step integration routines described in Section 6.2. Such a solution is described by Harrison (1985).

8.2.4 Heave and surge motions of tensioned buoyant platforms

A tensioned buoyant platform (TBP) is a floating structure connected to a sea bed foundation by vertical tethers which are kept in tension by excess buoyancy over weight in the platform structure. The use of TBPs as alternatives to oil production from fixed or catenary moored semisubmersibles or other floating structures is now well established. A TBP possesses a combination of desirable characteristics for an offshore work platform. The absence of a fixed rigid structure from sea surface to sea bed, coupled with the excellent station keeping characteristics of tensioned vertical tethers, satisfy the two important demands of economy and safe operation. Additionally, a TBP exhibits very much smaller wave induced heave, roll and pitch motions due to the presence of the vertical mooring tethers. Large surge and sway offsets due to wind and current steady force components will, however, cause the platform to be drawn down lower in the water and exhibit a so called 'set down'. Nevertheless, the wave frequency surge and sway motions of a TBP are desirable since horizontal wave loads are not fully reacted by the platform structure.

Conventional TBPs have structural features which are similar to those of semisubmersibles. These include the presence of deeply submerged main buoyancy chambers which give rise to a significant reduction in wave induced forces due to the exponential decay of wave particle velocities and accelerations with depth below the surface. The combination of horizontal pontoons and vertical columns supporting the working deck produces wave force cancellation at certain frequencies. This occurs because the upward wave force on vertical columns due to buoyancy changes can be almost exactly cancelled out by the downward wave force on horizontal pontoons. Additionally, the stiffness of the tethers and their axial tensions will radically influence the dynamic response of the platform to waves.

Methods of analysis for a TBP are very similar to those outlined earlier for a semisubmersible structure, except for the insertion of a stiffness matrix induced by the presence of the relatively stiff platform tethers. Thus the equation of motion, the evaluation of the coefficient matrices and the solution of the equation are the same as in Section 8.2.2 with the tether stiffness matrix $\mathbf{C}_t$ included.

The governing equation for heave motion of a tensioned buoyant platform of the configuration shown in Figure 8.11 is derived below. The platform is square in plan view with transverse pontoons (marked A in Figure 8.11) of length $2L$ and cross-sectional area A_h. The equation for heave motions is

$$(m + A)\ddot{z} + (C + C_t)z = F_z(t) \tag{8.67}$$

where

$$m = 4\rho(A_v d + 2A_h L) - \frac{4T}{g}$$

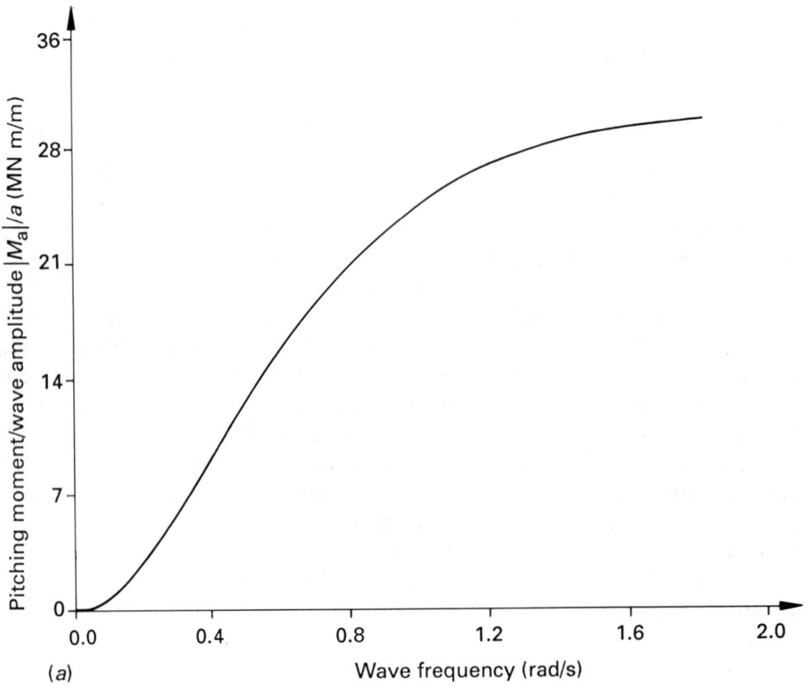
Pitching moment/wave amplitude |M_a|/a (MN m/m)
36
28
21
14
7
0
0.0
0.4
0.8
1.2
1.6
2.0
Wave frequency (rad/s)

(a)

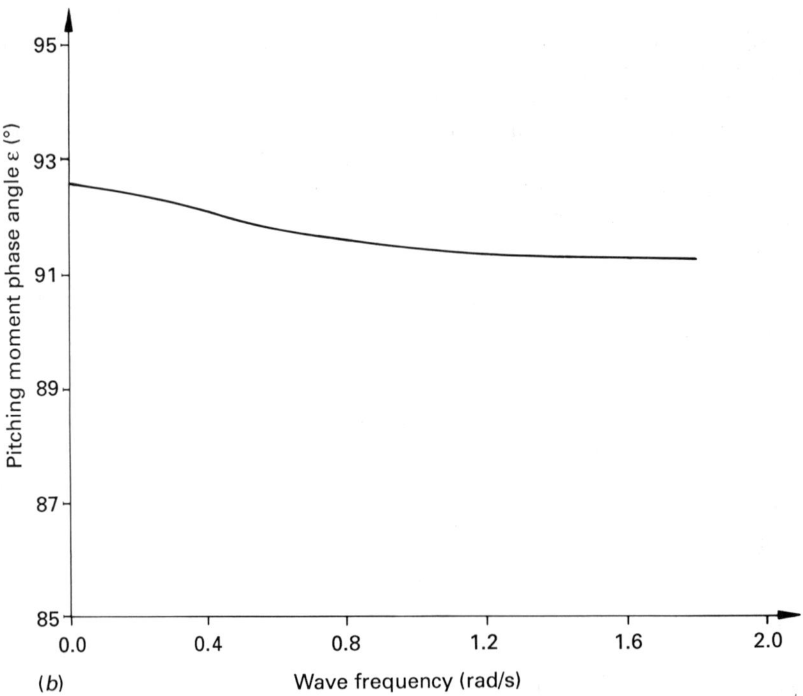
Pitching moment phase angle ε (°)
95
93
91
89
87
85
0.0
0.4
0.8
1.2
1.6
2.0
Wave frequency (rad/s)

(b)

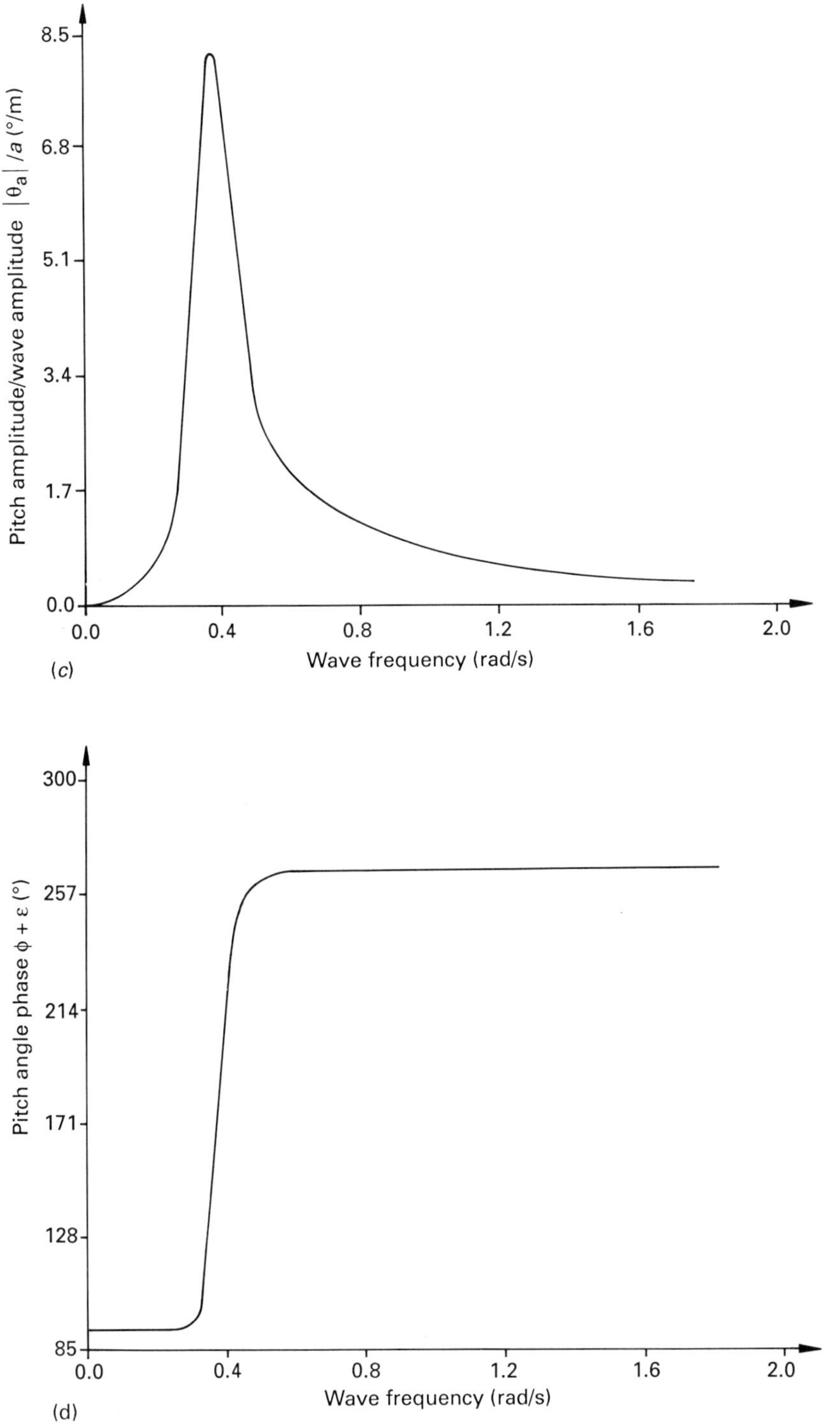

Figure 8.10. Response of an articulated column to waves

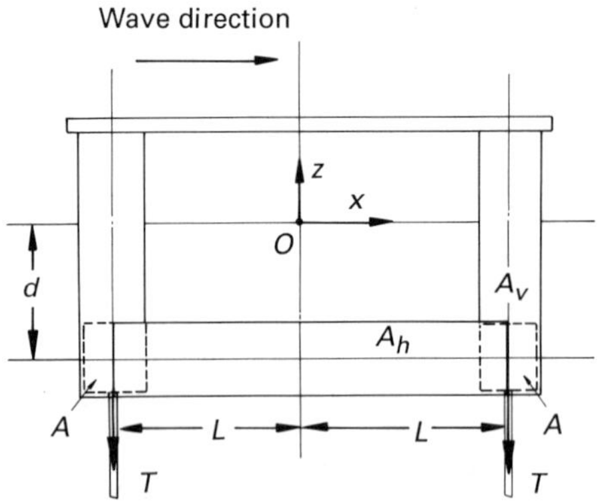

Figure 8.11. Side view of four-column tensioned buoyant platform. A are transverse pontoons of lengths $2L$ and area of cross-section A_h

with T being the tension in each of the tethers on the four corners of the platform. Also

$$\left.\begin{aligned} A &= 8\rho C_{\mathrm{mz}} A_h L \\ C &= 4\rho g A_v \\ &\text{and} \\ C_{\mathrm{t}} &= 4\lambda' \end{aligned}\right\} \tag{8.68}$$

where C_{mz} is the added mass coefficient for the pontoon cross-section in vertical motion and λ' is the axial stiffness of each of the tethers. In this analysis, the drag damping coefficient and wave induced drag forces are neglected for simplicity. The wave force in heave is

$$\begin{aligned} F_z(t) = &\int_{-L}^{L} 2\rho A_h(1 + C_{\mathrm{m}})\dot{w}\,\mathrm{d}x + 2A_v(p_f + p_r) \\ &+ 2\rho A_h L(1 + C_{\mathrm{m}})(\dot{w}_{\mathrm{f}} + \dot{w}_r) \end{aligned} \tag{8.69}$$

where $\dot{w}_f$ and $\dot{w}_r$ are the vertical wave particle accelerations at $z = -d$ and $x = +L$ and $-L$, respectively, with the remaining notation from Section 8.2.2. Then, using Equation (8.54),

$$\begin{aligned} F_z(t) = &\,4a\rho g \mathrm{e}^{-kd}[-A_v \cos kL + A_h(1 + C_{\mathrm{m}}) \sin kL \\ &+ A_h \cdot kL \cdot (1 + C_{\mathrm{m}}) \cos kL] \sin \omega t \end{aligned} \tag{8.70}$$

These equations have most of the characteristics of the governing equation for semisubmersibles except for the presence of stiffness term C_{t}, which is large and tends to substantially reduce the heave natural periods of such platforms to the range of 3–5 s from the usual semisubmersible heave resonance period of from 18–25 s. Using the definition of Equation (8.56), Figure 8.12 presents variations of the wave force and motion transfer functions with frequency for a TBP with $d = 35$ m, $L = 50$ m, $A_h = 100$ m^2 and $A_v = 150$ m^2. The total tether pre-tension and stiffness are taken to be 3.75 MN and 56.53 MN/m respectively. The added mass coefficient for vertical motion of the rectangular pontoons is taken to be 1.2. The wave force is slightly modified due to the additional transverse pontoons but the motion response exhibits a substantial reduction in magnitude due to the shift in resonant frequency.

A similar analysis to that above can be carried out for surge motion of a TBP with a governing equation of the form

$$(m + A')\ddot{x} + (C' + C_t')x = F_x(t) \tag{8.71}$$

The additional terms in the above equation are

$$\left.\begin{aligned} &A' = 4\rho C_{\text{mx}} A_h L + 4\rho A_v d \\ &C' = 0 \qquad C_t' = \frac{4T}{s} \end{aligned}\right\} \tag{8.72}$$

where C_{mx} and 1.0 are the added mass coefficients for horizontal flow normal to the pontoons and vertical columns respectively. The hydrostatic stiffness term C' is zero and the tether stiffness is expressed in terms of tension and tether length. An expression for $F_x(t)$ can be derived in the same way as for heave motion to yield

$$\begin{aligned} F_x(t) = &\int_{-d}^{0} 4\rho A_v(\dot{u}_f + \dot{u}_r)\,\mathrm{d}z \\ &+ \rho A_h L(1 + C_{\text{mx}})(\dot{u}_f + \dot{u}_r)_{z=-d} \\ &- 2A_h(p_f - p_r)_{z=-d} \end{aligned} \tag{8.73}$$

where $\dot{u}_f$, p_f and $\dot{u}_r$, p_r denote horizontal wave particle accelerations and wave pressures at $x = +L$ and $x = -L$ respectively and are given by

$$\dot{u} = -a\omega^2 \mathrm{e}^{+kzx}\cos(kx - \omega t) \tag{8.74}$$

and Equation (8.54) with C_{mx} and 1.0 taken as the added mass coefficients for horizontal flow normal to the pontoons and vertical columns respectively. Then, the wave force $F_x(t)$ can be written as

$$\begin{aligned} F_x(t) = &-\left[\frac{8\rho a\omega^2 A_v}{k}\left(1 - \mathrm{e}^{-kd}\right)\right. \\ &\left. + 2\rho a\omega^2 A_h L(1 + C_{\text{mx}})\mathrm{e}^{-kd}\right]\cos kL\cos\omega t \\ &- 4\rho g a A_h \mathrm{e}^{-kd}\sin kL\cos\omega t \end{aligned} \tag{8.75}$$

Note that the only force cancellations occurring in Equation (8.75) are due to zeroes of the combined $\cos kL$ and $\sin kL$ terms.

Figure 8.13 presents the surge force and motion response transfer function calculated from these equations. The tether length s is taken to be 450 m and $C_{\text{mx}} = 1.2$ with all other parameters being as in the calculation of heave response described above. Again the effect of tether stiffness dominates the motion transfer function with a low natural period and relatively large surge motions at predominant wave periods.

In the calculations described above, the added mass and drag coefficients for members of circular cross-section can be obtained from the graphs presented in Chapter 5. For non circular shapes, recommendations made by Det Norske Veritas (1982), and presented in Appendix G, may be used.

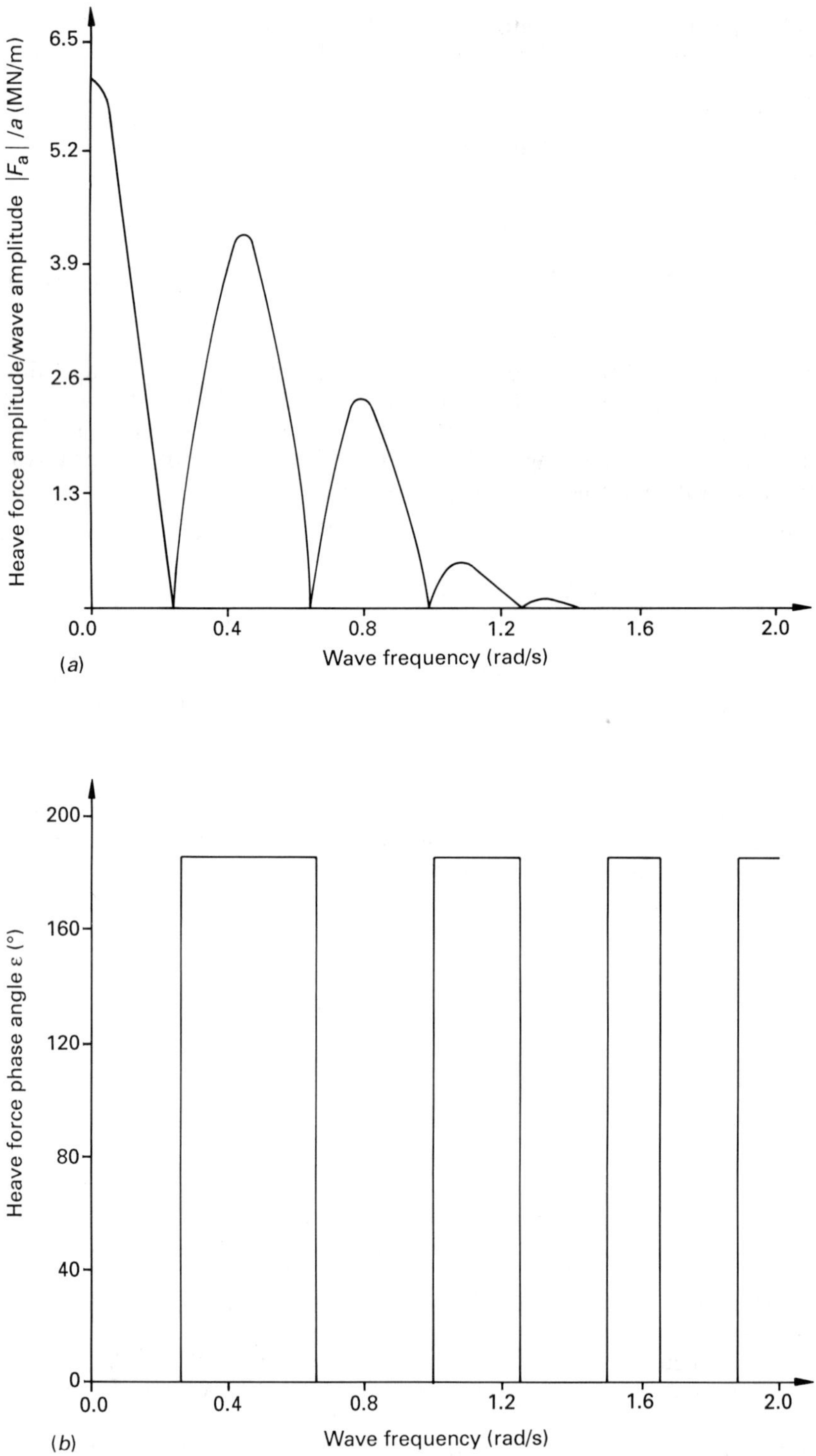
6.5
5.2
3.9
2.6
1.3
Heave force amplitude/wave amplitude $|F_a|/a$ (MN/m)
0.0
0.4
0.8
1.2
1.6
2.0
Wave frequency (rad/s)
(a)
200
160
120
80
40
0
Heave force phase angle ε (°)
0.0
0.4
0.8
1.2
1.6
2.0
Wave frequency (rad/s)
(b)

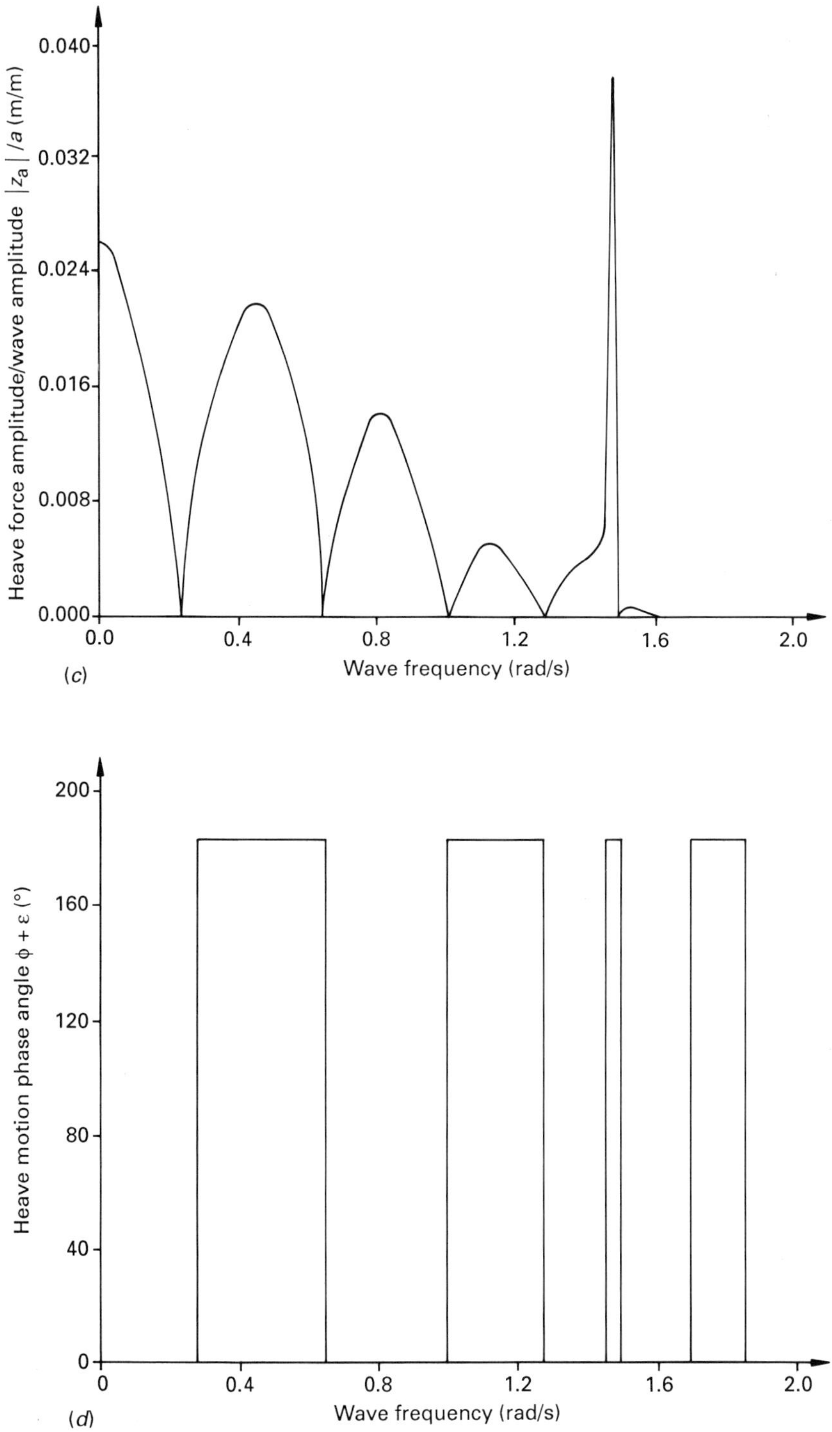

Figure 8.12. Heave force and motion response of four-column tensioned buoyant platform

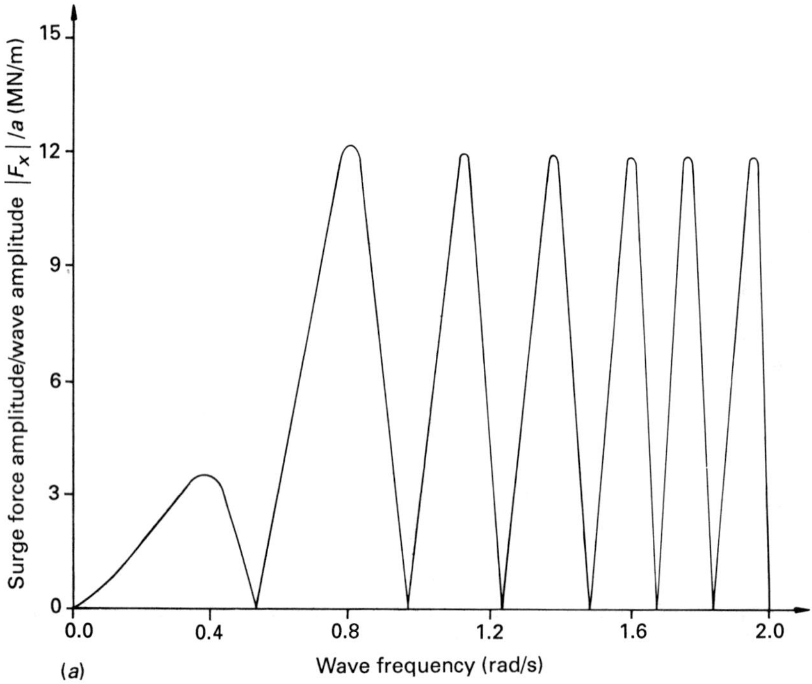
Surge force amplitude/wave amplitude $|F_x|/a$ (MN/m)
15
12
9
6
3
0
0.0
0.4
0.8
1.2
1.6
2.0
Wave frequency (rad/s)

(a)

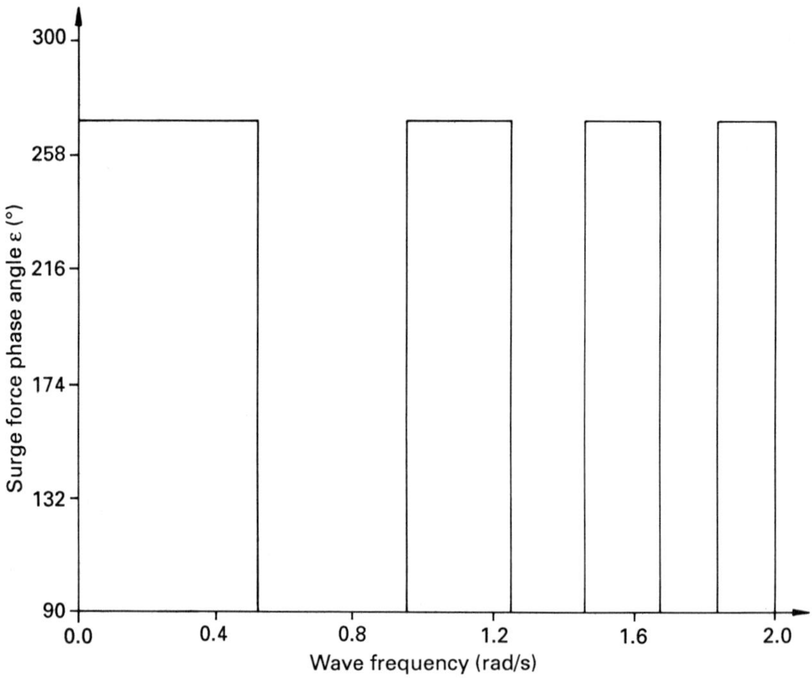
Surge force phase angle ε (°)
300
258
216
174
132
90
0.0
0.4
0.8
1.2
1.6
2.0
Wave frequency (rad/s)

(b)

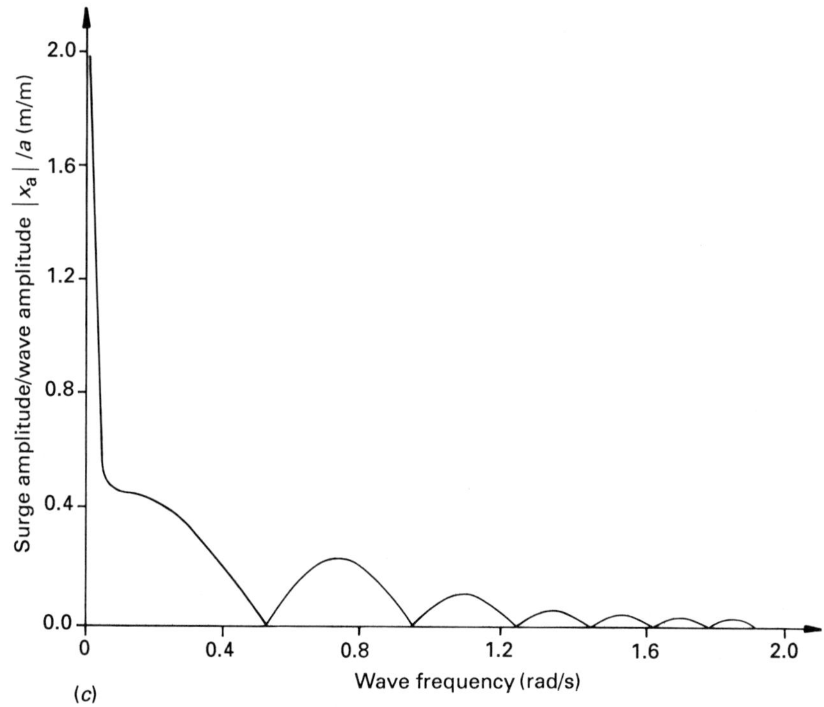

(c)

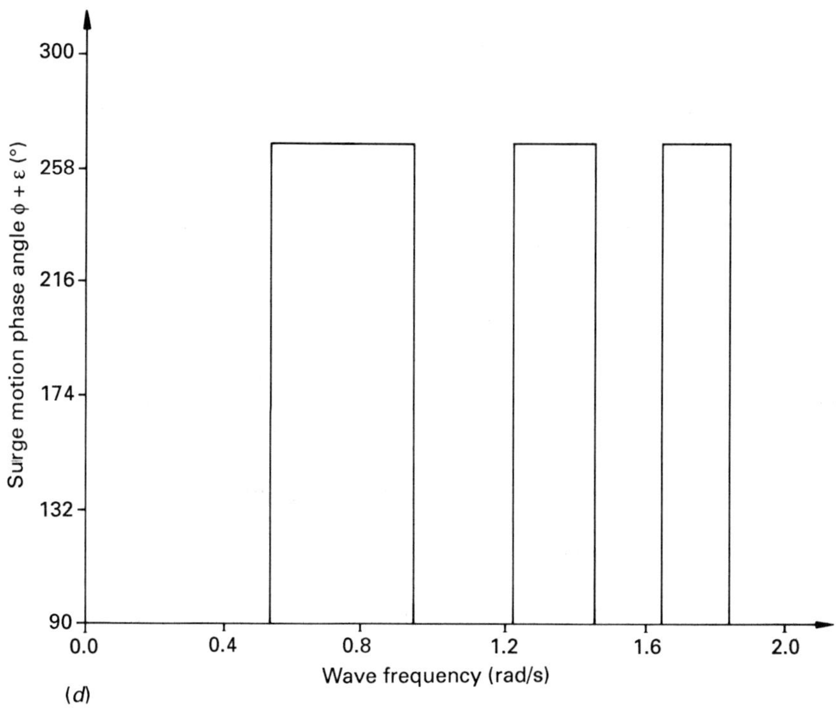

(d)

Figure 8.13. Surge force and motion response of four-column tensioned buoyant platform

8.3 Effects of moorings

As illustrated in Section 8.2.4, mooring systems modify the governing equations of a floating platform by introducing stiffness terms which alter the natural periods of the combined vessel and moorings to a lesser or greater extent. There are two kinds of mooring systems in common use for floating offshore structures – catenary moorings and tensioned vertical tethers. The mechanics of both of these are described in this section.

8.3.1 Catenary moorings

A catenary mooring system consists of freely hanging wire or chain lines connecting a surface platform to anchors on the sea bed some distance from the platform. The mooring lines are laid out, usually symmetrically in plan view, around the surface platform with a significant length laying on the sea bed and terminated by anchors. Each of the lines relies on an increase or decrease in line tension as it lifts off or settles on the sea bed to generate a restoring force on the surface platform. A spread of catenary lines thus generates a non-linear restoring force that increases with vessel horizontal offset and resists steady environmental forces. The equivalent spring stiffness imposed on horizontal vessel motions is generally too small to significantly influence wave frequency motions, although excitation by low frequency drift forces can induce dynamic magnification in platform horizontal motions and lead to high peak line tensions. The longitudinal and transverse motion dynamics of the mooring lines themselves can influence their behaviour particularly for long lengths in deep water. The role of thrusters within a thruster assisted mooring system also needs to be considered during analysis and design.

The basic theory underlying the behaviour of catenary lines is well known – see, for examle, O'Brien and Francis (1964). It is, however, instructive to review the basic mechanics of a catenary and to derive its governing equations.

Figure 8.14 shows a catenary mooring line deployed from points A on the submerged hull of a floating vessel to an anchor B on the sea bed. Note that part of the line between A and B is resting on the sea bed and that the horizontal dimension a is usually 5–20 times larger than the vertical dimension b. As the line mounting point on the vessel is shifted horizontally from point A_1, through A_2, A, A_3 to A_4, the catenary line laying on the sea bed varies from a significant length at A_1, to none at A_4. The cable tension in the vicinity of points A is due to the total weight in sea water of the suspended line length. The progressive effect of line lift from the sea bed due to horizontal vessel movement from A_1 to A_4, increases line tension in the vicinity of points A. This feature, coupled with the simultaneous decrease in line angle to the horizontal, causes the horizontal restoring force on the vessel to increase with vessel offset in a non-linear manner.

This behaviour can be described by the catenary equations which are

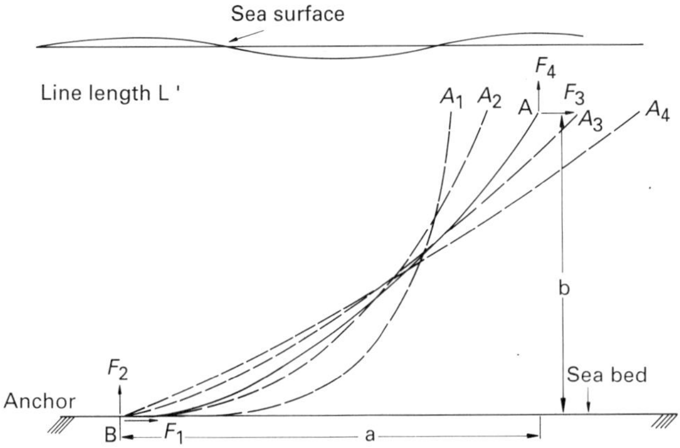

Figure 8.14. Catenary mooring line

derived below. Figure 8.15(b) shows an element A of a line of weight per unit length w hanging freely as shown in Figure 8.15(a). The element weight of $w \cdot \delta s$ is supported by a variation in tension along the line. A differential equation for this variation is obtained by taking vertical and horizontal equilibrium of the forces acting on the line element of Figure 8.15(b). Thus

$$w\delta s + \frac{\mathrm{d}}{\mathrm{d}s}(T\sin\theta)\delta s = 0 \tag{8.76}$$

$$\frac{\mathrm{d}}{\mathrm{d}s}(T\cos\theta) = 0 \tag{8.77}$$

Taking $\cos\theta = \mathrm{d}x/\mathrm{d}s$ and $\sin\theta = \mathrm{d}y/\mathrm{d}s$, integration of Equation (8.77) gives

$$T\frac{\mathrm{d}x}{\mathrm{d}s} = X \tag{8.78}$$

where X is a constant horizontal force component in the line due to the fact that no external horizontal forces act on the line in still water.

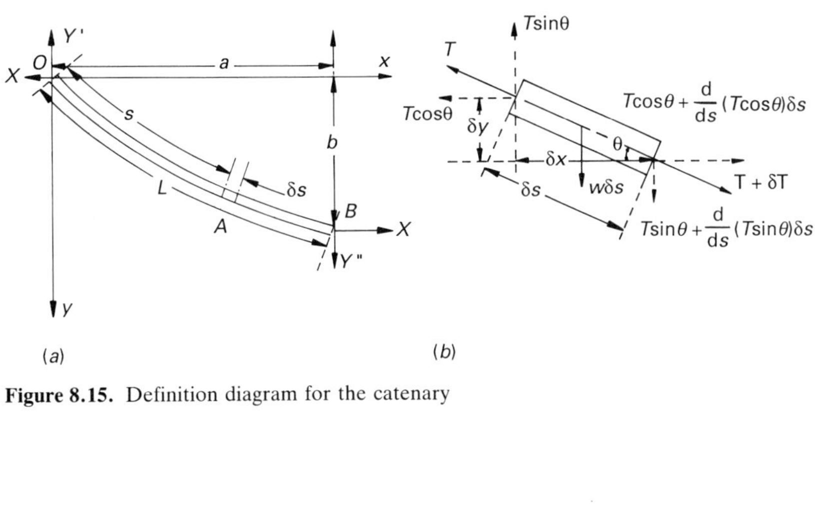

Figure 8.15. Definition diagram for the catenary

Substituting Equation (8.78) into Equation (8.76) then yields an equation of the form

$$X\frac{\mathrm{d}^2y}{\mathrm{d}x^2} + w\frac{\mathrm{d}s}{\mathrm{d}x} = 0 \tag{8.79}$$

which can also be written in terms of x and y only as

$$X\frac{\mathrm{d}^2y}{\mathrm{d}x^2} + w\left[1 + \left(\frac{\mathrm{d}y}{\mathrm{d}x}\right)^2\right]^{1/2} = 0. \tag{8.80}$$

These are both different forms of the differential equation for a simple catenary. The equations may be readily integrated to give, first

$$\frac{\mathrm{d}y}{\mathrm{d}x} = -\sinh\left(\frac{wx}{X} - \phi\right) \tag{8.81}$$

where the constant of integration ϕ is given by using boundary conditions which constrain the line to go through points $x = 0$, $y = 0$ and $x = a$, $y = b$. With this condition

$$\left.\begin{aligned} \phi &= \sinh^{-1}\left\{\frac{\lambda(b/a)}{\sinh\lambda}\right\} + \lambda \\ &\text{and} \\ \lambda &= \frac{wa}{2X} \end{aligned}\right\} \tag{8.82}$$

Integrating Equation (8.81) again and using the above boundary conditions gives

$$y = \frac{X}{w}\left[\cosh\phi - \cosh\left(\frac{2\lambda x}{a} - \phi\right)\right] \tag{8.83}$$

A number of worth while results can be derived from these equations. The length L of segment OB can be obtained as

$$\begin{aligned} L &= \int_0^a \frac{\mathrm{d}s}{\mathrm{d}x}\,\mathrm{d}x \\ &= \int_0^a \cosh\left(\frac{wx}{X} - \phi\right)\mathrm{d}x \\ &= \frac{2X}{w}\sinh\lambda\cosh(\phi - \lambda) \end{aligned} \tag{8.84}$$

Substituting $x = a$ and $y = b$ in Equation (8.83) and simplifying also gives

$$b = \frac{2X}{w}\sinh\lambda\sinh(\phi - \lambda) \tag{8.85}$$

Squaring Equations (8.84), (8.85) and subtracting yields the relationship

$$L^2 = b^2 + \frac{a^2\sinh^2\lambda}{\lambda^2} \tag{8.86}$$

which is useful for determining L from information on variables a, b, w and X. The vertical components at each end of the line segment are obtained as follows. First, at the left-hand end

$$Y' = X\left(\frac{\mathrm{d}y}{\mathrm{d}x}\right)_{x=0} = X\sinh\phi \tag{8.87}$$

Now from Equation (8.82)

$$\sinh\phi = \sinh\left\{\sinh^{-1}\left[\frac{\lambda(b/a)}{\sinh\lambda}\right] + \lambda\right\}$$

$$= \frac{\lambda b}{a}\coth\lambda + M\sinh\lambda \tag{8.88}$$

where

$$M = \cosh\left\{\sinh^{-1}\left[\frac{\lambda(b/a)}{\sinh\lambda}\right]\right\}$$

Taking hyperbolic cosine of Equation (8.82) gives

$$\cosh\phi = M\cosh\lambda + \frac{\lambda b}{a} \tag{8.89}$$

Squaring Equations (8.88), (8.89) and substracting gives

$$M^2 = 1 + \frac{\lambda^2 b^2}{a^2}(\coth^2\lambda - 1)$$

$$= \frac{L^2\lambda^2}{a^2\sinh^2\lambda}$$

after using Equation (8.86). Thus

$$M = \frac{wL}{2X\sinh\lambda} \tag{8.90}$$

and substituting into Equations (8.88) and (8.87) yields

$$Y' = \frac{w}{2}(b\coth\lambda + L) \tag{8.91}$$

Also the vertical component of tension at the right-hand end is given by

$$Y'' = X\left(\frac{\mathrm{d}y}{\mathrm{d}x}\right)_{x=a}$$

$$= \frac{w}{2}(b\coth\lambda - L) \tag{8.92}$$

after simplifying using Equation (8.84). Equations (8.91) and (8.92) yield the expected result that

$$Y' - Y'' = wL \tag{8.93}$$

Similarly, the tensions at each end of the line segment can be written as

$$T' = X\left(\frac{\mathrm{d}s}{\mathrm{d}x}\right)_{x=0}$$

$$= X\cosh\phi$$

$$= \frac{w}{2}(L\coth\lambda + b) \tag{8.94}$$

and

$$T'' = X\left(\frac{ds}{dx}\right)_{x=a}$$

$$= X\cosh(2\lambda - \phi)$$

$$= \frac{w}{2}(L\coth\lambda - b) \qquad (8.95)$$

Equations (8.91), (8.92), (8.94) and (8.95) can be combined to yield the simple result

$$\frac{Y' + Y''}{T' + T''} = \frac{b}{L} \qquad (8.96)$$

That is, the ratio of average vertical force component to average tension at the line ends equals the ratio of the vertical projection of the line to total line length.

The above equations can be used to derive line tensions and shape for any single line of a mooring pattern. However, a typical mooring analysis requires summation of the effects of up to 12 or more mooring lines with the co-ordinates of the surface vessel position on the water plane introducing two further variables. The complexity of this calculation makes it suitable for implementing it within a computer program. However, the catenary equations set out above are not entirely suitable for the basis of an efficient algorithm. The mathematical developments presented by O'Brien and Francis (1964) and Jennings (1965) give rise to an alternative set of equations describing the free hanging catenary which are much more suitable for automated calculations. Jennings (1965) has also extended the equations to apply for elastic lines and to account for temperature expansion effects. Peyrot (1980) has converted these mathematical details into a computer algorithm to derive the end forces and tension distribution in a catenary from a knowledge of its end co-ordinates, line elasticity and unstretched length. The equations underlying this procedure are repeated here for completeness but utilize slightly different variables which are defined by Figure 8.16. Each cable element is described by the catenary equations.

$$\left.\begin{aligned} L^2 &= b^2 + \frac{a^2\sinh^2\lambda}{\lambda^2} \\ Y' &= \frac{w}{2}\left(-b\,\frac{\cosh\lambda}{\sinh\lambda} + L\right) \end{aligned}\right\} \qquad (8.97)$$

where

$$\lambda = \frac{wa}{2X}$$

w is the cable weight per unit length in water and $X = |X_1| = |X_2|$ is the horizontal component of cable tension. Cable forces and tensions are related by the following equations:

$$\left.\begin{aligned} &Y'' = -Y' + wL_u, \qquad X_2 = -X_1, \\ &T_I = \sqrt{(X_1^2 + Y'^2)} \quad \text{and } T_J = \sqrt{(X_2^2 + Y''^2)} \end{aligned}\right\} \qquad (8.98)$$

where L_u is the unstressed cable length. By integrating along the cable length, the following relations are found

$$\left.\begin{aligned} a &= -X_1\left(\frac{L_u}{EA} + \frac{1}{w}\ln\frac{Y'' + T_J}{T_I - Y'}\right) \\ b &= \frac{1}{2EAw}\,(T_J^2 - T_I^2) + \frac{T_J - T_I}{w} \\ L &= L_u + \frac{1}{2EAw}\left(Y_4''T_J + Y'T_I + X_1^2\ln\frac{Y'' + T_J}{T_I - Y'}\right) \end{aligned}\right\} \quad (8.99)$$

where A is the area of cross-section of the line, E is its material modulus of elasticity and T_I, T_J are tensions at cable ends I and J respectively.

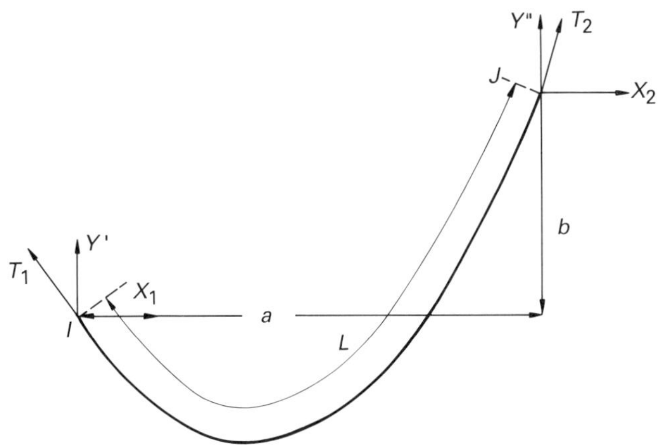

Figure 8.16. Mooring line element variables

As can be seen from this set of equations a, b and L can be written as functions of X_1 and Y' alone. An iteration procedure is then set up in a program subroutine with the starting values of X_1 and Y' found from Equations (8.97) and (8.98) with only the first term in an expansion of $[\sinh^2 \lambda/\lambda^2]$ being used. Hence initial values of a and b are found, and linear corrections δa and δb (which depend on a and b) are then applied to X_1 and Y' until δa and δb are less than a small imposed tolerance level. Hence the forces X_1, Y', X_2, and Y'' and cable tensions are found for any catenary. The line co-ordinates can be determined by replacing L_u by any fraction of L_u in Equations (8.99).

For mooring lines lying partially on the sea bed the above analysis is modified using an iteration procedure, so that additional increments of line are progressively laid on the bed until the suspended line is in equilibrium. Usually four or five iterations are sufficient to locate the final contact point. In many situations, multi-element lines made up of varying lengths and physical properties are used to increase the line restoring force. Such lines may be analysed by the algorithm above, where the analysis is performed on each cable element, and the imbalance in force at connecting points between elements is used to establish displacements through which these points must be moved to obtain equilibrium.

A typical mooring analysis is carried out by utilizing the above algorithm to calculate the forces exerted on the vessel due to each catenary line given the line end point co-ordinates on the surface vessel and sea bed together with line lengths and elasticities. These forces are then summed for all lines in the mooring spread to yield the resultant horizontal restoring force and vertical forces. The restoring force and tension in the most loaded line against vessel offset curve is then calculated by displacing the vessel through prescribed horizontal distances from its initial position.

The results of a typical analysis are presented in Figure 8.17. The steady component of environmental force is applied to the vertical axis of this diagram to obtain the resultant steady component of vessel offset from the horizontal axis. The slope of the force curve at this offset gives an equivalent linear stiffness C_t of the mooring system in the relevant direction for use in an equation of the form

$$(m + A)\ddot{x} + B_v|\dot{x}|\dot{x} + C_t x = F_x(t) \tag{8.100}$$

where co-ordinate x refers to a horizontal degree of freedom and the hydrotatic stiffness is zero.

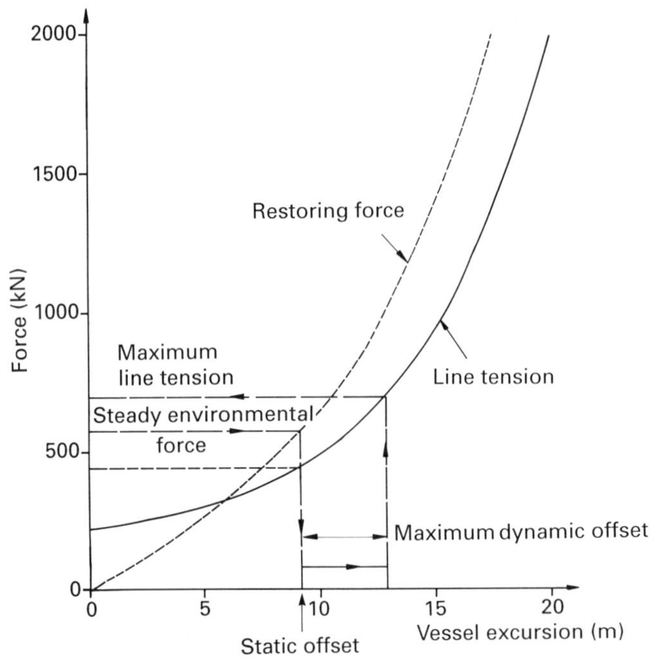

Figure 8.17. Restoring force and mast loaded line tension against vessel excursion for a typical catenary mooring system

The wave amplitude calculated from this equation, together with any offsets due to low frequency drift, are then applied on the horizontal axis of Figure 8.17 on the right-hand side of the static offset component so as to read off the maximum dynamic line tension.

It is also common practice to carry out time domain numerical solutions of a hybrid equation of motion representing the vessel, environmental and mooring system forces. The equation of motion is of the form

$$\mathbf{M}\ddot{\mathbf{X}} + \mathbf{B}\dot{\mathbf{X}} + \mathbf{B}_{\mathrm{v}}|\dot{\mathbf{X}}|\dot{\mathbf{X}} + \mathbf{K}\mathbf{X} = \mathbf{F}_0(t) + \mathbf{F}_{\mathrm{w}} + \mathbf{F}_{\mathrm{c}} + \mathbf{T} + \mathbf{F}_{\mathrm{m}}\{\mathbf{X}(t)\} \tag{8.101}$$

in matrix notation for the horizontal degrees of freedom of a floating vessel described through column vector $\mathbf{X}$; $\mathbf{M}$, $\mathbf{B}_{\mathrm{v}}$ and $\mathbf{K}$ are the total mass, drag damping and stiffness (excluding moorings) matrices of the vessel; $\mathbf{B}$ is a linear damping matrix contributed by damping due to waves radiating away from the vessel; $\mathbf{F}_0(t)$ is the wave exciting force, $\mathbf{F}_{\mathrm{W}}$ is the wind force, $\mathbf{F}_{\mathrm{c}}$ is the current force, $\mathbf{T}$ is the thruster force including interaction effects, and $\mathbf{F}_{\mathrm{m}}$ is the force exerted on the vessel by the moorings. The term $\mathbf{F}_{\mathrm{m}}\{\mathbf{X}(t)\}$ is evaluated using the static mooring analysis calculation described above – hence this analysis is referred to as hybrid since the dynamics of the surface vessel are calculated but the mooring lines are treated quasistatically. Equation (8.101) can be used to examine the dynamics of both intact and damaged (one line broken) mooring systems through the resultant offsets and line tensions obtained.

The wave exciting force term $\mathbf{F}_0(t)$ in the above equation can include low frequency or second-order wave drift forces. These second-order wave forces arise when linearized, or first-order wave theory is extended to account for finite wave heights and used to calculate wave loading. Although the resultant second-order wave forces are strictly only a correction to first-order wave forces, these corrections have a significant effect on mooring system performance for two reasons. Firstly second-order forces include a mean force component (wave drift force) which increases the total steady force to be reacted by the mooring system. Furthermore, second-order wave forces are usually subharmonic or superharmonic in nature, that is, they have frequencies which are lower or higher than the frequencies of predominant wave action.

It is of particular interest when an irregular wave incident on a moored structure induces second-order wave forces with components at low frequencies. Due to the low stiffness in surge, sway and yaw of catenary mooring systems, their natural frequencies are also low and mooring system resonance in surge, sway and yaw can be induced by second-order wave forces. The resultant large resonant motions are only limited by fluid induced damping forces and the peak mooring line tensions due to this phenomenon have to be considered in any comprehensive mooring analysis.

Mean and low frequency forces due to second-order wave loading also exhibit the non-linear property that for increasing wave amplitudes, second-order wave forces increase in proportion to the square of wave amplitude whereas first-order wave forces increase linearly with wave amplitude.

The earliest analysis of mean drift forces was made by Havelock (1940, 1942) for the mean drift force on a surface piercing body at high wave frequencies. Maruo (1960) offers an extended analysis to obtain the mean drift force on an arbitrary body from the mean rate of change of momentum in a control volume around the body.

More recently, Faltinsen and Michelsen (1974) have provided a modern numerical analysis method that obtains the mean horizontal drift force and yawing moment on an arbitrary geometry using momentum conservation. Pinkster and van Oortmorssen (1976), on the other hand, compute drift forces by direct integration of fluid pressures. A systematic body surface integral equation formulation for second-order forces has also been given by Standing *et al.* (1981) and others. Numerical solutions for low frequency drift forces for an arbitrary body in three dimensions have been given by Pinkster (1980) and by Standing *et al.* (1981).

However, low frequency drift forces calculated by using the above methods can only be satisfactorily applied in a design calculation through a time domain analysis which combines a drift force time history with the non-linear restoring force behaviour of a typical catenary mooring spread. The selection of floating platform added mass properties and of irregular wave spectra to be used for such design calculations pose additional problems.

One technique for integrating drift force calculations with a mooring analysis utilizes a slowly varying drift force time history generated by an inverse Fourier transform of a drift force spectrum derived from the methods outlined above. Alternatively, more approximate methods given by Oppenheim and Wilson (1980) or by Clauss (1982) may be used. Yet another approach to the problem is given by Hsu and Blankarn (1970) and Remery and Hermans (1971) with a semianalytical calculation of drift force time history directly from the wave elevation time history at the vessel position – this can itself be derived using an inverse Fourier transform of the wave elevation spectrum.

The solution of Equation (8.101) in the time domain with low frequency drift forces included is critically dependent on a good estimate of the damping force acting on the vessel at low frequencies. The level of damping force governs the maximum line tensions obtained and with the present state of the art, model tests are likely to yield the most accurate estimates of damping forces at low frequencies.

8.3.2 Tensioned moorings

The stiffness contribution of a taut mooring tether between a point on a surface platform and a sea bed foundation can be derived by idealizing the tether as weightless and perfectly elastic with a known constant tension and axial stiffness.

The stiffness contributions to all six rigid body degreees of freedom of the surface platform are derived for completeness with six elements of the restoring force vector $\mathbf{F}_{\mathrm{m}}$ written as

$$F_i = k_{ij}X_j$$

with k_{ij} being the (6 × 6) stiffness matrix with i and j denoting the row and column numbers respectively; X_j is a (6 × 1) column vector where i and j denote the six rigid degrees of freedom of the surface platform in surge, sway, heave, roll, pitch and yaw respectively.

Since the tether is assumed weightless, it is taken to lie along a straight line joining the two end co-ordinates. The resultant stiffness matrix is evaluated assuming that the surface platform moves through small displacements relative to the tether lengths involved.

Each tether has a constant tension T, an elastic stiffness λ' and is strung between co-ordinates (x_1, y_1, z_1) on the sea bed and (x_2, y_2, z_2) at the surface platform, both relative to the platform's principal axes system $Oxyz$, as shown in Figure 8.18.

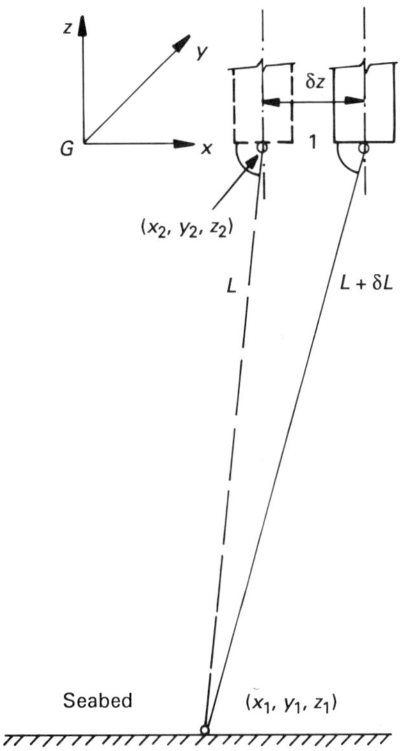

Figure 8.18. Notation for tether stiffness

The direction cosines are defined as follows:

$$\left.\begin{aligned} &\cos\alpha = \frac{x_2 - x_1}{L}, \cos\beta = \frac{y_2 - y_1}{L}, \cos\gamma = \frac{z_2 - z_1}{L} \\ &\text{where the tether length is} \\ &L = \surd[(x_2 - x_1)^2 + (y_2 - y_1)^2 + (z_2 - z_1)^2] \end{aligned}\right\} \quad (8.102)$$

Consider that the surface platform translates in the positive x direction through a small distance δx. Let $a = x_2 - x_1$, $b = y_2 - y_1$, and $c = z_2 - z_1$. Then to first order, the new tether length becomes

$$L + \delta L = L + \frac{a}{L}\delta x$$

Then the additional tension δT in the cable due to extension δx is

$$\delta T = \lambda' \delta L = \frac{\lambda'(x_2 - x_1)}{L}\delta x \tag{8.103}$$

The resulting restoring force component along the x axis can be written as

$$\delta T_x = (T + \delta T)\cos\alpha' - \mathrm{T}\cos\alpha$$

where T is the cable tension and α' is given by

$$\cos\alpha' = \frac{x_2 + \delta x - x_1}{L + \delta L} \tag{8.104}$$

Thus

$$\delta T_x = (T + \delta T)\left[\frac{x_2 - x_1}{L}\right]\left[\frac{1 + [(\delta x)/(x_2 - x_1)]}{1 + (\delta L/L)}\right] - \frac{T(x_2 - x_1)}{\mathrm{L}}$$

$$= T\cos\alpha\left\{\frac{\delta x}{x_2 - x_1} - \frac{\delta L}{L} + \frac{\delta T}{T}\right\} \tag{8.105}$$

to first order. Since $\delta L = \delta x \cos\alpha$;

$$\delta T_x = \frac{T}{L}\cos\alpha\left[\frac{L\delta x}{x_2 - x_1} - \cos\alpha\delta x + \frac{\lambda(x_2 - x_1)}{T}\delta x\right]$$

$$= \left(\lambda\cos^2\alpha + \frac{T}{L}\sin^2\alpha\right)\delta x \tag{8.106}$$

In the limit as $\delta x \to 0$

$$\frac{\delta T_x}{\delta x} = \frac{\partial T_x}{\partial x} = k_{11} = \lambda\cos^2\alpha + \frac{T}{L}\sin^2\alpha \tag{8.107}$$

which is the first term in the 6×6 stiffness matrix.

The remaining terms corresponding to the restoring forces due to translations can be derived in a similar fashion in terms of k_{ij} to be

$$\left.\begin{aligned}
k_{21} &= [\lambda + (T/L)]\cos\alpha\cos\beta = k_{12}\\
k_{31} &= [\lambda + (T/L)]\cos\alpha\cos\gamma = k_{13}\\
k_{22} &= \lambda\cos^2\beta + (T/L)\sin^2\beta\\
k_{32} &= [\lambda + (T/L)]\cos\beta\cos\gamma = k_{23}\\
k_{33} &= \lambda\cos^2\gamma + (T/L)\sin^2\gamma
\end{aligned}\right\} \tag{8.108}$$

The remaining 27 terms corresponding to restoring moments due to translations, restoring forces due to rotations and restoring moments due to rotations can all be expressed in terms of the first nine terms. These are given below for one-half of the symmetric matrix and include the leading diagonal.

$$\left.\begin{aligned}
k_{41} &= k_{31}y_2 - k_{21}z_2 \\
k_{51} &= k_{11}z_2 - k_{31}x_2 \\
k_{61} &= k_{21}x_2 - k_{11}y_2 \\
k_{42} &= k_{32}y_2 - k_{22}z_2 \\
k_{52} &= k_{21}z_2 - k_{32}x_2 \\
k_{62} &= k_{22}x_2 - k_{21}y_2 \\
k_{43} &= k_{33}y_2 - k_{32}z_2 \\
k_{53} &= k_{31}z_2 - k_{33}x_2 \\
k_{63} &= k_{32}x_2 - k_{31}y_2 \\
k_{44} &= k_{33}y_2^2 - 2k_{32}y_2z_2 + k_{22}z_2^2 \\
k_{54} &= k_{31}y_2z_2 - k_{21}z_2^2 - k_{33}y_2x_2 + k_{32}x_2z_2 \\
k_{64} &= k_{32}x_2y_2 - k_{22}x_2z_2 - k_{31}y_2^2 + k_{21}y_2z_2 \\
k_{55} &= k_{11}z_2^2 - 2k_{31}x_2z_2 + k_{33}x_2^2 \\
k_{65} &= k_{21}x_2z_2 - k_{32}x_2^2 - k_{11}y_2z_2 + k_{31}x_2y_2 \\
k_{66} &= k_{22}x_2^2 - 2k_{21}x_2y_2 + k_{11}y_1^2
\end{aligned}\right\} \quad (8.109)$$

The stiffness matrix is computed for each tether in a mooring system and summed to yield at total mooring stiffness matrix.

In some situations such as for deep water or low displacement TBPs in shallow water, it may become necessary to calculate the dynamic interaction between the platform and its tethers with their distributed added mass and damping forces accounted for. In these cases, a mathematical model of tether vibrations needs to be coupled with the dynamics of the surface platform in order to derive more accurate estimates of platform motions, tether vibration amplitudes and tether bending moments. Patel and Lynch (1983) present an example of such work.

8.4 Analysis with multiple degrees of freedom

8.4.1 Basic equations

The general multiple degree of freedom equations of motions for a stationary floating body in gravity waves can be formulated using an approach similar to that used for single degrees of freedom in Section 8.2.1. It is usual to formulate these matrix equations in the six rigid body degrees of surge, sway, heave, roll, pitch and yaw described by a column vector **X**, although additional degrees of freedom (such as structural deformations) can be incorporated, if required.

Now the force on a general floating body of semisubmersible space frame configuration can be described by the equation

$$\mathbf{M}\ddot{\mathbf{X}} = \sum \mathbf{M}_{\mathrm{A}}(\ddot{\boldsymbol{\eta}} - \ddot{\mathbf{X}}) + \sum \mathbf{M}_{\mathrm{FK}}\ddot{\boldsymbol{\eta}}$$

$$+ \sum \mathbf{B}_{\mathrm{v}}|\dot{\boldsymbol{\eta}} - \dot{\mathbf{X}}|(\dot{\boldsymbol{\eta}} - \dot{\mathbf{X}}) - \mathbf{K}\mathbf{X} - \mathbf{K}_{\mathrm{m}}\mathbf{X} \tag{8.110}$$

where $\mathbf{M}$, $\mathbf{M}_{\mathrm{A}}$ and $\mathbf{M}_{\mathrm{FK}}$ are the (6×6) coefficient matices quantifying structure physical mass, added mass and the Froude–Krylov added inertia respectively, $\mathbf{B}_{\mathrm{v}}$ is a (6×6) matrix representing the non-linear drag force contribution, while $\mathbf{K}$ and $\mathbf{K}_{\mathrm{m}}$ are (6×6) stiffness matrices contributed by the hydrostatic and mooring restoring forces acting on the vessel.

Wave particle velocities and accelerations are denoted by vectors $\dot{\boldsymbol{\eta}}$ and $\ddot{\boldsymbol{\eta}}$, respectively, with the summation signs in the wave force terms indicating a numerical integration over vessel submerged members to take account of spatial variations in wave properties. The contribution of linear wave radiation damping is neglected in this formulation. Furthermore, the above equation is strictly valid only for offshore structures with slender members. Section 8.4.2(b) examines the governing equation for monolithic structures of larger cross-sectional dimensions.

Equation (8.110) can be rewritten to give

$$(\mathbf{M} + \mathbf{M}_{\mathrm{A}})\ddot{\mathbf{X}} + \mathbf{B}_{\mathrm{v}}|\dot{\mathbf{X}}|\dot{\mathbf{X}} + (\mathbf{K} + \mathbf{K}_{\mathrm{m}})\mathbf{X}$$

$$= \sum (\mathbf{M}_{\mathrm{A}} + \mathbf{M}_{\mathrm{FK}})\ddot{\boldsymbol{\eta}} + \sum \mathbf{B}_{\mathrm{v}}|\dot{\boldsymbol{\eta}} - \dot{\mathbf{X}}|(\dot{\boldsymbol{\eta}} - \dot{\mathbf{X}}) + \mathbf{B}_{\mathrm{v}}|\dot{\mathbf{X}}|\dot{\mathbf{X}} \tag{8.111}$$

with a $\mathbf{B}_{\mathrm{v}}|\dot{\mathbf{X}}|\dot{\mathbf{X}}$ term added to both sides of the equation to obtain a conventional form for the left-hand side. Equation (8.111) would be difficult to solve in its present form due to the presence of $\dot{\mathbf{X}}$ in the wave force on the right-hand side of the equation. However, two features of this wave force expression suggest a useful simplification. These are concerned with the relatively large magnitude of the inertia term in the wave force and the decay with depth of wave particle velocities which makes the $\Sigma\, \mathbf{B}_{\mathrm{v}}|\dot{\boldsymbol{\eta}} - \dot{\mathbf{X}}|(\dot{\boldsymbol{\eta}} - \dot{\mathbf{X}}) + \mathbf{B}_{\mathrm{v}}|\dot{\mathbf{X}}|\dot{\mathbf{X}}$ term small for members with significant submergence below still water level.

With these approximations in mind, Equation (8.111) can be simplified to

$$(\mathbf{M} + \mathbf{M}_{\mathrm{A}})\ddot{\mathbf{X}} + \mathbf{B}_{\mathrm{v}}|\dot{\mathbf{X}}|\dot{\mathbf{X}} + (\mathbf{K} + \mathbf{K}_{\mathrm{m}})\mathbf{X}$$

$$= \sum (\mathbf{M}_{\mathrm{A}} + \mathbf{M}_{\mathrm{FK}})\ddot{\boldsymbol{\eta}} + \sum \mathbf{B}_{\mathrm{v}}|\dot{\boldsymbol{\eta}}|\dot{\boldsymbol{\eta}} \tag{8.112}$$

with the $\Sigma\, \mathbf{B}_{\mathrm{v}}|\dot{\boldsymbol{\eta}} - \dot{\mathbf{X}}|(\dot{\boldsymbol{\eta}} - \dot{\mathbf{X}}) + \mathbf{B}_{\mathrm{v}}|\dot{\mathbf{X}}|\dot{\mathbf{X}}$ term replaced by a $\Sigma\, \mathbf{B}_{\mathrm{v}}|\dot{\boldsymbol{\eta}}|\dot{\boldsymbol{\eta}}$ term. In order to explore the effects of this simplication, a residue wave force fraction R_i for a particular motion (heave, say) can be defined as

$$R_i = \frac{[\Sigma\, \mathbf{B}_{\mathrm{v}i}|\dot{\boldsymbol{\eta}}_i - \dot{\mathbf{X}}_i|(\dot{\boldsymbol{\eta}}_i - \dot{\mathbf{X}}_i) + \mathbf{B}_{\mathrm{v}i}|\dot{\mathbf{X}}_i|\dot{\mathbf{X}}_i - \Sigma\, \mathbf{B}_{\mathrm{v}i}|\dot{\boldsymbol{\eta}}_i|\dot{\boldsymbol{\eta}}_i]}{(K_i H/2)} \tag{8.113}$$

where the i refers to diagonal coefficients for heave motion. The residue force is non-dimensionalized with respect to the wave force amplitude at zero frequency by using the hydrostatic stiffness K_i in the mode of motion being considered and the wave height H.

Results presented later for semisubmersibles demonstrate that, after drag force linearization, the residue wave force that remains unaccounted for due to the simplification is negligible at wave frequencies even for large wave heights where the drag forces will be at their greatest. The drag force linearization used here is based on equal energy dissipation at resonance for the non-linear damping coefficients and their equivalent linear values. However, it should be noted that this linearization is a fundamentally different assumption from that implied in the reformulation of Equation (8.110) into Equation (8.112) which is quantified by residue factor R_i.

The formulation of the equation of motion for a monolithic ship shape hull form differs from that used for a semisubmersible since, for ship shapes, the diffraction of incident waves due to the large size of the hull form has to be accounted for. For these hull forms, the basic equation of motion can be written simply as

$$(\mathbf{M} + \mathbf{M}_A)\ddot{\mathbf{X}} + \mathbf{B}_p\dot{\mathbf{X}} + (\mathbf{K} + \mathbf{K}_m)\mathbf{X} = \mathbf{F}e^{i\omega t} \tag{8.114}$$

where $\mathbf{M}$, $\mathbf{M}_A$, $\mathbf{K}$ and $\mathbf{K}_m$ are defined above, $\mathbf{B}_p$ is the matrix of potential damping coefficients and $\mathbf{F}e^{i\omega t}$ is a complex vector of wave forces obtained by summing hull surface pressures from a diffraction analysis.

The above basic equations of motion are solved using different approaches for slender space framed structures and for large hull forms.

8.4.2 Coefficient matrices and wave forces

(a) *Morison equation approach*

This is applicable to slender space framed structures with characteristic cross-sectional dimensions that are less than 20% of incident wave lengths. The following additional assumptions are implied in the formulation of the equation of motion (8.112) above:

1. The vessel structure is assumed to be an assembly of cylindrical elements and horizontal rectangular pontoons. The cylindrical members and rectangular pontoons are assumed to have large ratios of length to cross-sectional dimension.
2. The motion amplitudes of the platform and wave motions are assumed to be small. As a consequence linear wave theory is used.
3. Wave forces on individual elements of the structure are computed as though other members were not present, that is, hydrodynamic interference between members is ignored.
4. The forces associated with sinusoidal wave motions are computed independently of the forces associated with absolute motions of the structure.
5. The non-linear drag damping term is linearized by assuming an equivalent linear damping which would dissipate the same energy at resonance as the non-linear damping. The contribution of wave radiation effects to the damping terms is assumed to be negligibly small.

The floating vessel is described by orthogonal co-ordinate axes with origin at the centre of gravity, x pointing forwards, y to port and z vertically upwards. The physical mass matrix is computed using the mass distribution of the vessel and is diagonal if the reference axes are principal axes of the vessel. The added mass matrix for the structure is obtained using a summation of the added mass matrix for each individual cylinder. Details of this approah are given in Section 3.11. These equations can readily be modified to account for rectangular cylinders.

The fluid damping matrix $\mathbf{B}_v$ for the structure is evaluated in a similar manner to the added mass matrix – as a sum of the contributions from each individual members. The derivation of the generalized damping matrix for an arbitrarily orientated circular cylinder can be derived using the method in Section 3.11 with the assumption that only drag forces normal to the cylinder axis are significant. Unlike the added mass matrix, the non-linear velocity square proportionality generates an asymmetric damping matrix. The corresponding matrices for the rectangular cylinder and non-elongated body are evaluated as special cases of the circular cylinder.

Contributions to the hydrostatic stiffness matrix $\mathbf{K}$ will only arise in the heave, roll and pitch degrees of freedom due to buoyancy forces in the water plane cutting members of the hull. If, for a member number n, A_{wn} denotes the water plane area (x_{wn}, y_{wn}) are the co-ordinates of the centroid of this water plane area, then the hydrostatic stiffness matrix elements k_{ij} can be written as the summations for all n

$$\left.\begin{aligned} k_{33} &= \rho g \sum A_{wn} \\ k_{43} &= \rho g \sum y_{wn} A_{wn} \\ k_{53} &= -\rho g \sum x_{wn} A_{wn} \\ k_{54} &= -\rho g \sum A_{wn} x_{wn} y_{wn} \\ k_{44} &= \rho g V (GM)_{\mathrm{P}} \\ k_{55} &= \rho g V (GM)_{\mathrm{R}} \end{aligned}\right\} \qquad (8.115)$$

where V is the vessel displacement by volume and $(GM)_{\mathrm{R}}$, $(GM)_{\mathrm{P}}$ are metacentric heights in roll and pitch respectively, while all other stiffness terms are zero. A linearized mooring system stiffness matrix $\mathbf{K}_{\mathrm{m}}$ can be obtained from the calculations described in Section 8.3 and inserted in the equation of motion to obtain motion responses of the moored rather than the free vessel.

The wave force calculation is based on the reduction of the terms on the right-hand side of Equation (8.112) into an oscillating force column vector by summing the effects of wave pressures, particle velocities and accelerations on all structural members of the semisubmersibe. Conventional linear wave theory equations are presented in a compact complex number form here which is readily amenable to inclusion in an efficient computer program.

The wave surface elevation η can be written as

$$\left.\begin{aligned}&\eta = \eta_0\beta e^{-i\omega t}\\ &\text{where}\\ &\beta = \exp[i(kx\cos\alpha + ky\sin\alpha)]\end{aligned}\right\} \quad (8.116)$$

where η_0 is the wave amplitude and α is the wave direction measured clockwise positive from the x axis. The components of velocity and acceleration in the x, y and z directions are

$$\left.\begin{aligned}&u = \omega\beta v_2\cos\alpha\eta_0\, e^{-i\omega t} \qquad \dot{u} = -i\omega t^2\beta v_2\cos\alpha\eta_0 e^{-i\omega t}\\ &v = \omega\beta v_2\sin\alpha\eta_0\, e^{-i\omega t} \qquad \dot{v} = -i\omega^2\beta v_2\sin\alpha\eta_0 e^{-i\omega t}\\ &w = -i\omega\beta v_1\eta_0\, e^{-i\omega t} \qquad \dot{w} = -\omega^2\beta v_1\eta_0 e^{-i\omega t}\\ &\text{where}\\ &v_1 = \frac{\sinh k(d - s' + z)}{\sinh kd},\; v_2 = \frac{\cosh k(d - s' + z)}{\sinh kd}\end{aligned}\right\} \quad (8.117)$$

d is the water depth, and s' is the z co-ordinate of still water level relative to reference axes at the centre of gravity. The wave dynamic pressure can be written as

$$p = \rho g\beta v_2\eta_0 e^{-i\omega t} \quad (8.118)$$

The three components of wave velocity and acceleration as well as wave elevation and wave pressure are complex quantities in these equations.

The above equations are used to efficiently compute wave velocities, accelerations and pressures at any point around the platform submerged members so that wave force calculations can be carried out. Normal forces on elongated cylindrical or rectangular members are computed by subdividing the members into elements, computing local wave properties and summing the consequential normal loading for all the elements. Axial forces due to wave dynamic pressures are also computed with due account being taken of surface piercing members where the dynamic pressure acts on one member end only.

The vessel response to an applied oscillating wave force vector is then computed by using an iterative technique to account for the non-linear drag damping force. A first approximation diagonal linear damping coefficient matrix is obtained by ignoring all non-diagonal terms in the total mass and stiffness matrices and assuming linear damping to be 10% of critical. The equations of motion are solved with this first approximation to the damping value. The column vector of these velocities is then substituted into the modulus sign in the damping term such that the equation

$$(\mathbf{M} + \mathbf{M}_A)\ddot{\mathbf{X}} + \mathbf{B}_v \frac{8}{3\pi} |\dot{\mathbf{X}}_{\text{last approx}}|\dot{\mathbf{X}} + (\mathbf{K} + \mathbf{K}_m)\mathbf{X} = \mathbf{F}(t) \quad (8.119)$$

is solved to obtain a better approximation for the column vector $\mathbf{X}$. The equivalent damping matrix $\mathbf{B}_{eq}$

$$\mathbf{B}_{eq} = \mathbf{B}_v \frac{8}{3\pi} |\dot{\mathbf{X}}| \quad (8.120)$$

is a standard result obtained by assuming equal work done at resonance by the non-linear and the equivalent linear damping terms (see Section 8.2.2). The iteration is continued until a specific tolerance (such as 1%) between successive approximations is achieved.

It is of interest to explore the effects of the drag force nonlinearity in Equation (8.112); both for the wave force excitation on the right-hand side and for the damping term on the left-hand side. Figures 8.19–8.21 display the vertical wave forces and heave motion responses as functions of wave period for a typical twin pontooned semisubmersible at drilling draught. Vessel heave forces and motions are presented for two values of incident wave height and the data presented are nondimesionalized with respect to the total heave wave force amplitude at zero frequency for heave forces and incident wave amplitude for heave motions.

Figure 8.19 separately displays the added mass and Froude–Krylov force on the horizontal pontoons, the Froude–Krylov force on the lower ends of the vertical columns and drag force on the horizontal pontoons for 2 m and 12 m wave heights. It needs to be pointed out that the vertical force on the horizontal pontoons and the vertical force on the column are 180° out of phase so that a cancellation effect occurs at a wave period of 18.7 s for the total wave force. Figure 8.19 illustrates how the drag force (curve C for 2 m wave height and curve D for 12 m wave height) is a relatively small fraction of the other force components on the structure.

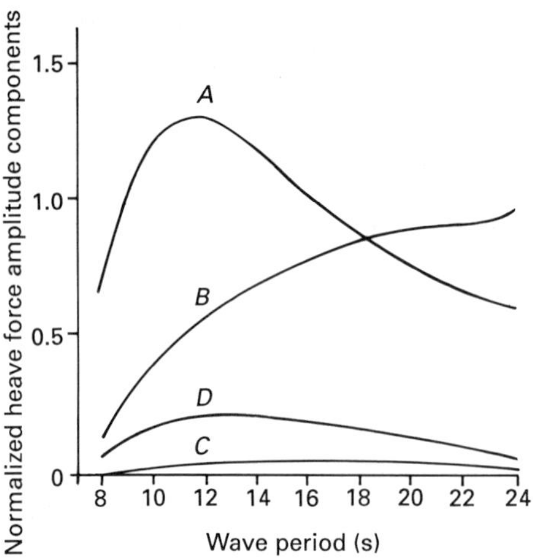

Figure 8.19. Heave force amplitude components as functions of wave period for 2 m and 12 m wave heights. Curve A denotes added mass and wave pressure force on pontoons. Curve B denotes wave pressure force on vertical columns, for both 2 m and 12 m wave heights. Curves C and D denote drag force for wave heights of 2 m and 12 m respectively

Figure 8.20 displays the total vertical wave force for 2 m and 12 m wave heights. The cancellation effect described above is not quite exact because of the small drag force contribution. In order to quantify the difference in wave force computed by the formulation of Equations (8.110) and (8.112) the unaccounted drag force is computed as a fractional residue term R_i as defined in Equation (8.113) and presented in Figure 8.20. It is apparent that the residue is negligible for wave periods up to 17 s and rises to a maximum of 0.15 for 12 m wave height and 20 s period. The small overall level of the residue R_i justifies the simplification made in Equation (8.112) from Equation (8.110).

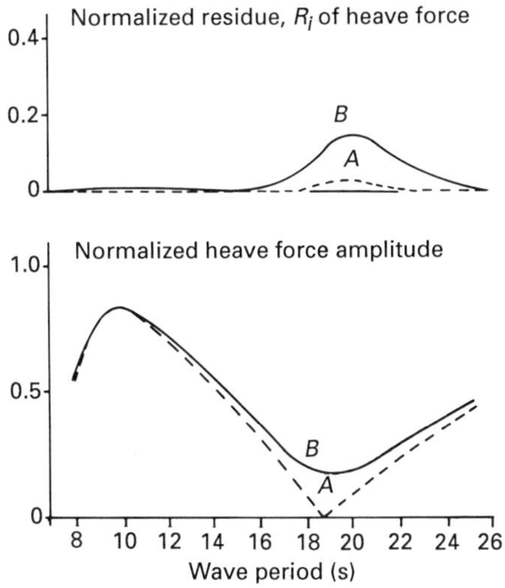

Figure 8.20. Total heave force amplitude and fractional residue of wave force as functions of wave period for 2 m (curves A) and 12 m (curves B) wave heights

The heave motion response is presented in Figure 8.21 and shows a strong influence of the drag force non-linearity around resonance with the heave response amplitude per unit wave amplitude reducing from 4.88 m/m at 1 m wave amplitude to 1.26 m/m a 6 m wave amplitude. The vessel motion response away from resonance is not significantly affected although there is some increase in response around 16–19 s due to the corresponding increase in wave force amplitude at these periods. The large change in the unit heave response at and around resonance is to be expected since the damping force in a vibratory system is dominant at resonance.

Morison's equation formulation of wave force on slender space frame structure such as semisubmersibles requires the choice of applicable added mass and drag coefficients. It is common practice to use drag and added mass coefficients from recommendations made by classification societies (see Appendix G, for example). This procedure has been followed for these calculations with an added mass coefficient of 1.24 and a drag coefficient of 1.34 used for vertical motion of the twin horizontal rectangular pontoons. In order to explore the sensitivity of the calculations to reasonable variations in these coefficients, the changes in the heave transfer function due to some of the combinations

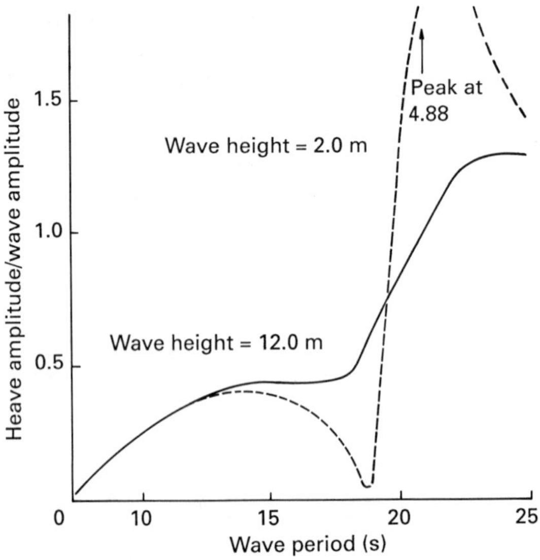

Figure 8.21. Heave motion response of semisubmersible vessel in head seas

of the added mass coefficient from 1.00 to 1.50 and of the drag coefficient from 1.00 to 2.00 are presented in Figure 8.22 for 6 m wave height. It is evident that the relatively large variation in the drag coefficient does not alter the heave response at wave frequencies to a significant extent. This is to be expected since variations in the drag force are small in comparison with inertia forces on the vessel. The small drag force amplitude does have some influence on the heave response amplitude at the cancellation frequency and at resonance. The

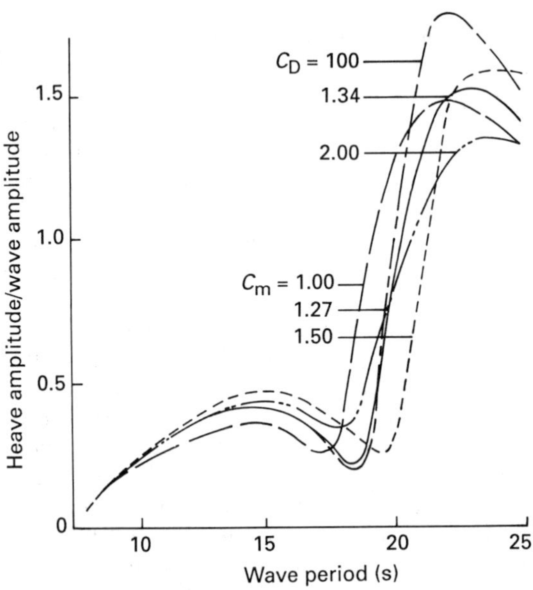

Figure 8.22. Sensitivity of heave motion response to drag and inertia coefficient variations

difference of two large inertia related forces around the cancellation frequency highlights the effect of the drag coefficient changes. The alteration of response at resonance is due to the change in damping level induced through the drag coefficient. It is also clear from Figure 8.22 that the relatively large variation in pontoon added mass coefficients serves primarily to alter the heave natural period with consequential changes in the heave response at wave frequencies. These results underline the importance of selecting the added mass coefficients carefully; perhaps by reference to potential flow diffraction theory results for isolated body cross-sectional shapes.

Nevertheless, this Morison equation calculation shows good agreement with both model test data and full scale measurements. Patel (1983) gives further information on the analyses and Lyons and Patel (1984) presents some comparison of calculations with test data.

(b) *Diffraction theory*

Calculations of wave induced motions of a large non-space frame structure in gravity waves requires a solution of the wave problem with no flow boundary conditions at the moving body surface in addition to the free surface and sea bed boundary conditions. The solution can be split into two related problems – the scattering wave problem defines wave forces on a floating body when fixed in space and with waves incident on it in an identical manner to the technique described in Section 5.6 for computing wave forces on a fixed body. The radiation wave problem is concerned with defining forces on the body (added mass and damping) due to its oscillation in otherwise still water. These oscillations will induce wave potentials such that the total wave potential in the fluid is the sum of the incident ϕ_w, scattered ϕ_s, and forced wave potentials ϕ_f, so that

$$\phi = \phi_w + \phi_s + \phi_f \tag{8.121}$$

and these must satisfy the boundary conditions at the body surface given by

$$\frac{\partial \phi_w}{\partial n} + \frac{\partial \phi_s}{\partial n} + \frac{\partial \phi_f}{\partial n} = V_n \tag{8.122}$$

where V_n is the velocity of the body surface in the direction n normal to the surface. This boundary condition can be applied at the mean body surface since the theory is applied for small motions; ϕ, together with its three components must also satisfy the Laplace equation and the free surface and sea bed boundary conditions. Furthermore, ϕ_s and ϕ_f must satisfy the radiation conditions.

Boundary conditions for the scattering and radiation wave problems can be split up from Equation (8.122) as

$$\left.\begin{aligned} &\frac{\partial \phi_w}{\partial n} + \frac{\partial \phi_s}{\partial n} = 0 \\ &\text{and} \\ &\frac{\partial \phi_f}{\partial n} = V_n \end{aligned}\right\} \tag{8.123}$$

respectively, both being applied on the body surface. The scattering problem is identical to the application of diffraction theory on fixed structures, as described in Section 5.6. The radiation problem can also be solved by using either boundary integral or boundary element techniques. Only the solution using boundary integral techniques is described here for brevity. As in Section 5.6, the analysis assumes inviscid, irrotational flow and that wave amplitudes are small. The unsteady flow around the floating vessel is calculated by introducing oscillating sources of unknown velocity potential on the vessel submerged surface that is discretized by a mesh of facets with an oscillating source on the surface of each facet.

A Green's function is used to represent the velocity potential of each source, which, because of the form of the Green's function, satisfies Laplace's equation, zero flow at the horizontal sea bed, the free surface and radiation boundary conditions. Now, the solution for scattered wave potential due to the stationary floating body, subjected to incident waves of potential, ϕ_w, is identical to that described in Section 5.6 for fixed structures. A set of linear simultaneous equations are obtained by equating the flow due to the local source plus the additional flow due to all other souces to the negative of the flow due to the undisturbed wave for each facet on the body surface. Solution of the resulting set of algebraic equations yields the unknown source strengths and, therefore, the velocity potential, ϕ_s, which is used to derive pressures and wave forces by integration over the body surface. Thus the wave force vector **F** of Equation (8.114) may be obtained for an incident wave of specified frequency and direction.

The velocity potential, ϕ_f, is obtained in a similar way to that above except for the use of a different boundary condition which reflects the fact that ϕ_f arises from body motions in otherwise still water. Thus, at all facets, the source strengths, ϕ_{fi}, are such that the flow due to the local source plus the flow due to all other sources equals the velocity component of the body along the facet normal. This velocity component will depend on the mode of motion (surge, sway, heave, and so on) in which the body is moving. All this can be represented by equating the normal velocity of the fluid and of the jth facet for the vessel moving in its kth mode of motion. This yields the equation

$$\sum_{\text{All } i} \frac{\partial \phi_{fi}}{\partial n_j} \sigma_{ik} = V_{jk} \tag{8.24}$$

where V_{jk} is the normal velocity of the jth facet with the vessel moving in its kth mode of motion. Furthermore, n_j is the normal to the jth facet, $\partial\phi_i/\partial n_j$ is the normal fluid velocity at the jth facet due to a unit source at the ith facet, and σ_{ik} are the unknown source strengths required in the kth mode. Application of Equation (8.124) for all facets produces a system of complex equations to be solved for the source strengths. Once these are known, the pressures at the facets are evaluated and their effects integrated over the vessel surface to yield forces in each mode of motion to unit motion in the kth mode.

These forces may be written as a complex square matrix, $G(\omega)$ which can be decomposed into its real and imaginary parts through the equation

$$\mathbf{G}(\omega) = \omega^2\mathbf{M}_A(\omega) - i\omega\mathbf{B}_P(\omega) \tag{8.125}$$

to yield frequency dependent added mass and damping matrices $\mathbf{M}_A(\omega)$ and $\mathbf{B}_P(\omega)$ which are required for Equation (8.114).

The inclusion of physical mass, hydrostatic and mooring stiffness matrices **M**, **K** and $\mathbf{K}_m$ completes derivation of all the coefficient matrices of Equation (8.114). The hydrodynamic coefficient matrices are, however, frequency dependent and require a diffraction analysis at all frequencies at which motions are required. Equation (8.114) is linear and can readily be solved to yield the displacement vector **X**. The exciting force vector $\mathbf{F}(\omega)$ and the coefficient matrices $\mathbf{M}_A(\omega)$ and $\mathbf{B}_P(\omega)$ can also be derived using finite element methods in an analogous way to that described for the boundary integral approach described above.

There is one further point of interest regarding the relationship between the scattered and forced wave potentials (ϕ_s and ϕ_f) for a floating vessel problem. The use of equations called Haskind relations (see Newman, 1962) enables the scattered wave potential ϕ_s to be expressed in terms of the incident and forced wave potentials ϕ_w and ϕ_f. Thus, once ϕ_f is calculated, ϕ_s need not be computed by diffraction analysis but can instead be derived using the Haskind relations.

Typical results of a boundary integral diffraction analysis for a ship shaped hull are shown in Figures 8.23 to 8.25. The discretization of the submerged hull geometry is shown in figure 8.21 using 277 triangular facets on the ship half hull. The vessel is of 263.7 m overall length, 40.8 m beam and 145 937 t displacement with 14.80 m draught floating in deep water. Figure 8.24 presents the variation of added mass and radiation damping coefficients with frequency for heave and pitch motions. Note that the variation in added mass is relatively small but the radiation damping shows large changes with very small values at some wave periods. Wave induced heave force and pitching moments and the resultant motion responses for head seas are presented in Figure 8.25.

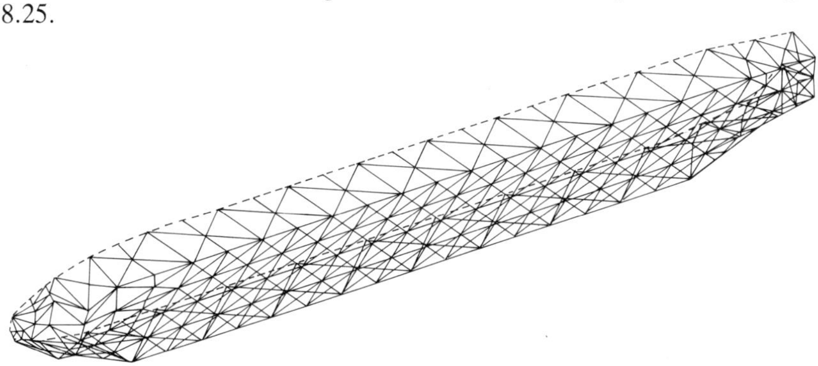

Figure 8.23. Facet discretization of submerged ship hull for diffraction theory

There is generally good agreement between model test data, full scale results and the results of diffraction analysis for wave induced motion

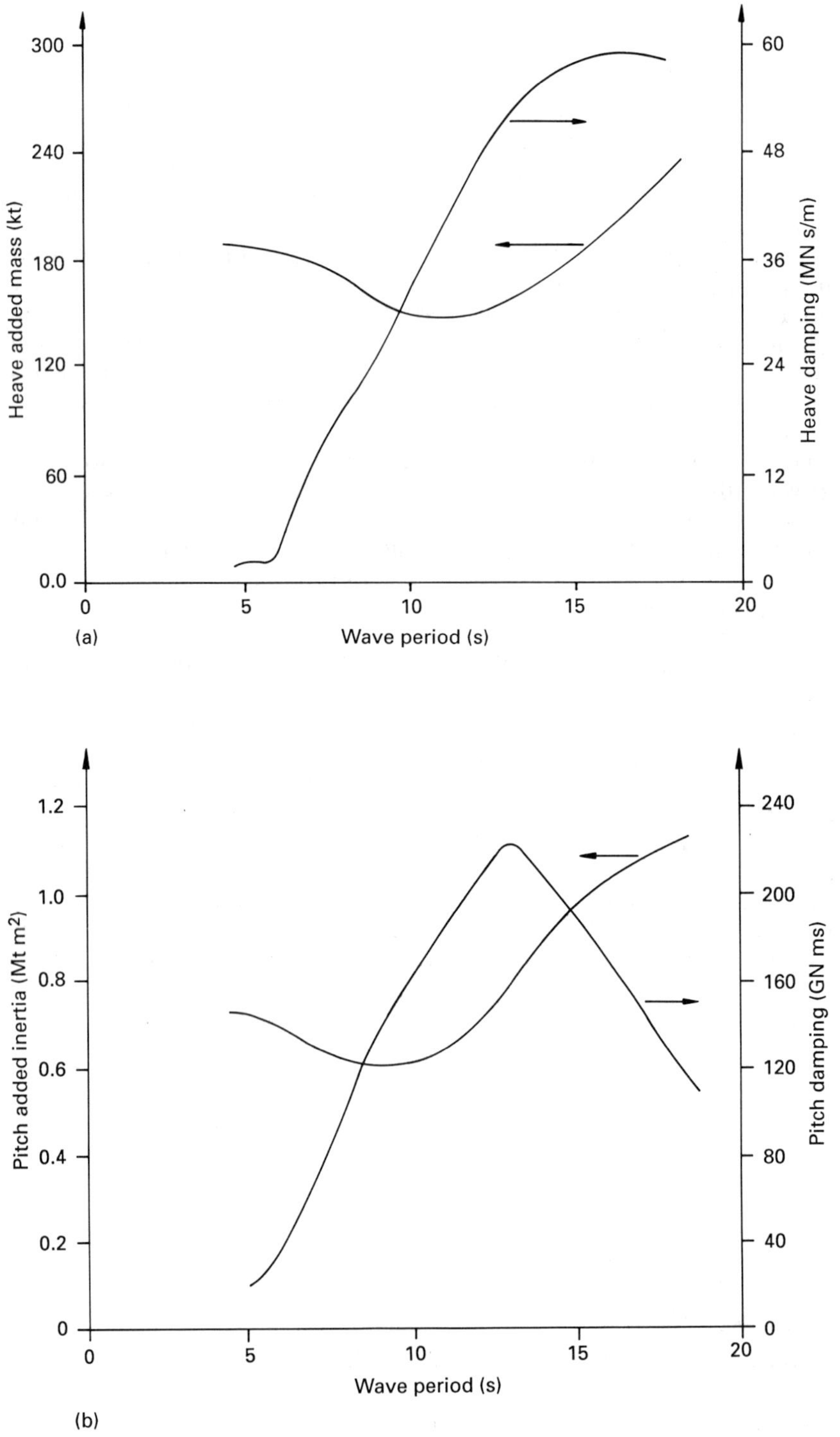

Figure 8.24. Added mass and damping coefficient variations with wave period for ship hull

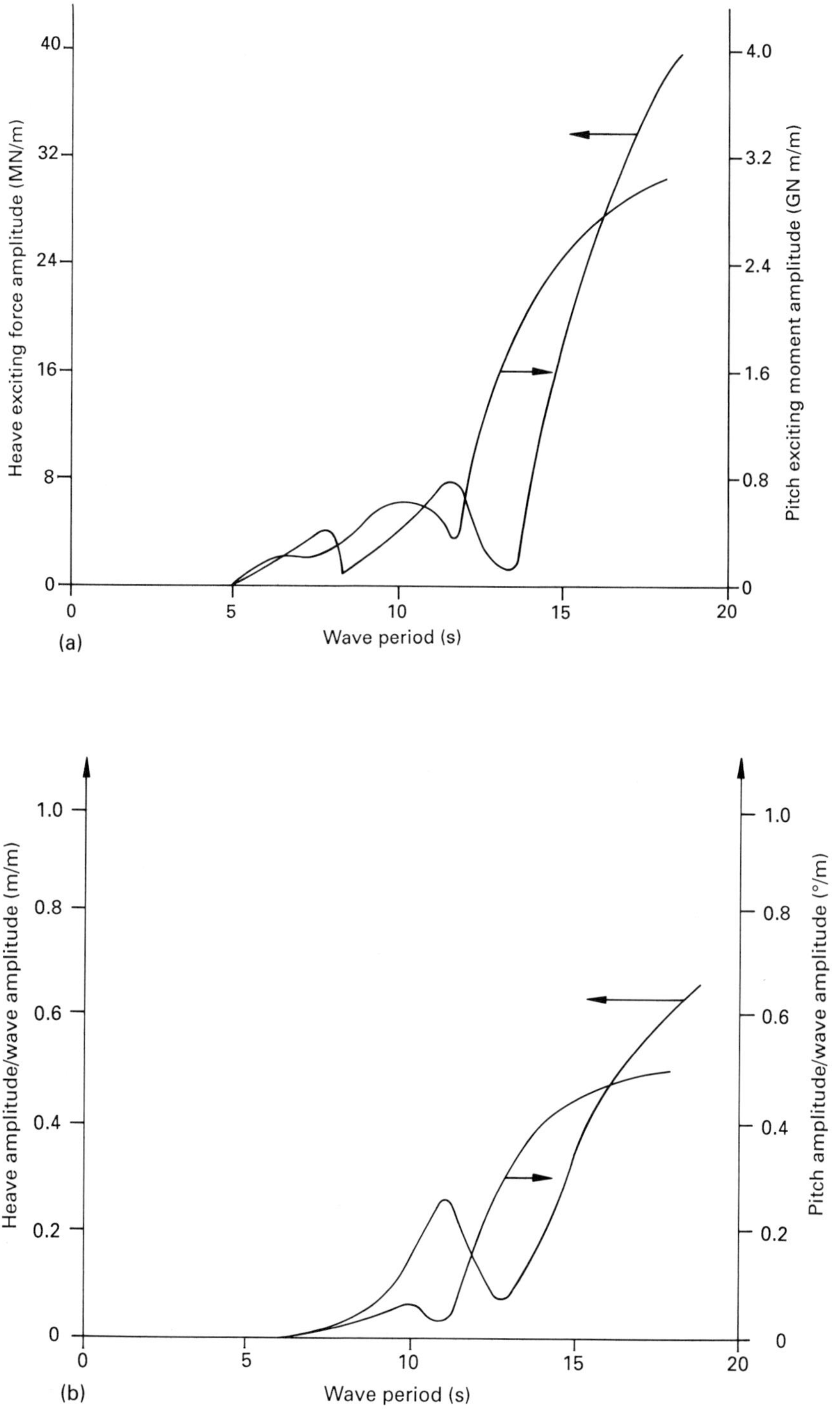

Figure 8.25. Variations of heave and pitch wave excitation forces and motion responses with wave period for ship hull

response – the one exception is the roll degree of freedom near resonance where non-linear potential and viscous effects cause some discrepancy. Brown *et al.* (1983) present a validation of diffraction theory predictions using model tests at two scale factors. Further work by Brown and Patel (1986) examines the reasons behind discrepancies between predictions and model tests for roll motions.

8.5 Analysis of continuously flexible structures

The operation of floating or compliant offshore structures requires the use of flexible tubular or tether connections between the surface platform and sea bed. The catenary mooring lines or taut tethers described in Section 8.3 are two examples. Others are offered by vertical marine risers which are compliant single or multiple steel pipe connections between the surface vessel and subsea well heads. In recent years, the offshore industry has pioneered the use of composite steel and elastomer construction for these tubular connections. These flexible risers can be arranged in catenary shapes from the surface platform and are often supported by an intermediate buoy. Flexible risers are able to operate with much larger surface platform offset (from above the subsea well head) than is permissible with more rigid vertical steel risers which are generally limited to offsets of between 7% and 10% of water depth.

All these continuously flexible structures require analysis during design to ensure that they have acceptable levels of deformations, stresses and fatigue lives due to forces induced by currents, waves and surface vessel motions. The presence of internal and external hydrostatic pressures has a fundamental effect on the governing equation for these tubular structures, as does the influence of the momentum of fluid flow within the pipe. This section is concerned with deriving the governing equations and developing solution procedures for the dynamics of marine risers of vertical or catenary configuration.

A vertical marine riser may be regarded as a hollow 'beam column'. The difference between a column subjected to lateral loading and a marine riser is that the riser is subjected to both internal and external hydrostatic pressure as well as axial and lateral loadings.

If the riser is simply approximated to a beam column then the governing differential equation for lateral static deflection is

$$\frac{\mathrm{d}^2}{\mathrm{d}y^2}\left(EI\,\frac{\mathrm{d}^2x}{\mathrm{d}y^2}\right) - T(y)\,\frac{\mathrm{d}^2x}{\mathrm{d}y^2} - w\,\frac{\mathrm{d}x}{\mathrm{d}y} = f \tag{8.126}$$

where EI is the riser bending stiffness, T is axial tension in the riser pipe wall, w is the weight per unit length of riser and contents and f is the lateral force per unit length. The co-ordinate system used is shown in Figure 8.26 with x measured from the bottom of the riser and positive upwards while y denotes horizontal riser deflection from a vertical through the riser base.

If, however, the hydrostatic pressure is included in the analysis, a slightly different form of Equation (8.126) is arrived at. The force due to the hydrostatic external pressure distribution which exists around the riser, and also the force due to internal pressure (which is related to well head pressure) are resolved into horizontal and vertical force components and incorporated into the governing equation for static deflection to give

$$\frac{d^2}{dy^2}\left(EI\frac{d^2x}{dy^2}\right) - [T(y) + A_o p_o - A_i p_i]\frac{d^2x}{dy^2}$$
$$- (\gamma_s A_s + \gamma_i A_i - \gamma_o A_o)\frac{dx}{dy} = f \qquad (8.127)$$

where the additional terms are the external hydrostatic pressure around the riser p_o, the internal hydrostatic pressure, p_i, with A_o being the cross-sectional area of riser bore and wall, A_i, the cross-sectional area of riser bore only and A_s, the cross-sectional area of riser wall; γ_i is the specific weight of fluid in the riser bore, γ_o is the specific weight of fluid surrounding the riser tube (sea water) and γ_s is the specific weight of riser pipe wall material.

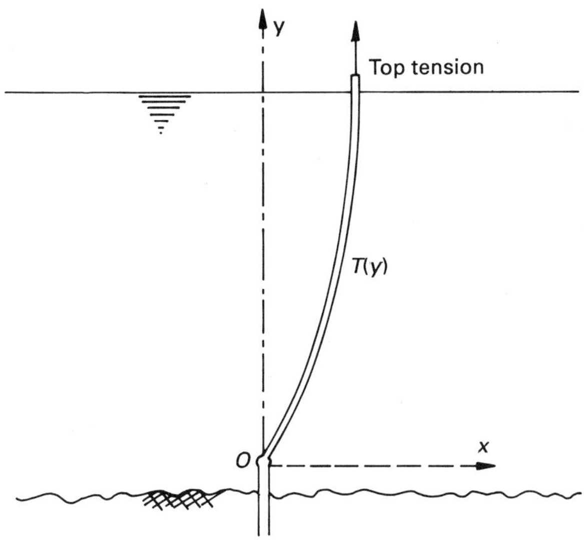

Figure 8.26. Conventional vertical riser notation

Equation (8.127) will be derived later but is valid for small deflections only, that is, for offset angles less than 10° from the vertical, and, therefore, the error in applying this equation to a vertical steel riser is usually negligible. Some interesting points concerning the effects of pressure on the riser may be deduced by further consideration of the second term in Equation (8.127). The $(A_o p_o - A_i p_i)$ term comes from the lateral effect of external and internal hydrostatic pressure. It's effect is similar to that of the actual tension in the riser wall since this term also multiplies the second derivative of displacement y. The pressure term does not modify the actual riser axial tension or the resultant direct stress in the riser wall. For this reason, the collection of parameters that multiply the second derivative is sometimes called 'effective tension' T_e, given by

$$T_e = T_o + A_o p_o - A_i p_i$$

The concept of effective tension is a convenient mathematical grouping of parameters that have a similar effect. Equation (8.127) demonstrates that the effect of external hydrostatic pressure is similar to that of a tensile axial force, while the internal pressure influences riser behaviour as would a compressive force; $(\gamma_s A_s + \gamma_i A_i - \gamma_o A_o)$ is equivalent to the corresponding term w in Equation (8.126).

Now, the differential equation describing the static behaviour of a marine riser of arbitrary geometry is derived using the notation of Figure 8.27 and the element of Figure 8.28. The analysis is restricted to two dimensions for simplicity. The static forces acting on the pipe element of Figure 8.28 can be listed as follows:

1. an axial tension and shear force within the pipe wall material;
2. a horizontal force due to the resultant of external and internal hydrostatic pressures, called $(F_{xo} + F_{xi})$;
3. a vertical force due to the resultant of external and internal hydrostatic pressures, called $(F_{yo} + F_{yi})$;
4. a drag force due to steady current. The velocity vector is resolved into components normal and tangential to the element, with only the normal component assumed to exert a distributed force of N per unit length.
5. the weight of the element (W_R) acting vertically downwards.

Summing components of force in the y direction for the element in Figure 8.28 yields the equation

$$(T + \mathrm{d}T)\sin(\theta + \mathrm{d}\theta) - T\sin\theta - (V + \mathrm{d}V)\cos(\theta + \mathrm{d}\theta)$$
$$+ V\cos\theta + (F_{yo} + F_{yi}) - W_R - N\cos\theta\, r\, \mathrm{d}\theta = 0 \qquad (8.128)$$

where W_R is the element weight. Similarly, summing forces in the x direction yields

$$(T + \mathrm{d}T)\cos(\theta + \mathrm{d}\theta) - T\cos\theta + (V + \mathrm{d}V)\sin(\theta + \mathrm{d}\theta)$$
$$- V\sin\theta + (F_{xo} + F_{xi}) + N\sin\theta\, r\, \mathrm{d}\theta = 0 \qquad (8.129)$$

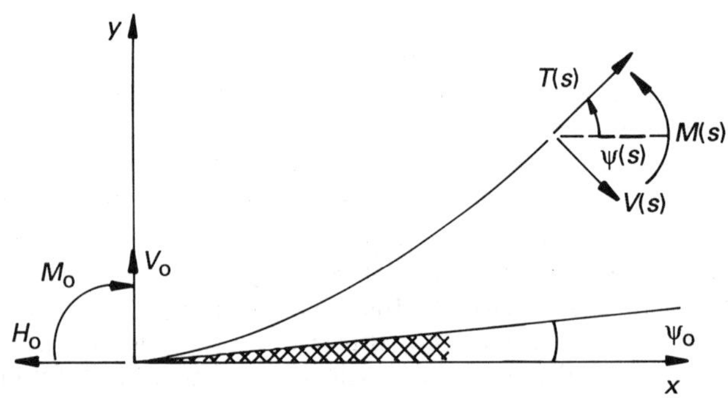

Figure 8.27. Notation for a riser of arbitrary geometry

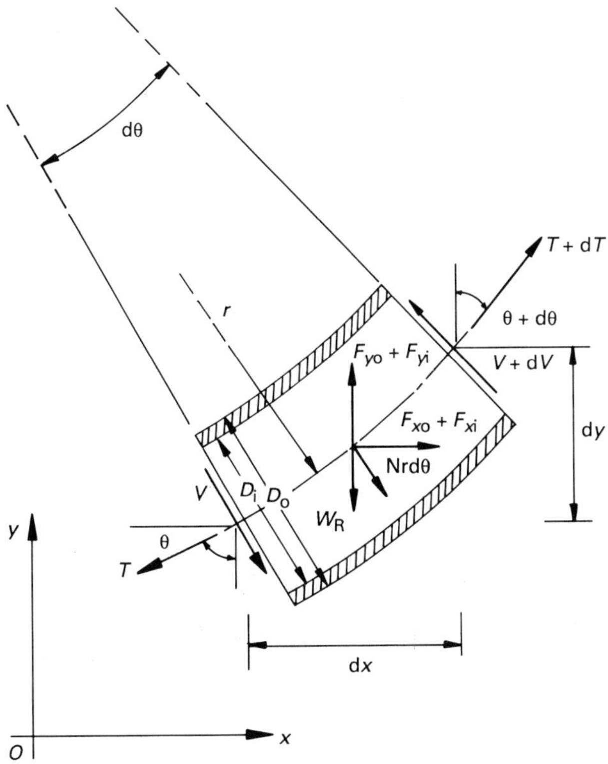

Figure 8.28. Riser element

These equations can be simplified for small $d\theta$ to

$$\left.\begin{aligned}&(T\cos\theta + V\sin\theta)\,d\theta + dT\sin\theta - dV\cos\theta + (F_{yo} + F_{yi})\\&- W_R - N\cos\theta\, r\, d\theta = 0\\&\text{and}\\&-(T\sin\theta - V\cos\theta)\,d\theta + dT\cos\theta + dV\sin\theta\\&+ (F_{xo} + F_{xi}) + N\sin\theta\, r\, d\theta = 0\end{aligned}\right\}\quad(8.130)$$

Combining these expressions gives

$$Td\theta - dV + (F_{yi} + F_{yo} - W_R)\cos\theta - (F_{xo} + F_{xi})\sin\theta - Nr\,d\theta = 0 \quad (8.131)$$

Appendix F presents a derivation of expressions for the resultant horizontal $(F_{xo} + F_{xi})$, and vertical forces $(F_{yo} + F_{yi})$ due to internal and external hydrostatic pressure. Substituting these expressions together with equations for the element weight and drag force into Equation (8.131) yields

$$T \cdot \mathrm{d}\theta - \mathrm{d}V - [(p_iA_i - p_oA_o) + r(\gamma_iA_i - \gamma_oA_o)(\cos\theta - \sin\theta\,\mathrm{d}\theta)]\cos^2\theta\,\mathrm{d}\theta - W_R\cos\theta - [(p_iA_i - p_oA_o) + r(\gamma_iA_i - \gamma_oA_o)(\cos\theta - \sin\theta\,\mathrm{d}\theta)]\sin^2\theta\,\mathrm{d}\theta - Nr\,\mathrm{d}\theta = 0. \tag{8.132}$$

and after simplication this becomes:

$$[T + p_oA_o - p_iA_i]\,\mathrm{d}\theta - \mathrm{d}V + \{(\cos\theta - \sin\theta\,\mathrm{d}\theta)(\gamma_oA_o - \gamma_iA_i) - \gamma_sA_s\cos\theta - N\}\,r\,\mathrm{d}\theta = 0 \tag{8.133}$$

with $W_R = \gamma_sA_s\,r\mathrm{d}\theta$ where γ_s is the weight per unit volume of the pipe material and A_s is the pipe wall area of cross-section. It is of interest at this stage to rewrite Equation (8.133) for a nearly vertical pipe. This can be done by using ϕ as the angle between the pipe element and the vertical such that $\phi = \pi/2 - \theta$ and $\mathrm{d}\phi = -\mathrm{d}\theta$.

Then, Equation (8.133) can be rewritten in terms of ϕ as

$$-[T + p_oA_o - p_iA_i]\,\mathrm{d}\phi - \mathrm{d}V - \{(\sin\phi + \cos\phi\,\mathrm{d}\phi)\cdot(\gamma_oA_o - \gamma_iA_i) - \gamma_sA_s\sin\phi - N\}\,r\,\mathrm{d}\phi = 0 \tag{8.134}$$

Dividing throughout by ds and using the following identities:

$$\frac{\mathrm{d}V}{\mathrm{d}s} = \frac{\mathrm{d}V}{\mathrm{d}y}\frac{\mathrm{d}y}{\mathrm{d}s} = \frac{\mathrm{d}V}{\mathrm{d}y}\left[1 + \left(\frac{\mathrm{d}x}{\mathrm{d}y}\right)^2\right]^{-1/2} = \frac{\mathrm{d}^2}{\mathrm{d}y^2}\left[-EI\frac{\mathrm{d}^2x}{\mathrm{d}y^2}\right]\left[1 + \left(\frac{\mathrm{d}x}{\mathrm{d}y}\right)^2\right]^{-1/2}$$

$$r\,\mathrm{d}\phi = -\mathrm{d}s$$

$$\sin\phi = \frac{\mathrm{d}x}{\mathrm{d}y}\left[1 + \left(\frac{\mathrm{d}x}{\mathrm{d}y}\right)^2\right]^{-1/2}$$

$$\cos\phi = \left[1 + \left(\frac{\mathrm{d}x}{\mathrm{d}y}\right)^2\right]^{-1/2}, \quad \frac{\mathrm{d}\phi}{\mathrm{d}y} = \frac{\mathrm{d}^2x}{\mathrm{d}y^2}\left[1 + \left(\frac{\mathrm{d}x}{\mathrm{d}y}\right)^2\right]^{-3/2}$$

and ignoring the $\cos\phi\,\mathrm{d}\phi$ term, we get:

$$\frac{\mathrm{d}^2}{\mathrm{d}x^2}\left[EI\frac{\mathrm{d}^2x}{\mathrm{d}y^2}\right] - \underbrace{(T + p_oA_o - P_iA_i)\frac{\mathrm{d}^2x}{\mathrm{d}y^2}}_{A}\ \underbrace{\left[1 + \left(\frac{\mathrm{d}x}{\mathrm{d}y}\right)^2\right]^{-1}}_{B} - (\gamma_sA_s - \gamma_oA_o + \gamma_iA_i)\frac{\mathrm{d}x}{\mathrm{d}y} = N\underbrace{\left[1 + \left(\frac{\mathrm{d}x}{\mathrm{d}y}\right)^2\right]^{1/2}}_{C} \tag{8.135}$$

Note that the term A in the above equation arises from the lateral effects of internal and external pressure and is the source of the concept of effective tension outlined earlier. Terms B and C, on the other hand, are due to the effects of riser orientation. Because small deflections are assumed in vertical riser analysis, terms B and C in Equation (8.135) may be to give

$$\frac{\mathrm{d}^2}{\mathrm{d}y^2}\left(EI\frac{\mathrm{d}^2x}{\mathrm{d}y^2}\right) - (T + p_oA_o - p_iA_i)\frac{\mathrm{d}^2x}{\mathrm{d}y^2} - (\gamma_sA_s - \gamma_oA_o + \gamma_iA_i)\frac{\mathrm{d}x}{\mathrm{d}y} = N$$

which is of similar form to Equation (8.127).

For a flexible riser, however, the approximation leading to elimination of terms B and C is not applicable. A more general approach is thus required. Now using the expression

$$\frac{d\theta}{ds} = \frac{1}{r} = \frac{d^2y/dx^2}{[1 + (dy/dx)^2]^{3/2}} \tag{8.136}$$

in Equation (8.133), dividing by (ds) and using

$$\cos\theta = \frac{dx}{ds} = \frac{1}{[1 + (dy/dx)^2]^{1/2}}, \; \sin\theta = \frac{dy}{ds} = \frac{dy/dx}{[1 + (dy/dx)^2]^{1/2}} \tag{8.137}$$

$$\frac{dV}{ds} = \frac{dV}{dx}\cdot\frac{dx}{ds} = \frac{dV}{dx}[1 + (dy/dx)^2]^{-1/2}$$

reduces term of zeroth order to

$$(T + p_oA_o - p_iA_i)\frac{d^2y}{dx^2}\cdot[1 + (dy/dx)^2]^{-1} - \frac{dV}{dx}$$
$$+ (\gamma_oA_o - \gamma_iA_i - \gamma_sA_s) - N[1 + (dy/dx)^2]^{1/2} = 0 \tag{8.138}$$

Taking

$$T_e = T + p_oA_o - p_iA_i$$
$$w_e = \gamma_oA_o - \gamma_iA_i - \gamma_sA_s \tag{8.139}$$

gives

$$T_e\frac{d^2y}{dx^2}\left[1 + \left(\frac{dy}{dx}\right)^2\right]^{-1} + w_e - \frac{dV}{dx}$$
$$- N\left[1 + \left(\frac{dy}{dx}\right)^2\right] = 0 \tag{8.140}$$

Equation (8.140) is the final form of the governing equation for a flexible riser. A comparison of Equation (8.140) and Equation (8.127) shows that some of the terms are common to both equations. For example, the effective tension T_e and effective weight w_e occur in both equations. The additional terms now account for both the horizontal and vertical effects of hydrostatic pressure.

Equation (8.140) may also be rewritten in terms of ψ and s (see Figure 8.27) by using the following expressions:

$$\frac{dy}{dx} = \tan\psi, \left[1 + \left(\frac{dy}{dx}\right)^2\right] = (1 + \tan^2\psi) = \sec^2\psi$$

and

$$\frac{d^2y}{dx^2} = \sec^3\psi\,\frac{d\psi}{ds}$$

to yield

$$T_e\frac{d\psi}{ds} + w_e\cos\psi - \left(N + \frac{dV}{ds}\right) = 0 \tag{8.141}$$

It is informative at this stage to transform Equation (8.141) into that for a simple catenary in vacuo so as to check its validity. Thus if,

$$\left.\begin{aligned} &p_o = p_i = 0,\ \gamma_o = \gamma_i = 0, \\ &T_e = T = H_0 \sec\psi,\ w_e = -\gamma_s A_s = -w \\ &\text{then} \\ &T\frac{d\psi}{ds} - w\cos\psi - \left(N + \frac{dV}{ds}\right) = 0 \end{aligned}\right\} \tag{8.142}$$

For zero drag force and zero shear variation, we get

$$H_0 \sec^2\psi \frac{d\psi}{ds} = w \tag{8.143}$$

which is the equation of the simple catenary. The complexity of the governing equations for a flexible riser requires that the solutions be obtained through a numerical finite element or finite difference computation. The finite element analysis uses the methods that have already been outlined in Section 6.1 and 6.2. Patel *et al.* (1984) gives further details of a technique that is used for vertical steel risers. A typical beam element and riser idealisation is given in Figure 8.29 with typical

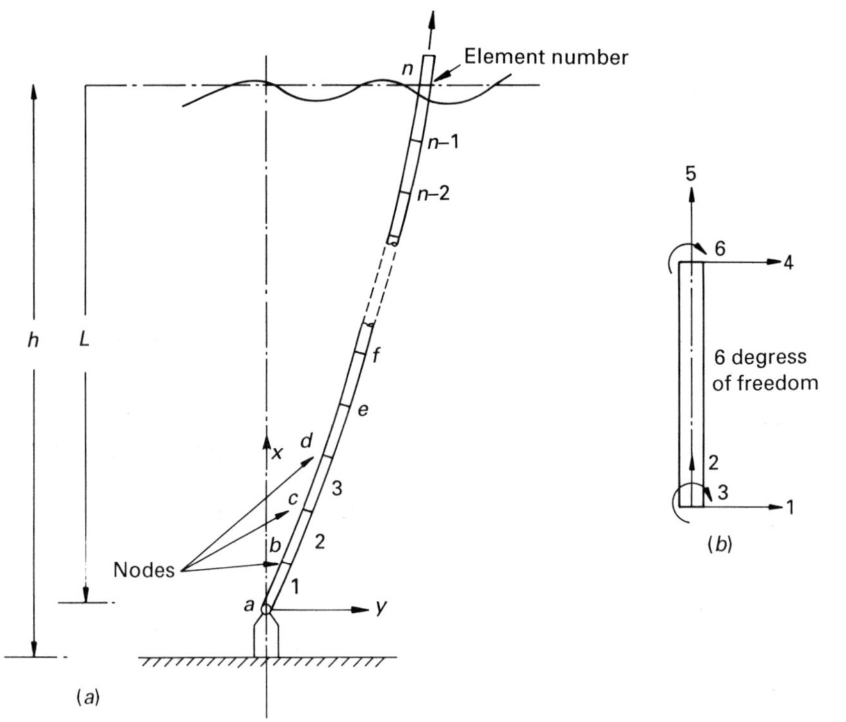

Figure 8.29. Element and global node description for a finite element idealization of a vertical riser: (*a*) riser element nodes; (*b*) single beam element

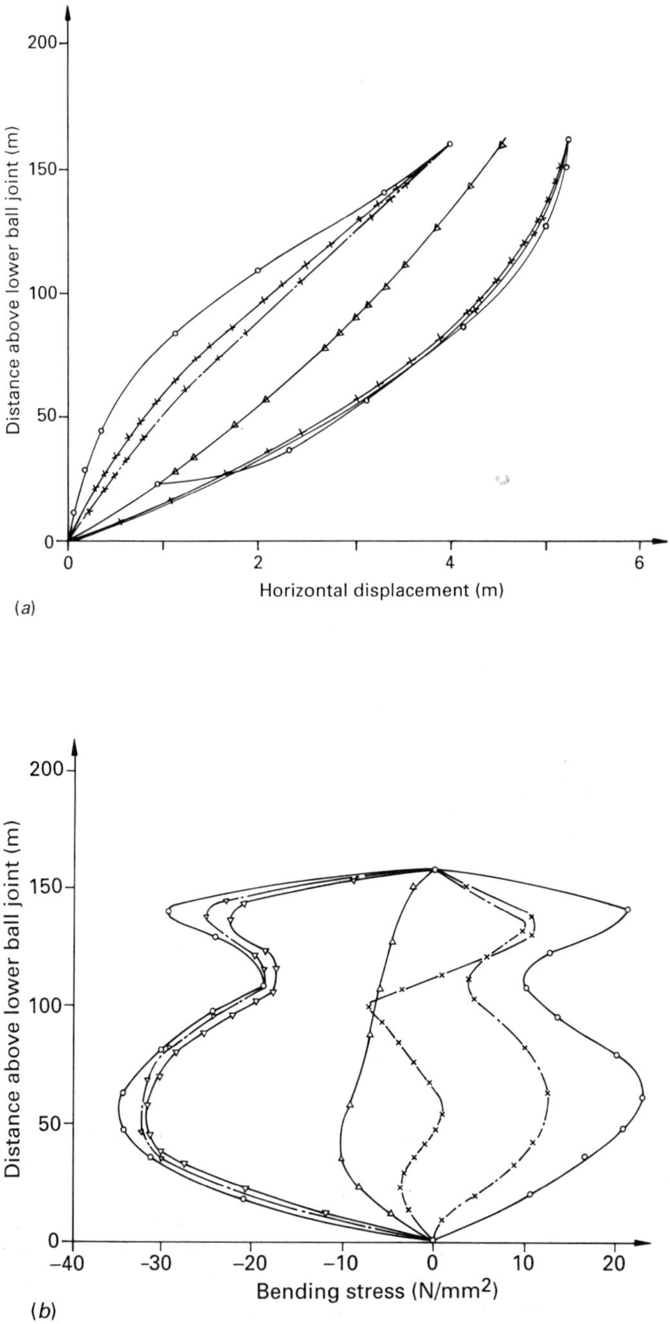

Figure 8.30. Displacement and stresses for a marine riser compared with API case 500-20-1D. *Key:* △ – static value; ○ – API results; ▽ – static and dynamic minimum; + – static and dynamic maximum. Full lines denote frequency domain analysis results, dashed lines denote time domain analysis results

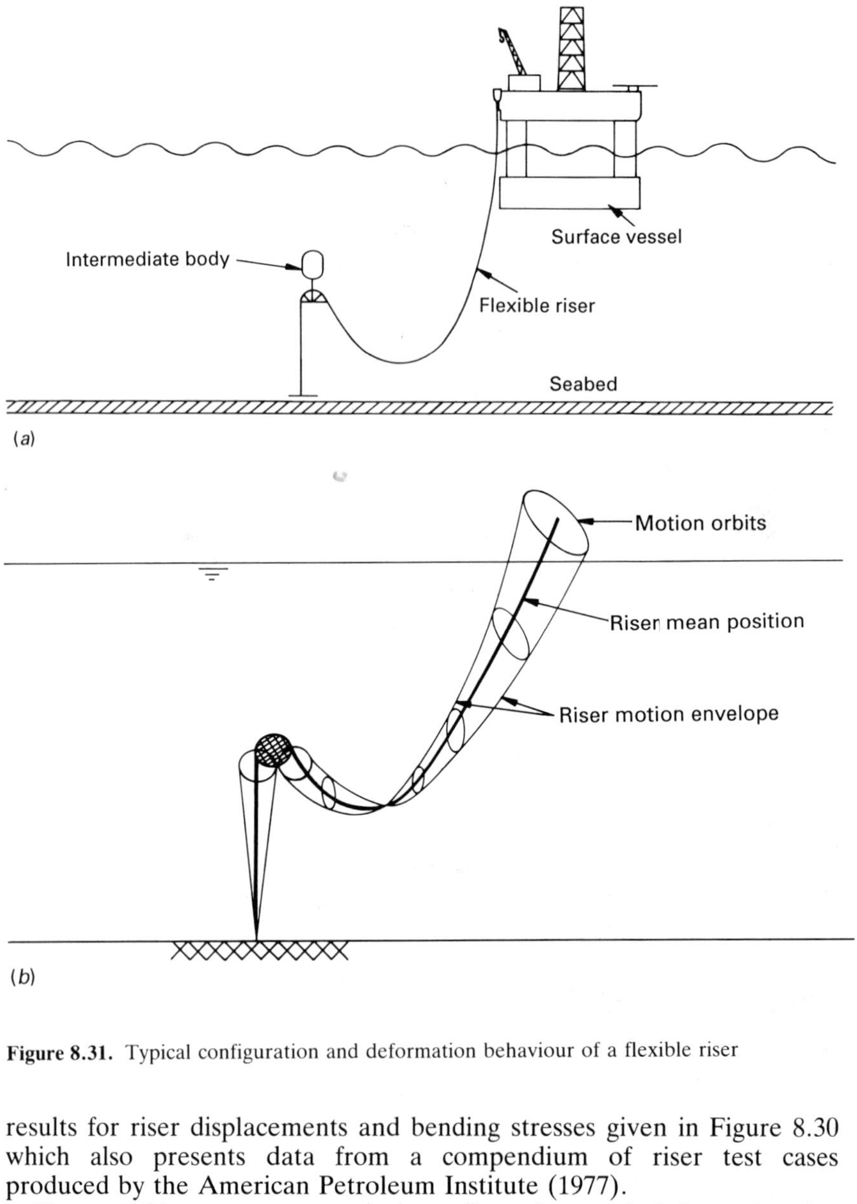

Figure 8.31. Typical configuration and deformation behaviour of a flexible riser

results for riser displacements and bending stresses given in Figure 8.30 which also presents data from a compendium of riser test cases produced by the American Petroleum Institute (1977).

Figure 8.31 presents a typical configuration and deformation behaviour of a flexible riser with an intermediate midwater buoy. The displacement envelope for the riser is also calculated using a finite element analysis that is an extension of the technique used for vertical risers.

There are also examples of essentially horizontal flexible pipe lines that arise when subsea pipe lines are fabricated at an on-shore location and towed out to an offshore site for installation. Such lines can be from

1 to 5 km in length and from 0.3 to 1.0 m in diameter. These extremely slender and flexible structures have to be kept in tension during the towing process and their weight per unit length, wave induced motions and stresses as well as installation procedures have to be carefully designed to prevent catastrophic loss of the line during installation. Patel and Seyed (1987) present typical calculations that can be carried out for towed pipe lines.

References

American Petroleum Institute, (1977), *Comparison of Marine Drilling Riser Analyses, American Petroleum Institute*, Bulletin 2 J, 1st ed, January.

Brown, D. T., Eatock-Taylor, R. and Patel, M. H. (1983), Barge motions in random seas – a comparison of theory and experiment, *Journal of Fluid Mechanics*, **129**, April, 385–407.

Brown, D. T. and Patel, M. H. (1986), On predictions of resonant roll motions for flat bottomed barges, *Transactions of the Royal Institution of Naval Architects*, **128**, 235–245.

Clauss, G. F. (1982), Slow drift forces in a barge-type structure – comparing model tests with calculated results, *Proceedings of the Offshore Technology Conference*, OTC 4435.

Det Norske Veritas (1982), *Rules for the Design, Construction and Inspection of Offshore Structures*, Appendix B, Loads, Published 1977, reprinted with corrections 1982, Det Norske Veritas, PO Box 300, N-1322, Hovik, Norway.

Faltinsen, O. M. and Michelsen, F. C. (1974), Motions of large structures in waves at zero Froude number, *Proceedings of the International Symposium on Dynamics of Marine Structures in Waves*, University College London, 99.

Harrison, J. H. (1985), On the hydrodynamics of compliant semisubmersibles, PhD thesis *Faculty of Engineering, University of London*.

Havelock, T. (1940), The pressure of water waves upon a fixed obstacle, *Proceedings of the Royal Society*, **175**, 409.

Havelock, T. (1942), The drifting force on a ship among waves, *Philosophical Magazine*, **33**, 467–475.

Hsu, F. H. and Blankarn, K. A. (1970), Analysis of peak mooring forces caused by slow drift oscillations in random seas *Proceedings of the Offshore Technology Conference*, OTC Paper 1159.

Jennings, A. (1965), Cable movements under two dimensional loads – A discussion, *Journal of the Structural Division, American Society of Civil Engineers*, **ST1**, February, 307–311.

Lyons, G. J. and Patel, M. H. (1984), Comparisons of theory with model test data for tensioned buoyant platforms, *Transactions of the ASME, Journal of Energy Resources Technology*, **106**, December, 426–436.

Maruo, H. (1960), The drift of a body floating in waves, *Journal of Ship Research*, **4**(3), 1.

Newman, J. N. (1962), The exiciting forces on fixed bodies in waves, *Journal of Ship Research*, **6**, 10–17.

O'Brien, W. T. and Francis, A. J. (1964), Cable movements under two dimensional loads, *Journal of the Structural Division, American Society of Civil Engineers*, **ST3**, June, 89–123.

Oppenheim, B. W. and Wilson, P. A. (1980), Continuous digital simulation of the second order slowly varying wave drift force, *Journal of Ship Research*, **24**, (3), 181–189.

Patel, M. H. (1983), On the wave induced motion response of semisubmersibles, *Transactions of the Royal Institution of Naval Architectures*, **125**, July, 221–228.

Patel, M. H. and Lynch, E. J. (1983), The coupled dynamics of tensioned buoyant platforms and mooring tethers, *Journal of Engineering Structures*, **5**, October, 299–308.

Patel, M. H., Sarohia, S. and Ng, K. F. (1984), Finite element analysis of the marine riser, *Journal of Engineering Structures*, **6**, July (3) 161–170.

Patel, M. H. and Seyed, F. B. (1987), Analysis and design considerations for towed pipe lines, *Proceedings of the 6th International Symposium on Offshore Mechanics, Offshore and Arctic Pipelines*, ASME, pp. 17–28.

Peyrot, A. H. (1980), Marine cable structures, *Journal of the Structural Division, American Society of Civil Engineers*, December, pp. 2391–2404.

Pinkster, J. A. (1980), *Low Frequency Second Order Wave Exciting Forces on Floating Structures*, Publication number 650, Netherlands Ship Model Basin, Wageningen, Netherlands.

Pinkster, J. A. and Oortmerssen, van G. (1976), Computation of the first and second order wave forces on bodies in regular waves, *Proceedings of the Second International Conference on Numerical Ship Hydrodynamics*, p. 136.

Remery, G. F. M. and Hermans, A. J. (1971), The slow drift oscillations of a moored object in random seas, *Proceedings of the Offshore Technology Conference*, OTC 1500.

Standing, R. G., DaCunha, N. M. C. and Matten, R. B. (1981), *Slowly Varying Second Order Wave Forces, Theory and Experiment*, National Maritime Institute, Report No R138.

Warburton, G. B. (1976) *The Dynamical behaviour of structures*, 2nd ed, Pergamon Press Ltd, Oxford.

Chapter 9

Model testing of offshore structures

Model tests of ships in still water and in waves have historically played an important role in the design of the vessel's propulsion system, the determination of its sea keeping characteristics and its maneouvrability. In recent years, developments in the offshore industry have prompted the use of wave tanks for the testing of a variety of floating, fixed and compliant offshore structures. These model tests are generally aimed at simulating the response of these structures to the harsh wave environments they will encounter in service and also to verify hydrodynamic theories used for the design of such structures. These theories invariably employ some degree of assumptions or empiricisms which cast doubt on the quality of their predictions, whereas model tests have the potential, if scaling laws are satisfied, of faithfully reproducing all the physics of a full scale condition.

The primary purpose of model testing facilities is to simulate, as accurately as possible at model scale, the environmental phenomena encountered by offshore structures during both normal operating and extreme conditions. Thus, the environmental effects of wind, current and waves must be simulated. For offshore structures operating in Arctic conditions or in earthquake zones, special purpose facilities must be capable of simulating icing phenomena and horizontal foundation accelerations analogous to those encountered during earthquakes. Since ocean wave loading, in most cases, is the largest environmental force acting on offshore structures, the majority of available facilities are designed for testing offshore structures in waves.

Due to this requirement for model testing in waves, there has been a continuing evolution of wave tank facilities with the capacity to generate a variety of wave conditions ranging from simple small amplitude sinusoidal waves to high amplitude random multi-directional seas with specified wave elevation power spectral densities and known directionality characteristics. The latter wave climate poses difficulties associated with generating the waves, analysing test data such as for wave properties and structural response and in interpreting resultant information. Nevertheless, such a wave climate closely resembles that encountered by offshore structures.

There are a large number of wave tanks in national research laboratories and universities that can generate long-crested regular and irregular waves – the majority of these tanks are also capable of generating a current flow and possess a travelling carriage above the tank for the testing of ships with forward speed. Some of these tanks are able to simulate wind loading by using banks of ducted air propellors. There are only a handful of wave tanks in the world that can generate short-crested (that is, directional) irregular waves. These tanks are generally square or of aspect ratio where the tank side along which wave generators are mounted is long compared to the other side. This is in contrast to most long-crested wave tanks where the tank length along the direction of wave propagation is very long compared to the wave crest length (or width) of the tank. These dimensions have developed historically because the latter long-crested wave tanks have evolved from ship testing practice where the tank length is necessary to get a reasonably long ship foward speed run.

Wind tunnels, either on their own, or in some rare cases built into the air space above a wave tank are used to simulate the atmospheric boundary layer flowing over the top sides of an offshore structure. This is a classic example of a situation where a model test can assist in quantifying wind forces which can otherwise only be obtained from calculation by employing extremely crude empirical relationships and estimated coefficients.

Some special purpose testing tanks are used for ship design. Cavitation tunnels are used to investigate ship propellar characteristics. They are much like wind tunnels but contain water at reduced pressure to simulate full scale cavitation effects at model scale. Testing of ship maneouvring is also carried out by using radio controlled free running models in large rectangular basins with or without waves. For offshore applications, tanks with refrigeration or with specially formulated ice-simulating chemical compounds are used to test the behaviour of offshore structures (or ship ice breakers) in ice bound waters. Other special purpose testing facilities for deep water or high current areas have also been constructed. However, the ultimate test of the effectiveness of an offshore structure design or of the theoretical predictions contributing to the design process is to make measurements of both the environment and the structure at full scale. Such measurements are increasingly being carried out as part of the structural monitoring and in-service maintenance of large fixed and floating offshore structures. In some cases, fundamental data from a full scale testing facility has been provided by specifically designed and installed structures in shallow water – the Christchurch Bay tower in the UK and an Ocean Test Structure in the USA are two examples. Interpretation of the data from such full scale studies suffers from the fact that environmental perturbations are uncontrolled and the complex mechanism governing the combined loading of wind, current and waves from different directions cannot be readily unravelled from measured data.

9.1 Similarity considerations

The results of dimensional analysis or of physical considerations of the magnitudes of different force mechanisms can be used to show that the behaviour of wind, current and wave loading on offshore structures is governed by non-dimensional numbers that are independent of the magnitudes of physical quantities within them. Thus, similarity of non-dimensional numbers between full scale and model scale serves as a measure of the similarity of the relative effects of physical phenomena governing the behaviour of both systems. This similarity can be applied to making measurements at model scale and applying appropriate scale factors to obtain the magnitude of the corresponding parameters at full scale.

Section 5.2.1 demonstrates the derivation of a non-dimensional Reynolds number using both dimensional analysis and also physical arguments based on the order of magnitudes of inertia and viscous forces within a flow field. Reynolds number similarity between model and full scale would, therefore, be confirmation of the relative forces from inertia and viscous effects being identical at both scales. A non-dimensional Froude number is also derived in Section 5.2.1 as a measure of the ratio of inertia to gravitational forces.

These concepts of dimensional analysis can be generalized in the following manner. If the force F on an offshore structure is influenced by the magnitude of several contributory physical quantities, some of which could be sea water density ρ, acceleration due to gravity g, wave height H, and period T, then F can be written as

$$F = f(\rho, g, H, T, \ldots) \qquad (9.1)$$

A dimensional analysis of all the physical quantities in Equation (9.1) can then be used to generate groups of these quantities in the form of independent dimensionless numbers ($N, N_1, N_2, \ldots$). Then Equation (9.1) can be written as

$$N = F(N_1, N_2, N_3, \ldots) \qquad (9.2)$$

where the dependent parameter F in Equation (9.1) appears in the number N only. Then, for the physical process described by Equation (9.1) at any scale, the non-dimensional numbers are equal so that

$$(N_1)_\mathrm{f} = (N_1)_\mathrm{m},\ (N_2)_\mathrm{f} = (\mathrm{N}_2)_\mathrm{m} \qquad (9.3)$$

and so on where subscripts f and m refer to two different scales such as full scale and model scale respectively. The derivation of Reynolds number in Section 5.2.1 illustrates an application of the above approach.

Physical arguments based on the levels of various force mechanisms may also be used to examine the role of some non-dimensional numbers. Thus, as in Section 5.2.1, with the notation used there

$$\text{Reynolds number} = \frac{\text{inertia force}}{\text{viscous force}} = \frac{\rho U^2 L^2}{\mu U L} = \frac{UL}{\nu} \qquad (9.4)$$

and

$$\text{Froude number} = \sqrt{\left(\frac{\text{inertia force}}{\text{gravitational force}}\right)} = \sqrt{\left(\frac{\rho U^2 L^2}{\rho L^3 g}\right)} = \frac{U}{\sqrt{(gL)}} \qquad (9.5)$$

These equations indicate that Reynolds number similarity is important for all flows whereas Froude number similarity is applicable to flows where the gravitational acceleration contributes to the governing mechanisms – free surface gravity waves being a classical example.

Other non-dimensional numbers become significant where elastic, pressure or capillary forces may contribute to the behaviour of a flow and the structure within it. Using the physical arguments described above, these non-dimensional numbers are

$$\text{Weber number} = \sqrt{\left(\frac{\text{inertia force}}{\text{elastic force}}\right)} = \sqrt{\left(\frac{\rho U^2 L^2}{E L^2}\right)} = \frac{U}{\sqrt{(E/\rho)}} \qquad (9.6)$$

$$\text{Euler number} = \sqrt{\left(\frac{\text{inertia force}}{\text{pressure force}}\right)} = \sqrt{\left(\frac{\rho U^2 L^2}{2p' L^2}\right)} = \frac{U}{\sqrt{(2p'/\rho)}} \qquad (9.7)$$

and

$$\text{Cauchy number} = \sqrt{\left(\frac{\text{inertia force}}{\text{capillary force}}\right)} = \sqrt{\frac{\rho U^2 L^2}{\sigma L}} = \frac{U}{\sqrt{(\sigma/\rho L)}} \qquad (9.8)$$

where E is Young's modulus for the structure immersed in the fluid field, p' is strictly a pressure difference and σ is the surface tension coefficient. The ratio of inertia force to the force due to the elasticity of the fluid itself can be defined as the Mach number which is

$$\text{Mach number} = \sqrt{\left(\frac{\text{inertia force}}{\text{fluid elasticity force}}\right)} = \sqrt{\left(\frac{\rho U^2 L^2}{p L^2}\right)}$$
$$= \frac{U}{\sqrt{(p/\rho)}} = \frac{U}{c} \qquad (9.9)$$

where p is the absolute pressure and c is the speed of sound within the fluid. The Mach number is only of concern in compressible fluids (that is, gases) where the flow speed U begins to approach the speed of sound c. In addition, there are several non-dimensional numbers which reflect the thermal conduction and convection behaviour of fluids. These are not mentioned since they are not considered during model testing of offshore structures.

Now the majority of model testing on offshore structures is carried out using fixed or floating rigid models representing structures with elastic deformation velocities that are negligible compared to wind, current or wave velocities in the surrounding fluid. For freely floating structures, however, their rigid body velocities themselves need no longer be small. In the case of wind tunnel testing for wind forces on structure top sides, similarity between model scale and full scale requires Reynolds number equivalence. This is also true for model tests of a structure in current flow as long as the current velocity is not large enough to induce wave making resistance on the structure. Model tests for wave effects on offshore structures require that Froude number equivalence be obtained since the gravitational acceleration plays a key role in wave action. These arguments indicate that both Reynolds and Froude number similarity is required for model testing of structurally rigid offshore structures. Now for Reynolds number similarity,

$$\frac{U_m L_m}{\nu_m} = \frac{U_f L_f}{\nu_f} \tag{9.10}$$

where a linear scale factor α (= L_f/L_m) usually takes values between 30 and 200. Now with model tests for wind forces in air and for current forces in water, with negligible wavemaking, Froude number similarity does not have to hold. For both cases, the ratio of model scale to full scale kinematic viscosities remains at unity because of the difficulty of finding suitable fluids that have sufficiently different kinematic viscosities for scaling purposes. Thus for Reynolds number similarity, the ratio of model to full scale velocities must be of value α. This is generally impossible to achieve in both wind tunnels and wave tanks to the extent that a Reynolds number disparity of one or two orders of magnitude between full scale and model scale is inevitable. This problem is alleviated by the fact that many physical quantities such as drag force on a circular cylinder vary slowly with Reynolds number except in critical regions (see Figure 5.5, for example). For circular cylinders, model scale Reynolds numbers are invariably in the pre-critical regime whereas full scale numbers are in the post-critical region. Some work in aeronautics has shown that using turbulence strips or trip wires to trigger transition can assist in simulating high Reynolds number flow in a low Reynolds number test but only in conditions with non-reversing incident flows.

Tests in waves require Froude number as well as Reynolds number similarity. The Froude number is such, however, that the similarity expressed by the equation

$$\frac{U_m}{\sqrt{(g_m L_m)}} = \frac{U_f}{\sqrt{(g_f L_f)}} \tag{9.11}$$

is achievable. Now the linear scale factor α is L_f/L_m and unless model test facilities are established on other planets, $g_m/g_f = 1$. Then for Froude number similarity

$$\frac{U_m}{U_f} = \sqrt{\left(\frac{1}{\alpha}\right)} \tag{9.12}$$

But the velocities can be written in terms of lengths and time scale factors T_m and T_f. This reduces the requirement for Froude number similarity to

$$\frac{T_f}{T_m} = \sqrt{\alpha} \tag{9.13}$$

which simply imposes a requirement to consider equivalent full scale time periods (such as wave period) to be equal to model scale time periods times the square root of the linear scale factor. Equation (9.13) can be easily satisfied in the data analysis procedures for a model test.

It is interesting to digress here to examine the non-dimensional parameters that govern similarity of a free gravity wave of height H, period T, wave length L, in water depth d, with gravitational acceleration g also playing a part in the physics of wave behaviour. Now a wave property, such as crest velocity U_c can be written as a function

$$U_c = f(g, H, T, L, d) \tag{9.14}$$

which after dimensional analysis reduces to a non-dimensional equation of the form

$$\left(\frac{U_c}{H/T}\right) = f\left(\frac{H}{gT^2}, \frac{H}{L}, \frac{d}{L}\right) \tag{9.15}$$

where for equivalence between model and full scale, the three non-dimensional numbers on the right-hand side must remain invariant. Taking the first number, this means

$$\frac{H_m}{g_m T_m^2} = \frac{H_f}{g_f T_f^2} \tag{9.16}$$

and with $g_m/g_f = 1$ and $\alpha = H_f/H_m$, the time scale factor becomes the same as in Equation (9.13).

Using Froude number scaling, the scale factor for various derived quantities is set out in Table 9.1 for the modelling of floating offshore structures in waves. Note that the acceleration scale factor remains 1 and that in order to permit full scale values in sea water to be translated to model scale for fresh water testing, a density scale factor $\bar{\lambda}$ is incorporated.

Table 9.1 Scale factors for Froude number scaling

Parameter	*Ratio of full scale to model scale parameter value*
Length	α
Mass	$\bar{\lambda}\alpha^3$
Density	$\bar{\lambda}$
Time	$\sqrt{(\alpha)}$
Volume	α^3
Moment of inertia	$\bar{\lambda}\alpha^5$
Force	$\bar{\lambda}\alpha^3$
Area	α^2
Velocity	$\sqrt{(\alpha)}$
Acceleration	1
Moment	$\bar{\lambda}\alpha^4$

Notes:

$\alpha = \dfrac{\text{Full scale length}}{\text{Model scale length}}$, $\bar{\lambda} = \dfrac{\text{Full scale water density}}{\text{Model scale water density}}$

It may be recalled that the Keulegan–Carpenter number

$$K_C = \frac{U_m T}{D} \tag{9.17}$$

where U_m is maximum velocity, T is wave period and D is a characteristic body cross-sectional dimension significant in wave loading. It can be readily seen that, with Froude scaling, the Keulegan–Carpenter number also remains invariant. For monolithic offshore structures where diffraction induced wave forces are dominant the absence of Reynolds number equivalence is not significant since viscous forces are negligible anyway.

The effect of Reynolds number miscaling becomes more important for slender structural elements where drag forces play a greater part.

Equivalence of the Euler or Cauchy numbers is not of concern in the testing of offshore structures since differential pressure or capillary effects are negligible. However, model testing of offshore structures where the effects of elastic structural deformations are included require that additional non-dimensional similarity conditions be satisfied. Examples of such structures are continuously flexible tensioned and catenary marine risers, guyed towers and the tethers of tensioned buoyant structures. Any fixed deep water structure with sufficient flexibility to have significant elastic deformation in gravity waves should also be included within this classification. For model testing of these structures, it is necessary that the mass distribution, structural elasticity and damping forces are correctly scaled between full scale and model scale. This requires that the non-dimensional parameters of Weber number, a density ratio ρ_s/ρ and the damping ratio, ξ, must remain invariant. Here ρ_s is the density of the structure and ρ of the wave fluid with the damping ratio being defined as

$$\xi = \frac{b}{\sqrt{(2mk)}} \tag{9.18}$$

where m, b and k are the structural mass, damping and stiffness. Equivalence of these numbers between model scale and full scale is difficult to achieve and requires the use of special materials to vary Young's modulus as well as sectional design to separate the mass and stiffness contributing elements of the test model. The damping ratio is difficult to scale, simply because the level of damping in the full scale structure is rarely quantified reliably. The construction and testing of model offshore structures with correctly scaled dynamic properties poses a challenging design problem. Typical model design and testing procedures for elastic structures in wind tunnels and wave tanks are described more fully by Pankhurst and Holder (1952) and Lyons and Patel (1986) respectively.

9.2 Test facilities

Wave tanks generating long- and short-crested waves and wind tunnels for wind force measurements form the three most widely used model testing facilities for offshore structures. A typical example of each of these installations is described here.

Worldwide there are only a few major wave tank facilities capable of generating irregular directional waves that are representative of open ocean waves. However, there are many more test tanks capable of generating long-crested, two-dimensional waves. These latter facilities are still valuable from two points of view. The primary advantage of using uni-directional 'long-crested' waves is that theoretical understanding of such waves is sufficiently far advanced to enable test data on the waves and their consequential structure loading to be meaningfully interpreted. Furthermore, nearly uni-directional sea waves are prevalent in many full scale situations where they can impose much more severe coherent loadings on an ocean structure compared to an equivalent multi-directional sea with the same energy. Therefore, long-crested waves and the associated test facilities are still routinely used for design work.

One of the most important fixtures on a long- or short-crested wave basin or tank is the wavemaker together with its control and actuating mechanism. In general, six kinds of wavemakers (illustrated in Figure 9.1) are used in wave tanks. These are:

1. The single or multiple flap oscillating horizontally about a pivot at the basin bottom. A multiple flap has an additional hinge above the base which allows the two segments of the flap to rotate independently.
2. A flap or flaps as in 1 but of reduced draught with a raised pivot.
3. A piston oscillating horizontally.
4. A wedge sliding on an incline.
5. A plunging wedge sliding up and down.
6. A pneumatic wavemaker operating by varying the air pressure above an enclosed cavity.

Single or multiple flap wave generators generate waves with less excitation at the tank bottom than at the surface and for this reason are not very suitable for generating waves in shallow water. The single or multiple plunger wavemaker, on the other hand, is very effective in generating shallow water waves, since the water motion is uniform from the free surface to the bottom of the tank. However, very high power is required because of the large moving mass.

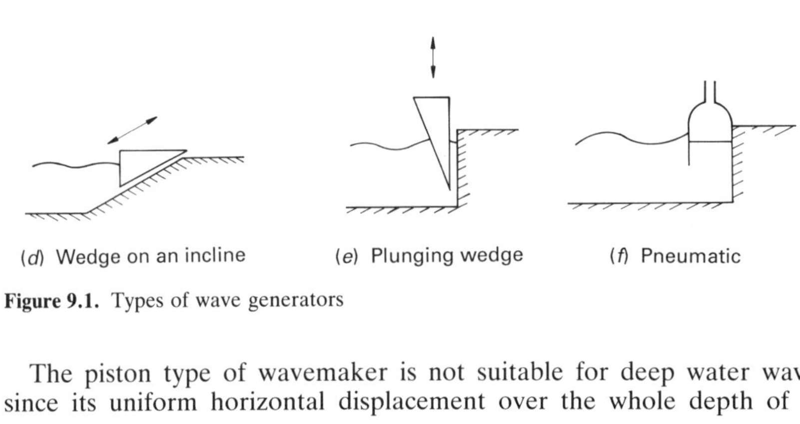

Figure 9.1. Types of wave generators

The piston type of wavemaker is not suitable for deep water waves, since its uniform horizontal displacement over the whole depth of the tank does not represent actual deep water waves for which the wave motion decreases exponentially with depth. Piston-type wavemakers also dissipate power from their rear surfaces. The wedge sliding on an incline of 30° to the horizontal is an efficient from of wave generation from the point of view of actuating power because of the absence of a mass of water behind the front vertical face of the generator.

To ensure that the generated wave form will not be unnecessarily distorted by reflections from the end of the tank, wave absorbers, known as beaches, are installed to absorb waves and minimize undesirable reflections. The beach normally consists of horizontal wooden slats parallel to incident crest lines, porous hessian material blankets or slightly consolidated highly porous metallic 'swarf' particles or strips. Beach structures are designed to absorb wave energy as efficiently as possible over a wide range of wave frequencies, although because of imperfect absorbtion, the reflected wave can still be of the order 5–10% of the incident wave, particularly at long wave periods.

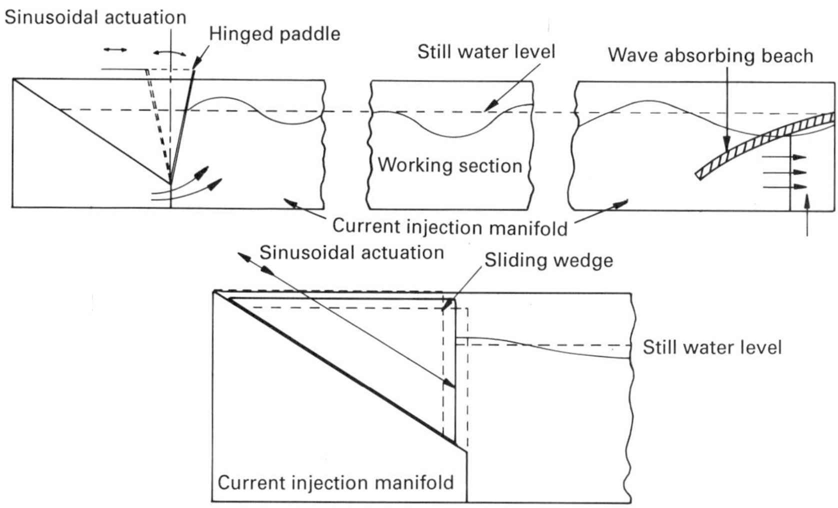

Figure 9.2. Typical long-crested wave tank with hinged paddle or sliding wedge-type wave generator

A typical wave tank using a wedge type wavemaker is described here. Such tanks have lengths of between 7 and 50 times their width with water depths of from 2 m to 12 m (see Figure 9.2). Glass observation panels are often fitted to the working section length of the tank. A facility for generating current flow, either with or against the direction of wave propagation, may be provided by two current injection manifolds at each end of the tank. At the wave generator end, the manifold is usually designed to direct flow into or out of the tank by a set of deflector vanes. The current manifold at the other end of the tank can be sited under a porous wave-absorbing beach.

Figure 9.2 shows a paddle and wedge-type wave generator mounted on the tank. The wave generating surface is often mounted with thin baffle plates at right-angles to their surface to eliminate cross-waves developing across the tank. Gaps between the wave generating surface and the sides and bottoms of the tanks are sealed as far as possible to reduce leakage and the consequent loss of wavemaking efficiency. The wave generators are sometimes split up into several segments along the width of the tank and are usually actuated by electro-hydraulic servo controlled rams which derive their feedback signal from a displacement transducer mounted on the ram. The ram control circuits require an input voltage varying with time to specify the variation of ram displacement with time. Movement of the front surface of the generator immersed in the tank contents induces gravity waves which propagate down the tank towards the wave absorbing beach. The required input control signal for ram displacement is a complicated function of the dynamic response of all the elements between the control voltage and the desired wave climate at the working section. A micro-computer operating in real time is used to obtain a demanded wave condition without time consuming manual preprocessing and signal storage.

The relationship between wave amplitude and generator stroke amplitude can be obtained from calculations. Havelock (1929) initially presented a general theory for wave generation on a water surface which has been a basis for much subsequent research. Theoretical results specifically applicable to the piston and paddle-type wavemakers were later obtained for the linear wave by Biesel (1951). These equations relate wave amplitude a to wave generator displacement amplitude r to yield

$$\frac{a}{r} = \frac{2\sinh^2 kd}{(kd + \sinh kd \cosh kd)} \tag{9.19}$$

for a piston and

$$\frac{a}{r} = \frac{2\sinh(kd)(1 - \cosh kd + kd \sinh kd)}{kd(kd + \sinh kd \cosh kd)} \tag{9.20}$$

for a paddle with stroke r measured horizontally at still water level.

However, the case of the hinged paddle type wavemaker with a draught less than the water depth has been analysed in detail by Hyun (1976). The theory describes the two-dimensional irrotational motion of an inviscid incompressible fluid in a semi-infinite tank with a rectangular cross-section and a constant water depth d. The wave amplitude some distance away from the wavemaker ($x > 3d$) is given in terms of the wave length and the geometry of the wavemaker. An equivalent theoretical analysis for the wedge-type wave generator has been developed by Gilbert *et al.* (1970). Both these calculations for the paddle and wedge-type wave generators are based on the assumption that the test tank is semi-infinite or that the absorption of waves by the beach at the opposite end is complete and that leakage between the wave generator and tank sides is negligible. In practice, even with the best-known wave absorbers, there are some reflected wave components present that must be accounted for.

At this stage, it is useful to consider how reflections modify the generated wave in a test tank of finite length with a wave absorber at the opposite end to the wave generator. Some time after the wave generator is first set into motion, a steady state is eventually reached due to successive partial reflections of the progressive wave off the wave absorber and the generator in turn. The initially generated wave train, called the primary incident wave, is partially reflected from the opposite end to form a primary reflected wave component which is then reflected from the wave generator surface to form a secondary reflected wave and so on. These higher reflections have progressively smaller amplitudes such that only the primary reflected wave is significant relative to the primary incident wave. The theoretical results given above only predict the initially generated primary incident wave; it is the properties of this wave that need to be extracted from measurements.

The required correction for reflection effects can be made by using a simple reflection theory presented by Ursell *et al.* (1960). This theory is based on the fact that a combination of the primary incident wave and its first partial reflection from the wave absorber will be a partial standing wave system along the tank length. Therefore, if a partial standing wave component is incorporated into the velocity potential solution for the theoretical primary incident wave, an expression for the measured wave amplitude a is given by Ursell *et al.* to be

$$a^2 = a_1^2[1 + 2e\cos\delta + 2e\cos(2kx + \delta) + 2e^2\cos 2kx + 2e^2] \tag{9.21}$$

where a_1 is the amplitude of the primary incident wave of wave number k, and e and δ are unknown parameters, depending on wave absorber characteristics and tank length. When e is small, Equation (9.21) becomes

$$a = a_1[1 + e\cos(2kx + \delta) + e\cos\delta + O(e^2)] \tag{9.22}$$

Thus the wave amplitude varies sinusoidally from point to point along the wave tank length between upper and lower amplitudes limits of a_{max} and a_{min}, respectively, which can be written as

$$a_{max} = a_1(1 + e + e\cos\delta) \tag{9.23}$$

and

$$a_{min} = a_1(1 - e + e\cos\delta) \tag{9.24}$$

If the wave absorber reflection coefficient ε is defined as the amplitude ratio of the primary reflected wave to the primary incident wave, it can be expressed in terms of the preceding parameters as

$$\varepsilon = \frac{a_{max} - a_{min}}{a_{max} + a_{min}} \tag{9.25}$$

$$= \frac{e}{I + e\cos\delta} = e + O(e^2) \tag{9.26}$$

Equation (9.22) predicts a sinusoidal variation of amplitude along the tank of wavelength equal to half the wavelength of the primary incident wave.

By recording the variation of measured wave amplitude a with x, both the reflection coefficient ε and the amplitude of the primary incident wave a_1 can be deduced in the following way. At a particular point along the wave tank, for a certain value of x, the amplitude a of the combined incident plus reflected wave is a maximum. This implies that $\cos(2kx + \delta) = 1$ and hence $(2kx + \delta) = 0$. Since a, x and k are known from the measurements, δ can be evaluated. On the other hand, when the measured wave amplitude is a minimum $(2kx + \delta) = \pi$ and a value for δ can be determined. An average value of δ is then derived by following this procedure for successive minima and maxima. The reflection coefficient ε can be calculated from Equation (9.25) using values of $a_{\max}$ and $a_{\min}$. Furthermore, Equation (9.26) and the known value of δ enables the parameter e to be deduced so that the amplitude of the primary incident wave can finally be obtained from Equations (9.23) or (9.24). It is to be noted that this method neglects the effect of the e^2 term in Equation (9.21) and, therefore, can only be regarded as accurate for values of ε ($\simeq e$, approximately) less than 10%.

For higher values of the reflection coefficient, however, an alternative method for determining the primary incident wave amplitude is available. This method is based on the property of Equation (9.21) that $a = a_1$ when $\cos kx = 0$, which occurs at possible values of x given by the equation

$$x = \left(\frac{m}{2} + \frac{1}{4}\right)\lambda \tag{9.27}$$

where $m = 0, 1, 2, 3, \ldots$. Measurements of wave amplitude at selected points along the tank length, satisfying the preceding condition, enable an accurate value of the primary incident wave amplitude to be obtained.

If this reflected wave correction procedure is applied to the generated wave, close agreement is obtained between theory and experiment for both paddle and wedge-type generators. This is illustrated by comparisons of theory and test data in Figure 9.3 obtained from Patel and Ionnaou (1980). The comparison is plotted in terms of parameter hf^2/g where h is water depth and f is wave frequency in Hz. There is good agreement between theory and tests but this is dependent on ensuring that edge leakage effects are reduced to a negligible level and on taking account of the wave reflected from the beach.

A straightforward comparison of the paddle and wedge wave generating performance (Figure 9.3(*b*)) indicates that the paddle produces wave heights approximately 30% below those produced by the wedge for the same generator stroke. This is to be expected since the wave energy input due to the to-and-fro motion of the wedge surface will be greater than that due to the hinged motion of the paddle. A further advantage of the wedge-type wave generator over the paddle is the absence of fluid on the wedge rear surface, which greatly reduces the total actuating power required, despite the fact that frictional forces in the wedge sliding mounts are likely to be large.

It is possible to completely eliminate edge leakage effects by the installation of flexible watertight membranes across gaps in the wave generator system, such that its motion is not restricted. However, the lack of an efficient wave absorber still imposes a significant limitation on

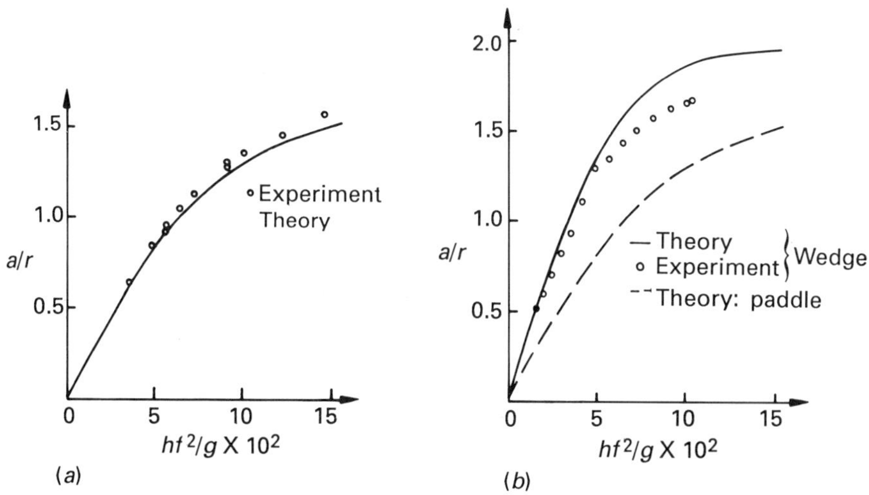

Figure 9.3. Comparison of theory and experiment for waves generated by paddle and wedge-type wave generators: (*a*) paddle type; (*b*) wedge type

the accuracy attainable in model testing. Measured reflection coefficients are usually less than 10% for most frequencies, but can be as high as 20–30% for low frequency long waves. Good quality water waves can only be effectively used when these reflections are avoided – either by the use of very long tanks which provide a reasonable length of testing time before reflections travel back from the opposite end, or by using computer-assisted experimental techniques which allow fast acquisition of data before reflections mar the quality of waves. The latter alternative is not very satisfactory when long testing times are essential for studies associated with low frequency phenomena such as slow drift oscillations.

The solution to this dilemma lies in two possible directions. It is feasible to improve the performance of absorber beaches by using additional absorbent material in configurations that extend further along the wave tank length. A second approach is to take advantage of recent developments on the use of a wave generator as both a generator and absorber of waves (Salter, 1974) which can substantially reduce the content of reflected waves within a tank. In this mode of operation, the wave generator is driven through a servo feedback with the error signal provided by the difference of the demanded and actual force exerted by the wave fluid on the generator. This type of control system enables a conventional wave generator (paddle, piston, wedge, or other) to be used both as a generator and an absorber of wave energy through a closed-loop control system that will generate only the wave train that is required, while at the same time absorbing unwanted reflection components returning to the wave generator and preventing the setting up of a partial standing wave system. Thus, it is feasible to generate a train of progressive waves even if no wave absorption occurs at the opposite end of the wave tank. However, this mode of operation requires a more sophisticated control system and actuating mechanism. In the majority of cases for small to medium size tanks, it is preferable to use the conventional 'open loop' system with a wave generator at one end of the tank and a good wave absorber at the other.

Figure 9.4 shows the plan and side views of a typical directional wave testing facility. Such basins can be up to 50 m by 80 m in plan with water depths of up to 10 m. Directional waves are generated by segmented paddle-type wave generators of from 30 up to 100 in number along one side of the basin. These segments can each be from 0.2 m to 0.8 m in width and are independently driven by electro-hydraulic servo-controlled rams or by servo-controlled electric d.c. motors. It is possible to drive the paddles electrically because the power requirement for each small segment of the generator is within the range deliverable by small electric motors. Many directional wave basins also have an additional long-crested wave generator along the adjacent side to enable the two wave generators to produce crossing seas with the auxiliary wave generator being able to produce higher waves. Wave absorbing beaches are installed on the other two sides of the basin.

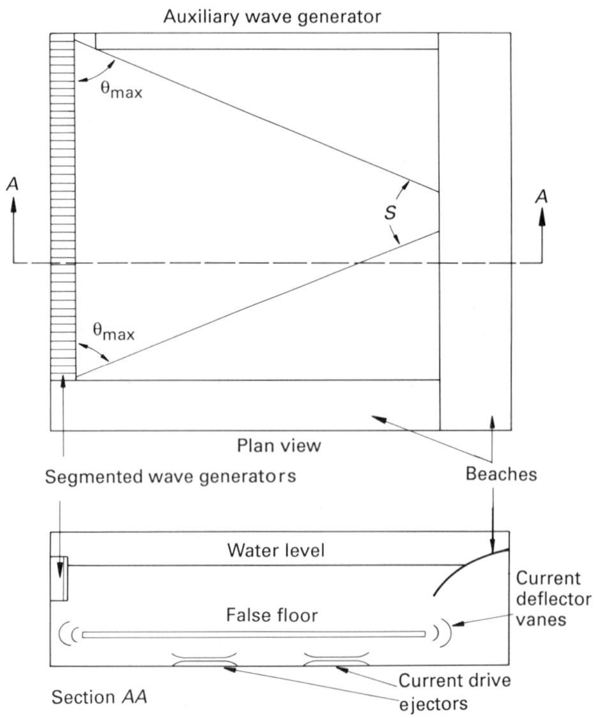

Figure 9.4. Typical directional wave facility

A current facility can be provided in various ways, with Figure 9.4 illustrating one technique where a movable false floor provides variable water depth together with a return path around which ejector pumps can be used to drive a current circulation. Both long- and short-crested wave tanks can also be installed with pits which are used to model deep water structures such as guyed towers and tensioned buoyant platforms. The design of the wave generators, their control and drive systems, beaches and associated equipment for directional wave basins are similar to those employed in long-crested wave tanks, Huse and Torum (1981) and Salter (1984) present descriptions of typical directional wave basins.

A wind tunnel simulating the atmospheric boundary layer is used for obtaining more accurate estimates of wind forces and overturning moments on fixed and floating offshore structures. The overturning moments are of particular interest for assessing the hydrostatic stability of floating structures. Figure 9.5 illustrates a typical return flow wind tunnel with a drive motor and turning vanes used to blow air to a settling chamber. The screens preceding the settling chamber reduce the level of turbulence in the free stream approaching the working section. The atmospheric boundary layer is generated by screens made up of graduated slats or cylinders which are carefully arranged to generate a representative boundary layer with the desired velocity profile and turbulence intensity. This boundary layer is allowed to develop for some distance before passing the test model which is mounted on a force dynamometer and often installed with flow visualization tracers. Further practical details of wind tunnel testing may be obtained from Pankhurst and Holder (1952).

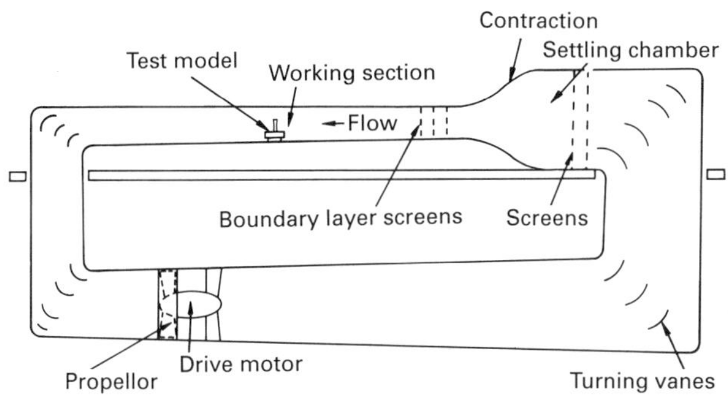

Figure 9.5. Typical wind tunnel facility

The design and construction of effective test models is an essential part of the model testing process. Conventional model construction techniques are used with wax, wood, glass fibre, aluminium or plastic as basic building materials. Particular care has to be taken to ensure that the model dimensions, mass distribution and elasticities correctly meet the scaling requirements outlined in Section 9.1. In general, it is more difficult to satisfy these requirements accurately for small models, that is, of large scale factor. The majority of test models for wave testing of offshore structures are constructed using from 20th to 80th scale although wind tunnel test models tend to be from 70th up to 150th scale. The model design also has to incorporate the instrumentation required and this can pose payload problems for small model sizes. A tilt table is used for ballasting the models to specific centre of gravity positions and for measuring the model moments of inertias.

Model test facilities have a large associated range of measuring equipment for sensing wave elevation, water particle velocities, model motions, forces, moments and pressures. All these instruments are designed to generate a direct voltage or current signal which is linearly proportional to the measured parameter in order to provide a signal that is easily converted into digital form by an analogue to digital converter.

The type of transducers used are illustrated by brief descriptions of two of these. The wave elevation sensor consists of two parallel vertical rods a small distance apart and partially immersed in the tank. Wave elevation is sensed by detecting the reaction to an applied alternating voltage across the rods. The supply frequency is in the kHz range and the resultant small current is demodulated and amplified by an associated circuit to give a d.c. voltage proportional to wave elevation. Arrays of three or more of these sensors are used to obtain estimates of directional distribution in short-crested seas. Early versions of motion sensing transducers are based around low friction potentiometers with conductive plastic tracks. Nylon lines mounted on the model pass around pulleys on the potentiometer spindles and are tensioned by small weights at the free ends of the lines. The various modes of motion of a model can, therefore, be directly transduced into a voltage signal by an appropriate combination of potentiometers. A more modern technique utilizes television cameras focused on light emitting diodes or contrast zones on the moving model. A digital circuit is then used to convert the resultant video signal into voltage outputs proportional to the motions and rotations of diode pairs or contrast zones. The majority of force sensing transducers used on models are based on strain gauged or displacement sensing load cell principles.

The design of test models and their instrumentation is continuously evolving to meet the requirements of modern deep water structures and to use the latest developments in electronic transducers and computing technology. It is common practice nowadays to use two or more independent computer systems for operating a wave tank facility – one machine may operate in real time to generate control signals for the wave generator, whereas the second and additional machines can be used for the data acquisition and analysis role. Patel (1984) describes an example of a typical computer and software system.

9.3 Mathematical basis for wave generation

It is necessary to develop a basis for generating control signals to one or more wave generators to enable a desired wave elevation to be obtained over the working area of a wave tank. This wave elevation is a combination of the transfer functions of the control signal conditioning filters, the electro-hydraulic or electric generator actuator, the wave elevation to stroke ratio of the wave generator and the motion of the wave down the tank.

Taking the case of long-crested waves first, the generation of regular waves poses no difficulties. A desired sinusoidal wave elevation time history can be readily worked back through all the system transfer functions in the frequency domain to yield the necessary control signal amplitude. For the generation of random waves, however, the requirement is to repeat the above procedure for a stationary, random time history having a predetermined power spectral density and a Gaussian probability distribution.

A brief review of alternative methods for filling this requirement is appropriate here. A range of analogue methods have employed thyratrons or temperature limited diodes to produce a white noise signal (a time series having a constant spectral density over the frequency range of interest). This signal has then been shaped to give the required spectral density by using a set of analogue band pass filters with their outputs summed together. These devices, however, suffer from temperature induced amplitude drifts and have to be separately set up for each spectral density required.

An alternative approach which can be implemented in analogue or digital hardware idealizes the required continuous spectrum into a series of constant bandwidth sections. The control signal is then derived by summing the sine waves at the centre frequencies of the constant bandwidth bars, the amplitudes of the sine waves being obtained from the area in each bandwidth section. Alternatively, the spectrum may be approximated to a number of constant amplitude sine waves at unequally spaced frequencies. This method is more complicated since the sine wave frequencies tend to cluster around the peak frequency of the desired spectrum and need to be carefully evaluated. In all these methods, a randomly selected phase difference is imposed on each frequency component to eliminate any coherent features developing in the summed waveform. This superposition of sine waves has the advantage of relative simplicity. The summing procedure can be expressed by the equation for wave elevation (η) of

$$\left.\begin{aligned} \eta &= \sum_{i=1}^{N} a_i \cos(\omega_i t + \phi_i) \\ &\text{where} \\ a_i &= \sqrt{[2S(\omega_i)(\omega_{i+1} - \omega_i)]} \end{aligned}\right\} \quad (9.28)$$

where $S(\omega)$ is the spectrum to be generated and ϕ_i are randomly selected phase angles between 0 and 2π. The values for a_i and ω_i can be selected for constant amplitude or constant frequency steps. However, the resulting time sequence is made up of discrete frequency component spikes. These can have the effect of artificially exaggerating or suppressing key sections of measured dynamic response functions from test models.

This major disadvantage of the superposition approach to building up a time series with a desired spectrum makes it essential that a time series with a continuous spectrum be used. Such a time series can be generated digitally by a 'pseudo-random' shift register with logical feedback. The signal produced is actually completely deterministic and periodic but the period is adjustable and capable of being extremely large. Then, within one complete cycle, the output time series has enough of the properties of white noise to be treated as such. The time series has a constant power spectrum over a certain frequency range and suffers no amplitude or zero drift. Golomb (1967) sets out the mathematical basis of the above characteristics.

Figure 9.6(a) shows a shift register consisting of N binary stages, each having a bipolar state of 1 or 0. Theoretically, the shift resistor states must be 1 or -1 but the 0 is used instead of -1 for programming convenience. At every clock period Δt, the values of each stage are shifted one place to the right. The first stage then takes a value determined by modulo 2 addition of the values at any two chosen stages r_1 and r_2. It can be shown that this 'feedback' or 'regenerative' part of the register (stages 1–n) goes through a sequence of states which is periodic and of maximum period $2^n - 1$. Golomb (1967) and Watson (1962) tabulate vaules of r_1 and r_2 which enable these maximum periods to be realized.

The random sequence of plus and minus 1s (with the -1s replaced by 0s, here) generated by a shift register at clock intervals of Δt can each be regarded as having an equal probability of occurrence. It can be shown (see Newland, 1975) that this time series will have a one-sided power spectral density $S(\omega)$ of

$$S(\omega) = \frac{1}{\pi f}\left[\frac{\sin(\omega/2f)}{(\omega/2f)}\right]^2 \tag{9.29}$$

where $f = 1/\Delta t$ Hz. This function is shown plotted in Figure 9.6(b). It has an approximately constant region up to a frequency of $f/2$ Hz. However, since the shift register sequence is periodic (with period $T = [2^n - 1]\Delta t$), the power spectral density must consist of a series of delta functions, spaced $2\pi/T$ radians apart, with the areas under each delta function given by Equation (9.29). Since T can be made very large for most shift register sequences, the spectral density can be regarded as being continuous.

Clearly, the shift register sequence described here has neither the required spectral density nor a Gaussian probability distribution. However, both can be achieved by weighting each stage of the register and summing to give an output after each clock period Δt of

$$\sum_{k=1}^{N} A_k a_k \tag{9.30}$$

where A_k are the shift register stage values and a_k are the corresponding weightings. Figure 9.7 shows a schematic diagram of the process. Since the summed output above is derived from a large number of essentially random variables, it can be regarded as Gaussian from the central limit theorem.

In order to impose a required spectral density onto the summed shift register output, the signal path transfer functions are converted to a combined transfer function $H_1(\omega)$, such that if an output wave spectral density of $S_w(\omega)$ is desired, the input control signal must have a spectral density of $S_c(\omega)$ given by

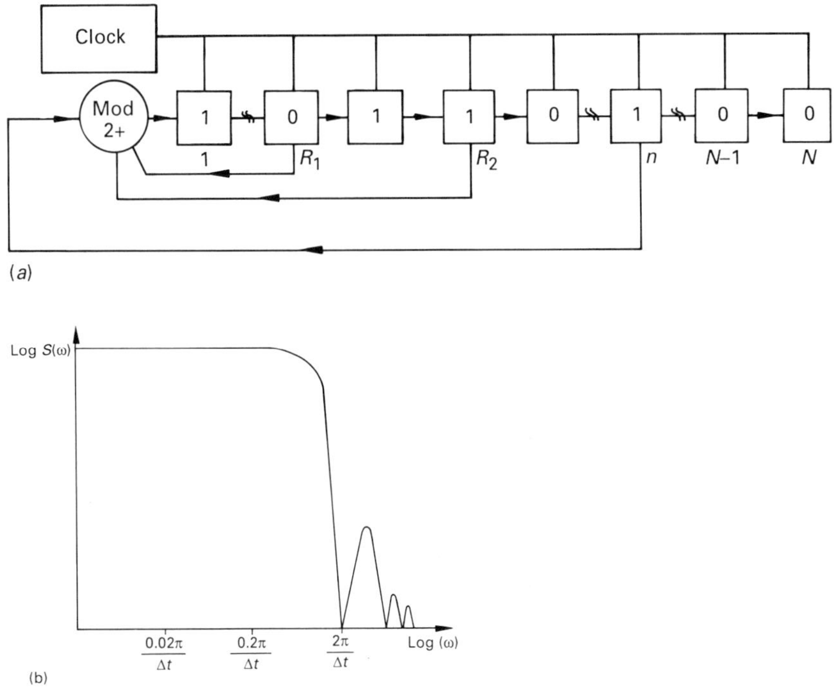

Figure 9.6. Shift register and its spectrum: (*a*) the shift register; (*b*) white noise spectrum

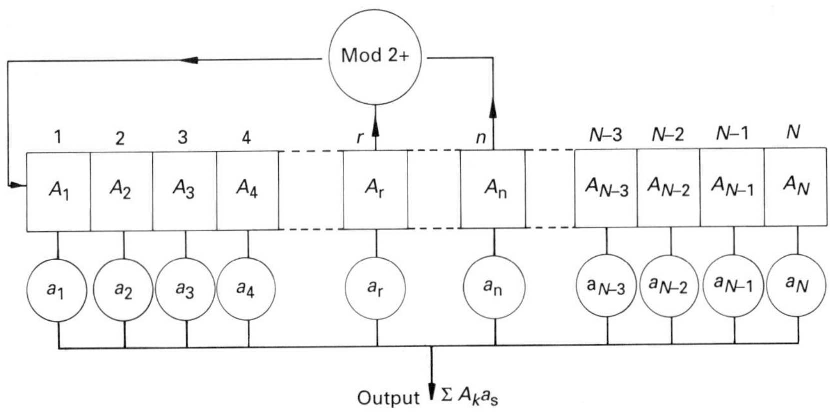

Figure 9.7. The shift register weightings

$$S_c(\omega) = \frac{S_w(\omega)}{|H_1(\omega)|^2} \tag{9.31}$$

In order to obtain the required control signal spectrum $S_c(\omega)$, the white noise spectrum $S(\omega)$ of the shift register sequence outputs has to be filtered digitally through a transfer function $H_2(\omega)$ such that

$$H_2(\omega) = +\sqrt{\left(\frac{S_c(\omega)}{S(\omega)}\right)} \tag{9.32}$$

This is done by applying the weightings, as illustrated in Equation (9.30), which are obtained as impulse response functions corresponding to the desired transfer function $H_2(\omega)$. The discretized transform can be defined by the equation

$$a_{k+1} = h(r\cdot\Delta t) = \sum_{k=0}^{N-1} H_2(k\cdot\Delta\omega)\exp[i(2\pi kr/N)] \tag{9.33}$$

for $r = 0, 1, 2, \ldots, N$ Δt is the shift register clock interval; $\Delta\omega$ is the spectrum frequency interval; and N are the number of desired discretizations. It can readily be shown that

$$N = \frac{2\pi}{\Delta t\cdot\Delta\omega} \tag{9.34}$$

It has to be noted that only the modulus of $H_2(\omega)$ is used here. $H_2(\omega)$ has been made imaginary and antisymmetric in Equation (9.33) giving a real and antisymmetric $h(t)$ which also yields an output signal with a zero mean value. To avoid aliasing, the output spectrum is well separated from the first high frequency end of the spectral density, $S_c(\omega)$.

The above approach to generating random waves has been formulated and implemented previously on an analogue system (see Fryer *et al.*, 1973). However, the set-up procedure for the analogue system is tedious. The impulse response transfer function weightings have to be precomputed for every spectral density required and then inserted into the shift registers through a set of potentiometers. The further development described by Patel (1981) automates the complete procedure through a micro-computer with the control signal path transfer functions incorporated within the software. The most important of these is the wave generator transfer function which is strongly frequency dependent (see Figure 9.3) and needs to be accurately defined in order to obtain good agreement between desired and actual wave heights. The transfer function can be conveniently incorporated in the software by means of a curve fit to a hyperbolic tangent function. The dynamic responses of the analogue filter, the electro-hydraulic servo-mechanism and phase effects due to wave propagation down the tank should also be accounted for in the software.

Figure 9.8 shows a comparison between demanded and measured spectra in the wave tank using this procedure. There are several advantages to implementing the random wave generation procedure digitally. The numerical time series output by the computer is of a 'pseudo-random' nature but at the same time it is actually completely deterministic and precisely repeatable. Thus the power spectral density derived from the signal remains constant, suffers no level or zero drift and can be repeatably specified via the control program input. On the other hand, a different 'ensemble' of the same power spectral density can be obtained for an experiment by either using a different segment of the control signal output or by altering the feedback paths in the shift register. The repeatable nature of the random wave can be used to advantage with a trigger signal facility. For large test models which significantly distort their local wave environment, an initial wave run can be made in the absence of the model to measure the wave only. A second run with the identical random wave and the data acquisition triggered at precisely the same instant can then be used to record model responses.

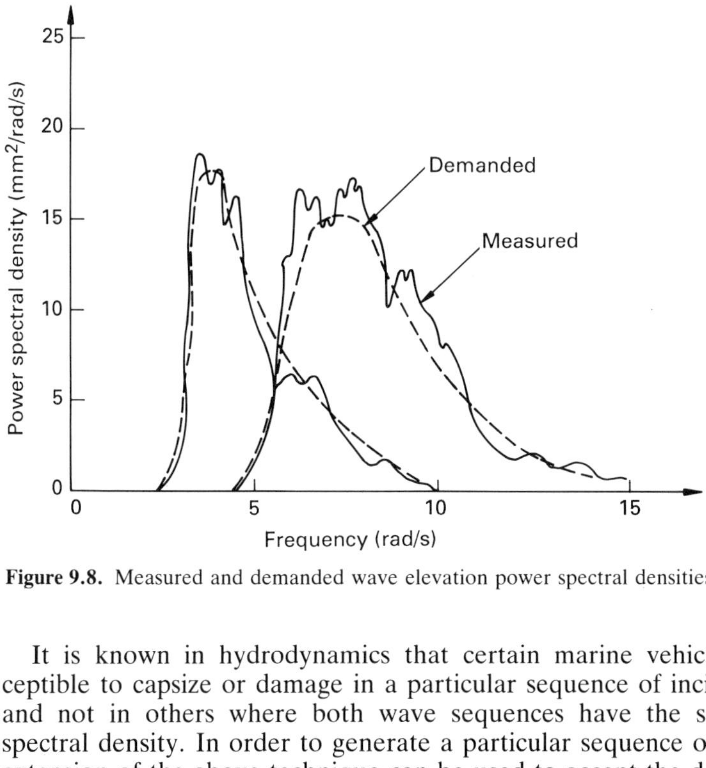

Figure 9.8. Measured and demanded wave elevation power spectral densities

It is known in hydrodynamics that certain marine vehicles are susceptible to capsize or damage in a particular sequence of incident waves and not in others where both wave sequences have the same power spectral density. In order to generate a particular sequence of waves, an extension of the above technique can be used to accept the desired wave sequence as input, to convert it to the frequency domain, pass the resultant spectrum through the necessary amplitude and phase transfer functions and to regenerate a control signal in the time domain which has, therefore, been adjusted for both the amplitude and phase of the transfer functions in the control signal path. This technique not only has the capacity to reproduce certain wave sequences but it can also generate periodic waves with desired shapes such as may be typical for wind generated water waves. Patel and Korloo (1985) give further details of this technique.

The generation of short-crested waves using the segmented paddles shown in Figure 9.4 adds another dimension to the problem of defining wave generator control signals. Each paddle segment has to be simultaneously driven in order to generate waves with a prespecified elevation spectrum and directional distribution over the working area of the basin. Simultaneous control signal generation is only possible with the use of powerful digital computers able to generate analogue outputs from digital values simultaneously for a large number of cycles.

There are three ways of using segmented wave paddles to generate directional seas. One of the most elegant of these is sometimes called the snake method of wave generation. If the segmented paddles are oscillated such that a travelling sinusoidal displacement of wave length L and period T is propagated along the line of wave makers then the resultant water wave will propagate at a direction θ to the line of wavemakers such that

$$\sin\theta = \frac{\lambda}{L} \tag{9.35}$$

where λ is the wavelength of the wave at period T. Figure 9.9 illustrates this generation of an angled wave using a plan view of a wave basin with reference axis Oxy and a line of wave makers with mean position of $y = 0$. A travelling wave of length L in wave maker displacement along the positive x axis generates waves with crest lines shown, at an angle θ to axis $y = 0$. This form of wave generation was first suggested by Biesel (1954).

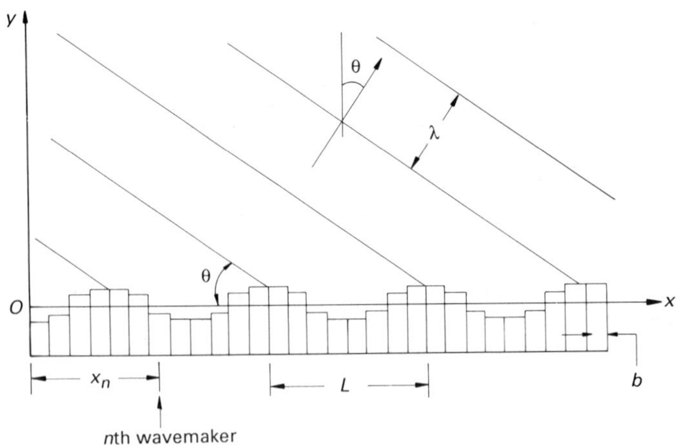

Figure 9.9. Generation of directional seas

There is a lower limit to the wave period achievable with these 'snake' wavemakers due to the finite width b of individual wavemaker segments. In practical terms, the snake motion wavelenth needs to be greater than $3b$ and it is suggested by Biesel (1954) that a practical lower limit to the wave period that can be generated without extraneous disturbances to the water surface is given by the equation

$$T \geqslant 0.53\sqrt{(\sqrt{2} + \sin\theta)} \tag{9.36}$$

Above this limit of wave period, however, a wave elevation can be generated by a summation of sinusoidal wave components at different frequencies and directions as illustrated by Equation (4.124) and rewritten here as

$$\left.\begin{aligned}\eta &= \sum_{i=1}^{N}\sum_{j=1}^{M} a_{ij}\cos[k_i(x\cos\theta_j + y\sin\theta_j) - \omega_i t + \phi_{ij}]\\ &\text{with}\\ a_{ij} &= \sqrt{[2S(\omega_i,\theta_j)(\omega_{i+1}-\omega_i)(\theta_{j+1}-\theta_j)]}\end{aligned}\right\} \quad (9.37)$$

using the notation of Figure 9.9. Here $S(\omega_i, \theta_j)$ is the directional wave spectrum discretized into N frequency components and M direction components with ϕ_{ij} being randomly selected phase angles between 0 and 2π. Using the snake wave-making method of organizing the directionality of waves, the displacement r_n of the nth wave paddle is given by the equation

$$r_n = \sum_{i=1}^{N}\sum_{j=1}^{M} H_i a_{ij}\cos[k_i \sin\theta_j x_n - \omega_i t + \varepsilon_{ij}] \quad (9.38)$$

where H_i is the transfer function at frequency ω_i between paddle and generated wave amplitude. Equation (9.38) is analogous to (9.37) for long-crested wave generation.

This technique of wave generation is implemented by using a large number of frequency and directional components to make the spectrum as continuous as possible, although it has to be recognized that the spectral energy is generated at a series of discrete frequencies and directions. This may have an influence on interpretation of results from experiments where the test structure may have an extremely narrow range of resonant excitation or other frequency dependent phenomena. However, the frequency component superposition technique does allow generation of very unusual wave forms by appropriate selection of the matrix of starting phases ε_{ij}. Artificial sea states with prescribed patterns, freak waves, and isolated high plunging breakers can be generated for closer examination of the physics of such extreme conditions.

The snake wavemaker technique can also be used in a second method to generate a directionality distribution, but by using a wave elevation spectrum obtained by digitally filtering a pseudo-random shift register sequence. The wave elevation spectrum is first converted to a wave generator displacement spectrum through the generator transfer function. The spectrum is then split up into the required directional spread by using wavemaker displacements obtained from summing travelling waves corresponding to frequency and wave direction. This approach is used less frequently than the frequency superposition technique described above and is, therefore, not followed up in detail here.

The third approach to directional wave generation using segmented flaps exploits the diffraction pattern set up by each oscillating flap to build up a continuous variation of wave energy with direction. It has been shown by Naeser (1979) that the diffracted wave radiating out from oscillatory motion of an isolated paddle of width b (Figure 9.10) has a directional distribution that can be approximated by

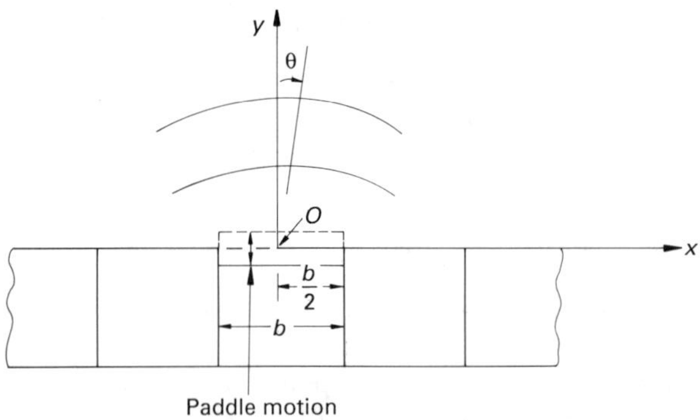

Figure 9.10. Wave diffraction from a paddle moving in isolation

$$\left.\begin{aligned} a &= \frac{a_0 b}{\sqrt{[\lambda(y-b)]}} \cos^{s/2}\theta, \\ &\text{where} \\ \theta &= \arctan(x/y) \end{aligned}\right\} \quad (9.39)$$

for $y > 3b$, and $y > b|x|/\lambda$ where a is the wave amplitude at point (x, y) within the range of validity defined above and a_0 is the wave amplitude generated by the paddle if it was confined in a channel of width b. The index s is given by

$$\left.\begin{aligned} s &= 2\left(\frac{b}{\lambda} + 0.6\right)^3 \\ &\text{for } b \geqslant 0.2\lambda. \end{aligned}\right\} \quad (9.40)$$

The validity of these equations has been tested through comparison with experiment by Naeser (1979). Assuming that the diffracted wave pattern from all the segments in a wavemaker array can be superimposed and that the diffraction pattern from one paddle is not influenced by the irregular wavemaker shape due to the motion of the other paddles, a desired directionality distribution can be generated by summing the diffraction patterns from each of the paddle segments. In defining the algorithm for wavemaker control, the value of b and thus s can be varied by running two or more flaps with the same demanded displacement time history. It is then possible to superimpose wave directions to obtain a directional distribution which is continuous but limited to be of the cosine form as shown in Equation (9.39). Further details of this procedure are given by Naeser (1979).

One key difference between snake and diffraction based wave generation is the area over which the resultant directional distribution is uniform. For the snake technique, the directional distribution will only be produced in a triangle region defined by the maximum inclining angle θ_{max} that can be generated by the paddles – this is defined by lines S in Figure 9.4. The diffraction based technique will, on the other hand, produce the required directional distribution everywhere in the tank due to reflections from side walls. For the snake wave techniques, it is desirable to mount models closer to the paddles, whereas for the diffraction wave techniques, better results are obtained by mounting models well away from the wave generator. In both cases, test models should not be mounted close to side walls to reduce the effects of multiple reflections between model and adjacent side wall.

9.4 Data analysis procedures

Analysis procedures for data acquired from tank tests are based on numerical algorithms derived from the deterministic and statistical descriptions of waves' phenomena presented in Chapter 4. Test data from wave elevations and model responses are recorded by analogue to digital conversion in a computer using a specified sampling interval. A variety of data analysis procedures are implemented in computer programs to process this time series data. Examples of typical analyses are listed below.

(a) *General data statistics*
A typical statistical analysis implements calculations of the following quantities:

1. mean value, mean square value, standard deviation and variance;
2. normalized data with respect to the mean value;
3. probability and cumulative probability density function and their moments;
4. the Rayleigh distribution function of peaks, significant signal height; and
5. the skewness and kurtosis;

(b) *Analysis of regular wave free and forced responses.*
The natural period and logarithmic decrement of a damped oscillatory time history are determined by using at least five cycles of a signal to determine the locations of peaks and troughs in the data enabling the frequency to be determined. The logarithmic decrement of the response is determined from the gradient of the appropriate straight line using a least squares approach (see Section 8.1.1 for further details).

Forced periodic responses are quantified by a numerical harmonic analysis. The frequency of a reference signal channel is determined by a simple peak and trough searching routine. The numerical harmonic analysis then extracts the amplitude components of all the other channels at this frequency together with the phase difference between each channel and the reference signal.

(c) *Spectral analysis*
The mathematical procedures for auto- and cross-spectral estimates outlined in Section 4.7.1 are implemented in computer programs. The raw data are usually conditioned before input to the spectral analysis – typically by applying a $\frac{1}{10}$th cosine tapered data window to improve the effective filter shape and to remove zero drift by averaging the first and

last $\frac{1}{3}$ of the records and adjusting all the records on the assumptions that this drift is linear. The spectral analysis technique also generates estimates of coherence between the signal and reference channels.

(d) *Analysis of directional wave data*
Wave data from directional wave basins needs to be acquired from multiple sensors and to be analysed by the discrete Fourier transform method or the maximum likelihood technique. Software implementing the procedures set out in Section 4.7.5 is used for this purpose.

9.5 Full-scale measurements

There is no doubt that the ultimate model test would be at full scale with a scale factor of one, realistic environmental conditions and no concerns related to scaling laws. It is becoming increasingly common to supplement design studies by using the results of full scale measurements on specifically designed test structures or from data obtained during monitoring of existing structures.

The Christchurch Bay Tower is a typical facility specially designed to obtain measurements on wave loading and foundation response at full scale conditions. Figure 9.11 shows a side view of the facility and the instruments mounted on it. The structure is situated in Christchurch Bay on the south coast of Great Britain in water with a mean depth of 8.4 m. It has 113 channels of data recorded on site with statistical information calculated at the tower and telemetered back at regular intervals.

Figure 9.11 shows the sensors mounted on the tower. There are ten channels of data on wave elevation, tide level, wind speed and direction, current speed and direction, atmospheric pressure and sea temperature. A further 20 channels are dedicated to measuring current velocity components below the water surface from both ball-type and electromagnetic current sensors. There are 27 channels of strain gauge transducer data, together with 28 channels of water pressure data. The tower foundation response was monitored by 28 channels of pore water and earth pressure transducers as well as soil shear stress and tower tilt sensors.

The Christchurch Bay Tower was first installed in December 1975, but suffered a foundation failure in January 1976. It was lifted and re-sited in mid 1976 but suffered further storm damage in late 1976 after which it was lifted, re-built and re-placed on site in 1978. The tower has enabled acquisition of valuable data that continues to provide further information and insight on the mechanics of wave loading on an offshore structure. Bishop (1978) and Bishop *et al.* (1980) describe the facility and review some of the data measured from it.

A typical structural monitoring system on an existing production structure is of more engineering interest because it provides fundamental data as well as in-service wave and structural loading data to the platform owner to enable the structure to be maintained both economically and effectively. Figure 9.12 shows a typical jacket platform and the

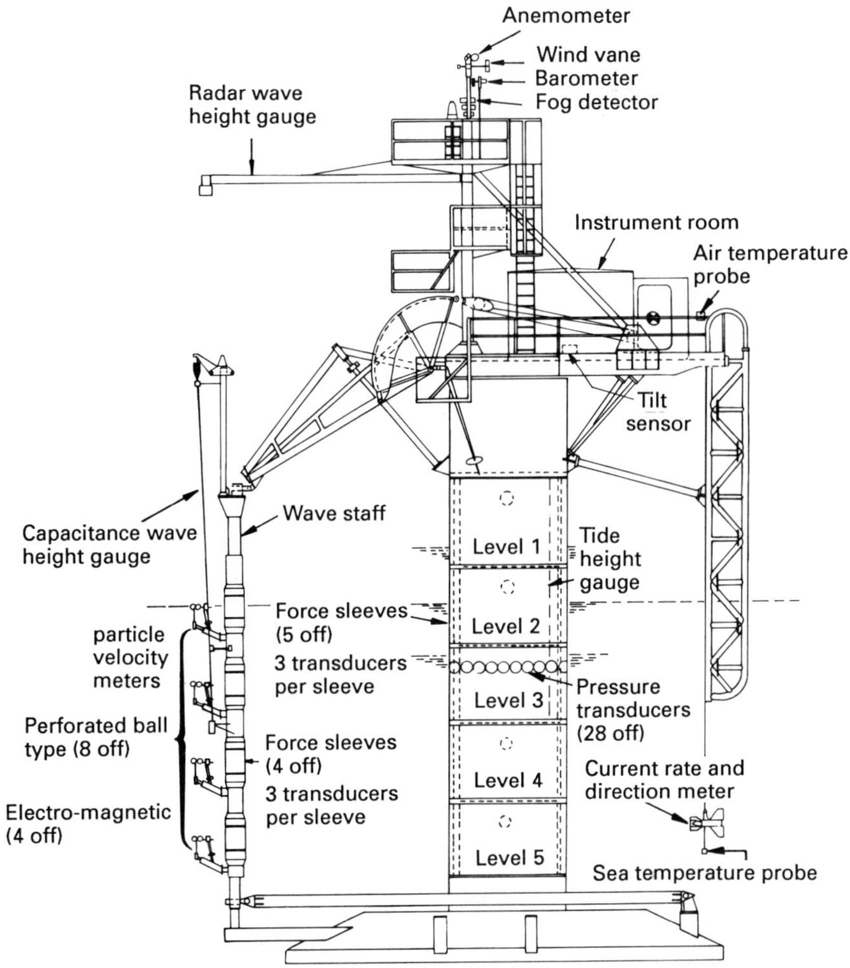

Figure 9.11. Christchurch Bay Tower test facility (from Bishop, 1978)

environmental condition and structural response data that would be collected over the life of the structure. Typical structural monitoring projects can acquire over 100 channels of data which are usually analysed and reduced to a more manageable size on site before being returned to the design office for further scrutiny.

Effective analysis of the data from these full scale measurements poses problems because of the uncontrolled incident environment being a combination of wind, current and wave effects. Research is continuing into how the data from full scale measurements can be analysed and interpreted to lead to improved theories for use in design or to a better understanding of the physical processes that are taking place. Full scale structural monitoring also provides a means by which in-service behaviour and structure maintenance programmes can be planned more effectively. There is an increasing acceptance of the fact that the initial

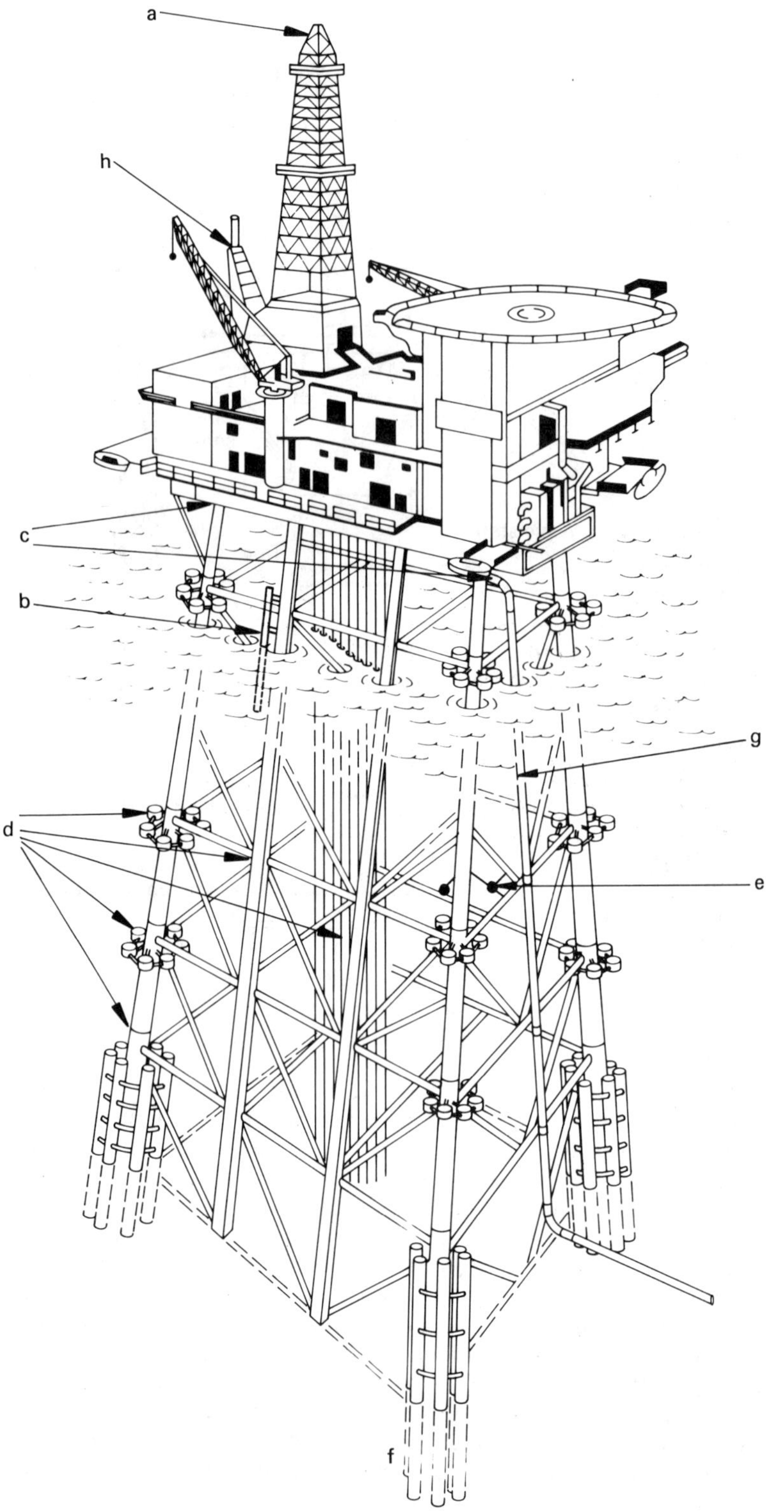
a
h
c
b
g
d
e
f

capital cost of a structural monitoring system is likely to be small compared to the savings in maintenance costs that can be achieved by effective use of the data arising from the system.

References

Biesel, F. (1951), Les appareils generateures de houle en laboratoire, *La Houille Blanche*, **6**, 147–165

Biesel, F. (1954), Wave machines, *Ships and Waves*, Stevens Institute of Technology, Hoboken, NJ, USA, pp. 288–304

Bishop, J. R. (1978), *A Description of the Christchurch Bay Wave Force Project*, National Maritime Institute, Report No. OT-R-7804, NMI R33, April

Bishop, J. R., Tickell, R. G. and Gallagher, K. A. (1980), The UK Christchurch Bay Report Project, *Proceedings of the 12th Annual Offshore Technology Conference*, 5–8 May, Paper OTC 3796

Fryer, D. K., Gilbert, G. and Wilkie, A. (1973), A wave spectrum synthesizer, *Journal of Hydraulics Research*, **11**, (3)

Gilbert, G., Thompson, D. M. and Brewer, A. J. (1970), Design curves for regular and random wave generators, *Journal of Hydraulic Research*, **9**, (2), 163–172

Golomb, S. (1967), *Shift register sequences*, Holden-Day

Havelock, T. H. (1929), Forced surface waves on water, *Philosophical Magazine*, Ser. 7, **8**, 569–576

Huse, E. and Torum, A. (1981), Norwegian Hydrodynamics Laboratory – ocean basin design philosophy, *Proceedings of the International Symposium on Hydrodynamics in Ocean Engineering*, 24–28 August, Volume 2, pp. 1133–1170

Hyun, J. M. (1976), Theory of hinged wavemakers of finite draft in water of constant depth, *Journal of Hydronautics*, **10**, (1), January, 2–7

Lyons, G. J. and Patel, M. H. (1986), Theory and model tests for the vortex shedding induced transverse response of marine risers, *Journal of Sound and Vibration*, **111**, (2), December, 467–487

Naeser, H. (1979), Generation of uniform directional spectra in a wave basin using the natural diffraction of waves, *Proceedings of Port and Ocean Engineering Under Arctic Conditions* POAC, pp. 621–632

Newland, D. E. (1975), *Random Vibrations and Spectral Analysis*, Longman, London

Pankhurst, R. C. and Holder, D. W. (1952), *Wind Tunnel Techniques*, Pitman, London

Patel, M. H. (1984), A software system for a wave tank facility. *Journal of Advances in Engineering Software*, **6**, (4), 214–221

Patel, M. H. (1981), A real time control system for the generation of water waves in a test tank. *Proceedings of the 2nd International Conference on Engineering Software*, CML Publications, Southampton, pp. 539–553

Patel, M. H. and Ionnaou, P. A. (1980), Comparative performance of paddle and wedge type wave generators, *Journal of Hydronautics*, **14**, (1), January, 5–9

Patel, M. N. and Korloo, J. (1985), On the generation of phase controlled water waves. *The Naval Architect*, November, E458–60

Salter, S. H. (1974), Wave power, *Nature*, **249**, 720–724

Salter, S. H. (1984), Physical modelling of directional seas, *Proceedings of the Symposium on Description and Modelling of Directional Seas*, Technical University, Denmark 18–20 June, Paper B1

Figure 9.12. Typical data sources for structural monitoring of a jacket. *Key*: a – wind vane and aneomometer; b – wave elevation sensing staff; c – horizontal acceleration sensors at each corner; d – strain gauges; e – water particle velocity meters; f – soil settlement gauges; g – riser surface temperature and strain gauges; h – flare tower acceleration sensors

Ursell, F., Dean, R. G. and Yu, Y. S. (1960), Forced small amplitude water waves: comparison of theory and experiment, *Journal of Fluid Mechanics*, **7**, Pt 3, pp. 33–52

Watson, E. J. (1962), Primitive polynomials (mod 2). *Mathematics of Computation*, Volume 16, pp. 368–369

Appendix A

Physical quantities and units

Any physical quantity must be expressed with two factors – the numerical value of the quantity and the name of the unit that this value is expressed in. The offshore industry employs units based on both the Systeme International d'Unites (SI) and on Imperial units. These are described below together with conversion factors between them. Both systems consist of the following base units:

Quantity	*SI units*		*Imperial units*	
	Name of unit	*Unit symbol*	*Name of unit*	*Unit symbol*
length	metre	m	feet	ft
mass	kilogram	kg	pound	lb
time	second	s	second	s
electric current	ampere	A	ampere	A
thermodynamic temperature	kelvin	K	kelvin	K
luminous intensity	candela	cd	candela	cd
amount of substance	mole	mol	mole	mol

The term principal units is used to denote the above units or combinations derived from them. For example, m/s is the principal SI unit of velocity, whereas, ft/s is the principal Imperial unit of velocity. Some principal units also have special names and unit symbols. The SI also comprises units derived from the principal units by means of the prefixes listed below. These are auxiliary SI units.

Factor by which the unit is multiplied	*Prefix*	*Symbol*
$1\,000\,000\,000\,000 = 10^{12}$	tera	T
$1\,000\,000\,000 = 10^{9}$	giga	G
$1\,000\,000 = 10^{6}$	mega	M
$1\,000 = 10^{3}$	kilo	k
$100 = 10^{2}$	(hecto)	h

Factor by which the unit is multiplied	*Prefix*	*Symbol*
$10 = 10^{1}$	(deca)	da
$0.1 = 10^{-1}$	(deci)	d
$0.01 = 10^{-2}$	(centi)	c
$0.001 = 10^{-3}$	milli	m
$0.000\,001 = 10^{-6}$	micro	μ
$0.000\,000\,001 = 10^{-9}$	nano	n
$0.000\,000\,000\,001 = 10^{-12}$	pico	p
$= 10^{-15}$	femto	f
$= 10^{-18}$	atto	a

The use of prefixes expressing multiplication by an integral power of 1000 is preferred. A prefix is considered to be combined with the simple unit symbol to which it is directly attached. For example,

$$1\ \text{cm}^3 = (10^{-2}\ \text{m})^3 = 10^{-6}\ \text{m}^3$$

Prefixed are attached only to the base units or to those principal units that have special names. The factors in compound unit symbols should be separated. Thus mN means milliNewton whereas m N or m.N is metre Newton.

Definitions of physical quantities

The definitions of physical quantities in SI and Imperial units are presented below together with convenient conversion factors between them. The numbers given in each statement are exact unless otherwise indicated.

Plane angle

one radian	= 1 rad	
one degree	= 1°	= (π/180)rad
one grade	= 1 grade	= (π/200)rad

Solid angle

one steradian = 1 sr

Length

one metre	= 1 m (base unit)
one foot	= 0.3048 m
one inch	= 25.4 mm
one yard	= 0.9144 m
one mile	= 5280 ft
one mile	= 1.609 344 km

Volume

one cubic metre	= 1 m^3
one litre	= 0.001 m^3
one gallon	= 1 UK gal
	= 0.004 546 091 9 m^3, exactly
1 US gal	= 0.003 785 411 784 m^3, exactly
Thus	
1 UK gal	= 0.004 546 m^3, approximately
1 US gal	= 0.003 785 m^3, approximately

1 old litre	= 1.000 028 litre
1 barrel	= 42.0 US gal = 34.9723 UK gal = 0.158 987 m^3

Time

one second	= 1 s
one minute	= 1 min = 60 s
one hour	= 1 h = 60 min = 3600 s
one day	= 1 d = 24 h = ($\frac{1}{7}$) week
one year	= 1 a = 365.2425 × 24 × 3600 s (average calendar year)

Angular velocity (ω)

one radian per second = 1 rad/s

Rotational frequency ($\omega/2\pi$)

one revolution per second = 1 rev/s

Frequency

one hertz = 1 Hz = 1 s^{-1}

Velocity

one metre per second	= 1 m/s
one knot	= 1.852 km/h
one knot	= 0.514 444 m/s, approximately

Acceleration

one metre per second squared = 1 m/s^2
standard acceleration = g_n = 1 f
1 f = 9.806 16 m/s^2 = (980.665/30.48) ft/s^2 = 32.172 ft/s^2, approximately at 0° latitude and sea level. The local acceleration due to gravity is not a constant quantity. It is approximately 1 f.

Mass

one kilogramme	= 1 kg
one gramme	= 0.001 kg
one tonne	= 1 t = 1000 kg = 1 Mg
one pound	= 1 lb = 0.453 592 37 kg
one ton	= 1 UK ton (long ton) = 2240 lb = 1.016 046 9 tonne
one American ton	= 1 US ton (short ton) = 2000 lb = 0.907 184 74 tonne
one slug	= 1 slug = 1 lbf s^2/ft = 32.172 lb

Force

one newton	= 1 N = 1 kg m/s^2
one dyne	= 1 dyn = 1 g cm/s^2 = 10^{-5} N
one poundal	= 1 pdl = 1 lb ft/s^2
one pound-force	= 1 lbf = 1 slug ft/s^2 = 4.448 N
one kilogramme-force	= 1 kgf
one ton-force	= 1 tonf
where lbf/lb	= kgf/kg = tonf/ton = 1 f

Pressure (p), *stress* (σ, τ)

one pascal = one newton per square metre = 1 Pa = 1 N/m^2
one bar = 1 bar = 10^5 N/m^2 = 1 $Mdyn/cm^2$
one standard atmosphere = 1 atm = 1.013 25 bar

1 mmHg = 13.5951 × 0.1 × 980.665 dyn/cm^2
(This is a fixed unit of pressure. The actual pressure difference corresponding to a column of mercury of height 1 mm will depend on the density of the mercury and on local gravity.)
one pound force per square inch = 6.894 76 N/m^2, approximately

Gauge pressure
May mean *either* the excess of a true (absolute) pressure over the actual atmospheric pressure or the excess of a true (absolute) pressure over a pressure of 1 atmosphere

Viscosity (dynamic viscosity)
one newton second per metre squared = 1 N s/m^2 = 1 kg/m s
= one pascal second = 1 Pa s

one poise = 1 P = 0.1 kg/m s
1 cP = 1 mPa s

Energy, work, heating

one joule	= 1 J	= 1 N m
one calorie	= 1 cal	= 4.1868 J
one British Thermal Unit	= 1 Btu	= 2.326 J lb/g = 778 ft lbf, approximately
one thermochemical calorie	= 4.184 J	

Power

one watt	= 1 W	= 1 J/s
one horse-power	= 1 hp	= 550 ft lbf/s = 745.7 W, approximately

Thermodynamic temperature
This quantity is defined in accordance with the laws of thermodynamics.
one kelvin = 1 K (base unit)
The kelvin is, by definition, equal to $\frac{100}{27316}$ times the thermodynamic temperature of the triple point of ordinary water substance.
one rankine = 1 R = $(\frac{5}{9})$ K

Any unit of thermodynamic temperature may be used as a unit of temperature difference or as a unit of increment of temperature.

Celsius temperature θ

definition: $\theta = T - T_C$, where $T_C = 273.15$ K

Fahrenheit temperature ϕ

definition: $\phi = T - T_F$, where $T_F = 459.67$ R

Any unit of thermodynamic temperature may be used as a unit for a quantity defined to be equal to a temperature difference.

The temperature level of a system may be indicated by the equivalent sentences:

'The temperature is t_K K'

'The temperature is t_C °C'

'The temperature is t_R °R'

'The temperature is t_F °F'

wherein appear numbers given by

$$t_K = T/K,\ t_C = (T - T_C)/K,$$

$$t_R = T/R,\ t_F = (T - T_F)/R,$$

whence

$$t_C = \theta/K$$

and

$$t_F = \phi/R$$

In these sentences no adjective qualifies 'temperature'. The information omitted at the point is later implied by the temperature-level indicators:

K, called 'degree Kelvin'

°C, called 'degree Celsius'

°R, called 'degree Rankine'

°F, called 'degree Fahrenheit'

These are not pure unit symbols. They are suitable for use in sentences and on instruments. They should not be used in equations and must not be used in compound unit symbols. The quantities T, $T_2 - T_1$ and ΔT are physically significant. The quantities θ, ϕ are physically insignificant.

When the word 'temperature' is used alone (without qualification, implication of statement) it must be taken to refer to T. For example, the body temperature of a healthy human may be written as

$$T = 310 \text{ K} = 558 \text{ R}$$

$$\theta = 36.85 \text{ K} = 66.33 \text{ R}$$

$$\phi = 54.627 \text{ K} = 98.33 \text{ R},$$

or expressed by the sentences

'The temperature is 310 K'

'The temperature is 558 °R'

'The temperature is 36.85 °C'

'The temperature is 98.33 °F'

and is often express the same fact by

'The temperature is 310 K'

'The temperature is 558 R'

etc.

Electric flux ψ
coulomb C = A.s

Electric charge Q

Electric potential difference
volt V = Wb/s

Electric flux density D
C/m^2

Magnetic flux ϕ
weber = Wb = V.s

Electric current I

Magnetic potential difference
ampere A = C/s

Magnetic flux density B
tesla = T = Wb/m^2

Electric field strength E
V/m

Magnetic field strength H
A/m

Capacitance C
farad F = C/V

Inductance L
henry H = Wb/A

Permittivity $\varepsilon = D/E$
C/V m = F/m

Permeability $\mu = B/H$
Wb/A m = H/m

V·A = W   C·V = J = Wb·A   F·H = s²

Resistance R

Conductance
siemens, S = A/V

Reactance X

Impedance Z
ohm, Ω = V/A

In terms of the base unit A other named units are given by

C = A s

V = W/A   W = J/A

F = C²/J   H = J/A²

Ω = W/A²   S = A²/W

T = J/A m²

Amount of substance

The amount of particular substance is proportional to the number of particles of type specified or implied by the designation of the particular substance.

one mole = 1 mol   (base unit)

The mole is the amount of substance in a system containing as many particles of type specified or implied by the designation of the particular substance as there are atoms of carbon-12 in a system containing an amount of carbon-12 having mass 0.012 kilogramme.

Universal physical quantities

permeability of a vacuum $\mu_0 = 4\pi \times 10^{-7}$ H/m

$\varepsilon_0 = 1/(\mu_0 c_0^2)$ exactly

$= 8.85 \times 10^{-12}$ F/m, approximately

Measured constant physical quantities:
speed of light $c_0 = 2.99792 \times 10^8$ m/s
charge of proton $e = 1.6021 \times 10^{-19}$ C
mass of electron $m_e = 9.109 \times 10^{-31}$ kg
molar number (Avogadro's constant) $N_n = 6.023 \times 10^{23}$ mol^{-1}
molar ideal-gas constant $R_n = 8.314$ J/mol K
Boltzmann constant, molecular ideal-gas constant $k = 1.3805 \times 10^{-23}$ J/K
Planck's constant $h = 6.626 \times 10^{-34}$ J s
Stefan–Boltzmann constant $\sigma = 5.670 \times 10^{-8}$ J/m² K⁴ s
gravitational constant $G = 6.67 \times 10^{-11}$ N m²/kg

Table A1. Summary of units for physical quantities

Quantity	*Measure formula*	*Quantity*	*Measure formula*
		Mass per unit area	ML^{-2}
Fundamental Units		Viscosity	$ML^{-1}T^{-1}$
Length	L	Kinematic viscosity	L^2T^{-1}
Mass	M	Diffusivity	L^2T^{-1}
Time	T	Stress, pressure	$ML^{-1}T^{-2}$
		Strain	$M^\circ L^\circ T^\circ$
Speed or velocity	LT^{-1}	Coefficient of elasticity,	$ML^{-1}T^{-2}$
Acceleration	LT^{-2}	elastic modulus.	
Momentum	MLT^{-1}	Linear stiffness	MT^{-2}
Force	MLT^{-2}	Angular stiffness	ML^2T^{-2}
Energy, work	ML^2T^{-2}	Mass moment	ML
Power	ML^2T^{-3}	Moment of inertia, product	ML^2
Moment of Force	ML^2T^{-2}	of inertia	
Angular momentum	ML^2T^{-1}	First moment of area	L^3
Angle	$M^\circ L^\circ T^\circ$	Second moment of area	L^4
Angular velocity	T^{-1}	Vorticity	T^{-1}
Angular acceleration	T^{-2}	Circulation	L^2T^{-1}
Area	L^2	Coefficient of friction	$M^\circ L^\circ T^\circ$
Volume	L^3	Coefficient of restitution	$M^\circ L^\circ T^\circ$
Density	ML^{-3}	Flexural rigidity *EI* of a beam	ML^3T^{-2}

Appendix B

Tensor notation

Tensor notation is an abbreviated form of expression for algebraic and differential equations with two, three or more independent variables. The general theory of tensor calculus is also used in continuum fluid mechanics and in other physical sciences to formulate governing equations which are valid in every co-ordinate system and simultaneously contain the transformation properties of individual quantities. These general properties of tensors are not considered here. A description of the notation for Cartesian tensors is set out to assist in reading the tensorially expressed equations in the text.

Cartesian tensors are restricted to transformations from one homogenous co-ordinate system to another. In three-dimensional space, the number of components of a tensor equals 3^N where N is the order of the tensor. A tensor of zero order is a scalar quantity, whereas a tensor of order one has three co-ordinate components in physical space and is a vector. Tensors of second and third order also arise in continuum mechanics.

The components of a tensor of any order and the tensor itself can be represented by indicial notation. A generic letter denotes the variable of interest with the indices written typically as subscript letters.

Under the rules of indicial notation, a letter index may occur once or twice in a given term. When an index occurs in a term but is not repeated, that index is taken to have values 1, 2, 3 to N where N is an integer denoting the range of the index. Unrepeated indices are known as free indices and the tensorial rank of a term is equal to the number of free indices appearing in that term. When an index appears twice in a term, that index takes on all the values in its range and the resulting terms are summed. Repeated indices are called dummy indices in this summation convention.

Thus the tensor a_i is of first order with one free index and denotes a vector quantity with the number of components equal to the range of i (usually three in continuum mechanics). The tensor $a_{ij}b_j$ with one free index is also a first order tensor. If j has a range of 3 then

$$a_{ij}b_j = a_{i1}b_1 + a_{i2}b_2 + a_{i3}b_3.$$

Second-order tensors have two free indices such as D_{ij} or in equation form

$$D_{ij} = a_{ijk}c_k.$$

Tensor equations provide highly compact notation for algebraic equations. Thus the tensor equation

$$Y_i = a_{ij}x_j$$

with a range of three on both indices represents the equations

$$Y_1 = a_{11}x_1 + a_{12}x_2 + a_{13}x_3$$

$$Y_2 = a_{21}x_1 + a_{22}x_2 + a_{23}x_3$$

$$Y_3 = a_{31}x_1 + a_{32}x_2 + a_{33}x_3$$

and the tensor equation

$$P_{ij} = q_{ip}r_{jq}s_{pq}$$

with a range of 2 for i and j denotes four equations, the first two of which are

$$p_{11} = q_{11}r_{11}s_{11} + q_{11}r_{12}s_{12} + q_{12}r_{11}s_{21} + q_{12}r_{12}s_{22}$$

$$p_{12} = q_{11}r_{21}s_{11} + q_{11}r_{22}s_{12} + q_{12}r_{21}s_{21} + q_{12}r_{22}s_{22}$$

For tensors where only terms correponding to equal indices are non-zero, the Kronecker-delta may be used. This is defined as

$$\delta_{ij} = 1 \text{ if } i = j$$

$$= 0 \text{ if } i \neq j$$

Appendix C

Complex number theory

The main results of complex number theory are summarised below. For a complex number $Z = x + \mathrm{i}y$, x is the real part and y is the imaginary part with $\mathrm{i} = \sqrt{-1}$. The complex number can also be represented in modulus and argument form as

$$
\begin{aligned}
Z &= r\exp[\mathrm{i}(\theta + 2n\pi)\\
&= r[\cos(\theta + 2n\pi) + \mathrm{i}\sin(\theta + 2n\pi)]
\end{aligned}
$$

where n is any integer, r is the argument and θ is the modulus with a principal value in the range of from $-\pi$ to π; Z can be represented on an Oxy or $Or\theta$ plane (the Argand diagram) by the co-ordinates (x, y) or (r, θ). Thus

$$r = |Z| = \sqrt{(x^2 + y^2)}$$

and

$$\theta = \arctan\left(\frac{y}{x}\right) \text{ with } -\pi < \theta \leq \pi$$

also

$$x = r\cos\theta \text{ and } y = r\sin\theta$$

For two complex numbers

$$Z_1 = x_1 + \mathrm{i}y_1 = Z_1\mathrm{e}^{\mathrm{i}\theta_1} \text{ and } Z_2 = x_2 + \mathrm{i}y_2 = r_2\exp(\mathrm{i}\theta_2)$$

the usual arithmetic operations can be written as

$$
\begin{aligned}
Z_1 \pm Z_2 &= (x_1 \pm x_2) + \mathrm{i}(y_1 \pm y_2)\\
Z_1Z_2 &= (x_1x_2 - y_1y_2) + \mathrm{i}(x_1y_2 + x_2y_1)\\
&= r_1r_2\exp[\mathrm{i}(\theta_1 + \theta_2)]\\
\frac{Z_1}{Z_2} &= \frac{(x_1 + \mathrm{i}y_1)(x_2 - \mathrm{i}y_2)}{(x_2 + \mathrm{i}y_2)(x_2 - \mathrm{i}y_2)}\\
&= \frac{(x_1x_2 + y_1y_2) + \mathrm{i}(x_2y_1 - x_1y_2)}{x_2^2 + y_2^2}\\
&= \frac{r_1}{r_2}\exp(\theta_1 - \theta_2)
\end{aligned}
$$

De Moivre's theorem

$$e^{i\theta} = \cos\theta + i\sin\theta$$

$$e^{in\theta} = \cos n\theta + i\sin n\theta$$

$$= (\cos\theta + i\sin\theta)^n$$

$\bar{Z} = (x - iy)$ is the complex conjugate of $Z = x + iy$ $\quad |Z|^2 = Z\cdot\bar{Z}$

$$x = \frac{(Z + \bar{Z})}{2}, \quad y = \frac{(Z - \bar{Z})}{2}$$

Powers of Z

$$Z^n = (r\cos\theta + ir\sin\theta)^n$$

$$= r^n(\cos n\theta + i\sin n\theta)$$

Roots of *Z*

$$Z^{1/n} = (r\cos\theta + r\sin\theta)^{1/n}$$

$$= r^{1/n}\left[\cos\left(\frac{\theta + 2\pi m}{n}\right) + i\sin\left(\frac{\theta + 2\pi m}{n}\right)\right]$$

where m = 0, 1, 2, . . .$(n - 1)$.

Hyperbolic functions

$$e^{i\theta} = \cos\theta + i\sin\theta$$

$$e^{-i\theta} = \cos\theta - i\sin\theta$$

$$\cos\theta = \frac{e^{i\theta} + e^{-i\theta}}{2} = \cosh i\theta$$

$$\sin\theta = \frac{e^{i\theta} - e^{-i\theta}}{2i} = \frac{1}{i}\sinh i\theta = -i\sinh i\theta$$

If θ replaced by $i\theta$, then

$$\cos i\theta = \cosh\theta \quad \sin i\theta = i\sinh\theta$$

Appendix D

The divergence and Green's theorems

The divergence (or Gauss's) theorem

If three functions $P(x, y, z)$, $Q(x, y, z)$ and $R(x, y, z)$ have single valued, continuous first derivatives in a three-dimensional volume of space V bounded by a closed regular surface, S then the divergence (or Gauss) theorem yields the equation

$$\int_V \left(\frac{\partial P}{\partial x} + \frac{\partial Q}{\partial y} + \frac{\partial R}{\partial z}\right) \mathrm{d}V = \oint_S (P\,\mathrm{d}S_x + Q\,\mathrm{d}S_y + R\,\mathrm{d}S_z) \tag{D1}$$

where $\mathrm{d}S_x$, $\mathrm{d}S_y$ and $\mathrm{d}S_z$ are projections of an element of the surface S in the x, y and z direction respectively pointing outwards from the surface of volume V.

Green's theorem

Green's theorem can be derived from the divergence theorem and is usually expressed in three forms. If scalar functions $P(x, y, z)$ and $Q(x, y, z)$ and their partial derivatives are continuous within a three dimensional volume V bounded by a regular closed surface S, then, the first form of Green's theorem can be expressed as

$$\int_V \left(\frac{\partial P}{\partial x}\frac{\partial Q}{\partial x} + \frac{\partial P}{\partial y}\frac{\partial Q}{\partial y} + \frac{\partial P}{\partial z}\frac{\partial Q}{\partial z}\right) \mathrm{d}V + \int_V P\left(\frac{\partial^2 Q}{\partial x^2} + \frac{\partial^2 Q}{\partial y^2} + \frac{\partial^2 Q}{\partial z^2}\right) \mathrm{d}V$$

$$= -\oint_S P\frac{\partial Q}{\partial n}\,\mathrm{d}S \tag{D2}$$

where n denotes the normal to the closed surface S pointing in to the volume V.

A second form of Green's theorem is obtained by putting $P = Q$ to give

$$\int_V \left[\left(\frac{\partial P}{\partial x}\right)^2 + \left(\frac{\partial P}{\partial y}\right)^2 + \left(\frac{\partial P}{\partial z}\right)^2\right] \mathrm{d}V + \int_V P\left(\frac{\partial^2 P}{\partial x^2} + \frac{\partial^2 P}{\partial y^2} + \frac{\partial^2 P}{\partial z^2}\right) \mathrm{d}V$$

$$= -\oint_S P\frac{\partial P}{\partial n}\,\mathrm{d}S$$

A third form of Green's theorem is obtained by interchanging P and Q in Equation (D2) and subtracting the resultant equation from (D2) to yield

$$= \int_V \left[P\left(\frac{\partial^2 Q}{\partial x^2} + \frac{\partial^2 Q}{\partial y^2} + \frac{\partial^2 Q}{\partial z^2} \right) - Q\left(\frac{\partial^2 P}{\partial x^2} + \frac{\partial^2 P}{\partial y^2} + \frac{\partial^2 P}{\partial z^2} \right) \right] \mathrm{d}V$$

$$= - \oint_S \left(P\frac{\partial Q}{\partial n} - Q\frac{\partial P}{\partial n} \right) \mathrm{d}S$$

Green's function may be applied to a velocity potential ϕ, which will satisfy the Laplace equation

$$\frac{\partial^2 \phi}{\partial x^2} + \frac{\partial^2 \phi}{\partial y^2} + \frac{\partial^2 \phi}{\partial z^2} = 0$$

The three forms of Green's theorem thus simplify further when applied to velocity potential functions.

Appendix E

Lagrange's equation of motion

An equation of motion for any system can be derived from first principles by making use of scalar energy quantities in a variational form through Hamilton's principle. This principle states that for any system, the variation of its kinetic and potential energy plus the variation of the work done by non-conservative forces over any time interval must sum to zero. This can be written mathematically as

$$\int_{t_1}^{t_2} \delta(T - V)\,\mathrm{d}t + \int_{t_1}^{t_2} \delta W_{\mathrm{nc}}\,\mathrm{d}t = 0 \tag{E1}$$

where T is the total kinetic energy of the system, V is the system potential energy including both strain energy and gravitationl or other potential energy from conservative external forces, W_{nc} is the work done by non-conservative external forces such as damping, δ is an operator denoting variation of the quantity following it and the integration is carried out from time t_1 to t_2.

For a static system, $T = 0$ and Hamilton's principle reduces to

$$\delta(V - W_{\mathrm{nc}}) = 0 \tag{E2}$$

due to the invariance with time for a statical condition.

Hamilton's principle can be used to derive a generalized equation of motion for multiple degrees of freedom by writing T, V and δW_{nc} in terms of a set of generalized coordinates

$$q_1, q_2, \ldots, q_n.$$

T, V and δW_{nc} can be expressed in terms of these generalized co-ordinates or their time derivatives for the dynamics of a mechanical or structural system. For any general case these functions can be written as

$$T = T(q_1, q_2, \ldots, q_N, \dot{q}_1, \dot{q}_2, \ldots, \dot{q}_N) \tag{E3}$$

$$V = V(q_1, q_2, \ldots, q_N) \tag{E4}$$

and

$$\delta W_{\mathrm{nc}} = Q_1\delta q_1 + Q_2\delta q_2 + \ldots + Q_n\delta q_n \tag{E5}$$

The kinetic energy T will be a function of the generalized co-ordinates and their time derivatives, whereas the potential energy V will be a function of the generalized co-ordinates only. Coefficients $Q_1, Q_2, \ldots, Q_n$ are generalized forcing functions applied to the co-ordinates $q_1, q_2, \ldots, q_n$.

Substituting Equations (E3)–(E5) in (E1) and expanding the first term in a variation form yields

$$\left.\begin{aligned}&\int_{t_1}^{t_2}\left(\frac{\partial T}{\partial q_1}\delta q_1+\frac{\partial T}{\partial q_2}\delta q_2+\ldots+\frac{\partial T}{\partial q_n}\delta q_n+\frac{\partial T}{\partial \dot q_1}\delta \dot q_1\right.\\&+\frac{\partial T}{\partial \dot q_2}\delta \dot q_2+\ldots+\frac{\partial T}{\partial \dot q_n}\delta \dot q_n-\frac{\partial V}{\partial q_1}\delta q_1-\frac{\partial V}{\partial q_2}\delta q_2\\&\left.\ldots-\frac{\partial V}{\partial q_n}\delta q_n+Q_1\delta q_1+Q_2\delta q_2+\ldots+Q_n\delta q_n\right)\mathrm{d}t=0\end{aligned}\right\}\tag{E6}$$

where $\delta q_1, \delta q_2, \ldots, \delta q_n$ are arbitrary small changes in co-ordinates $q_1, q_2, \ldots, q_n$.

This equation can be generalized in terms of an ith term to yield

$$\int_{t_1}^{t_2}\sum_{i=1}^{n}\left(\frac{\partial T}{\partial q_i}\delta q_i+\frac{\partial T}{\partial \dot q_i}\delta \dot q_i-\frac{\partial V}{\partial q_i}\delta q_i+Q_i\delta q_i\right)\mathrm{d}t=0\tag{E7}$$

Integrating the velocity dependent second term yields

$$\int_{t_1}^{t_2}\frac{\partial T}{\partial \dot q_i}\delta \dot q_i\,\mathrm{d}t=\left[\frac{\mathrm{d}T}{\delta \dot q_i}\delta q_i\right]_{t_1}^{t_2}-\int_{t_1}^{t_2}\frac{\partial}{\partial t}\left(\frac{\partial T}{\partial \dot q_i}\right)\delta q_i\,\mathrm{d}t=-\int_{t_1}^{t_2}\frac{\partial}{\partial t}\left(\frac{\partial T}{\partial \dot q_i}\right)\partial q_i\,\mathrm{d}t\tag{E8}$$

with $\delta q_i(t_1)=\delta q_i(t_2)=0$ since operator δ denotes variation in q_i over the time interval from t_1 to t_2. Then Equation (E7) can be written as

$$\int_{t_1}^{t_2}\left\{\sum_{i=1}^{n}\left[-\frac{\partial}{\partial t}\left(\frac{\partial T}{\partial \dot q_i}\right)+\frac{\partial T}{\partial q_i}-\frac{\partial V}{\partial q_i}+Q_i\right]\delta q_i\right\}\mathrm{d}t=0\tag{E9}$$

Since all the δq_i for $i = 1$ to n are arbitrary, the term in square brackets must vanish separately for all is.

Thus after rearrangement of this term, we get

$$\frac{\partial}{\partial T}\left(\frac{\partial T}{\partial \dot q_i}\right)-\frac{\partial T}{\partial q_i}+\frac{\partial V}{\partial q_i}=Q_i$$

for any of the generalized co-ordinates q_i. This is Lagrange's equation of motion.

Appendix F

Forces on a cylindrical element due to internal and external pressure

A hollow cylindrical member submerged in a fluid and containing a fluid within itself will experience a force due to the external and internal hydrostatic pressures of both fluids acting on the surfaces of the cylinder. An element of the cylinder is shown in Figure F1. The resultant force is obtained by finding the force on an arbitrary section of the element, (shaded portion of Figure F1), and resolving it into components before integrating to obtain the total force on the element. Note that only the force on the curved walls of the cylinder due to hydrostatic pressure is evaluated. The force on the end cross-sections is not considered here since the cylinder is taken to be very long and the end cross-section will usually terminate to a coupling such that hydrostatic pressure will not act on the cylinder cross-sections. Furthermore, the axes system used in Figure F1 is such that the hydrostatic pressure is taken to increase linearly along the vertical axis.

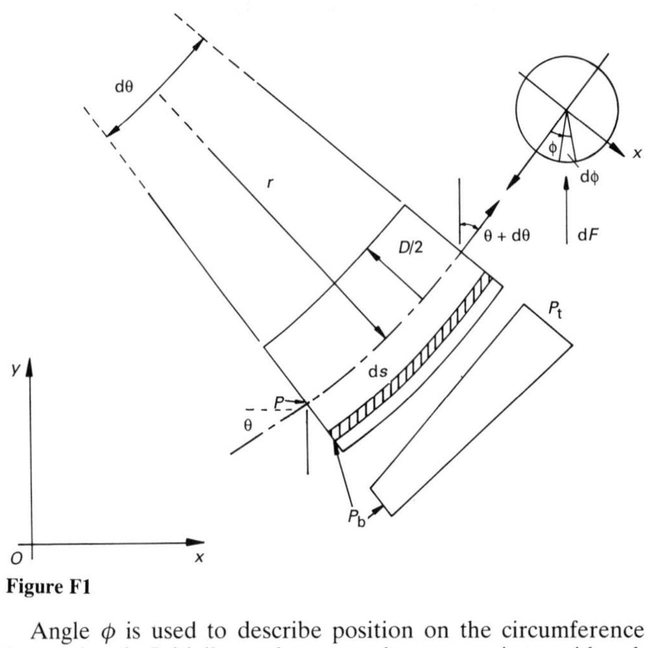

Figure F1

Angle ϕ is used to describe position on the circumference of the element to be analysed. Initially, only external pressure is considered in the derivation. Forces due to internal pressure can be readily deduced from those due to external pressure by a simple reversal of signs and change of diameter.

As shown in Figure F1, the length ds of any strip on the cylinder circumference parallel to its axis is given by

$$\mathrm{d}s = (r + \tfrac{1}{2}D\cos\phi)\,\mathrm{d}\theta \tag{F1}$$

where r is the element radius of curvature and D is the diameter of the pressure bearing surface; θ and ϕ are defined in Figure F1. If the hydrostatic pressure on the centre line of the element at its lower end is p, then the pressure p_{b} at various levels along the bottom surface is given by

$$p_{\mathrm{b}} = p - \tfrac{1}{2}\gamma D\cos\theta\cdot\sin\phi \tag{F2}$$

where γ is the weight per unit volume of the fluid medium. Also, the corresponding pressure p_{t} at the top of the element is hydrostatic, and given by the equation

$$p_{\mathrm{t}} = p_{\mathrm{b}} - \gamma\cdot\cos\theta\,\mathrm{d}s \tag{F3}$$

since

$$\cos\theta = \frac{\mathrm{d}y}{\mathrm{d}s}$$

The area of section of element described by arc dϕ is given by

$$\mathrm{d}A = \tfrac{1}{2}D\cdot\mathrm{d}s\cdot\mathrm{d}\phi \tag{F4}$$

The force which acts on this section of the element is then

$$\mathrm{d}F = \tfrac{1}{2}(p_{\mathrm{b}} + p_{\mathrm{t}})\,\mathrm{d}A \tag{F5}$$

Substituting the expressions derived for dA, p_{b}, p_{t} into (F5) gives

$$\mathrm{d}F = \tfrac{1}{2}(p_{\mathrm{b}} + p_{\mathrm{b}} - \gamma\,\mathrm{d}s\cos\theta)(\tfrac{1}{2}D\cdot\mathrm{d}s\cdot\mathrm{d}\phi) \tag{F6}$$

$$\mathrm{d}F = (p - \tfrac{1}{2}\gamma D\cos\theta\cdot\sin\phi - \tfrac{1}{2}\gamma\,\mathrm{d}s\sin\theta)(\tfrac{1}{2}D\,\mathrm{d}s\,\mathrm{d}\phi) \tag{F7}$$

Replacing (F1) for ds leads to

$$\mathrm{d}F = [p - \tfrac{1}{2}\gamma D\cos\theta\cdot\sin\phi - \tfrac{1}{2}\gamma(r + \tfrac{1}{2}D\cos\phi)\,\mathrm{d}\theta\sin\theta]$$
$$\cdot(\tfrac{1}{2}r + \tfrac{1}{4}D\cos\phi)D\cdot\mathrm{d}\theta\cdot\mathrm{d}\phi. \tag{F8}$$

Expanding the individual terms gives

$$\mathrm{d}F = [\tfrac{1}{2}pDr\,\mathrm{d}\theta - \tfrac{1}{4}\gamma Dr^2\sin\theta(\mathrm{d}\theta)^2]\,\mathrm{d}\phi$$
$$+ [\tfrac{1}{4}pD^2\,\mathrm{d}\theta + \tfrac{1}{4}\gamma D^2 r\cos\theta\,\mathrm{d}\theta - \tfrac{1}{4}\gamma D^2 r\sin\theta(\mathrm{d}\theta)^2]\sin\phi\,\mathrm{d}\phi$$
$$+ [\tfrac{1}{8}\gamma D^3\cos\theta\,\mathrm{d}\theta - \tfrac{1}{8}\gamma D^3\sin\theta(\mathrm{d}\theta)^2]\sin^2\phi\,\mathrm{d}\phi \tag{F9}$$

The differential force may be resolved into it's three directional components – F_x, F_y and F_z along the x, y and z axes respectively. In this case, the analysis is restricted to two dimensions and since there is no deformation out of the x–y plane, the resultant force in the z direction is taken as zero, i.e.

$$F_z = 0.0$$

$$\mathrm{d}F_x = -\mathrm{d}F\sin\theta\sin\phi \tag{F10}$$

$$\mathrm{d}F_y = \mathrm{d}F\cos\theta\sin\phi \tag{F11}$$

$$\mathrm{d}F_x = \int_{\phi=0}^{\phi=2\pi} -\mathrm{d}F\sin\theta\sin\phi \tag{F12}$$

$$\mathrm{d}F_y = \int_{\phi=\pi}^{\phi=2\pi} \mathrm{d}F\cos\theta\sin\phi \tag{F13}$$

Therefore

$$F_x = -\,[\tfrac{1}{2}pDr\sin\theta\,\mathrm{d}\theta - \tfrac{1}{4}\gamma r^2 D\sin^2\theta(\mathrm{d}\theta)^2]\int_{\phi=0}^{2\pi}\sin\phi\,\mathrm{d}\phi$$

$$-\,(\tfrac{1}{4}pD^2\sin\theta\,\mathrm{d}\theta + \tfrac{1}{4}\gamma D^2 r\sin\theta\cos\theta\,\mathrm{d}\theta - \tfrac{1}{4}\gamma D^2 r\sin^2\theta(\mathrm{d}\theta)^2)$$

$$\int_{\phi=0}^{2\pi}\sin^2\phi\,\mathrm{d}\phi - [\tfrac{1}{8}\gamma D^3\cos\theta\sin\theta\,\mathrm{d}\theta - \tfrac{1}{16}\gamma D^3\sin\theta(\mathrm{d}\theta)^2]\int_{\phi=0}^{2\pi}\sin^3\phi\,\mathrm{d}\phi$$

Using

$$\int_0^{2\pi}\sin\phi\cdot\mathrm{d}\phi = 0, \quad \int_0^{2\pi}\sin^2\phi\cdot\mathrm{d}\phi = \pi,$$

$$\int_0^{2\pi}\sin^3\phi\cdot\mathrm{d}\phi = 0 \quad \text{and} \quad A = \pi D^2/4,$$

gives the force in the x direction on a curved element as

$$F_x = -[pA + r\gamma A(\cos\theta - \sin\theta\,\mathrm{d}\theta)]\sin\theta\,\mathrm{d}\theta \tag{F15}$$

When the force due to internal pressure is considered, it's form will be the same but of opposite sign. Combining the effects of internal and external pressure for the most general case gives

$$(F_{xo} + F_{xi}) = [(p_i A_i - p_o A_o) + r(\gamma_i A_i - \gamma_o A_o)$$
$$(\cos\theta - \sin\theta\,\mathrm{d}\theta)]\sin\theta\,\mathrm{d}\theta \tag{F16}$$

where p_o, p_i are the external and internal pressures respectively at the level of the bottom of the element centre line.

The vertical force F_y is obtained in a similar way. Before the integration is performed, the expression for the force in the vertical direction appears as

$$F_y = [\tfrac{1}{2}pDr\,\mathrm{d}\theta - \tfrac{1}{4}\gamma Dr^2\sin\theta(\mathrm{d}\theta)^2]\cos\theta\left(\int_0^{2\pi}\sin\phi\,\mathrm{d}\phi\right)$$

$$+\,(\tfrac{1}{4}pD^2\,\mathrm{d}\theta + \tfrac{1}{4}\gamma D^2 r\cos\theta\,\mathrm{d}\theta - \tfrac{1}{4}\gamma D^2 r\sin\theta(\mathrm{d}\theta)^2)\cos\theta$$

$$\times\left(\int_0^{2\pi}\sin^2\phi\,\mathrm{d}\phi\right) + [\tfrac{1}{8}\gamma D^3\cos\theta\,\mathrm{d}\theta$$

$$-\,\tfrac{1}{8}\gamma D^3\sin\theta(\mathrm{d}\theta)^2]\cos\theta\int_0^{2\pi}\sin^3\phi\,\mathrm{d}\phi \tag{F17}$$

In evaluating this integral the following are used:

$$\int_0^{2\pi}\sin\phi\,\mathrm{d}\phi = 0$$

$$\int_0^{2\pi}\sin^3\phi\,\mathrm{d}\phi = 0, \int_0^{2\pi}\sin^2\phi\cdot\mathrm{d}\phi = \pi.$$

Thus Equation (F17) becomes

$$F_y = [pA + r\gamma A(\cos\theta - \sin\theta\,\mathrm{d}\theta)]\cos\theta\,\mathrm{d}\theta \tag{F18}$$

As before the effect of including the internal pressure acting on the element can be done quite easily. The final expression for the vertical force on a curved inclined element due to both internal and external pressure is given by

$$(F_{yo} + F_{yi}) = [(p_o A_o - p_i A_i) + r(\gamma_o A_o - \gamma_i A_i)$$
$$(\cos\theta - \sin\theta\,\mathrm{d}\theta)]\cos\theta\,\mathrm{d}\theta \tag{F19}$$

Summary of hydrodynamic coefficients

This section presents tables for added mass coefficients, drag coefficients and Strouhal number for the variety of typical shapes.

Table G1 Added mass coefficients (C_m) for two-dimensional bodies with cross-sections shown.

Section through body	C_m	Section through body	C_m
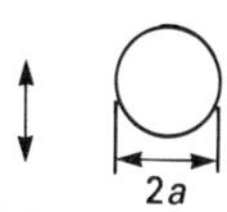	1.0	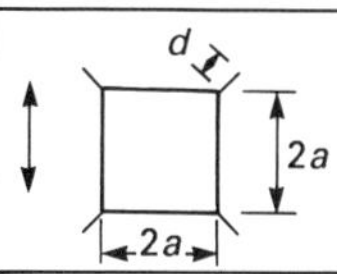 $d/a = 0.05$ $d/a = 0.10$ $d/a = 0.25$	1.61 1.72 2.19
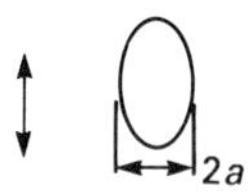	1.0	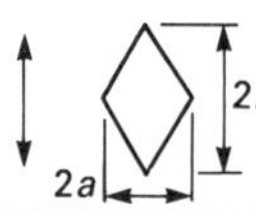 $a/b = 2$ $a/b = 1$ $a/b = 1/2$ $a/b = 1/5$	0.89 0.76 0.67 0.61
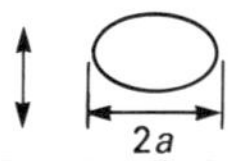	1.0	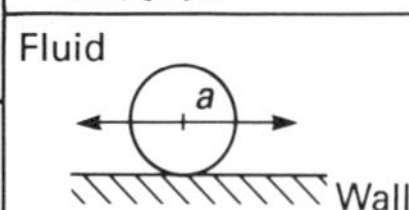	2.29
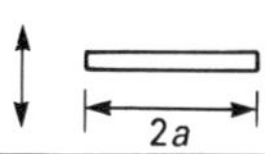	1.0		
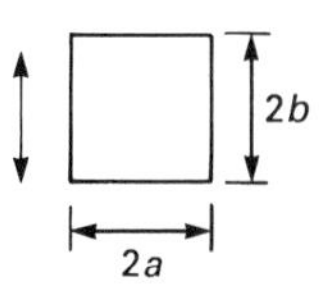 $a/b = \infty$ $a/b = 10$ $a/b = 5$ $a/b = 2$ $a/b = 1$ $a/b = 1/2$ $a/b = 1/5$ $a/b = 1/10$	1.0 1.14 1.21 1.36 1.51 1.70 1.98 2.23	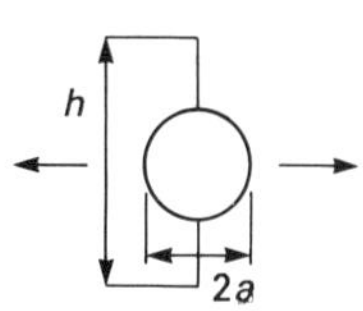	$1 + \left(\frac{h}{2a} - \frac{2a}{h}\right)^2$

The directions of motion are indicated by arrows and the values given are based on potential flow theory only. Added mass force, F, is expressed as

$$F = \rho C_m V \dot{U}$$

where ρ is fluid density, $V(=\pi a^2)$ is a 'reference' volume and $\dot{U}$ is body acceleration (from Det Norske Veritas, 1982).

Table G2 Added mass coefficients C_m for three-dimensional bodies with the shapes shown

Body shape	b/a	C_m		V
1. Flat plates Circular disc 		0.64		$\frac{4}{3}\pi a^2$
Eliptical disc 	b/a	C_m		$\frac{\pi}{6}a^2b$
	∞	1.0		
	12.75	0.99		
	7.0	0.97		
	3.0	0.90		
	1.5	0.76		
	1.0	0.64		
Rectangular plates 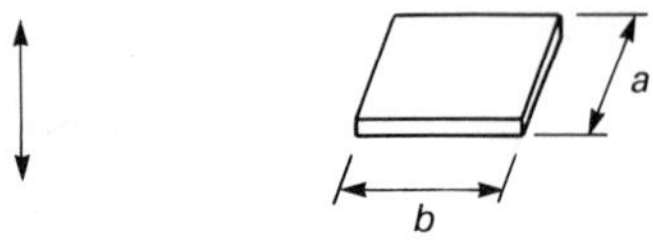	b/a	C_m		$\frac{\pi}{4}a^2b$
	1.0	0.58		
	1.5	0.69		
	2.0	0.79		
	3.0	0.83		
	∞	1.00		
Triangular plates 		$\frac{1}{\pi}(\tan\theta)^{3/2}$		$\frac{a^3}{3}$
2. Bodies of revolution Spheres 		0.5		$\frac{4}{3}\pi a^2$
Ellipsoids	b/a	C_m		$\frac{4}{3}\pi a^2 b$
Lateral 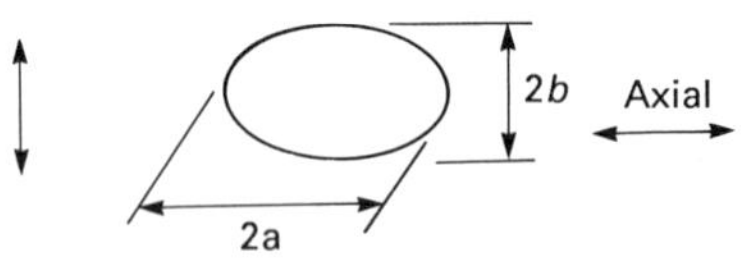		Axial	Lateral	
	1.5	0.30	0.62	
	2.0	0.21	0.70	
	2.51	0.16	0.76	
	3.99	0.08	0.86	
	6.97	0.04	0.93	
	9.97	0.02	0.96	
3. Square prisms 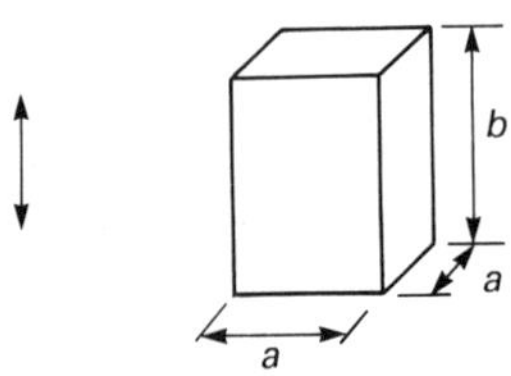	b/a	C_m		$a^2 b$
	1	0.68		
	2	0.36		
	3	0.24		
	4	0.19		
	5	0.15		
	6	0.13		
	7	0.11		
	10	0.08		

The directions of motion are indicated by arrows. Added mass force F is expressed as

$$F = \rho C_m V \dot{U}$$

where ρ is fluid density, $\dot{U}$ is body acceleration in direction shown and V is a 'reference' volume tabulated above (from Det Norske Veritas, 1982).

Table G3 Drag coefficients C_D for three-dimensional bodies with cross-sections shown

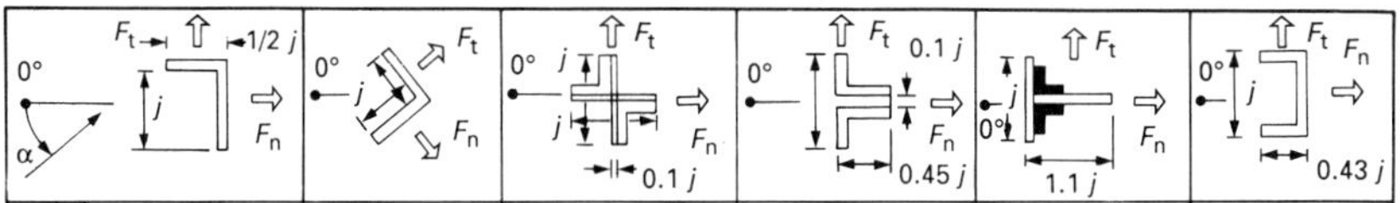

°	C_{Dn}	C_{Dt}	C_{Dn}	C_{Dt}	C_{Dn}	C_{Dt}	C_{Dn}	C_{Dt}	C_{Dn}	C_{Dt}	C_{Dn}	C_{Dt}
dg												
0	+1.9	+0.95	+1.8	+1.8	+1.75	+0.1	+1.6	0	+2.0	0	+2.05	0
45	+1.8	+0.8	+2.1	+1.0	+0.85	+0.85	+1.5	−0.1	+2.1	+0.9	+1.85	+0.6
90	+2.0	+1.7	−1.9	−1.0	+0.1	+1.75	−0.95	+0.7	−1.6	+2.15	0	+0.6
135	−1.8	−0.1	−2.0	+0.3	−0.75	+0.75	−0.5	+1.05	−1.1	+2.4	−1.6	+0.4
180	−2.0	+0.1	−1.4	−1.4	−1.75	−0.1	−1.5	0	−1.7	±2.1	−1.8	0

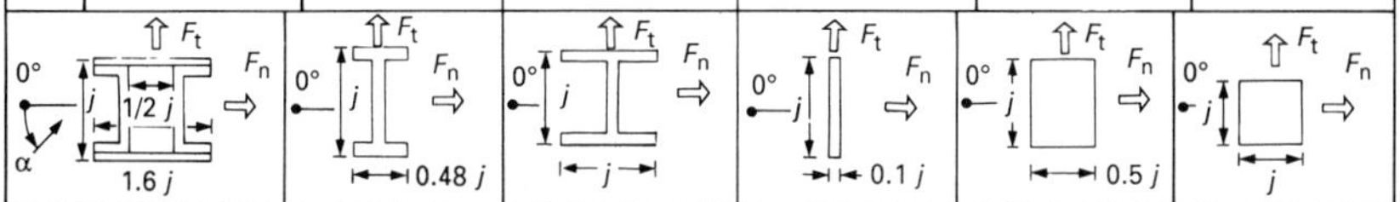

°	C_{Dn}	C_{Dt}	C_{Dn}	C_{Dt}	C_{Dn}	C_{Dt}	C_{Dn}	C_{Dt}	C_{Dn}	C_{Dt}	C_{Dn}	C_{Dt}
dg												
0	+1.4	0	+2.05	0	+1.6	0	+2.0	0	+2.1	0	+2.0	0
45	+1.2	+1.6	+1.95	+0.6	+1.5	+0.5	+1.8	+0.1	+1.4	+0.7	+1.55	+1.55
90	0	+2.2	0	+0.9	0	+1.9	0	+0.1	0	+0.75	0	+2.0

The wind direction is indicated by arrows on the left. The in line and transverse drag forces (F_n and F_t) and corresponding coefficients (C_{Dn} and C_{Dt}) are formulated as

$$F_n = \tfrac{1}{2}\rho U^2 dj^2 C_{Dn}$$

$$F_t = \tfrac{1}{2}\rho U^2 dj^2 C_{Dt}$$

where ρ is fluid density, U is velocity and d is the dimension shown in the table (from Det Norske Veritas, 1982).

Table G4 Drag coefficients C_D for three-dimensional bodies on a horizontal surface

Plan shapes	$\frac{d}{b}$	C_D for h/b: Up to 1/2	1	2	4	6
Wind 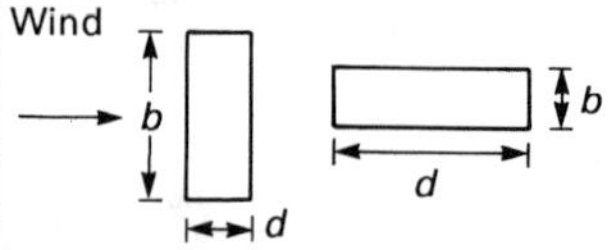	⩾ 4	1.2	1.3	1.4	1.5	1.6
	⩾1/4	0.7	0.7	0.75	0.75	0.75
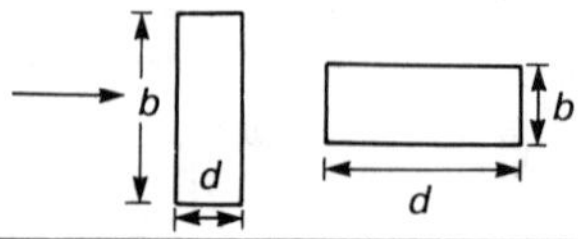	3	1.1	1.2	1.25	1.35	1.4
	1/2	0.7	0.75	0.75	0.75	0.8
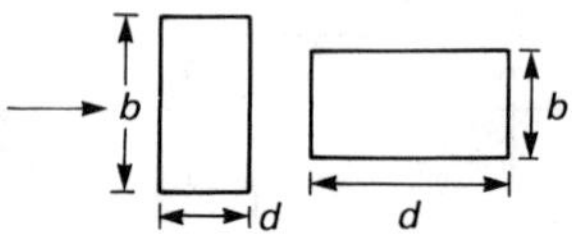	2	1.0	1.05	1.1	1.15	1.2
	1/2	0.75	0.75	0.8	0.85	0.9
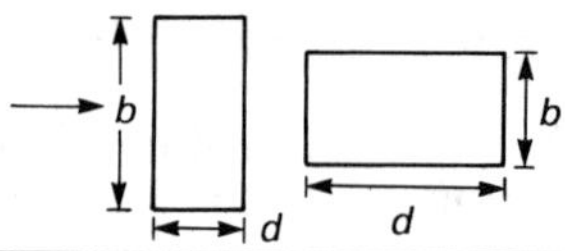	1 1/2	0.95	1.0	1.05	1.1	1.15
	2/3	0.8	0.85	0.9	0.95	1.0

Plan shape	$\frac{d}{b}$	C_D for h/b: Up to 1/2	1	2	4	6	10	20
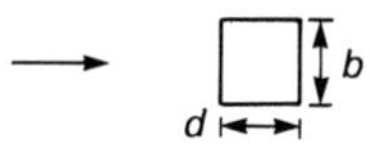	1	0.9	0.95	1.0	1.05	1.1	1.2	1.4

b is the dimension normal to the wind direction, d is the dimension along the wind direction and h is the height. The drag force F is formulated as

$$F = \tfrac{1}{2}\rho U^2 A C_D$$

where ρ is fluid density, U is velocity, and A is projected member area in the direction of incident wind (from Det Norske Veritas, 1982).

Table G5 Strouhal number *S* for members of various cross-sectional shapes

Flow dirn	Member shape	S	Flow dirn	Member shape	S
→	*t* = 2.0; 50; *t*; 50	0.120	↓	*t* = 1.0; 12.5; 12.5; 25; 50	0.147
↓		0.137			
→	*t* = 0.5; 25; 25	0.120	↓	*t* = 1.0; 12.5; 12.5; 12.5; 50	0.150
↓	*t* = 1.0; 25; 50	0.144	←	*t* = 1.0; 50; 50	0.145
			↑		0.142
			↙		0.147
↓	*t* = 1.5; 12.5; 50	0.145	←	*t* = 1.0; 25; 25	0.131
			↑		0.134
			↙		0.137
↓	*t* = 1.0; 25; 50	0.140	→	t = 1.0; 25; 25 25 25	0.121
↑		0.153	↓		0.143
↓	*t* = 1.0	0.145	→	*t* = 1.0; 25; 25 12.5 25	0.135
↑		0.168			
→	*t* = 1.5; 50	0.156	→	*t* = 1.0; 50; 100	0.160
↓		0.145			
Cylinder	25; 11.800 < Re < 19.100	0.200	→	*t* = 1.0; 25; 50	0.114
			↑		0.145

Member dimensions are in millimetres. Strouhal number S is defined as

$$S = \frac{Df}{U}$$

where D is the member cross-sectional direction perpendicular to the member or fluid direction of motion, f is the vortex shedding frequency in Hz and U is the flow velocity (from Det Norske Veritas, 1982).

Reference

Det Norske Veritas, (1982), Rules for the design, construction and inspection of offshore structures, 1977, Reprinted with corrections, 1982. Det Norske Veritas, PO Box 300, N-1322, Hovik, Norway.

Index

Page numbers in *italics* refer to figures and tables, with *t* denoting tables. Textual matter may occur on the same pages.